T0338186

WiMAX Evolution

WiMAX Evolution

Emerging Technologies and Applications

Marcos D. Katz

VTT, Finland

Frank H.P. Fitzek

Aalborg University, Denmark

A John Wiley and Sons, Ltd, Publication

Library of Congress Cataloging-in-Publication Data

Katz, Marcos D.
    WiMAX evolution : emerging technologies and applications / Marcos Katz, Frank Fitzek.
        p. cm.
    Includes bibliographical references and index.
    ISBN 978-0-470-69680-4 (cloth)
1. Wireless communication systems. 2. Broadband communication systems. 3. Mobile communication
systems. 4. Wireless LANs. 5. IEEE 802.16 (Standard) I. Fitzek, Frank H. P. II. Title.
    TK5103.2.K36 2009
    621.384–dc22   2008038550

A catalogue record for this book is available from the British Library.

ISBN 9780470696804 (H/B)

Set in 10/12pt Times by Sunrise Setting Ltd, Torquay, UK.
Printed in Great Britain by CPI Antony Rowe.

Contents

IV   Advanced WiMAX Architectures   85

5   WiMAX Femtocells   87
Chris Smart, Clare Somerville and Doug Pulley

6   Cooperative Principles in WiMAX   105
Qi Zhang, Frank H.P. Fitzek and Marcos D. Katz

VI WiMAX Evolution and Future Developments   305

16 MIMO Technologies for WiMAX Systems: Present and Future   307
Chan-Byoung Chae, Kaibin Huang and Takao Inoue

17 Hybrid Strategies for Link Adaptation Exploiting Several Degrees of Freedom in WiMAX Systems   335
Suvra Sekhar Das, Muhammad Imadur Rahman and Yuanye Wang

18 Applying WiMAX in New Scenarios: Limitations of the Physical Layer
    and Possible Solutions   367
    Ilkka Harjula, Paola Cardamone, Matti Weissenfelt, Mika Lasanen,
    Sandrine Boumard, Aaron Byman and Marcos D. Katz

19 Application of Radio-over-Fiber in WiMAX: Results and Prospects   385
    Juan Luis Corral, Roberto Llorente, Valentín Polo, Borja Vidal, Javier Martí,
    Jonás Porcar, David Zorrilla and Antonio José Ramírez

List of Contributors

Sassan Ahmadi
Intel Corporation
Mail Stop: JF3-336
2111 NE 25th Avenue
Hillsboro
OR 97124
USA
`sassan.ahmadi@intel.com`

Alexander Bachmutsky
Nokia Siemens Networks
313 Fairchild Drive
Mountain View
CA 94043
USA
`alexander.bachmutsky@nsn.com`

Thomas Michael Bohnert
SAP Research CEC Zurich
Kreuzstrasse 20
8008 Zurich
Switzerland
`thomas.michael.bohnert@sap.com`
and `tmb@nginet.de`

Sandrine Boumard
VTT Technical Research Centre of Finland
Kaitoväylä 1
FI-90571 Oulu
Finland
`sandrine.boumard@vtt.fi`

Aaron Byman
EB Corp.
Tutkijantie 7
90570 Oulu
Finland
`Aaron.Byman@elektrobit.com`

Paola Cardamone
THALES Security Solutions and Services S.p.A.
via Provinciale Lucchese, 33
50019 Sesto Fiorentino
Firenze
Italy
`paola.cardamone@gmail.com`

Thomas Casey
Elektrobit
Keilasatama 5
02150 Espoo
Finland
`thomas.casey@elektrobit.com`

Chan-Byoung Chae
Wireless Networking and Communications
    Group
Department of Electrical and Computer
    Engineering
The University of Texas at Austin
Austin, TX
USA
`cbchae@ece.utexas.edu`

Francesco Chiti
Department of Electronics and
    Telecommunications
University of Florence
via di S. Marta 3
I-50139 Florence
Italy
`francesco.chiti@unifi.it`

Juan Luis Corral
Nanophotonics Technology Center
Universidad Politécnica de Valencia
Camino de Vera s/n
46022 Valencia
Spain
`jlcorral@ntc.upv.es`

Marília Curado
DEI-CISUC
University of Coimbra
Polo II, Pinhal de Marrocos
3030-290 Coimbra
Portugal
marilia@dei.uc.pt

Suvra Sekhar Das Ph.D
Tata Consultancy Services
Innovation Lab, Convergence Practice,
    Tata Consultancy Services
Kolkata
India
suvra.das@tcs.com

Michael Devetsikiotis
Electrical and Computer Engineering
North Carolina State University
Raleigh
NC 27695-7911
USA
mdevets@ncsu.edu

Romano Fantacci
Department of Electronics and
    Telecommunications
University of Florence
via di S. Marta 3
I-50139 Florence
Italy
romano.fantacci@unifi.it

Frank H.P. Fitzek
Electronic Systems – Mobile Device Group
Aalborg University
Denmark
ff@es.aau.dk

Francisco Fontes
Portugal Telecom Inovação
R. Eng. José Ferreira Pinto Basto
3810-106 Aveiro
Portugal
fontes@ptinovacao.pt

Avraham Freedman
Hexagon System Engineering Ltd
P.O. Box 10149
14 Imber Street
Suite 51
Petach Tikva 49001
Israel
avif@hexagonltd.com

Ilkka Harjula
VTT Technical Research Centre of Finland
Kaitoväylä 1
FI-90571 Oulu
Finland
ilkka.harjula@vtt.fi

Matthias Hollick
Multimedia Communications Lab (KOM)
TU Darmstadt
Merckstr. 25
64283 Darmstadt
Germany
matthias.hollick@kom.tu-darmstadt.de

Kaibin Huang
Department of Electrical and Electronic
    Engineering
Hong Kong University of Science and
    Technology
Hong Kong
huangkb@ieee.org

Jie Hui
Intel Communication Technology Lab
Portland, Oregon
USA
Jie.Hui@intel.com

Jyrki Huusko
VTT Technical Research Centre of Finland
Kaitoväylä 1
FI-90571 Oulu
Finland
Jyrki.Huusko@vtt.fi

Takao Inoue
Wireless Networking and Communications
    Group
Department of Electrical and Computer
    Engineering
The University of Texas at Austin
Austin, TX
USA
inoue@ece.utexas.edu

Riku Jäntti
Department of Communications and Networking
Helsinki University of Technology
PL 3000
02015 TKK Espoo
Finland
riku.jantti@tkk.fi

Marcos D. Katz
VTT Technical Research Centre of Finland
Kaitoväylä 1
FI-90571 Oulu
Finland
Marcos.Katz@vtt.fi

Giada Landi
Nextworks
Via Turati, 43
56125 Pisa
Italy
g.landi@nextworks.it

Mika Lasanen
VTT Technical Research Centre of Finland
Kaitoväylä 1
FI-90571 Oulu
Finland
mika.lasanen@vtt.fi

Moshe Levin
Hexagon System Engineering Ltd
P.O. Box 10149
14 Imber Street, Suite 51
Petach Tikva 49001
Israel
moshe@hexagonltd.com

Roberto Llorente
Nanophotonics Technology Center
Universidad Politécnica de Valencia
Camino de Vera s/n
46022 Valencia
Spain
jlcorral@ntc.upv.es

Leonardo Maccari
Department of Electronics and
   Telecommunications
University of Florence
via di S. Marta 3
I-50139 Florence
Italy
leonardo.maccari@unifi.it

Dania Marabissi
Department of Electronics and
   Telecommunications
University of Florence
via di S. Marta 3
I-50139 Florence
Italy
dania.marabissi@unifi.it

Javier Martí
Nanophotonics Technology Center
Universidad Politécnica de Valencia
Camino de Vera s/n
46022 Valencia
Spain
jmarti@ntc.upv.es

Ricardo Matos
IT/UA Telecommunications Institute/University
   of Aveiro
Campus Universitário de Santiago
3810-193 Aveiro
Portugal
ricardo.matos@ua.pt

Parag S. Mogre
Multimedia Communications Lab (KOM)
TU Darmstadt
Merckstr. 25
64283 Darmstadt
Germany
parag.mogre@kom.tu-darmstadt.de

Edmundo Monteiro
University of Coimbra
Pinhal de Marrocos, Polo II
3030 Coimbra
Portugal
edmundo@dei.uc.pt

Pedro Neves
Portugal Telecom Inovação
R. Eng. José Ferreira Pinto Basto
3810-106 Aveiro
Portugal
pedro-m-neves@ptinovacao.pt

Jari Nurmi
Elektrobit
Kehräämöntie 5
87400 Kajaani
Finland
jari.nurmi@elektrobit.com

Ioannis Papapanagiotou
Electrical and Computer Engineering
North Carolina State University
Raleigh
NC 27695-7911
USA
ipapapa@ncsu.edu

Kostas Pentikousis
VTT Technical Research Centre of Finland
Kaitoväylä 1
FI-90571 Oulu
Finland
kostas.pentikousis@vtt.fi

Jarno Pinola
VTT Technical Research Centre of Finland
Kaitoväylä 1
FI-90571 Oulu
Finland
jarno.pinola@vtt.fi

Esa Piri
VTT Technical Research Centre of Finland
Kaitoväylä 1
FI-90571 Oulu
Finland
esa.piri@vtt.fi

Valentín Polo
AIMPLAS
València Parc Tecnològic
C/ Gustave Eiffel, 4
46980 Paterna
Spain
vpolo@aimplas.es

Jonás Porcar
DAS Photonics S.L.
Camino de Vera s/n
Building 8F
46022 Valencia
Spain
jporcar@dasphotonics.com

Doug Pulley
picoChip
Riverside Buildings
108 Walcot Street
Bath BA1 5BG
UK
doug.pulley@picochip.com

Muhammad Imadur Rahman Ph.D
Center for TeleInFrastrutur (CTIF)
Department of Electronic Systems
Aalborg University
Denmark
imr@ieee.org

Antonio José Ramírez
DAS Photonics S.L.
Camino de Vera s/n
Building 8F
46022 Valencia
Spain
aramirez@dasphotonics.com

Wonil Roh
Samsung Electronic Corp., Ltd
416 Maetan-3dong
Yeongtong-gu
Suwon-city
Gyeonggi-do, 443-742
Korea
wonil.roh@samsung.com

Susana Sargento
IT/UA Telecommunications Institute/University
   of Aveiro
Campus Universitário de Santiago
3810-193 Aveiro
Portugal
susana@ua.pt

Gerrit Schulte
acticom
Am Borsigturm 42
13507 Berlin
Germany

Christian Schwingenschloegl
Siemens AG
Corporate Technology, Information and
   Communication
Otto-Hahn-Ring 6
81730 Munich
Germany
chris.schwingenschloegl@siemens.com

Patrick Seeling
Department of Computing and
   New Media Technologies
University of Wisconsin - Stevens Point
Science Building, Room B243
Stevens Point
WI 54481
USA
pseeling@uwsp.edu

Paulo Simões
DEI-CISUC
University of Coimbra
Polo II, Pinhal de Marrocos
3030-290, Coimbra
Portugal
psimoes@dei.uc.pt

Chris Smart
picoChip
Riverside Buildings
108 Walcot Street
Bath BA1 5BG
UK
chris.smart@picochip.com

Clare Somerville
picoChip
Riverside Buildings
108 Walcot Street
Bath BA1 5BG
UK
clare.somerville@picochip.com

Roshni Srinivasan
Intel Corporation
2200 Mission College Boulevard RNB 5-123
Santa Clara
CA 95052
USA
roshni.srinivasan@intel.com

Dirk Staehle
University of Wuerzburg
Institute of Computer Science
Chair of Distributed Systems
Am Hubland
D-97074 Wuerzburg
Germany
dstaehle@informatik.uni-wuerzburg.de

Ralf Steinmetz
Multimedia Communications Lab (KOM)
TU Darmstadt
Merckstr. 25
64283 Darmstadt
Germany
ralf.steinmetz@kom.tu-darmstadt.de

Daniele Tarchi
Department of Electronics and
    Telecommunications
University of Florence
via di S. Marta 3
I-50139 Florence
Italy
daniele.tarchi@unifi.it

Rath Vannithamby
Intel Corporation
2111 NE 25th Avenue
Mail Stop JF3-206
Hillsboro
OR 97124
USA
rath.vannithamby@intel.com

Borja Vidal
Nanophotonics Technology Center
Universidad Politécnica de Valencia
Camino de Vera s/n
46022 Valencia
Spain
borvirod@ntc.upv.es

Nenad Veselinovic
Elektrobit
Keilasatama 5
02150 Espoo
Finland
nenad.veselinovic@elektrobit.com

Yuanye Wang M.Sc
Aalborg University
Radio Access Technology Section
Department of Electronic Systems
Aalborg University
Denmark
ywa@es.aau.dk

Matti Weissenfelt
VTT Technical Research Centre of Finland
Kaitoväylä 1
FI-90571 Oulu
Finland
matti.weissenfelt@vtt.fi

Vladimir Yanover
Alvarion Ltd
11/4 Nahshon Str.
Kfar Saba 44447
Israel
vladimir.yanover@alvarion.com

Qi Zhang
Department of Communications,
    Optics and Materials
Technical University of Denmark
Denmark
qz@com.dtu.dk

Xiongwen Zhao
Elektrobit
Keilasatama 5
02150 Espoo
Finland
xiongwen.zhao@elektrobit.com

Andreas Ziller
Siemens AG
Corporate Technology, Information and
    Communication
Otto-Hahn-Ring 6
81730 Munich
Germany
andreas.ziller@siemens.com

David Zorrilla
DAS Photonics S.L.
Camino de Vera s/n
Building 8F
46022 Valencia
Spain
dzorrilla@dasphotonics.com

Foreword

Mobile WiMAX: the Enabler for the Mobile Internet Revolution

The Internet has become one the most important assets for the growth of economies across the globe. More than a billion people use the Internet at their workplace and in their daily lives for business interactions, social interactions and entertainment. The Internet has had a profound effect on the economy of developed and developing nations having made economic activity more efficient, accessible and affordable. Most of the productivity gains in today's economies are thanks to the Internet and ecommerce. There have been profound social impacts from increased the access to valuable information and social interaction between the masses. The impact is at many socioeconomic levels: business productivity, energy savings, healthcare delivery, improved government functions, education, improved citizen interactions (locally and globally), etc. Despite the benefits of the Internet, today only about 20% of the World's population have access to the Internet. In particular, the emerging countries that could benefit greatly are seriously deprived of this valuable asset. There are a number of reasons for the small number of users in the emerging countries: lack of infrastructure, affordability of personal computers, unaffordable access fees, etc.

The next big step in the evolution of the Internet is ubiquitous availability enabled through mobile Internet. This revolutionary step is poised to increase the value of the Internet enormously as it will create a fundamental shift in the use of the Internet by bringing the Internet to the users as opposed to users having to go to the Internet. For this vision to become a reality, a number of requirements need to be met. First and foremost, affordable and ubiquitous mobile Internet access needs to be provided using the mobile cellular concept. This is poised to be fulfilled thanks to mobile WiMAX. Secondly, affordable and low-power mobile Internet devices and mobile PCs are needed. This is also happening with the computer industry making huge strides in making these devices more affordable. The low-cost netbook category with examples such as the ASUS Eee PC and variety of small mobile PCs or Mobile Internet Devices (MIDs) are now available and will undoubtedly become even more affordable in the near future.

Mobile WiMAX has been designed with the purpose of enabling mobile Internet from the physical layer to the network layer. The physical layer design relies on Orthogonal Frequency Division Multiple Access (OFDMA) and Multiple Input Multiple Output (MIMO) as the two key technologies to optimize coverage and spectral efficiency. In addition, sophisticated techniques for link adaptation and error control provide improved performance and robustness. Mobile WiMAX technology includes many other important aspects such as security and power-saving methods, provisions for location-based services, support for hierarchical deployments, quality-of-service, and open Internet user and network management schemes, which are essential in enabling deployment and consumer adoption of the technology.

The Internet is dynamic by nature and is evolving rapidly on the application level and creating ever-increasing demands on connectivity. Studies indicate that Internet traffic has been doubling roughly every two years. Mobile Internet will undoubtedly change the Internet as we know it today and may create even more traffic than ever anticipated. Mobile WiMAX needs to evolve constantly to keep up with the growth of mobile Internet. The WiMAX industry has already been working on the next technology in IEEE 802.16m to build the basis for the next generation of mobile Internet.

This book provides the material that is essential to understand the underlying concepts for mobile WiMAX and it also provides an overview of technologies that will enable the evolution of the technology in the future. I sincerely hope that the book will further motivate researchers and developers to create innovative ideas and techniques that will help fulfill the promise of the new era of mobile Internet.

Siavash M. Alamouti, Intel Fellow
Chief Technology Officer, Mobile Wireless Group

Preface

The remarkable development of wireless and mobile communications in the last two decades is a unique phenomenon in the history of technology. Even the most optimistic predictions on penetration of mobile subscribers and capabilities of wireless devices have been surpassed by reality. In a quarter of century the number of mobile subscribers soared from a few to half the world population (in 2008), and according to some forecasts by 2010 the number of mobile users will exceed the number of toothbrush users (four billion). The Wireless World Research Forum (WWRF) envisions that by year 2017 there will be seven trillion wireless devices serving seven billion people. Two main development directions in untethered communications can be identified, wide-area communications, with the omnipresent cellular systems as the most representative example, and short-range communications, involving an array of networking technologies for providing wireless connectivity over short distances, for instance Wireless Local Area Networks (WLANs), Wireless Personal Area Networks (WPANs), Wireless Body Area Networks (WBANs), Wireless Sensor Networks (WSNs), Bluetooth, etc. Recent years have witnessed an enormous growth in interest in the metropolitan wireless networks. This should not be a surprise, as in 2008, for the first time in history more than half of the world population lives in urban areas, according to the United Nations Population Fund. WiMAX (Worldwide Interoperability for Microwave Access) is the most representative worldwide initiative focusing on metropolitan communications. WiMAX, based on the IEEE 802.16 standard, defines wireless networks combining key characteristics of wide-area cellular networks as well as short-range networks, namely mobility and high data throughput. IEEE 802.16 is a very active and rapidly evolving standard that serves as the fundamental basis for WiMAX systems. Several amendments are currently being developed addressing particular technical aspects or capabilities, including 802.16g, 802.16h, 802.16i, 802.16j, 802.16k and 802.16m. There are already several books dealing with WiMAX technology, describing mostly the basic operating principles, current standards and associated technical solutions. The current vertiginous developments in the WiMAX arena have lead the Editors to conceive of this book, taking over where most of the published WiMAX volumes left off, that is, looking in future directions. Leading research scientists and engineers from key WiMAX industry, academia and research centers worldwide have contributed to this book with their ideas, concepts, concrete technical suggestions and visions.

As WiMAX as a whole encompasses a very broad area, it is impossible to find a single author able to write in detail about a large array of advanced concepts and solutions applicable at different system levels of WiMAX: the Editors have thus invited specialists in the field to contribute with their ideas in different chapters. The goal of this book is

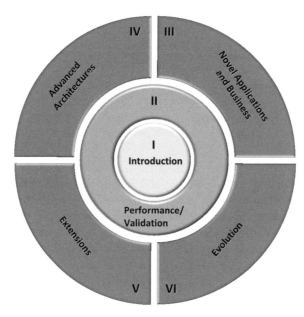

Figure 1 WiMAX evolution: organization of the book.

to create concrete supportive links between the presented concepts and future metropolitan communication systems, discussing technical solutions as well as novel identified scenarios, business applications and visions that are likely to become integral parts of the future WiMAX. Thus, this book tries to answer questions including the following. Which are the emerging WiMAX technologies that are being developed? What are the new scenarios for deploying WiMAX? What are the most promising WiMAX applications and business? How are standards evolving? What are the visions of industry? What are the capabilities and measured performance of real (commercial) WiMAX systems?

As shown in Figure 1, this book has been organized into six independent parts, covering different aspects of WiMAX technology and its evolution. Part One overview of the current state of WiMAX technology, serving as an introduction to WiMAX. Part Two presents measurements and validation results carried out on real state-of-the-art WiMAX testbeds (fixed and mobile), providing unique results on the achievable capabilities of commercial equipment operating in real scenarios. Novel scenarios and business cases for WiMAX are considered in Part Three. In Part Four new promising architectures for WiMAX are discussed, including wireless sensor networks, mesh and cooperative networking as well as femtocells. Part Five discusses several extensions to the current WiMAX, that is, new solutions that can be used in conjunction with the current WiMAX standard. Finally, Part Six looks into technical developments beyond the immediate WiMAX future, including PHY and MAC evolution, prospects and visions, emerging technologies, evolution of standards, etc.

WiMAX Evolution: Emerging Technologies and Applications is a book intended for research, development and standardization engineers working in industry, as well as for scientists in academic and research institutes. Graduate students conducting research in

WiMAX and next generation mobile communications will also find in this book relevant material for further research. The Editors think that this book provides novel views and detailed technical solutions, foreseeing future WiMAX while being a stimulating source of inspiration for further advanced research in the field.

The Editors welcome any suggestions, comments or constructive criticism on this book. Such feedback will be used to improve forthcoming editions. The Editors can be contacted at wimaxeditor@es.aau.dk.

Marcos D. Katz
VTT (Technical Research Centre of Finland), Finland

Frank H.P. Fitzek
Aalborg University, Denmark

September 2008

Acknowledgements

At times, our own light goes out, and is rekindled by a spark
from another person. Each of us has cause to think with deep
gratitude of those who have lighted the flame within us.

Albert Schweitzer

The Editors are deeply indebted to each and every contributor to this book. Without the valuable contributions and enthusiastic participation of specialists around the globe this book would have never been possible. We wish to place on record our deep appreciation to all of the authors of the chapters, who are, in alphabetical order:

Sassan Ahmadi, Alexander Bachmutsky, Sandrine Boumard, Aaron Byman, Thomas Bohnert, Paola Cardamone, Thomas Casey, Chan-Byoung Chae, Francesco Chiti, Juan Luis Corral, Marília Curado, Suvra Sekhar Das, Michael Devetsikiotis, Romano Fantacci, Francisco Fontes, Avraham Freedman, Ilkka Harjula, Matthias Hollick, Kaibin Huang, Jie Hui, Jyrki Huusko, Takao Inoue, Riku Jäntti, Giada Landi, Mika Lasanen, Moshe Levin, Roberto Llorente, Leonardo Maccari, Dania Marabissi, Javier Martí, Ricardo Matos, Parag S. Mogre, Edmundo Monteiro, Pedro Neves, Jari Nurmi, Ioannis Papanagiotou, Kostas Pentikousis, Jarno Pinola, Esa Piri, Valentín Polo, Jonás Porcar, Doug Pulley, Muhammad Imadur Rahman, Antonio Ramírez, Wonil Roh, Susana Sargento, Gerrit Schulte, Christian Schwingenschloegl, Patrick Seeling, Paulo Simões, Chris Smart, Clare Somerville, Roshni Srinivasan, Dirk Staehle, Ralf Steinmetz, Daniele Tarchi, Rath Vannithamby, Nenad Veselinovic, Borja Vidal, Yuanye Wang, Matti Weissenfelt, Vladimir Yanover, Qi Zhang, Xiongwen Zhao, Andreas Ziller and David Zorrilla.

We would like to express our gratitude to several people and organizations that supported this book. First, we are grateful to Mr Siavash Alamouti, Intel Fellow and CTO of the Mobile Wireless Group, for his motivating and enlightening foreword.

VTT, the Technical Research Centre of Finland, provided financial and logistical support for the preparation of this book. We are grateful to Technology Director Dr Jussi Paakkari, Technology Manager Kyösti Rautiola and Research Professor Dr Aarne Mämmelä for their unconditional support during this initiative. We also thank our research colleagues at VTT (Communications Platform Group, and in particular the Cooperative and Cognitive Networks Team) for their technical contributions, motivating discussions and for creating a truly pleasant working atmosphere. Our colleagues from the Converging Networks Laboratory (CNL) also deserve our deep appreciation, particularly Dr Marko Jurvansuu, Jyrki Huusko, Marko Palola, Dr Kostas Pentikousis and Dr Martin Varela Rico.

The European Project WEIRD (WiMAX Extension to Isolated Research Data Networks), coordinated and technically supervised by Enrico Angori (Datamat, Italy) and Marcos Katz, respectively, was the source of several chapters of this book. We are grateful to the WEIRD consortium and its people across Europe for the received support. For their support and enlightening discussions, we are also grateful to Gerrit Schulte (acticom, Germany), Kari Horneman (Nokia Siemens Networks, Finland), Dr Wonil Roh (Samsung Electronic Corp., Ltd, Korea), Dr Jaakko Talvitie (Elektrobit, Finland) and Professor Garik Markarian (Lancaster University, UK).

Parts of the book were financed by the X3MP project granted by the Danish Ministry of Science, Technology and Innovation. Furthermore we would like to thank our colleagues from Aalborg University, Denmark for their support, namely Børge Lindberg, Ben Krøyer, Peter Boie Jensen, Bo Nygaard Bai, Henrik Benner, Finn Hybjerg Hansen and Svend Erik Volsgaard.

The Editors would like to thank Nokia for providing invaluable technical support as well as mobile devices for testing purposes. Special thanks go to Harri Pennanen, Nina Tammelin and Per Møller from Nokia. We are grateful to Jarmo Tikka (Nokia) who kindly provided the N810 wireless tablets that were used in the measurement setup of Chapter 6. Particular thanks go to Alberto Bestetti and Antonio Cimmino (Alcatel-Lucent, Italy) and Arto Grönholm (Alcatel-Lucent, Finland) for support with the WiMAX equipment used in some of the measurement test-beds.

We thank John Wiley & Sons Ltd, for their encouragement and support during the process of creating this book. Special thanks to Tiina Ruonamaa, Anna Smart and Sarah Tilley for their kindness, patience, flexibility and professionalism. Birgitta Henttunen from VTT, Finland is acknowledged for her support in many administrative issues.

Finally, the Editors would like to thank their respective families for their support and understanding during the entire process of creating this book.

List of Acronyms

μC MicroController

16-QAM 16 Quadrature Amplitude Modulation

2G 2nd Generation

3G 3rd Generation

3GPP 3rd Generation Partnership Project

3GPP2 3rd Generation Partnership Project 2

4G Fourth Generation

A/V Audio/Visual

AAA Authentication, Authorization and Accounting

AAS Adaptive Antenna System

AC Admission Control; Antenna Circulation

ACIR Adjacent Channel Interference Ratio

ACK Acknowledgement

ACR Absolute Category Rating

ADSL Asymmetric Digital Subscriber Line

AG Antenna Grouping

AMC adaptive modulation and coding

AMR Adaptive Multi-Rate

AMS Adaptive MIMO Switching

AP Access Point

APD Adaptive Power Distribution

APFR Adaptive Power Fixed Rate

API Application Programming Interface

APMC Adaptive Power, Modulation and Coding

AQ Assessed QoS

ARP Address Resolution Protocol

ARQ Automatic Repeat Request

AS Antenna Selection

ASN Access Service Network

ASN-GW Access Service Network Gateway

ATM Asynchronous Transfer Mode

AVC Advanced Video Coding

AWGN Additive White Gaussian Noise

BD Block Diagonalization

BE Best Effort

BER Bit Error Rate

BF Beamforming

BGP Border Gateway Protocol (routing)

BLER Block Error Rate

BOM Bill Off Materials

bps Bits Per Second

BPSK Binary Phase Shift Keying

BS Base Station

BSID Base Station Identifier

BWA Broadband Wireless Access

C/I Carrier to Interference Ratio

CAPEX Capital Expenditures

CATV Cable Television

CBC Cipher Block Chaining

CBF Coordinated Beamforming

CBR Constant Bit Rate

CC Chase Combining; Convolutional Code; Coordination Center

CCF Call Control Function

CCP2P Cellular Controlled Peer to Peer

CDF Cumulative Distribution Function

CDL Clustered Delay Line

CDMA Code Division Multiplex Access

CELP Code Excited Linear Prediction

CH Compressed Header

C/I Carrier-to-Interference Ratio

CID Connection Identifier

CI-STBC Coordinate Interleaved Space–Time Block Code

CMIP Client Mobile IP

CN Correspondent Node

CN Core Network

CNL VTT Converging Networks Laboratory

CNR Channel-to-Noise Ratio

CoA Care-of-Address

CODEC Compression/Decompression

COST European Cooperation in the Field of Scientific and Technical Research

COTS Commercial Off The Shelf

CP Cyclic Prefix

CPE Customer Premises Equipment

CPS Common Part Sublayer

CPU Central Processing Unit

CQI Channel Quality Indicator

CQICH Channel Quality Indicator Channel

CRC Cyclic Redundancy Check

CS Convergence Sublayer

C-SAP Control Service Access Point

CSG Closed Subscriber Group

CSI Channel State Information

CSN Connectivity Services Network

CTS Clear to Send

DAS Distributed Antenna System

DCA Dynamic Channel Allocation

DCD Downlink Channel Descriptor

DCF Discounted Cash Flow

DES Data Encryption Standard

DFB Distributed Feedback

DHCP Dynamic Host Configuration Protocol

DL Downlink

DMTBR Dynamic Multiple-Threshold Bandwidth Reservation

DNS Domain Name System

DNS-SD Dynamic Name System Service Discovery

DPT Dirty Paper Theory

DRR Deficit Round Robin

DRX Discontinuous Reception

DS-CDMA Direct Sequence Code Division Multiple Access

DSL Digital Subscriber Line

DSLAM Digital Subscriber Line Access Multiplexer

DWRR Deficit Weighed Round Robin

EAP Extensible Authentication Protocol

ECMP Equal Cost Multi-Path

EDF Earliest Deadline First

EpBR Energy per Bit Ratio

ertPS Extended Real-Time Polling Service

ERT-VR Extended Real-Time Variable Rate

ESP Encapsulating Security Payload

ETX Expected Transmission Count

EVD Eigenvalue Decomposition

EVRC Enhanced Variable Rate Codec

FA Foreign Agent

FBSS Fast Base Station Switching

FCH Frame Control Header

FDD Frequency-Division Duplex

FDM Frequency Division Multiplexing

FEC Forward Error Correction

FER Frame Error Rate

FFMS Forest Fire Monitoring Station

FFT Fast Fourier Transform

FIFO First In First Out

FP Framework Programme

FPAR Fixed Power Adaptive Rate

FPGA Field-programmable Gate Array

FTP File Transfer Protocol

FUSC Fully Used Subcarriers

GA Generic Adapter

GIS Geographic Information Systems

GIST General Internet Signaling Transport

GMH Generic MAC Header

GoS Grade of Service

GPRS General Packet Radio Service

GPS Global Positioning System

GRE Generic Routing Encapsulation

GSM Global System for Mobile Communications

GTP GPRS Tunneling Protocol

GUI Graphical User Interface

GW Gateway

HA High Availability; Home Agent

HARQ Hybrid Automatic Repeat Request

HD High Definition

HFC Hybrid Fiber Coaxial

HFDD Half-duplex Frequency Division Duplex

HFR Hybrid Fiber Radio

HHO Hard Handover

HO Handover

HSDPA High Speed Data Packet Access

HSPA High Speed Packet Access

HSRP Hot Standby Router Protocol

HTTP Hyper Text Transfer Protocol

HW Hardware

ICMP Internet Control Message Protocol

ICT Information and Communication Technologies

ID Identification

IETF Internet Engineering Task Force

IFFT Inverse Fast Fourier Transform

IMDD Intensity Modulation, Direct Detection

IMS IP Multimedia Subsystem

IMT International Mobile Telecommnications

IP Internet Protocol

Ipsec Internet Protocol Security

IPTV Internet Protocol Television

IPv4 Internet Protocol version 4

IPv6 Internet Protocol version 6

IQ Intrinsic QoS

IQA Instrumental Quality Assessment

IRR Internal Rate of Return

ISD Inter-site Distance

IST Information Society Technologies

ITU International Telecommunications Union

kbps kilobits per second (1000 bits s^{-1})

KPI Key Performance Indicator

L1 Layer 1 (Physical Layer)

L2 Layer 2 (Data Link Layer)

L2TP Layer 2 Tunneling Protocol

LA Link Adaptation

LACP Link Aggregation Control Protocol

LAG Ling Aggregation

LAN Local Area Network

LBC Load Balancing Cycle

LBS Location Based Services

LDAP Lightweight Directory Access Protocol

LLA Low Level Agent

LLL Lenstra–Lenstra–Lovász

LOS Line-of-Sight

LPC Linear Predictive Coding

LPM Loss Packet Matrix

LSB Least Significant Bit

LTE Long Term Evolution

LU Lenstra–Lenstra–Lovász

MAC Medium Access Control

MAN Metropolitan Area Network

MAP Medium Access Protocol; Mobile Application Part

MBAC Measurement Based Admission Control

MBB Make Before Break

MBMS Multimedia Broadcast Multicast Service

Mbps Megabits per second (1 000 000 bits s^{-1})

MBS Mesh Base Station; Multicast and Broadcast Service

MCBCS Multicast and Broadcast Service

MCS Modulation and Coding Scheme

MCW Multi Codeword

MDHO Macro Diversity Handover

MeSH IEEE 802.16-2004 Mesh Mode

MIB Management Information Base

MICS Media Independent Command Service

MIES Media Independent Event Service

MIH Media Independent Handover

MIHF Media Independent Handover Function

MIHO Mobile Initiated Handover

MIHU Media Independent Handover User

MIIS Media Independent Information Service

MIMO Multiple Input Multiple Output

MIP Mobile Internet Protocol

ML Maximum Latency

MLD Maximum Likelihood Decoder

MLI Modulation Level Information

MM Mobility Management

MMF Multimode Fiber

MMR Mobile Multihop Relay

MMSE Minimum Mean Square Error

MN Mobile Node

MOS Mean Opinion Score

MPEG Moving Picture Experts Group

MRC Maximum Ratio Combining

MRT Maximum Ratio Transmission

MRTR Minimum Reserved Traffic Rate

MS Mobile Station

M-SAP Management Service Access Point

MSB Most Significant Bit

MSDU MAC Service Data Unit

MSE Mean Square Error

MSID Mobile Subscriber ID

MSTR Maximum Sustained Traffic Rate

MTBF Mean Time Between Failures

MTU Maximum Transmission Unit

NACK Negative Acknowledgement

NAI Network Access Identifier

NC Network Coding

NCMS Network Control and Management System

NDCQ Nondegenerate Constraint Qualification

NE Network Element

NET Network Layer

NGMN Next-Generation Mobile Network

NGN Next Generation Network

NIHO Network Initiated Handover

NLOS Non-Line-of-Sight

NMS Network Management System

NPV Net Present Value

NRM Network Reference Model

nrt Non-real-time

nrtPS Non-real-time Polling Service

NSIS Next Steps in Signaling

NSLP NSIS Signaling Layer Protocol

NTLP NSIS Transport Layer Protocol

NTP Network Time Protocol

NWG Network Working Group

O&M Operations and Management

OFDM Orthogonal Frequency Division Multiplexing

OFDMA Orthogonal Frequency Division Multiple Access

OGBF Orthogonal Beamforming

OMC Operation and Maintenance Center

OMF Operation and Maintenance Function

OPEX Operational Expenditures

OSPF Open Shortest Path First

P2P Peer to Peer

PA ITU Pedestrian A

PB ITU Pedestrian B

PAN Personal Area Network

PAPR Peak to Average Power Ratio

PBE Perfect Bayesian Equilibrium

PC Paging Controller; Power Control

PCM Pulse Code Modulation

PDA Personal Digital Assistant

PDU Protocol Data Unit

PEP Performance Enhancing Proxy

PER Packet Error Rate

PHB Per Hop Behavior

PHY Physical Layer

PLC Packet Loss Concealment

PLR Packet Loss Rate

PMIP Proxy Mobile IP

PMP Point to Multipoint

PN Psedorondam Noise

POF Plastic Optical Fiber

PQ Perceived QoS

PSTN Public Switched Telephone Network

PTMP Point-to-Multipoint

PTP Precision Time Protocol

PTP Point-to-point

PU2RC Per-User Unitary and Rate Control

PUSC Partially Used Subcarrier; Partially Used Subchannelization

QAM Quadrature Amplitude Modulation

QoE Quality of Experience

QoS Quality of Service

QPSK Quadrature Phase-Shift Keying

RADIUS Remote Authentication Dial-In User Service

RAN Radio Access Network

RAU Remote Antenna Unit

RB Resource Block

RF Radiofrequency

RFC Request for Comments (IETF standard document)

RMF Resource Management Function

RMS Root Mean Square

RoF Radio-over-Fiber

ROHC Robust Header Compression

RRM Radio Resource Management

RS Relay Station

RSS Received Signal Strength

RSSI Received Signal Strength Indicator

rt real-time

RTP Real-time Transport Protocol

rtPS Real-Time Polling Service

RTS Request to Send

RTT Round Trip Time

RT-VR Real-Time Variable Rate

Rx Receive

SA Specific Adapter

SAF Service Availability Forum

SAMPDA Simple Adaptive Modulation and Power Adaptation Algorithm

SAP Service Access Point

SBS Serving Base Station

SC Serra do Carvalho

SCM Spatial Channel Model

SCR Spare Capacity Report

SCTP Stream Control Transmission Protocol

SCW Single Codeword

SDMA Spatial Division Multiple Access

SDU Service Data Unit

SE Spectral Efficiency

SF Service Flow

SFDR Spurious Free Dynamic Range

SFM Service Flow Management

SID Silent Insertion Descriptor

SINR Signal-to-Interference + Noise Ratio

SIP Session Initiation Protocol

SISO Single Input Single Output

SL Serra da Lousã

SLA Service Level Agreement

SM Spatial Multiplexing

SMF Singlemode Fiber

SMS Short Message Service

SNMP Simple Network Management Protocol

SNR Signal-to-Noise Ratio

S-OFDMA Scalable Orthogonal Frequency Division Multiple Access

SOHO Small Office/Home Office

SON Self-Organized Network

SP Synchronization Pattern

SRA Simple Rate Adaptation

SRD System Requirement Document

SS Subscriber Station

SSL Secure Socket Layer

STBC Space Time Block Coding

STC Space-Time Coding

SUI Standford University Interim

SW Software

TBS Target Base Station

TCP Transmission Control Protocol

TDD Time Division Duplex

TDM Time Division Multiplexing

TDMA Time Division Multiple Access

TEM Telecommunications Equipment Manufacturer

TETRA Terrestrial Trunked Radio

TTI Transmission Time Interval

TTP Trusted Third Party

TWG Technical Working Group

Tx Transmit

UC University of Coimbra

UCD Uplink Channel Descriptor

UDP User Datagram Protocol

UGS Unsolicited Grant Service

UL Uplink

UMB Ultra Mobile Broadband

UMTS Universal Mobile Telecommunications System

UMTS-LTE Universal Mobile Telecommunications Systems – Long Term Evolution

VAD Voice Activity Detection

VBR Variable Bit Rate

VCEG Video Coding Experts Group

VCSEL Vertical Cavity Surface Emitting Laser

VDT Virtual Drive Test

VLSI Very-Large-Scale Integration

VoD Video on Demand

VoIP Voice over Internet Protocol

VP Vector Perturbation

VR Virtual Router

VRRP Virtual Router Redundancy Protocol

W3GPP third generation partnership project

WAC Wireless Access Controller

WDM Wavelength Division Multiplexing

WEIRD WiMAX Extension to Isolated Research Data Networks

WEP Wired Equivalent Privacy

WiFi Wireless Fidelity

WiMAX Worldwide Interoperability for Microwave Acccess

WINNER Wireless World Initiative New Radio

WLAN Wireless Local Area Network

W-LSB Windowed Least Significant Bits

WMAN Wireless Metropolitan Area Network

WMN Wireless Mesh Network

WNC Wireless Network Coding

WNEA WiMAX Network Element Advertisement

WPAN Wireless Personal Area Network

WRR Weighted Round Robin

WSN Wireless Sensor Network

WT WiMAX Terminal

WWRF Wireless World Research Forum

ZFBF Zero-Forcing Beamforming

Part I

Introduction

1

Introduction to WiMAX Technology

Wonil Roh and Vladimir Yanover

WiMAX stands for Worldwide Interoperability for Microwave Access. WiMAX technology enables ubiquitous delivery of wireless broadband service for fixed and/or mobile users, and became a reality in 2006 when Korea Telecom started the deployment of a 2.3 GHz version of mobile WiMAX service called WiBRO in the Seoul metropolitan area to offer high performance for data and video. In a recent market forecast published in April 2008, WiMAX Forum Subscriber and User Forecast Study, the WiMAX Forum projects a rather aggressive forecast of more than 133 million WiMAX users globally by 2012 (WiMAX Forum, 2008c). The WiMAX Forum also claims that there are more than 250 trials and deployments worldwide.

The WiMAX Forum is an industry-led non-profit organization which, as of the 1st quarter of 2008, has more than 540 member companies including service providers, equipment vendors, chip vendors and content providers. Its primary mission is to ensure interoperability among IEEE 802.16 based products through its certification process.

The air interface of WiMAX technology is based on the IEEE 802.16 standards. In particular, the current Mobile WiMAX technology is mainly based on the IEEE 802.16e amendment (IEEE, 2006a), approved by the IEEE in December 2005, which specifies the Orthogonal Frequency Division Multiple Access (OFDMA) air interface and provides support for mobility.

The selection of features to be implemented in WiMAX systems and devices is presented in the mobile WiMAX System Profile Release 1.0 (WiMAX Forum, 2007) which was developed in early 2006 and is currently maintained by the WiMAX Forum (WiMAX Forum, 2008a). It is this very technology defined in WiMAX Forum (2007) that was adopted by International Telecommunications Union (ITU) as the 6th air interface of IMT-2000 family

WiMAX Evolution: Emerging Technologies and Applications   Edited by Marcos D. Katz and Frank H.P. Fitzek
© 2009 John Wiley & Sons, Ltd

(ITU, 2007). The flexible bandwidth allocation and multiple built-in types of Quality-of-Service (QoS) support in the WiMAX network allow the provision of high-speed Internet access, Voice Over IP (VoIP) and video calls, multimedia chats and mobile entertainment. In addition, the WiMAX connection can be used to deliver content to multimedia gadgets such as the iPod.

Since the completion of the Release 1.0 Mobile System Profile, the WiMAX Forum has been working on a certification program which is a critical step for the proliferation of any modern communication technology throughout the world. As the result, the first WiMAX Forum Certified Seal of Approval for the 2.3 GHz spectrum was awarded to four base stations and four mobile stations in April 2008 (WiMAX Forum, 2008d). In June 2008, another four base stations and six mobile stations were awarded the WiMAX Forum Certified Seal of Approval for the 2.5 GHz spectrum with advanced features such as Multiple Input Multiple Output (MIMO) in time for commercial deployments around the world (WiMAX Forum, 2008e).

This chapter is intended to provide a high-level overview of the current mobile WiMAX technology with an emphasis on the Physical (PHY) layer and Medium Access Control (MAC) layer features. Some recent discussions and developments of further WiMAX evolution path are also addressed briefly at the end of the chapter.

1.1   Overview of State-of-the-art WiMAX Technology

1.1.1   Structure of the System Profile

As stated earlier, mobile WiMAX products and certification follow the IEEE 802.16 air interface specifications. The network specifications of mobile WiMAX products, however, are being developed internally by the WiMAX Forum, which include the end-to-end networking specifications and network interoperability specifications. The Network Working Group (NWG) within the WiMAX Forum is responsible for these network specifications, some of which involve Access Service Network (ASN) control and data plane protocols, ASN profiles, Connectivity Services Network (CSN) mobility support, Authentication, Authorization and Accounting (AAA) interworking with other technologies, and various services such as Location-Based Service (LBS), Multicast and Broadcast Service (MCBCS) etc. In this chapter, however, we will focus on the overview of mobile WiMAX technology from the air interface perspective.

Figure 1.1 presents the aforementioned composition of the current mobile WiMAX technology, commonly referred to as Release 1.0 profile. Its air interface specifications consist of four related IEEE 802.16 Broadband Wireless Access Standards, that is, IEEE Standard 802.16-2004, IEEE Standard 802.16-2004/Cor.1-2005, IEEE Standard 802.16e-2005 and the IEEE Draft Standard P802.16-2004/Cor.2.

Not all of the optional features defined in these IEEE Standards are implemented in WiMAX products and tested for certifications. Through extensive technical investigation analysis to build up the best competitive products, the WiMAX Forum Technical Working Group (TWG) published the first version of mobile WiMAX System Profile Release 1 in early 2006 (WiMAX Forum, 2007). The latest published version to date (Release 10 rev. 1.6.1) incorporated error fixes and minor corrections without touching the main features selected in the first revision.

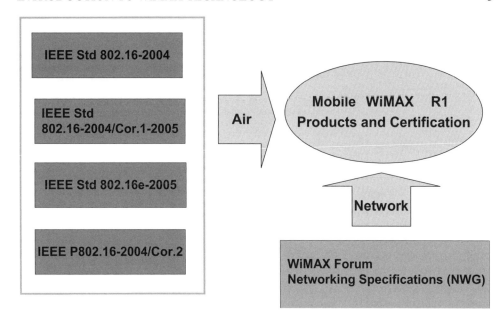

Figure 1.1 Mobile WiMAX Release 1.0 products and certification.

In Figure 1.2, a more detailed view of the construction of the mobile WiMAX system profile is presented from the air interface perspective.

The system profile is composed of five subprofiles, namely, PHY, MAC, radio, duplexing mode and power classes. Even though there are many different combinations of center frequencies and channel bandwidths accommodating different regional spectrum regulations, all Release 1 mobile WiMAX products share the same PHY and MAC features (profiles) and the same duplexing mode which is Time Division Duplex (TDD). In the following, some detailed descriptions of key PHY and MAC features in the mobile WiMAX system profile are offered.

1.1.2 Key PHY Features

In the following we give the key PHY features of mobile WiMAX technology and provide short descriptions.

1.1.2.1 Scalable OFDMA

OFDMA is the multiple access technique for mobile WiMAX. OFDMA is the Orthogonal Frequency Division Multiplexing (OFDM) based multiple access scheme and has become the *de-facto* single choice for modern broadband wireless technologies adopted in other competing technologies such as 3GPP's Long Term Evolution (LTE) and 3GPP2's Ultra Mobile Broadband (UMB). OFDMA demonstrates superior performance in non-line-of-sight (N-LOS) multi-path channels with its relatively simple transceiver structures and allows efficient use of the available spectrum resources by time and frequency subchannelization.

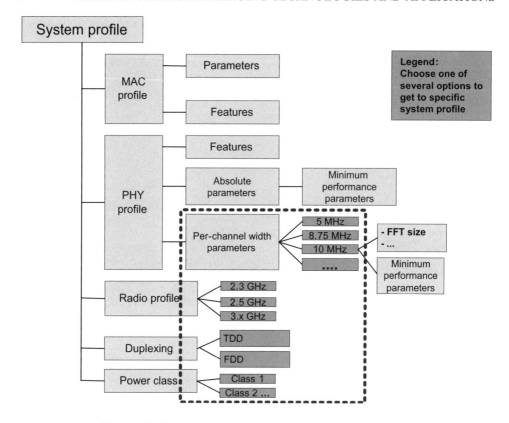

Figure 1.2 Structure of the mobile WiMAX system profile.

The simple transceiver structure of OFDMA also enables feasible implementation of advanced antenna techniques such as MIMO with reasonable complexity. Lastly, OFDMA employed in mobile WiMAX is scalable in the sense that by flexibly adjusting FFT sizes and channel bandwidths with fixed symbol duration and subcarrier spacing, it can address various spectrum needs in different regional regulations in a cost-competitive manner.

1.1.2.2   TDD

The mobile WiMAX Release 1 Profile has only TDD as the duplexing mode even though the baseline IEEE Standards contain both TDD and Frequency Division Duplex (FDD). Even though future WiMAX Releases will have FDD mode as well, TDD is in many ways better positioned for mobile Internet services than FDD.

First of all, Internet traffic is asymmetric typically with the amount of downlink traffic exceeding the amount of uplink traffic; thus, conventional FDD with the same downlink and uplink channel bandwidth does not provide the optimum use of resources. With TDD products, operators are capable of adjusting downlink and uplink ratios based on their service needs in the networks.

In addition, TDD is inherently better suited to more advanced antenna techniques such as Adaptive Antenna System (AAS) or Beamforming (BF) than FDD due to the channel reciprocity between the uplink and downlink. Mobile Internet with increased multimedia services naturally requires the use of advanced antenna techniques to improve capacity and coverage.

1.1.2.3 Advanced Antenna Techniques (MIMO and BF)

Various advanced antenna techniques have been implemented in the mobile WiMAX Release 1 profile to enable higher cell and user throughputs and improved coverage. As a matter of fact, mobile WiMAX was the first commercially available cellular technology that actually realized the benefits of MIMO techniques promised by academia for years. With its downlink and uplink MIMO features, both operators and end-users enjoy up to twice the data rates of Single-Input Single-Output (SISO) rate, resulting in up to 37 Mbps for downlink and 10 Mbps for uplink sector throughput using just 10 MHz TDD channel bandwidth.

Mobile WiMAX also enhances the cell coverage with its inherent BF techniques. Coupled with TDD operation, its powerful BF mechanism allows base stations to accurately form a channel matching beam to a terminal station so that uplink and downlink signals can reach reliably from and to terminals at the cell edge, thus effectively extending the cell range.

1.1.2.4 Full Mobility Support

Full mobility support is yet another strength of the mobile WiMAX products. The baseline standard of mobile WiMAX was designed to support vehicles at highway speed with appropriate pilot design and Hybrid Automatic Repeat Request (HARQ), which helps to mitigate the effect of fast channel and interference fluctuation. The systems can detect the mobile speed and automatically switch between different types of resource blocks, called subchannels, to optimally support the mobile user. Furthermore, HARQ helps to overcome the error of link adaptation in fast fading channels and to improve overall performance with its combined gain and time diversity.

1.1.2.5 Frequency Reuse One and Flexible Frequency Reuse

From the operators' perspective, securing greater frequency spectrum for their services is always costly. Naturally it is in their best interest if a technology allows decent performance in the highly interference-limited conditions with frequency reuse one. Mobile WiMAX technology was designed to meet this goal in a respectable way with its cell-specific subchannelization, low rate coding and power boosting and deboosting features. It also enables real-time application of flexible frequency reuse where frequency reuse one applied to terminals close to the cell center whereas a fraction of frequency is used for terminals at the cell edge, thereby reducing heavy co-channel interference.

1.1.3 Key MAC Features

The MAC layer of mobile WiMAX (802.16e) technology includes the following features which provide for high efficiency and flexibility.

1.1.3.1   Connection-based Data Transmission with Classification and QoS per Connection

The WiMAX technology provides an environment for connection-oriented services. For each service, certain classification rules are specified to define the category of traffic associated with the connection. For example, it could be Internet Protocol (IP) traffic destined for a specific IP address/port. For each connection, certain QoS parameters are defined, for example, the minimum reserved rate and maximum sustained rate. There are several types of scheduling such as real-time services that can be applied based on the application requirements. A special scheduling type (ertPS) is defined for the VoIP service with silence suppression and adaptive codecs.

1.1.3.2   Scheduled Transmissions and the Flexible Bandwidth Allocation Mechanism

Bandwidth allocation mechanism is based on real time bandwidth requests transmitted by the terminals, per connection. Bandwidth requests may be transmitted using a contention based mechanism or they can be piggybacked with the data messages. The Base Station executes resources allocation based on the requests and QoS parameters of the connection.

1.1.3.3   MAC Overhead Reduction

WiMAX technology includes support of the general Purpose Header Suppression (PHS) and IP Header Compression (ROHC). PHS can be used for packets of virtually any format such as IPv4 or IPv6 over Ethernet. It is beneficial if a considerable part of the traffic has identical headers which is typical for IP or Ethernet destination addresses. The PHS mechanism replaces the repeated part of the header with a short context identifier, thus reducing the overhead associated with headers.

ROHC is a highly efficient IETF standard for which WiMAX MAC has all necessary support.

1.1.3.4   Mobility Support: Handover

Handover procedures include numerous means of optimization. In particular, to reduce time expenses for the mobile to find the central frequency and acquire parameters of the neighbor base station, the mobile can apply a scanning process when the mobile is away from the serving base station to scan the wireless media for neighbor base stations. Information collected during scanning such as central frequencies of the neighbor base stations can then be used in actual handover. In some deployment scenarios, scanning can be performed without service interruption. For this purpose, information about the central frequency and parameters of the neighbor base stations is periodically advertised by the serving base station.

To shorten the time needed for the mobile to enroll into the new cell the network is capable of transferring the context associated with the mobile from the serving base station to the target base station.

All of these means provide a potential for high optimization in terms of handover latency. Under ideal conditions the interval of service interruption may be as short as several 5 ms frames. The specific handover optimization scheme used in a particular handover depends on the information available to the mobile.

1.1.3.5 Power Saving: Sleep Mode

Sleep mode is the primary procedure for power saving. In sleep mode the mobile is away from the base station for certain time intervals, normally of exponentially increasing size. During these intervals the mobile remains registered at the base station but can power down certain circuits to reduce power consumption.

1.1.3.6 Power Saving: Idle Mode

If the mobile has no traffic for a long time it can switch to idle mode in which it is no longer registered at any particular base station. To resume traffic between the network and the mobile, a paging procedure may be used by the network.

1.1.3.7 Security

The security sublayer provides Extensible Authentication Protocol (EAP)-based mutual authentication between the mobile and the network. It protects against unauthorized access to the transferred data by applying strong encryption of data blocks transferred over the air. To keep the encryption keys fresh, the security sublayer employs an authenticated client/server key management protocol which allows the base station to distribute keying material to mobiles. Basic security mechanisms are strengthened by adding digital-certificate-based Subscriber Station (SS) device authentication to the key management protocol.

1.1.3.8 MAC Layer Support for the Multicast and Broadcast Service

Multicast and Broadcast Services (MBSs) allow WiMAX mobile terminals to receive multicast data even when they are in idle mode. The most popular application of this feature is TV broadcasting to mobile terminals.

1.1.4 Advanced Networking Features

The WiMAX Forum developed specifications of the network infrastructure which complement the 802.16e specifications of the air interface (see WiMAX Forum (2008b)).

1.2 WiMAX Evolution Path

Figure 1.3 provides an overview of mobile WiMAX roadmaps for standards and products. The first release labeled as Release 1.0 is described earlier in this chapter. The other two, Releases 1.5 and 2.0, are short-term and long-term migration, respectively, and their brief summaries are provided in this section. The corresponding IEEE baseline standards for Releases 1.5 and 2.0 are IEEE 802.16 REV2 (IEEE, 2008) and IEEE 802.16m (IEEE, 2006b), respectively. Owing to the dependency on the ongoing IEEE standards, REV2 and 16m, the schedules of Releases 1.5 and 2.0 are projections by the authors and may change.

Each new generation of the technology needs changes in the profile and/or the standard itself.

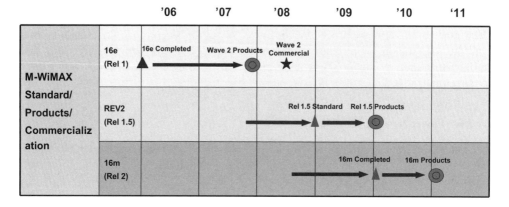

Figure 1.3 Roadmap of mobile WiMAX standards and products.

1.2.1   Release 1.5

The WiMAX Forum is currently working on the short-term migration of the profile called Release 1.5. Generally it is focused on optimization introducing optimized FDD/HFDD operations and features that can be added to Release 1.0 WiMAX devices through a software upgrade. This generation includes additional selection of 802.16 elements available through the ongoing IEEE 802.16 REV2 standard to address the following.

1.2.1.1   Efficient FDD/HFDD Operations

Optimization of FDD/HFDD operations is based on splitting the 802.16 frame into partitions to be used by two distinct groups of mobiles having separated the control channels such as downlink and uplink MAPs, fast feedback channels and HARQ ACK channels. Such a solution allows for the reuse of the design of Release 1.0 (TDD) chipsets while not compromising on the system performance in order to address FDD markets around the world.

1.2.1.2   New Band Classes

New band classes (introduced in WiMAX Forum Certification Profiles) will be added in Release 1.5, mostly to provide a solution for FDD bands.

1.2.1.3   Enhanced MIMO/BF Operations

Closed-loop operations for MIMO and BF are optionally considered to further improve the throughput and coverage beyond Release 1.0 which contains only open-loop MIMO and BF features.

1.2.1.4   Enhanced MAC Performance (Particularly Improved VoIP Capacity)

Release 1.0 is highly optimized to data communication such as TCP/IP. The nature of the data traffic implies 'bursty' transmission demand. To properly serve such a demand, Release 1.0

Table 1.1 Key Requirements of IEEE 802.16m.

Item	Requirements
Carrier frequency	Licensed band under 6 GHz
Operating bandwidth	5–20 MHz Other bandwidths can be considered as necessary
Duplex	Full-duplex FDD, Half-duplex FDD, TDD
Antenna Technique	Downlink $\geq$ (2Tx, 2Rx) Uplink $\geq$ (1Tx, 2Rx)

Peak data rate (peak spectral efficiency)	Type	Link direction	MIMO	Normalized peak data rate (bps/Hz)
	Baseline	Downlink	2×2	8.0
		Uplink	1×2	2.8
	Target	Downlink	4×4	15.0
		Uplink	2×4	5.6

Item	Requirements
Data latency	Downlink < 10 ms, Uplink < 10 ms
State transition latency	max 100 ms
Handover interruption time	Intra-frequency handover latency <30 ms Inter-frequency handover latency <100 ms

Throughput and VoIP capacity	Downlink	Uplink
Average sector throughput (bps/Hz/sector)	2.6	1.3
Average user throughput (bps/Hz)	0.26	0.13
Cell edge user throughput (bps/Hz)	0.09	0.05
VoIP capacity (active calls/MHz/sector)	30	30

Item	Requirements
MBS Spectral efficiency	Inter-base station distance 0.5 km > 4 bps/Hz Inter-base station distance 1.5 km > 2 bps/Hz
Enhanced MBS	Max MBS channel reselection interruption times: intra-frequency <1 s; inter-frequency <1.5 s
LBS Position accuracy	Handset-based: 50 m (67% of cdf), 150 m (95% of cdf) Network-based: 100 m (67% of cdf), 300 m (95% of cdf)

technology uses the mechanism of downlink and uplink MAPs which are control messages transmitted each frame, that is, every 5 ms. While this is perfect for bursty traffic, support for streaming (VoIP, video) data needs further optimization. The idea for optimization is to use persistent resource allocation so that a single MAP message provides information on periodic resources assignment matching the needs of a specific stream.

1.2.1.5   Extended and Enhanced Networking Features

Most of the extensions are related to MBSs. The Release 1.5 extension provides for more flexible allocation of MBS zones which is suitable also for small (micro and pico) cells. Another attractive part of Release 1.5 is the set of features supporting LBSs.

1.2.1.6   Support for WiMAX and WiFi: Bluetooth Coexistence in the Same Mobile

Special attention is paid to provide more efficient support to WiMAX terminals having additional wireless Local Area Network (LAN) and/or Personal Area Network (PAN) interfaces. As the timing of WiFi or Bluetooth interfaces does not match the timing of the WiMAX interface, special arrangements are needed to, for example, prevent the Bluetooth transmitter from interfering with the WiMAX receiver and vice versa.

1.2.2   Release 2.0

As was mentioned earlier, the long-term migration from Release 1.0 is known as Release 2.0 and the corresponding specification is being developed in the IEEE 802.16m project (IEEE, 2006b). Unlike Release 1.5, which focused on FDD/HFDD and software-based additions to Release 1.0, the goal of Release 2.0 is to meet International Mobile Telecommnications (IMT)-Advanced requirements for next-generation mobile networks which will be done by providing major improvements in all areas. The requirements for 16m can be found in (IEEE, 2007) and Table 1.1 summaries some of the key requirements of 16m.

References

IEEE (2006a) IEEE Std 802.16e-2005. Amendment to IEEE Standard for Local and Metropolitan Area Networks – Part 16: Air Interface for Fixed Broadband Wireless Access Systems – Physical and Medium Access Control Layers for Combined Fixed and Mobile Operation in Licensed Bands, February.

IEEE (2006b) IEEE P802.16. IEEE Standard for Local and Metropolitan Area Networks – Part 16: Air Interface for Fixed Broadband Wireless Access Systems – Amendment: Advanced Air Interface, December.

IEEE (2007) IEEE 802.16m. System Requirements, http://www.wirelessman.org/tgm/docs/80216m-07_002r4.pdf, October.

IEEE (2008) IEEE P802.16Rev2. Revision of IEEE Std 802.16-2004 and consolidates material from IEEE Std 802.16e-2005, IEEE Std 802.16-2004/Cor1-2005, IEEE Std 802.16f-2005 and IEEE Std802.16g-2007, June.

ITU (2007) Recommendation ITU-R M.1457. Detailed Specifications of the Radio Interfaces of International Mobile Telecommunications-2000 (IMT-2000).

WiMAX Forum (2007) WiMAX Forum Mobile System Profile Release 1.0 Approved Specification, Revision 1.4.0, May.

WiMAX Forum (2008a) www.wimaxforum.org.

WiMAX Forum (2008b) www.wimaxforum.org/technology/documents/WiMAX_Forum_Network_ Architecture_Stage_2-3_Rel_1v1.2.zip.

WiMAX Forum (2008c) WiMAX Technology Forecast, www.wimaxforum.org/technology/downloads/wimax_forum_wimax_forecasts_6_1_08.pdf.

WiMAX Forum (2008d) WiMAX Forum Announces First Mobile WiMAX Certified Products at WiMAX Forum Congress Asia 2008,
www.wimaxforum.org/news/pr/view?item_key=59390fb727bfa15b5b8d11bf9341b2b1176099f8.

WiMAX Forum (2008e) WiMAX Forum announces first certified MIMO 2.5 GHz Mobile WiMAX products,
www.wimaxforum.org/news/pr/view?item_key=cffca4e77e1900b83fa727fe754a60bc0db849e6.

Part II

WiMAX Validation: Validating Current Fixed and Mobile WiMAX through Advanced Testbeds

2

WiMAX Performance in Practice

Kostas Pentikousis, Esa Piri, Jarno Pinola and
Ilkka Harjula

According to some estimates, by 2010 WiMAX operators will cover areas inhabited by more than 650 million people; however, current deployment is not meeting previous expectations and predictions. Moreover, non-vendor, third-party empirical evaluations of the technology are far from common. In fact, most WiMAX studies have up to now solely employed simulation, modeling and analytical tools. It is not really well understood what is in practice possible with Commercial Off-The-Shelf (COTS) WiMAX equipment. This chapter presents results from our recently concluded baseline, VoIP and IPTV synthetic traffic measurement and analysis studies over WiMAX at the VTT Converging Networks Laboratory (CNL) in Oulu, Finland, and aims at contributing to our understanding about the potential of WiMAX in practice.

This chapter is organized as follows. Section 2.1 provides a short literature review of other empirical studies of WiMAX. Section 2.2 introduces our fixed WiMAX testbed and our methodology with respect to traffic generation, metrics and testbed host clock synchronization. It also provides baseline capacity measurements and relates them with previously published works. Section 2.3 presents our empirical evaluation of VoIP over fixed WiMAX. Here we explain the issues arising from header overhead in VoIP, discuss the performance of different VoIP codecs and highlight the performance gains that may be attained by employing VoIP aggregation. Section 2.4 presents our evaluation of emulated IPTV audio/video streaming over fixed WiMAX, for both line-of-sight (LOS) and non-line-of-sight (NLOS) conditions. Section 2.5 presents an empirical evaluation study of the baseline capacity of a mobile WiMAX testbed, which we consider to be the first publicly disclosed. Finally, Section 2.6 summarizes this chapter and Section 2.7 provides pointers for further reading.

WiMAX Evolution: Emerging Technologies and Applications   Edited by Marcos D. Katz and Frank H.P. Fitzek
© 2009 John Wiley & Sons, Ltd

Figure 2.1   WiMAX deployment and locations of testbeds and field trials with publicly reported results from third parties. (Source: Maravedis (2008).)

2.1   Empirical Evaluations of WiMAX

WiMAX has received a lot of attention during recent years. Figure 2.1 illustrates the worldwide coverage by the end of 2007 for fixed and mobile WiMAX based on data published by Maravedis (2008), a market research firm. A country is shaded in the map if either a commercial or a trial network is in operation. Clearly, there is a lot of interest for WiMAX across the globe already and, as we will see in Chapter 3, there are several different possibilities for the technology, which can be deployed in different topologies and various markets, supporting a diverse set of applications.

However, as Maravedis points out, the vast majority of current WiMAX deployments do not support mobility. In addition, a large proportion of all deployments are based on solutions uncertified by the WiMAX Forum. This is to some degree a result of the main use of fixed WiMAX up to now. That is, WiMAX has been primarily deployed as a last-mile broadband connectivity technology, in particular for sparsely populated rural areas. Such use is conducive to proprietary solutions. However, as more WiMAX Forum-certified equipment becomes available, WiMAX deployment is expected to proliferate. According to some estimates, by 2012 WiMAX networks could cover areas inhabited by more than a billion people.

Nevertheless, although there is significant interest in WiMAX technology (see Andrews *et al.* (2007), IEEE 802.16 Working Group (2004, 2005), Pinola and Pentikousis (2008), WiMAX Forum (2007a,b) and the references therein) and high expectations about future deployment (see http://www.wimaxforum.org), WiMAX equipment is yet to become readily available at affordable prices. This means that third-party empirical evaluations are not easy to find.

On the other hand, WiMAX network operators do not release information about the performance of their (operational) networks due to business concerns. At the same time, WiMAX equipment vendors tend to release figures that relate to the best capacity of their offerings. The popular press, as one would expect, usually reports only 'headline' figures, without specific technical details. This creates an information gap for researchers in the area of Broadband Wireless Access (BWA) as it is not very clear what current WiMAX technology can deliver in practice.

Given this background, it is not surprising that most WiMAX studies employ simulation and modeling. Before starting our fixed WiMAX empirical evaluation, we were able to find only a handful of publicly reported, peer-reviewed studies of WiMAX performance in practice, which we briefly discuss next. Figure 2.1 highlights the locations of the testbeds or field trials where empirical performance evaluation studies have been conducted by third parties and have been disclosed publicly.

Scalabrino *et al.* (2006, 2007) used a fixed WiMAX testbed deployed in Turin, Italy, to empirically evaluate Voice over IP (VoIP) performance over WiMAX. In particular, they focused on scenarios where service differentiation is employed in the presence of significant amounts of elastic background traffic. Unfortunately, although their testbed included three Subscriber Stations (SSs), the authors did not report any results from their simultaneous use. That is, their evaluation considers emulated VoIP calls over point-to-point links. Moreover, perhaps due to a different radio resource allocation, Scalabrino *et al.* (2007) reported that the bottleneck in their testbed proved to be the downlink, which is not the case in our own testbed, as we will see later in this chapter.

Grondalen *et al.* (2007) reported active traffic measurement results from a fixed WiMAX field trial near Oslo, Norway. They focused their evaluation on the performance of bulk Transmission Control Protocol (TCP) and User Datagram Protocol (UDP) transfers but did not empirically measure the performance of VoIP or other multimedia traffic. They did measure throughput under both LOS and NLOS conditions and correlated it with Received Signal Strength Indicator (RSSI) values at 15 distinct locations. Grondalen *et al.* (2007) showed that their WiMAX system, employing the same modulation and Forward Error Correction (FEC) as our testbed described in the following section, can deliver a throughput of 9.6 Mbps to a single flow in the downlink. As they note, this throughput level can be attained by SSs located at a distance of up to 5 km from the Base Station (BS).

Martufi *et al.* (2008) discussed the merits of using WiMAX for emergency services, such as environmental monitoring and fire prevention. Chapter 3 provides further details about this type of scenario. Martufi *et al.* (2008) describe a testbed deployed in the mountainous area near Coimbra, Portugal. They performed measurements for their video surveillance application and report throughput of 1.1 and 1.3 Mbps for the uplink and downlink, respectively. For this link, the BS–SS distance was nearly 23 km, with direct LOS, and a 3.5 MHz channel bandwidth. They also measured throughput when additional channel bandwidth is allocated. For a 7 MHz channel, downlink throughput reaches 2.7 Mbps and uplink throughput doubles to 2.2 Mbps. Finally, when a 14 MHz channel is used, the measured throughput is 5.6 and 4.5 Mbps in the downlink and uplink, respectively. Using the same equipment, Neves *et al.* (2007) reported that for another link with a BS–SS link span of 19 km in LOS, and 3.5 MHz channel bandwidth, a throughout of 2 and 2.4 Mbps in the uplink and downlink, respectively, can be sustained. Neves *et al.* (2007) did not study VoIP performance in their testbed.

Mignanti *et al.* (2008) also reported on FTP and VoIP performance over WiMAX in the Wind testbed in Ivrea, Italy. Their results indicate acceptable mean opinion scores for VoIP in a cell with a 2 km radius, but do not comment on overall (cumulative) throughput. Unfortunately, the results of Mignanti *et al.* (2008) are not directly comparable with ours due to differences at the physical layer.

In Section 2.3 we evaluate VoIP performance over fixed WiMAX using synthetic traffic generation. We measure the capacity of a fixed WiMAX testbed to carry three different types of VoIP traffic, namely G.723.1 (ITU-T, 1996), G.729.1 (ITU-T, 2006) and Speex, an open-source variable bit rate audio codec (Valin, 2008). We quantify both uplink and downlink performance using emulated VoIP traffic in terms of cumulative goodput, packet loss and mean opinion scores, based on the ITU-T E-Model (ITU-T, 1998, Sengupta *et al.*, 2006).

Moreover, we empirically investigate the benefits of employing application- and network-level VoIP aggregation when using fixed WiMAX as a backhaul. Aggregation appears as a promising approach for increasing the overall network efficiency and resource utilization. Hoene *et al.* (2006), for example, found that it is better to change the packet rate rather than the coding rate when a VoIP flow encounters limited capacity links.

There have been several proposals regarding VoIP flow aggregation for different wireless technologies ranging from Wireless Local Area Networks (WLANs) to cellular networks. We put these aggregation schemes to the test in our fixed WiMAX testbed and compare their performance with nonaggregated VoIP. As we will see, aggregation can more than double the effective VoIP backhaul capacity of our fixed WiMAX link in terms of the number of sustained flows.

2.2   Fixed WiMAX Testbed Evaluation

Figure 2.2 illustrates the experimental facility used in our empirical performance evaluation. The testbed is part of the VTT Converging Networks Laboratory and is located in Oulu, Finland. It comprises an Airspan MicroMAX-SoC fixed WiMAX BS, operating in the 3.5 GHz frequency band, two SSs and several PCs. One SS is an Airspan EasyST indoor unit and the other is an Airspan ProST.

Symmetrically on the BS and SS sides we connect GNU/Linux PCs that act as traffic sources/sinks. As shown in Figure 2.2, each PC has two network interfaces. All PCs use one of their two interfaces to connect with an IEEE 1588 Precision Time Protocol (PTP) synchronization server over Ethernet. PTP is specified in International Electrotechnical Commission (2004). In this way, the traffic under observation does not interfere with the synchronization message exchanges.

Table 2.1 summarizes our testbed configuration. The BS Medium Access Control (MAC) implementation, in practice, only supports best-effort scheduling. Since the testbed is deployed in an indoor laboratory environment with short link spans between the BS and SSs, the transmission power is set to only 1.0 dBm. Laboratory conditions are overall relatively static in nature and we thus measure fixed WiMAX performance in a best-case scenario under well-controlled conditions.

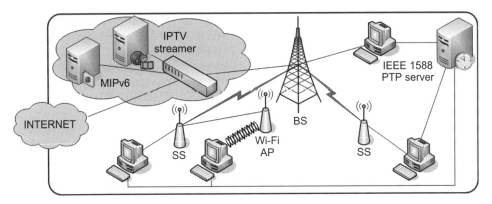

Figure 2.2 Schematic of the VTT CNL fixed WiMAX testbed.

Table 2.1 VTT CNL fixed WiMAX testbed configuration.

Base station	Airspan MicroMAX-SoC
Subscriber stations	EasyST and ProST
PHY	WiMAX 16d, 256 OFDM FDD
Frequency band	3.5 GHz
Channel bandwidth	3.5 MHz
BS and SS Tx power	1.0 dBm
Downlink modulation	16 QAM and 64 QAM
Uplink modulation	BPSK, QPSK, 16 QAM and 64 QAM
MAC scheduling	Best effort

The testbed equipment supports different FEC code rates as well as modulation schemes. Of course, different operational conditions have a direct effect on the maximum capacity of the WiMAX link. The downlink uses Quadrature Amplitude Modulation (QAM) only: either 64 QAM or 16 QAM. The uplink, in addition to QAM, supports Quadrature Phase Shift Keying (QPSK) and Binary Phase Shift Keying (BPSK) modulation as well.

2.2.1 Audio and Video Traffic over WiMAX

We are interested in actively measuring the performance of our fixed WiMAX testbed under different traffic loads. We are, in particular, seeking to quantify the capacity of our WiMAX equipment to carry packet streams corresponding to audio/visual (A/V) content in both the uplink and downlink.

Network traffic corresponding to A/V content has typically more stringent requirements than elastic flows. For example, according to ITU-T (2003), a VoIP call is to be considered of excellent quality if the one-way packet delay does not exceed 150 ms. Moreover, the ITU recommendation sets an upper limit (400 ms) on what is considered acceptable with respect to one-way packet delay for normal VoIP conversations (Sousa *et al.*, 2008). Although similar recommendations exist for other kinds of Internet traffic, dictating, for example, the maximum time necessary to load a Web page, VoIP is typically considered the application with the most stringent requirements with respect to one-way delay. On the other hand, video streaming typically demands considerable amounts of bandwidth.

With their increasing popularity, both VoIP and video streaming pose challenges to all network infrastructures. In particular, though, future wireless access networks, including those employing WiMAX, will have to cope with A/V content traffic while making the most of the limited radio resources, as discussed in Chapters 3 and 13. We expect that the empirical results from our testbed, reported later in this chapter, will be of interest to network practitioners and researchers alike as we transition towards an Internet where wireless access and A/V traffic become the norm, rather than the exception.

2.2.2   Traffic Generation

Our first realization is that we need to instantiate several flows in our testbed in a controlled manner so that we can empirically measure WiMAX performance under different traffic loads and patterns. Traffic generators are very handy in a laboratory environment as they can inject different types of traffic, with various profiles and characteristics, in a straightforward manner, while allowing for reproducible experiments.

We opted to use JTG, a simple and flexible traffic generator for GNU/Linux, available from Manner (2008). JTG can be configured and controlled through a command line interface and can emulate several packet flows in parallel. JTG allows us to experiment with Constant Bit Rate (CBR) traffic, such as VoIP without silence suppression, trace-driven or Variable Bit Rate (VBR) traffic, such as television streaming over IP (IPTV), and elastic traffic, as it can generate UDP and TCP traffic with configurable characteristics.

For example, we can use a JTG instance to quantify the WiMAX link capacity using a single TCP or UDP flow, as we do in Section 2.2.4. Or, we can employ several JTG instances to emulate dozens of VoIP conversations and IPTV streams. As we explain later in Section 2.3, we generate packet streams that emulate VoIP calls with payload sizes and packet transmission intervals according to the corresponding codec specifications. On the other hand, for the emulated IPTV traffic we employ trace-driven traffic generation based on an actual IPTV transmission.

2.2.3   Host Clock Synchronization

When evaluating the performance of multimedia traffic over the network, measuring one-way delay accurately is a major concern. To obtain accurate one-way delay measurements, synchronization inaccuracy, that is, the clock offset between the testbed hosts must be an order of magnitude smaller than the measured one-way delays. For small testbeds in a laboratory environment, this translates into synchronization accuracy in the range of a few hundred microseconds, as the theoretical Round Trip Time (RTT) link latencies of WiMAX are expected to be less than 50 ms.

To achieve this level of synchronization, one would typically rely on Global Positioning System (GPS) hardware to synchronize the clocks on every host machine in the testbed. This approach is followed by Prokkola *et al.* (2007). However, GPS-based clocks are expensive and their price can be a limiting factor in many testbed setups. There are limitations other than budgetary when considering GPS-based synchronized measurements. In particular, for indoor testbeds, the location of the laboratory in the building can make the use of GPS-based clock synchronization difficult or impractical.

Alternatively, one can use software-based host clock synchronization. The Network Time Protocol (NTP), described by Mills (1992), is well known and used widely to synchronize ordinary hosts in a network. However, NTP normally allows for a synchronization accuracy that is approximately of the order of tens of milliseconds, which is not sufficient for our experiments. Furthermore, we note that although GPS-based synchronization can provide timestamps that refer to a universal clock, in testbed measurements of the kind we are interested in, absolute synchronization with a global reference clock is not necessary. All that is needed for accurate one-way delay measurements in the lab is synchronization of all hosts with the clock of a single node, which defines the testbed 'master clock' for the purposes of our empirical evaluation.

IEEE and International Electrotechnical Commission (2004) have recently standardized a client–server-based synchronization protocol, PTP. PTP provides a promising alternative to GPS-based synchronization, as it can be implemented in software without the need for special (and expensive) hardware. For the experiments described in this chapter, we employ an open-source software-only implementation of PTP, which has been documented to achieve host synchronization accuracy of the order of tens of microseconds according to Correll *et al.* (2005). This solution is ideal for our setting as we can use as many PCs as necessary for our experiments and measure one-way delay accurately. The only requirement is that the testbed network has to be geographically constrained, so that the hosts can exchange PTP synchronization messages over a Local Area Network (LAN) with small end-to-end delays.

Before proceeding with the VoIP and A/V measurements presented in the remainder of this chapter, all dual-interface hosts in our WiMAX testbed were synchronized with a PC acting as the testbed master clock (marked as 'PTP server' in Figure 2.2). During the experiments, all hosts continued to exchange clock synchronization messages over the dedicated Ethernet interface.

The open-source PTP implementation employed in our testbed is called Precision Time Protocol Daemon (PTPd) and is freely available from Correll (2008). The distribution includes both PTP client and server functionalities and can be easily deployed inside a single LAN. According to the PTPd logs, the clocks of the COTS PCs in our testbed were synchronized with an accuracy of tens of microseconds (always less than 100 µs and typically in the range of 30–50 µs).

In order to validate the clock offset values reported by PTPd, we consider one-way delay measurements employing GPS-based synchronization as the state of the art. As detailed in Pentikousis *et al.* (2008a), we used QoSMeT, a highly-accurate proprietary GPS-based measurement tool developed by VTT (Prokkola *et al.*, 2007), to measure the one-way delay in our WiMAX testbed. QoSMeT reported that the median one-way delay is 8.7 ms in the downlink and 23.5 ms in the uplink.

We inject the same traffic patterns as QoSMeT into the uplink and downlink directions and measure the one-way packet delay while having all PCs synchronized using PTP. Figures 2.3 and 2.4 present the Cumulative Distribution Functions (CDFs) of the one-way delay as measured in the testbed when we use QoSMeT (i.e. GPS-based synchronization) and PTPd-based synchronization.

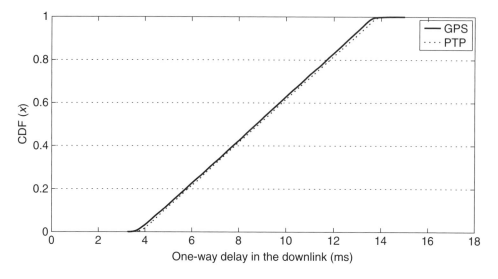

Figure 2.3 Comparison between GPS- and PTP-based one-way delay measurements for the BS-SS downlink.

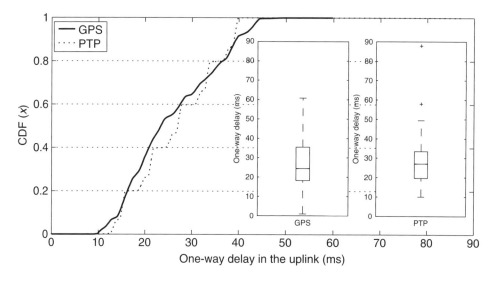

Figure 2.4 Comparison between GPS- and PTP-based one-way delay measurements for the BS–SS uplink.

Figure 2.3 illustrates that we can be confident about the one-way delay values reported later in this chapter for the downlink. In fact, we argue that the measurements are as accurate as they would have been had GPS-based synchronization been employed.

Table 2.2 Average throughput of fixed WiMAX link with different modulation schemes, at link saturation and with negligible packet loss (<0.1%).

Modulation	Uplink (Mbps)	Downlink (Mbps)
64 QAM FEC: 3/4	5.6	9.6
64-QAM FEC: 2/3	4.8	9.2
16-QAM FEC: 3/4	3.8	6.9
QPSK	1.5	N/A
BPSK	0.5	N/A

Figure 2.4 includes inset boxplots of the measured one-way delays in the uplink. With respect to the uplink one-way delay measurements, we note that the results reported in the rest of this chapter (using PTP-based clock synchronization) may actually be pessimistic on median when compared with GPS-synchronized active traffic measurements.

2.2.4 Baseline Capacity Measurements

In order to establish the upper bound of the empirically measured WiMAX capacity, we use a single greedy UDP flow. We stress test the WiMAX uplink and downlink separately, while making sure that there is negligible packet loss. All transmitted UDP packets are 1500 bytes long, including the UDP and IP headers. This packet size is equal to the Ethernet Maximum Transmission Unit (MTU) and it is the recommended packet size for WiMAX access networks. At link saturation and while employing the most efficient modulation scheme, 64 QAM and FEC 3/4, the average UDP throughput was measured to be 9.6 Mbps in the downlink and 5.6 Mbps in the uplink. We made sure that packet loss at link saturation was negligible (<0.1%).

Throughput is typically defined as the ratio of the total network traffic received over a certain amount of time. If we are interested in the application-layer information rate then we use goodput as our metric of choice. Goodput is defined as the ratio of application payload over the time needed to completely receive the payload at the destination. In other words, when calculating goodput, we exclude all transport and network layer headers and include only the application payload.

For packets carrying large application payloads, as is the case with Ethernet MTU-sized packets, the header overhead is rather small and thus the difference between the average throughput and goodput values is marginal. For example, the average goodput over UDP/IP at saturation point in our testbed is 9.4 Mbps in the downlink and 5.5 Mbps in the uplink, when 64 QAM with FEC 3/4 is used. However, as we will see later in the chapter, when header overhead is significant, the two metrics tell a different story about the access network performance.

Table 2.2 summarizes our baseline measurements with all supported modulation schemes. As one would expect, the lowest average throughput, of the order of 500 kbps, is recorded when BPSK is used in the uplink. This is an order of magnitude less than the average throughput recorded when 64 QAM with FEC 3/4 is used. In the downlink, we note that 64 QAM with FEC 3/4 outperforms 64 QAM with FEC 1/2 by 4.3%, and 16 QAM with FEC 3/4 by 39%.

To sum up, the modulation scheme and the FEC code rate play a critical role in the measured throughput (and goodput). Our testbed equipment employs mechanisms that automatically adapt modulation and FEC according to the operational conditions. For example, when the received Signal to Noise Ratio (SNR) deteriorates considerably, say, because the SS operates under NLOS conditions, the modulation scheme changes from QAM to QPSK or even to BPSK. Thus, the actual performance of individual SSs in a real deployment may vary greatly.

In the remainder of this chapter we only report empirical measurements for 64 QAM with FEC 3/4 employed, which provide an upper bound of what is possible in practice today with COTS fixed WiMAX equipment.

2.3   VoIP Over Fixed WiMAX

As mentioned earlier, there is great interest in using VoIP over packet-switched wireless networks, but there are few publicly available empirical studies of VoIP over fixed WiMAX. In this section we present our empirical evaluation study of three different VoIP codecs over WiMAX and discuss the benefits of application- and network-layer aggregation with respect to the effective capacity of our WiMAX testbed.

2.3.1   VoIP Overhead

Typically, VoIP calls do not place high demands on bandwidth. In fact, codec technology has been making great strides towards minimizing bandwidth demands, while delivering even better audio quality. As better audio compression techniques are developed, VoIP packets carrying voice samples tend to become smaller, which is reflected in the total bandwidth consumed by a VoIP call. For example, a codec standardized in the late 1980s, such as G.711, requires 100 kbps of available bandwidth ITU-T (1988). A recently developed codec, such as Speex, on the other hand, is highly configurable allowing for layered encoding. Speex supports 12 different payload bitrates ranging between 2.15–24.6 kbps and 4–44.2 kbps for the narrowband and wideband codecs, respectively (Valin, 2008). Nevertheless, codecs do have to sample voice conversations frequently, thus generating traffic flows with small packet sizes and short inter-packet times.

Raake (2006) pointed out that when considering most current VoIP codecs, each VoIP packet typically carries a voice sample representing 10–60 ms of audio. This roughly translates into voice frame sizes in the range of 20–100 bytes, depending on the codec employed. The Real-time Transport Protocol (RTP) (Schulzrinne *et al.*, 2003) is often used to encapsulate voice frames, which are then transmitted over UDP/IP. This, in turn, means that each VoIP packet comprises the voice sample and the RTP, UDP and IP headers. The RTP and UDP headers consume 12 and 8 bytes, respectively. The IPv4 standard header consumes another 20 bytes; the IPv6 fixed header part consumes 40 bytes. In other words, the total header overhead is 40 or 60 bytes, depending on whether IPv4 or IPv6 is used, respectively.

To sum up, due to the small size of a single voice frame, the header overhead when each voice sample is sent over RTP/UDP/IP can often exceed 100%. Compare this level of overhead with the case when the recommended packet size for WiMAX, the Ethernet MTU (1500 bytes), is used. In this case, the RTP/UDP/IP header overhead for an MTU-size packet is only 2.74%. The TCP/IP header overhead is also 2.74%, while the UDP/IP header overhead is only 1.9%. Of course, this header overhead is an issue that emerges in all networks, although it particularly hinders wireless access networks, including WiMAX. Several solutions have been proposed, including VoIP aggregation (empirically evaluated and discussed in Section 2.3.5) and Robust Header Compression (ROHC), which is evaluated in practice by Piri *et al.* (2008) and in Chapter 10.

In the following sections we empirically evaluate the performance of three commonly used VoIP codecs at the VTT CNL fixed WiMAX testbed (see Section 2.2) using synthetic traffic generation. First we consider two ITU-T standardized codecs, namely G.723.1 (ITU-T, 1996) and G.729.1 (ITU-T, 2006) and then an open-source VBR audio codec specifically designed for speech compression in VoIP applications over packet-switched networks (Valin, 2008).

2.3.2   Synthetic G.723.1 VoIP Over WiMAX

G.723.1 (ITU-T, 1996) is a low-rate, narrowband codec with small processing requirements for real-time encoding and decoding. We synthetically generate simultaneous unidirectional flows with JTG (see Section 2.2.2) stressing the WiMAX uplink and downlink separately.

The emulated G.723.1 codec information rate is assumed to be 6.4 kbps and thus the traffic generator injects a new packet carrying an emulated sample frame size of 24 bytes every 30 ms. In other words, 24 bytes carry 30 ms worth of speech. The application information rate in this case is 9.6 kbps (accounting for the codec payload and the RTP headers). The total network bitrate for each emulated VoIP flow is slightly over 17 kbps.

Recall that the average goodput attained with Ethernet MTU-sized packets, at link saturation with negligible loss (<0.1%), is 9.4 Mbps and the maximum throughput is 9.6 Mbps. In theory, a simple back-of-the-envelope calculation based on the maximum throughput measured would tell us that the fixed WiMAX downlink can sustain more than 560 G.723.1 emulated unidirectional flows. In practice, this is not the case. Our stress tests showed that the fixed WiMAX downlink can only carry up to 200 synthetic unidirectional VoIP flows with negligible loss (<0.1%). At link saturation, which is attained in the downlink with 400 VoIP flows, the packet loss rate exceeds 5%, which is unacceptable for G.723.1.

Similarly, in the uplink, the goodput attained with Ethernet MTU-sized packets, at link saturation with negligible loss (<0.1%), is 5.5 Mbps and the maximum throughput is 5.6 Mbps. Although these figures would predict that the uplink can sustain more than 320 flows, in practice, we can only have 150 G.723.1 emulated unidirectional flows without observing significant packet loss. If we inject 175 unidirectional flows in the uplink we exceed its saturation point and observe packet loss ratios greater than 5%.

2.3.3   Synthetic G.729.1 VoIP Over WiMAX

G.729.1 (ITU-T, 2006) is a sophisticated wideband codec, which is based on a layered structure. The layered encoding allows for 12 different payload bitrates ranging from 8 to 32 kbps. The core layer information rate is 8 kbps, carrying 20 ms worth of voice data. The core layer is the only layer necessary for a successful decoding of all received VoIP packets. The additional layers improve the received audio quality considerably, but are not as critical as the core layer.

Although G.729.1 defines sophisticated traffic adaptation features, we do not employ them in our empirical evaluation. Instead we choose a single operation mode with four additional layers, and synthetically generate emulated codec payloads of 40 bytes at a constant codec information rate of 16 kbps. The application information rate is 20.8 kbps and the total network bitrate for each emulated G.729.1 VoIP flow (including the RTP/UDP/IP headers) is 32 kbps.

In this set of experiments we consider the case of bidirectional VoIP flows which are terminated inside the same WiMAX cell. That is, we emulate G.729.1 VoIP calls by injecting the same number of flows in both uplink and downlink in parallel, modeling a more realistic VoIP call scenario. Moreover, for the tests described in the previous section we used only a single SS, while for the measurements described in the remainder we use two SSs.

Evidently, the bidirectional VoIP flows need to transit both the uplink and the downlink of the same BS. As we saw earlier, the uplink is the bottleneck in our testbed due the specific (fixed) radio resource allocation. Our experiments showed that the WiMAX BS can handle 65 bidirectional VoIP flows with an average loss rate of less than 5%, which is an acceptable sample loss threshold for G.729.1. Note that G.729.1 is designed to be more tolerant to packet loss than G.723.1.

Through separate experiments we found out that in our lab the central scheduling performed by the BS allows us to apportion a number of flows evenly between PCs connected to SS1 and SS2, split the same number of flows in any random manner between the two SSs or use only one of the SSs while observing negligible packet loss and similar performance.

Although the experiment setup is different from that described in the previous section, we could attempt a comparison. Effectively, the uplink can sustain 130 synthetic G.729.1 VoIP flows with acceptable packet loss (<5%). That translates into a average cumulative goodput of 2.7 Mbps (and a cumulative throughput of 4.16 Mbps). This is only 49% of the maximum attainable goodput with Ethernet MTU-sized packets.

2.3.4 Synthetic Speex VoIP over WiMAX

Speex is an open-source VBR audio codec specifically designed for speech compression in VoIP applications over packet-switched networks (Valin, 2008). Speex can be used with three different sampling rates (narrowband at 8 kHz, wideband at 16 kHz and ultra-wideband at 32 kHz), and has a large range of operational bitrates (2.15–44.2 kbps). Speex uses Code Excited Linear Prediction (CELP) for encoding voice samples and is robust to packet loss. Owing to its open source and good quality, it has been incorporated into several applications, including Microsoft's Xbox Live.

Similarly with the experiments in the previous section we consider the case of duplex, bidirectional parallel VoIP flows. We select the wideband codec variant with a codec information rate of 12.8 kbps. That is, for each VoIP flow, JTG generates 50 packets per second with 32 bytes of emulated codec payload. After including the necessary RTP, UDP and IP headers, each JTG instance injects 28.8 kbps of total emulated Speex CBR traffic into the testbed network.

In our testbed, we were able to have a total of 100 Speex flows in the uplink with negligible (<0.1%) average packet loss. Again, this is nearly half of what a back-of-the-envelope calculation would have predicted based on the maximum recorded goodput. Owing to the large header overhead, the average cumulative goodput in the uplink is only 1.76 Mbps. This is only 32% of the average goodput achieved by a single UDP flow transmitting Ethernet MTU-sized packets.

2.3.5 VoIP Aggregation

The empirical evaluations presented above make it clear that the header overhead, coupled with the processing delays associated with each packet, and the small inter-arrival packet times put a drag on VoIP performance over fixed WiMAX. A simple way to reduce header overhead is to pack more voice frames in a single packet before transmitting it over the WiMAX link. For example, a G.723.1 VoIP client can buffer two or three voice frames and encapsulate them together, using the same RTP header, before handing them over to UDP. Schulzrinne *et al.* (2003) explicitly allow voice frame aggregation. As we will see shortly, application-layer aggregation performs significantly better than trivial encapsulation, which places a single voice frame in each RTP/UDP/IP packet.

As mentioned earlier, VoIP packets usually carry voice data which corresponds to only a few tens of milliseconds. When voice frame aggregation is employed, audio samples need to be buffered at the sender before they are transmitted over the WiMAX access network. When a VoIP application bundles samples together, it will need to cope with increased end-to-end delays, over and above their normal level. The additional buffering delays caused by aggregation need to be considered carefully by application developers as they may actually worsen the quality of VoIP calls if they are excessive. According to Kitawaki and Itoh (1991), round-trip delays of 500 ms reduce conversational efficiency by approximately 20–30%. In addition, recall that ITU-T (2003) recommends that one-way packet delays should not exceed 400 ms, and should remain below 150 ms for optimal VoIP call quality. Thus, aggregation levels that introduce voice frame buffering delays that exceed 100 ms may be problematic for long-distance VoIP calls.

Next, we quantify the capacity of our fixed WiMAX testbed in terms of the number of synthetic G.723.1 VoIP flows it can sustain with negligible packet loss depending on whether aggregation is employed or not. Then we measure the testbed capacity to sustain VoIP flows using Mean Opinion Scores (MOSs) as the metric to determine the performance saturation point.

2.3.5.1 Application-layer VoIP Aggregation

As we reported earlier, the fixed WiMAX downlink can sustain 200 unidirectional synthetic G.723.1 flows with negligible loss. When JTG is configured to emulate the aggregation of two G.723.1 voice frames per packet, the WiMAX downlink can sustain 650 flows with negligible packet loss. This represents an impressive effective capacity increase of 225%. Note that in this case the buffering delay increases by 30 ms for every other packet, which can be easily handled by modern VoIP applications. If the VoIP sender bundles three voice frames per packet, then our testbed can sustain more than 800 synthetic unidirectional flows in the downlink while incurring negligible packet loss.

In the uplink, the testbed can sustain 150 synthetic unidirectional synthetic G.723.1 flows with negligible loss. When two voice samples are encapsulated per packet injected into the WiMAX uplink, nearly 300 flows can be sustained with negligible loss, effectively doubling the capacity of the testbed in terms of number of calls. We refer to this application-layer aggregation scheme as 'L1'. If three voice samples are aggregated, referred to as 'L2', then the WiMAX uplink can carry at least 400 unidirectional synthetic flows without any packet loss.

It is important to note that the two ends of the VoIP call may decide to use aggregation in an asymmetric manner. Given the relative uplink/downlink radio resource allocation in our testbed, which may be quite typical of future real-world deployments, the VoIP peers could adopt aggregation for the traffic traversing the uplink, but continue to use trivial encapsulation for the reverse traffic direction. Or, alternatively, given the higher one-way delay observed in the uplink, the two VoIP peers may decide to use trivial encapsulation in the uplink direction in order to avoid exacerbating the end-to-end one-way delay. Aggregation could be adopted in the downlink, on the other hand, sparing valuable access network resources for other traffic, such as IPTV or Video on Demand (VoD), which is unidirectional by nature and can use the leftover downlink bandwidth.

2.3.5.2   Network-layer VoIP Aggregation

Despite the impressive performance gains promised by application-layer aggregation, its adoption necessitates software updates across all end hosts. It also requires that hosts are aware that they are connected over WiMAX and they are willing to share this information with their peer.

An alternative to application-layer aggregation, which does not necessitate the cooperation of end users or the rollout of new or updated VoIP software, is network-layer aggregation. In network-layer aggregation a Performance Enhancing Proxy (PEP) is introduced at both the SS and the BS ends. The PEP can work in cooperation with other active network elements, such as routers and firewalls, in order to detect VoIP traffic and start aggregating trivially encapsulated voice samples. The PEP can bundle together several packets, possibly belonging to different VoIP calls, and transmit them over the fixed WiMAX link. Its peer on the other end of the WiMAX link will deaggregate the packets and route them to their final destination.

Similarly to application-layer aggregation, a PEP can bundle two or more complete VoIP packets into a new packet augmented with a PEP header. The term 'complete VoIP packet' here means the VoIP codec payload plus the RTP, UDP and IP headers. At the transmitting PEP side, multiple VoIP packets including their RTP/UDP/IP headers are bundled; at the receiving PEP, on the other end of the fixed WiMAX link, deaggregation takes place. Note that this is a rather trivial network-layer aggregation scheme that does not attain any gain in terms of cumulative goodput. In fact, this aggregation scheme actually increases the overall header overhead due to the introduction of the PEP header. Nevertheless, the PEPs make better use of the network resources simply by decreasing the number of packets that need to be transmitted over the WiMAX link.

Our testbed evaluation showed that when a simple network-layer aggregation scheme is adopted, more than 500 synthetic unidirectional G.723.1 flows can be sustained in the downlink while incurring negligible packet loss. In the uplink, nearly 300 flows can be sustained with negligible loss.

Table 2.3 summarizes our empirical effective capacity measurements over the WiMAX testbed.

Table 2.3 Number of emulated unidirectional VoIP flows sustained in the fixed WiMAX testbed with negligible packet loss (<0.1%).

Voice sample encapsulation	Downlink	Uplink
Trivial (one G.723.1 sample/packet)	200	150
Application-layer aggregation		
Two G.723.1 samples/packet	650	300
Three G.723.1 samples/packet	800	400
Network-layer aggregation		
Two G.723.1 samples/packet	500	300
Three G.723.1 samples/packet	550	300

2.3.5.3 Mean Opinion Scores

Up to now we considered negligible packet loss as the only metric with which we gauge the capacity of our testbed to deliver emulated VoIP calls. This approach has its limitations. First, modern VoIP applications can cope with moderate packet losses quite well. For example, sample loss ratios of up to 5% can be concealed by most modern codecs. Second, call quality, as perceived by the end users, does not depend only on packet loss, but also on one-way delay and other factors, including the codec employed.

ITU-T (1998) defines the R-factor, also known as the R-score, which combines different aspects of voice impairments. R-factor values range between 0 and 100; scores above 70 indicate VoIP streams of decent quality according to Sengupta *et al.* (2006).

The R-factor, in general, can be calculated using the following equation based on the observed packet loss and delay

$$R = R_0 - I_s - I_d - I_e + A, \tag{2.1}$$

where R_0 represents the basic SNR, including noise sources such as circuit noise and room noise, I_s factors in the effect of impairments to the voice signal, I_d captures the impairments due to delays, I_e takes into consideration the effects caused by the use of low-bitrate codecs and A is an expectation factor for compensating for the above impairments under various user conditions. For wired technologies A is set to zero.

We could further subdivide I_s, I_d, and I_e, but in our case we assume default values for speech transmission. After all, we are only emulating VoIP calls. Thus, we obtain the following equation from Equation (2.1):

$$R = 94.2 - I_d(d) - I_e(c, l) \tag{2.2}$$

in which I_e is a function of the codec (c) in use and the loss rate (l).

We can calculate $I_e(c, l)$ using

$$I_e(c, l) = \gamma_1 + \gamma_2 \times \ln(1 + \gamma_3 \times l), \tag{2.3}$$

where γ_1, γ_2, and γ_3 are codec-specific values. Since we are emulating traffic from a G.723.1 codec we have $\gamma_1 = 15$, $\gamma_2 = 90$, $\gamma_3 = 0.05$, as per Scalabrino *et al.* (2006).

I_d factors in delays (d) as detailed in Cole and Rosenbluth (2001) and ITU-T (1998):

$$I_d(d) = 0.024d + 0.11(d - 177.3)H(d - 177.3), \tag{2.4}$$

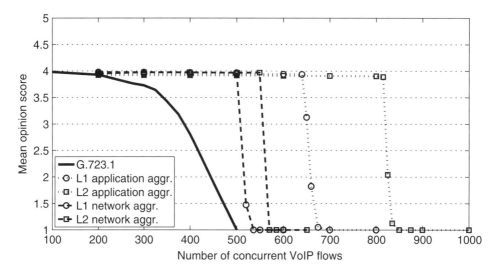

Figure 2.5   MOS for VoIP trivial encapsulation and application- and network-layer aggregation over the fixed WiMAX downlink.

where $H(x)$ is the step function ($H(x) = 1$ if $x \geq 0$ and 0 otherwise).

ITU-T (1998) specifies a nonlinear mapping from the R-score to the MOS scale which is based on subjective listening tests. MOS is a five-point Absolute Category Rating (ACR) quality scale where five (5) indicates excellent quality and one (1) the worst quality. Of course, there were no listening tests in our testbed evaluation. Instead, based on the ITU-T (1998) E-Model, we estimate the listening MOS using Equation (2.2) for the R-factor and the observed packet loss and delay, as follows

$$\text{MOS} = 1 + 0.035\, R + 7 \times 10^{-6} R(R - 60)(100 - R). \tag{2.5}$$

Figures 2.5 and 2.6 present the calculated MOS scores, based on Equation (2.5), for nonaggregated and aggregated synthetic G.723.1 VoIP flows when traversing the WiMAX downlink and uplink, respectively.

Figure 2.5 clearly illustrates that without aggregation, MOS deteriorates to unacceptable levels (<3.5) after injecting 350 parallel flows. Note that this capacity level is significantly higher than that estimated with negligible packet loss (200 flows) in Section 2.3.2. However, it is also significantly less than what one would expect based a back-of-the-envelope calculation (560 flows) using the maximum throughput measured and the total network bitrate of each emulated VoIP call.

When network-layer aggregation is employed, our empirical evaluation indicates that MOS remains at a high quality level (≥ 4) even when 500 synthetic unidirectional G.723.1 flows are injected in parallel. The maximum capacity in terms of the number of unidirectional synthetic G.723.1 flows with very high quality MOS score is attained with the L2 application-layer aggregation: more than 800 simultaneous flows can be sustained in our WiMAX testbed downlink.

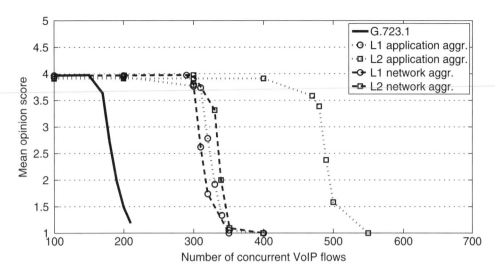

Figure 2.6   MOS for VoIP trivial encapsulation and application- and network-layer aggregation over the fixed WiMAX uplink.

In the uplink, the one-way delay remained in all cases under 100 ms, even when the voice frame loss rate exceeded 20%. Thus, the effect of the increase in one-way delay in the uplink MOS cannot be remarkable. Generally, though, the packet loss rate experienced by a VoIP flow plays a more critical role when calculating the R-score (and, thus, the equivalent MOS score) than delay deterioration, as Chatterjee and Sengupta (2007) explain.

Without aggregation, only 175 concurrent flows can be injected into the uplink before the calculated listening MOS deteriorates to an unacceptable level (<3). As Figure 2.6 illustrates, network-layer aggregation nearly doubles the effective capacity of our testbed in terms of VoIP flows. Interestingly, there is no significant difference in performance between L1 and L2 network-layer aggregation. Moreover, if the estimated listening MOS is our metric of choice, network-layer aggregation performs equally well with L1 application-layer aggregation. Finally, L2 application-layer aggregation allows our fixed WiMAX testbed to sustain up to 480 high-quality VoIP flows in the uplink. This is a remarkable increase in effective capacity of 174% when compared with the trivial encapsulation of one G.723.1 voice frame per packet.

2.3.5.4 Summary

Our empirical evaluation indicates that significant effective capacity gains can be achieved when voice frame aggregation is employed. The evidence from our testbed indicates that COTS fixed WiMAX equipment may be seriously underperforming when having to deal with large numbers of small and frequently arriving packets. We showed that the significant header overhead in a VoIP packet consumes a remarkable amount of resources and reduces the number of VoIP flows that can be properly accommodated considerably.

Simple aggregation schemes which bundle together multiple voice sample frames into one IP packet appear to be very promising. Our empirical evaluation indicates that application-layer aggregation may more than double the effective capacity of a WiMAX cell. Of course, VoIP application developers need always to consider the effect of aggregation on the end-to-end delay.

From an operator point of view, we found that the introduction of transparent network-layer VoIP aggregation can prove to be as effective as application-layer aggregation. Simple PEPs can effectively double the number of sustained VoIP flows and they may take advantage of information already available at other active network elements such as routers and firewalls. PEPs can also be installed, for example, at the traffic control systems residing at the ingress/egress of a backhaul WiMAX link.

2.4   IPTV over fixed WiMAX

Video streaming applications, including IPTV, VoD, and contemporary web- and P2P-based video services, require significantly more network resources than VoIP. Video is considered a premium service. Delivering high data rates and constraining jitter consistently, over both short and long periods, are difficult challenges to meet, in particular when wireless communication is involved. In order to deliver high-quality video over a network, support for high data rates, bounded latencies and low delay jitter are typically required. Of course, the exact bounds on all of these parameters depend on the specifics of the video streaming application.

Different video applications have slightly different demands with respect to the network capabilities. For example, live video streaming is the most demanding: the maximum acceptable end-to-end delay and jitter need to be low, so that the interactivity and real-time qualities of a live video feed are preserved and smooth playback of the video content is possible. From a football fan's point of view, there is nothing worse than hearing the neighbor celebrating the scoring of a goal during a football match while his own video feed has yet to play out the event. On the other hand, for VoD services, the bounds on end-to-end delay are not a critical factor, as extensive buffering can be employed. Larger delay jitter can also be tolerated, if pre-play buffering is used at the receiving end.

WiMAX and IEEE 802.16 are from the ground up designed to be able to handle different QoS requests. In particular, WiMAX can deal with different video streaming applications, as the QoS classes defined for the MAC layer take all of the necessary requirements into account. The Real-Time Variable Rate (RT-VR) and Extended Real-Time Variable Rate (ERT-VR) QoS classes are tailored for real-time applications such as video and audio streaming which use VBR in their transmissions. The RT-VR QoS class provides guaranteed bitrates and delay for all applications requiring it. If ERT-VR is chosen, low delay jitter, at least for the WiMAX access link, is guaranteed as well.

Unfortunately, though, not all currently available COTS WiMAX equipment support QoS provisioning. In fact, we expect that for many deployments best-effort delivery could be the norm.

Video packet streams comprise mainly large payloads of several hundreds of bytes. In contrast with VoIP, the total packet size can often be close to the Ethernet MTU and, therefore, header overhead is small. In other words, efficient bandwidth use is feasible for video streaming over a WiMAX air interface provided, of course, that the capacity of the link is sufficient for the target rate of the video streams. In fact, by employing state-of-the-art video codecs, such as the ITU-T H.264/Advanced Video Coding (AVC) (also known as ISO/IEC MPEG4 Part 10 International Telecommunication Union (2005)), the nominal data rate of each video stream can be relatively low, enabling a single WiMAX BS to deliver several video streams simultaneously to multiple users in its cell.

H.264/AVC is a video codec designed to emerge as the *de facto* codec for a variety of different multimedia devices operating over a variety of network technologies. This video coding standard is the result of the joint work of the ITU-T Video Coding Experts Group (VCEG) and the ISO/IEC Moving Picture Experts Group (MPEG). The improvements in coding efficiency through enhanced video compression and representation capabilities make H.264/AVC suitable also for use in wireless and mobile communication systems, as explained by Wiegand *et al.* (2003).

One of the most promising video applications using H.264/AVC is IPTV. There are several predictions that claim that IPTV and VoIP are to be the next 'big thing' in the IP-based multimedia services world. Many network service providers around the world anticipate that multimedia services will increase their revenue dramatically. As one would expect, the topic has attracted attention also in the WiMAX research community (see, for example, She *et al.* (2007), Retnasothie *et al.* (2006) and Uilecan *et al.* (2007), just to cite a few).

WiMAX, with the expected high data rates and forthcoming mobility support, poses as a technology over which new kinds of IPTV services can be deployed. For example, TV transmissions with High Definition (HD) picture quality can be delivered to multiple mobile users located in the same WiMAX cell. Such services could be offered, for example, to users of public transport by strategically locating WiMAX BSs along the main routes of the bus and train network. Unfortunately, once again due to the lack of affordably available WiMAX equipment, publicly disclosed performance evaluations are mainly based on theoretical analysis and simulations.

We fill this gap by empirically investigating two use cases for H.264/AVC-encoded IPTV over our fixed WiMAX testbed. In particular, we focus on measuring the capacity of current COTS WiMAX equipment to carry mixed multimedia traffic comprising VoIP and IPTV flows. Interested readers can find more details about this evaluation in Pentikousis *et al.* (2008a,b).

The first use case considers communication that takes place exclusively inside a single WiMAX cell. In the second use case, the hosts inside a single WiMAX cell communicate with remote peers located outside the cell. For both cases, we saturate the uplink WiMAX link with bidirectional synthetic Speex-encoded VoIP flows. As mentioned earlier, the uplink can sustain 100 synthetic Speex VoIP flows. Since the overall capacity of the fixed WiMAX downlink (9.6 Mbps) is larger than the uplink (5.6 Mbps), we introduce in the downlink, in addition to the Speex VoIP flows, several emulated IPTV streams. That is, although the uplink is saturated by the VoIP flows only, we let the IPTV flows effectively use the remaining network resources in the downlink.

In our experiments, the data rate of the transmissions is set to 512 kbps for the VBR video and to 192 kbps for the CBR audio, which together comprise the A/V signal of the IPTV channel. This setting corresponds to an actual transmission of an IPTV music channel configured at 360 × 288 pixels and 25 frames per second, which is suitable for display on portable/mobile devices equipped with small integrated screens.

Our testbed measurements indicate that even when all communications are between nodes inside a single WiMAX cell and, thus, through a single BS, it is possible to inject as many as five simultaneous H.264/AVC encoded IPTV streams into the downlink without observing service degradation. Note that in this case the WiMAX BS carries in addition to the five IPTV streams, 50 simultaneous bidirectional synthetic Speex VoIP 'calls', which originate and terminate inside the cell.

In the second use case, when the emulated VoIP calls involve peers outside the WiMAX cell, which may be a more realistic assumption for real-world deployments, the number of IPTV streams in the downlink can be increased to seven without service degradation. As a matter of fact, the downlink can sustain eight emulated IPTV transmissions with packet loss less than 5%, but with high one-way packet delays, which exceed 600 ms. If buffering can be employed, such as for VoD applications, the IPTV application can use pre-play buffering and deliver the content to the user in a seamless manner.

We also experimented with situations where the signal quality is decreased so that the modulation scheme used in the transmissions is forced to change from 64 QAM to 16 QAM. This would be the case in NLOS operation, for SSs located at cell edges, or in deep shadowing conditions, and of course the performance is naturally lower. In more densely populated suburban areas, where the cell size is smaller due to the large number of users and the height of the buildings is low, this should not be a problem. Difficulties with signal quality will be encountered more likely in vast rural areas with larger cells or in metropolitan city centres with tall buildings, as discussed in Chapter 3.

We find the results presented very promising for real WiMAX deployments. Recall that in our testbed experiments we were not able to test any other scheduling than best effort due to lack of support from the vendor equipment. Furthermore, we used the default radio resource allocation. Operators can configure different cells based on the anticipated traffic load. They can also increase the channel bandwidth allocation and use directional antennas. In the future, WiMAX equipment will also come with adaptive antenna techniques implemented such as beamforming and Multiple Input Multiple Output (MIMO), which effectively multiply the capacity of each WiMAX link. Traffic engineering techniques and teletraffic analysis and tools can also assist operators in accurately estimating and managing the user population and application usage in each WiMAX cell.

All in all, WiMAX has some highly attractive features considering IPTV applications, including the ability to serve point-to-multipoint topologies well, an all-IP system architecture, high data rates that are expected to keep increasing in the future, mobility support and QoS differentiation. These are all salient features of the WiMAX technology. Our testbed measurements, although conducted with rudimentary COTS WiMAX equipment, indicate that the technology's potential is not simply theoretical.

2.5   Mobile WiMAX Testbed Evaluation

An important difference between fixed WiMAX and mobile WiMAX is the physical layer. Mobile WiMAX uses Orthogonal Frequency Division Multiple Access (OFDMA) as its physical layer transmission scheme instead of plain Orthogonal Frequency Division Multiplexing (OFDM). OFDMA can also be used as a multiple-access mechanism when groups of data subcarriers, called subchannels, are allocated to different users. This kind

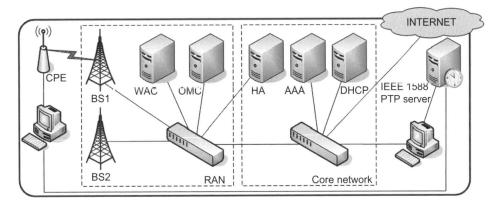

Figure 2.7 Schematic of the VTT CNL mobile WiMAX testbed.

of subchannelization can be used flexibly in mobile WiMAX for both the uplink and the downlink. This allows for resourceful dynamic bandwidth allocation, which can be taken advantage of, for example, to mitigate frequency selective channel fading due to mobility. To this end, mobile WiMAX specifies subchannel allocation schemes based on randomly allocated subcarriers from the available bandwidth.

Mobile WiMAX also introduces more scalability into the actual physical layer parameters. Multiple OFDMA profiles with varying amounts of subcarriers, cyclic prefix durations and channel bandwidths are supported, which allows the wireless link design to be optimized according to the environment where the system is deployed.

Pinola and Pentikousis (2008) overview several important aspects of mobile WiMAX, including the mobile WiMAX network reference model, node network entry and re-entry, and mobility support, which cannot be covered here due to space constraints.

2.5.1 The VTT CNL Mobile WiMAX Testbed

In this section we present our first evaluation of the recently commissioned VTT CNL mobile WiMAX testbed, which is illustrated in Figure 2.7. The testbed comprises two Alcatel-Lucent 9116 BSs operating in the 3.5 GHz frequency band. As shown in Figure 2.7, the Radio Access Network (RAN) consists of the BSs, the Wireless Access Controller (WAC), and the Operation and Maintenance Center (OMC). The core network includes a Mobile IP (MIP) Home Agent (HA), a Radius-based Authentication, Authorization, and Accounting (AAA) server, a Dynamic Host Configuration Protocol (DHCP) server, and a Domain Name System (DNS) server.

WAC is responsible for establishing and releasing data paths between the BS and the Core Network (CN) for a Mobile Node (MN) through the Call Control Function (CCF) subsystem. This includes, for example, MN authentication and authorization, IP connectivity establishment, QoS services and mobility inside the area of the current WAC and macro-mobility with the assistance of MIP. MN authentication is carried out by the AAA server. DHCP is used to provide IP addresses to MNs.

Table 2.4 VTT CNL mobile WiMAX testbed configuration.

Base station	Alcatel-Lucent 9116 BS
Subscriber stations	ZyXEL MAX-210M1 CPE
PHY	WiMAX 16e, 512 OFDMA TDD
Frequency band	3.5 GHz
Channel bandwidth	5 MHz
BS Tx power	35 dBm
Downlink modulation	64 QAM FEC: 1/2
Uplink modulation	16 QAM FEC: 1/2

Secondly, WAC provides functions which ensure the data transport, both traffic and control, between the WAC and other network elements. This is handled by the Data Transport Function (DTF) subsystem. Basically, all MN traffic is transparently handled and routed between the BS and CN by WAC. DTF intercepts, among others, authentication messages and DHCP packets and relays them to the corresponding entities.

Finally, another two functional subsystems of WAC are related with operation, maintenance and management. The Operation and Maintenance Function (OMF) subsystem is responsible for interfacing with OMC. OMC maintains state for the entire WAC cluster of RANs, which includes BSs, WAC, and RAN switch(es). OMF is in charge of configuration, performance counters, and software upgrades in the RAN entities. The Cluster Management Function (CLF) makes sure that all WAC cluster subsystems function correctly. CLF supervises, for example, cluster start-up and restart, and failover mechanisms against hardware and software failures.

2.5.2   Baseline Capacity Measurements

The measurements presented in this section are performed in a fashion similar to that described in Section 2.2. For example, all PCs are synchronized using PTP as discussed in Section 2.2.3. PCs acting as traffic generators/sinks are connected, on one end, with a Customer Premises Equipment (CPE) made by ZyXEL (MAX-210M1) and, on the other end, with the CN.

Table 2.4 summarizes the testbed configuration for the measurement results reported below. Our lab experience with the particular CPE model indicates that its performance may be the limiting factor for the overall measured performance. For example, among other things, the BS supports many more modulation and FEC schemes, for both uplink and downlink, such as 64 QAM, than those listed in Table 2.4, but the CPE does not.

As with the empirical evaluation of fixed WiMAX presented earlier, we evaluate the performance of the mobile WiMAX testbed by emulating different application traffic patterns. In this section we present the results for UDP traffic with varying packet sizes. The payload sizes chosen for the measurements are 100, 472 and 1372 bytes. The smallest payload size corresponds to ITU-T (1988) G.711 encoded voice samples. The largest payload represents, for example, video streaming applications. The payload size of 472 provides a reference middle point, mainly for comparing the testbed performance with the cases where very small or nearly Ethernet MTU-sized packets are injected into the network. However, it is important to highlight that the baseline results below are more in line with those presented

Table 2.5 Mobile WiMAX: average throughput with different packet sizes at link saturation.

MTU (bytes)	Downlink throughput (Mbps)	Uplink throughput (Mbps)
128	1.72	1.46
500	4.66	1.49
1400	5.13	1.53

in Section 2.2.4 than with those presented in Section 2.3. In short, in this baseline study we do not attempt to emulate VoIP or video traffic in a sophisticated manner. A more detailed evaluation of the capacity of our mobile WiMAX testbed with synthetic VoIP and IPTV traffic is part of our ongoing work.

Each of the payloads are generated using programs written in Perl, which add UDP/IP headers before each packet is injected into the network. Thus, at the network layer, the packets are sized 128, 500 and 1400 bytes, for the three measurement runs. The measured maximum throughput values for the three chosen packet sizes in downlink and uplink directions are given in Table 2.5.

As the throughput values presented in Table 2.5 include the UDP/IP headers, the larger overhead for the smaller packets is not taken into account. Just like the measurements with the fixed WiMAX tested, we note that the downlink performs clearly below par when small 128-byte packets are injected. The one-way delay in the downlink is just over 23 ms in the downlink and ranges between 25 and 30 ms in the uplink. However, at link saturation the one-way delay in the downlink exceeds 33 ms, and jumps to 180 ms in the uplink. With 128-byte packets, the mobile WiMAX downlink can deliver UDP throughput of approximately 1.7 Mbps with losses of the order of 1–2%; the uplink saturates in a similar fashion just above 1.4 Mbps.

When we inject 500-byte packets, UDP throughput improves significantly. At link saturation the mobile WiMAX testbed can deliver a UDP throughput of 4.7 and 1.5 Mbps in the downlink and uplink, respectively. With respect to one-way delay we once again observe a behavior similar to that described above. When the downlink handles a small number of packets the downlink and uplink one-way delay is approximately 30–34 ms. After the link saturation point is crossed, downlink one-way delay exceeds 43 ms while the uplink one-way delay increases dramatically to 600–700 ms. Of course, packet loss increases accordingly as well.

Finally, when we consider 1400-byte packets, the measured throughput values reach their maximum for our current testbed setup. Note that the MTU size used in the CPE configuration is 1400 bytes. With 1400-byte packets, the one-way delay in the downlink is 25–30 ms when the system is not congested. After the link saturation point, downlink one-way delay jumps to 85–90 ms, throughput saturates at 5.1 Mbps, and the packet loss ratio starts to increase. In the uplink, the one-way delay jumps from 35 ms to over 1.6 seconds after the link saturation point is reached. At the same point, the uplink UDP throughput saturates at 1.5 Mbps.

2.6 Summary

We presented a comprehensive evaluation of WiMAX performance in practice based on empirical investigations that employed two WiMAX testbeds located in Oulu, Finland. The first testbed uses Airspan fixed WiMAX equipment. The second testbed uses Alcatel-Lucent mobile WiMAX equipment.

Baseline capacity measurements showed that the fixed WiMAX testbed can deliver a throughput of 9.6 and 5.6 Mbps in the downlink and uplink, respectively, to a single UDP flow. Then, we considered the performance of aggregated and nonaggregated VoIP over our fixed WiMAX testbed. We measured the performance of different transmission schemes in terms of cumulative goodput, packet and sample loss rates, and calculated the objective MOSs using the R-Score specified by ITU. We found that VoIP flows carrying single sample payloads generated by the G.723.1 codec are clearly underperforming in both uplink and downlink and that simple voice sample aggregation schemes can more than double the effective capacity of the fixed WiMAX testbed when measured in terms of emulated VoIP calls with high-level MOSs.

We also evaluated VoIP and A/V streaming over a point-to-multipoint WiMAX testbed with one BS and two SSs. We measured the capacity of a single WiMAX cell under LOS and NLOS conditions to deliver tens of emulated bidirectional VoIP calls and a handful of emulated IPTV streams. Although our fixed WiMAX equipment does not have full support for QoS differentiation, we find the results presented very promising for real WiMAX deployments.

Last but not least, we presented a baseline capacity measurements study using our mobile WiMAX testbed, which, to the best of our knowledge, is the first of its kind provided by a third-party and disclosed publicly. We expect that our empirical evaluation results will be of great interest to other researchers in the BWA area, including simulationists and practitioners.

2.7   Further Reading

This chapter has been based on our experience with the fixed and mobile WiMAX testbeds of the VTT Converging Networks Laboratory. Our empirical evaluation methodology along with more detailed results can be found in Pentikousis *et al.* (2008a,b,c), Pinola (2008), Piri (2008) and Piri *et al.* (2008).

Readers interested in finding out more about the VTT Converging Networks Laboratory should visit http://cnl.willab.fi.

Acknowledgements

We would like to thank our colleagues Frerk Fitzek and Tuomas Nissilä for their invaluable work in earlier incarnations of this study; Mikko Hanski, Pekka Perälä, and Jarmo Prokkola for their assistance with the QoSMeT measurements; and Antti Heikkinen for his support while collecting the IPTV traces.

Part of this work was conducted at the VTT Converging Networks Laboratory within the framework of the IST 6th Framework Programme Integrated Project WEIRD (IST-034622), which was partially funded by the Commission of the European Union.

Study sponsors had no role in study design, data collection and analysis, interpretation or writing this report. The views expressed do not necessarily represent the views of VTT, the WEIRD project, or the Commission of the European Union.

References

Andrews, J.G., Ghosh, A. and Muhamed, R. (2007) *Fundamentals of WiMAX: Understanding Broadband Wireless Networking*. Prentice Hall, Englewood Cliffs, NJ.

Chatterjee, M. and Sengupta, S. (2007) VoIP over WiMAX. *WiMAX Applications* (eds Ahson, S. and Ilyas, M.). CRC Press, Boca Raton, FI., pp. 55–76.

Cole, R.G. and Rosenbluth, J.H. (2001) Voice over IP performance monitoring. *Proceedings of ACM SIGCOMM*, **31**, 9–24.

Correll, K. (2008) PTP daemon (PTPd). http://ptpd.sourceforge.net.

Correll, K., Barendt, N. and Branicky, M. (2005) Design considerations for software only implementations of the IEEE 1588 Precision Time Protocol. *Proceedings of the IEEE*, **1588**.

Grondalen, O., Gronsund, P., Breivik, T. and Engelstad, P. (2007) Fixed WiMAX field trial measurements and analyses. *Proceedings of the 16th IST Mobile and Wireless Communications Summit*, pp. 1–5.

Hoene, C., Karl, H. and Wolisz, A. (2006) A perceptual quality model intended for adaptive VoIP applications. *International Journal of Communication Systems*, **19**(3), 299–316.

IEEE 802.16 Working Group (2004) IEEE Standard for Local and Metropolitan Area Networks. Part 16: Air Interface for Fixed Broadband Wireless Access Systems. *IEEE Standard 802.16-2004*.

IEEE 802.16 Working Group (2005) IEEE Standard for Local and Metropolitan Area Networks. Part 16: Air Interface for Fixed Broadband Wireless Access Systems. Amendment 2: Physical and Medium Access Control Layer for Combined Fixed and Mobile Operation in Licensed Bands. *IEEE Standard 802.16e-2005*.

International Electrotechnical Commission (2004) Precision Clock Synchronization Protocol for Networked Measurement and Control Systems. *IEC 61588:2004(E), IEEE 1588-2002(E)*.

International Telecommunication Union (2005) Advanced video coding for generic audiovisual services. *ITU-T Recommendation H.264*.

ITU-T (1988) Pulse Code Modulation (PCM) of Voice Frequencies. *ITU-T Recommendation G.711*.

ITU-T (1996) Dual rate speech coder for multimedia communications transmitting at 5.3 and 6.3 kbs. *ITU-T Recommendation G.723.1*.

ITU-T (1998) The E-Model, a computational model for use in transmission planning. *ITU-T Recommendation G.107*.

ITU-T (2003) *ITU-T Recommendation G.114*.

ITU-T (2006) G.729 based Embedded Variable bit-rate coder: an 8-32 kbit/s scalable wideband coder bitstream interoperable with G.729. *ITU-T Recommendation G.729.1*.

Kitawaki, N. and Itoh, K. (1991) Pure delay effects on speech quality in telecommunications. *IEEE Journal on Selected Areas in Communications*, **9**, 586–593.

Manner, J. (2008) Jugi's Traffic Generator (JTG). http://hoslab.cs.helsinki.fi/savane/projects/jtg.

Maravedis (2008) http://www.maravedis-bwa.com.

Martufi, G., Katz, M., Neves, P.M., Curado, M., Castrucci, M., Simoes, P., Piri, E. and Pentikousis, K. (2008) Extending WiMAX to new scenarios: key results on system architecture and test-beds of the WEIRD project. *Proceedings of the Second European Symposium on Mobile Media Delivery (EUMOB)*, Oulu, Finland.

Mignanti, S., Tamea, G., Marchetti, I., Castellano, M., Cimmino, A., Andreotti, F., Spada, M., Neves, P.M., Landi, G., Simoes, P. and Pentikousis, K. (2008) WEIRD testbeds with fixed and mobile WiMAX technology for user applications, telemedicine and monitoring of impervious areas. *Proceedings of the Fourth International Conference on Testbeds and Research Infrastructures for the Development of Networks and Communities (TRIDENTCOM)*, Innsbruck, Austria.

Mills, D.L. (1992) Network Time Protocol (Version 3). *Request for Comments 1305*, IETF.

Neves, P., Simoes, P., Gomes, A., Mario, L., Sargento, S., Fontes, F., Monteiro, E. and Bohnert, T. (2007) WiMAX for emergency services: an empirical evaluation. *Proceedings of the International Conference Next Generation Mobile Applications, Services and Technologies*, pp. 340–345.

Pentikousis, K., Pinola, J., Piri, E. and Fitzek, F. (2008a) An experimental investigation of VoIP and video streaming over fixed WiMAX. *Proceedings of the Fourth International Workshop on Wireless Network Measurements (WiNMee)*, Berlin, Germany.

Pentikousis, K., Pinola, J., Piri, E. and Fitzek, F. (2008b) A measurement study of Speex VoIP and H.264/AVC video over IEEE 802.16d and IEEE 802.11g. *Proceedings of the Third Workshop on MultiMedia Applications over Wireless Networks (MediaWiN)*, Marrakech, Morocco.

Pentikousis, K., Piri, E., Pinola, J., Fitzek, F., Nissilä, T. and Harjula, I. (2008c) Empirical evaluation of VoIP aggregation over a fixed WiMAX testbed. *Proceedings of the Fourth International Conference on Testbeds and Research Infrastructures for the Development of Networks and Communities (TRIDENTCOM)*, Innsbruck, Austria.

Pinola, J. (2008) Mobility management in next generation wireless metropolitan area networks. *Master's Thesis*, The University of Oulu, Department of Electrical and Information Engineering, Oulu, Finland.

Pinola, J. and Pentikousis, K. (2008) Mobile WiMAX. *The Internet Protocol Journal*, **11**(2), 19–35.

Piri, E. (2008) Implementation of media independent handover for wireless broadband multimedia applications. *Master's Thesis*, The University of Oulu, Department of Electrical and Information Engineering, Oulu, Finland.

Piri, E., Pinola, J., Fitzek, F. and Pentikousis, K. (2008) ROHC and aggregated VoIP over fixed WiMAX: an empirical evaluation. *Proceedings of the 13th IEEE Symposium on Computers and Communications (ISCC)*, Marrakech, Morocco.

Prokkola, J., Hanski, M., Jurvansuu, M. and Immonen, M. (2007) Measuring WCDMA and HSDPA delay characteristics with QoSMeT. *Proceedings of the IEEE International Conference on Communications (ICC)*, Glasgow, Scotland, pp. 492–498.

Raake, A. (2006) *Speech Quality of VoIP: Assessment and Prediction*. John Wiley & Sons Ltd, Hoboken, NJ.

Retnasothie, F.E., Özdemir, M.K., Yücek, T., Celebi, H., Zhang, J. and Muththaiah, R. (2006) Wireless IPTV over WiMAX: challenges and applications. *Proceedings of the WAMICON*.

Scalabrino, N., Pelegrini, F.D., Chlamtac, I., Ghittino, A. and Pera, S. (2006) Performance evaluation of a WiMAX testbed under VoIP traffic. *Proceedings of the ACM WiNTECH*, pp. 97–98.

Scalabrino, N., Pellegrini, F.D., Riggio, R., Maestrini, A., Costa, C. and Chlamtac, I. (2007) Measuring the quality of VoIP traffic on a WiMAX testbed. *Proceedings of the Third International Conference on Testbeds and Research Infrastructures for the Development of Networks and Communities (TRIDENTCOM)*, pp. 1–10.

Schulzrinne, H., Casner, S., Frederick, R. and Jacobson, V. (2003) RTP: a transport protocol for real-time applications. *Request for Comments 3550*, IETF.

Sengupta, S., Chatterjee, M., Ganguly, S. and Izmailov, R. (2006) Improving *R*-Score of VoIP streams over WiMAX. *Proceedings of the IEEE International Conference on Communications (ICC)*, **2**, 866–871.

She, J., Hou, F., Ho, P.H. and Xie, L.L. (2007) IPTV over WiMAX: key success factors, challenges, and solutions. *IEEE Communications Magazine*, **45**(8), 87–93.

Sousa, B., Pentikousis, K. and Curado, M. (2008) Experimental evaluation of multimedia services in WiMAX. *Proceedings of the Fourth International Mobile Multimedia Communications Conference (MobiMedia)*, Oulu, Finland. ACM Press, New York.

Uilecan, IV., Zhou, C. and Atkin, G.E. (2007) Framework for Delivering IPTV Services over WiMAX Wireless Networks. *Proceedings of the EIT*.

Valin, J.M. (2008) Speex: a free codec for free speech. http://www.speex.org.

Wiegand, T., Sullivan, G.J., Bjøntegaard, G. and Luthra, A. (2003) Overview of the H.264/AVC Video Coding Standard. *IEEE Transactions on Circuits and Systems for Video Technology*, **13**(7), 560–576.

WiMAX Forum (2007a) WiMAX End-to-End Network Systems Architecture Stage 2: Architecture Tenets, Reference Model and Reference Points, *Release 1.1.0*, WiMAX Forum.

WiMAX Forum (2007b) WiMAX End-to-End Network Systems Architecture Stage 3: Detailed Protocols and Procedures, *Release 1.1.0*, WiMAX Forum.

Part III

Novel Scenarios

3

Novel WiMAX Scenarios for Future Broadband Wireless Access Networks

Pedro Neves, Kostas Pentikousis, Susana Sargento, Marília Curado, Paulo Simões and Francisco Fontes

3.1   Introduction

One of the concerns about the current Internet access technologies is that they cannot provide broadband access to all areas in a cost-effective manner. Current technologies either require a substantial investment in cabling and other infrastructure or cannot deliver broadband connections to several users round the clock. To address this issue, several proposals have been put forward that improve the efficiency of specific access technologies. For broadband wireless access, one promising option is based on the IEEE 802.16-2004 (IEEE, 2004) and IEEE 802.16e-2005 (IEEE, 2005) standards. Worldwide Interoperability for Microwave Access (WiMAX) (WiMAX Forum, 2008), is a broadband wireless access technology for local and metropolitan area networks (LANs/MANs), based on IEEE 802.16. WiMAX is an attractive broadband wireless alternative that can be used in urban and rural areas as well as in more demanding remote locations. By deploying WiMAX, broadband Internet access can be provided at only a fraction of the cost of wiring undeveloped areas.

Furthermore, IEEE 802.16 supports fixed subscriber stations, according to the IEEE 802.16-2004 (IEEE, 2004) standard, and mobile nodes, based on the IEEE 802.16e-2005 (IEEE, 2005) standard. The latter allows for node mobility in broadband wireless MAN scenarios. WiMAX is capable of supporting high-mobility nodes, with velocities exceeding 60 km/hour, while delivering application-layer throughput in excess of 10 Mbps. Using

WiMAX Evolution: Emerging Technologies and Applications   Edited by Marcos D. Katz and Frank H.P. Fitzek
© 2009 John Wiley & Sons, Ltd

different profiles, WiMAX can cover wide areas, which may reach 15 km in Non-Line-Of-Sight (NLOS) conditions and up to 50 km in Line-Of-Sight (LOS) environments, which is extremely important for rural areas. Using this access technology, operators can reach users distributed over large areas, with low installation costs when compared to fiber, cable or Digital Subscriber Line (DSL) deployments. Operational and management costs are also expected to be lower, which is an important factor especially when considering developing countries or rural areas. Another important factor is interoperability. Presently, the WiMAX Forum is leading the development of standardized system profiles that WiMAX Forum certified equipment must comply with. As an outcome of this effort, equipment prices are expected to decrease over time, and both users and operators will be able to avoid being locked in proprietary solutions by a single vendor.

In this chapter we present several scenarios for which we already envision the use of WiMAX technology ranging from fixed to mobile solutions, from backhaul for coverage extension to last mile connectivity, from business to residential and from urban to rural and impervious areas. Of course, this list of scenarios is not exhaustive and we expect that several more scenarios will also emerge. Let us also recall that the IEEE 802.16 working group is in fact developing its standards for both local and metropolitan wireless access. In the future, it may be the case that in-building connectivity could be provided solely with WiMAX or that, when prices become competitive, low-end profile equipment can be used as an alternative to WiFi (IEEE, 2007), thus taking advantage of the inherent Quality of Service (QoS) and security features of WiMAX.

This chapter is organized as follows. Section 3.2 describes possible scenarios for the use of WiMAX as last mile connectivity, including use cases where WiMAX is employed as a backhaul technology for coverage extension. Section 3.3 presents telemedicine-related scenarios, both for remote long-term patient monitoring and emergency services support. Section 3.4 presents several scenarios where WiMAX can prove to be valuable: environmental monitoring of volcanoes, forests and glaciers, to name a few. Finally, Section 3.5 summarizes and concludes this chapter.

3.2 WMAN Network Provider

In this section we present a very interesting set of application scenarios for WiMAX, related with its capacity to provide last mile connectivity to Internet users, based either on the fixed (IEEE, 2004) or on the mobile (IEEE, 2005) profile of the technology. Furthermore, we discuss developments initiated by newly established IEEE 802.16 working groups, which aim at enhancing WiMAX with relay and mesh networking functionalities. These working groups will also look into methods that will allow WiMAX to deliver even higher data rates, as required by the ITU IMT-Advanced (International Mobile Telecommunications – Advanced) (ITU, 2003b), for 4G compliant radio access technologies.

3.2.1 Broadband Wireless Access

IEEE 802.16 and WiMAX have evolved considerably over the years. Once considered as solely a fixed wireless local loop technology, for the most part suitable only for residential connectivity, WiMAX has emerged as a perfect candidate for providing broadband wireless access in numerous and quite diverse scenarios, as we explain in the remainder of this section.

WiMAX can backhaul traffic from other technologies, such as WiFi in Point-to-Point (PTP) and Point-to-Multipoint (PTMP) scenarios, replacing wires in remote areas, or even in business and residential urban areas.

Another likely scenario has one or more WiMAX Base Stations (BSs) providing access to several Subscriber Stations (SSs) connected through a mesh multi-hop network, extending WiMAX access coverage. This particular scenario is envisioned for urban residential areas. Furthermore, the mobility support of IEEE 802.16e-2005 opens up a different set of business opportunities for operators, turning IEEE 802.16 into a viable technology for the next generation of mobile communications, so-called 4G, as users will be able to access a wide range of services while moving.

Next, we review broadband access scenarios for rural and urban areas.

3.2.1.1 Rural Broadband Access

Delivering high-quality, wireless broadband access to rural communities is a challenging WiMAX deployment scenario for both developed and developing countries. This is mainly due to the large distances between the remote area and the operator infrastructure. As a rule, the small number of potential subscribers in rural areas does not entice operators to put in place the wired infrastructure necessary for broadband access. Even in developed countries, it is not always cost-effective to deploy wired solutions and public sector subsidies are often necessary. In developing countries, deploying wired alternatives is simply not viable. Wired solutions, such as DSL or cable, are only easily deployed in areas with existing infrastructure. As a result, rural areas are often left uncovered contributing to the Digital Divide (ITU, 2003a), or are covered with bare bones connectivity. In this case, WiMAX technology, particularly fixed WiMAX, can prove instrumental for providing broadband wireless access over large distances.

For example, in Finland, a country with low overall population density (about 16 people per square kilometer) but at the forefront of wireless communications developments, municipalities in the Kainuu Region have entered a joint venture with the local network operator (KPO) to deploy fixed WiMAX in suburban and rural areas. The population density in the Kainuu Region is only a quarter of the national population density (4 people per square kilometer). Until recently, residents in such areas had to contemplate narrowband-only and expensive connectivity options, as the economics of covering them with other broadband technologies are prohibitive.

Figure 3.1 illustrates a typical rural access connectivity scenario using fixed WiMAX technology. One or more WiMAX BSs are connected to the operator core network and provide wireless broadband connectivity to the rural area. Several WiMAX SSs, connected to the WiMAX BSs, are distributed over the remote area, including public institutions, residential areas, enterprises and healthcare centers, in a point to multipoint topology.

In these areas, nomadic scenarios are also an interesting application that can benefit WiMAX-based wireless access subscribers. The user can take his indoor WiMAX SS to another place and connect automatically to the WiMAX BS without requiring any technical support or contact between the customer and the network operator. For example, during construction projects, a company can benefit from the nomadic capability provided by WiMAX, connecting the construction head-office to the construction site temporarily, without having to subscribe for another access with the network provider.

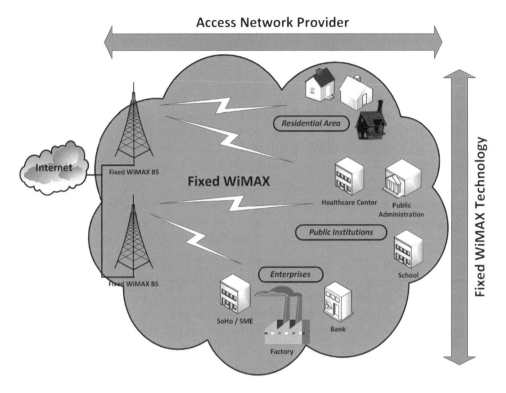

Figure 3.1 Rural point-to-multipoint fixed WiMAX access.

As mentioned above, fixed WiMAX can deliver capacities in the order of 10 Mbps with link spans that have been publicly demonstrated to exceed 20 km in LOS conditions as presented later in this section. Of course, application-level throughput depends on the network capacity, the channel bandwidth allocated and how sectors are defined. Operators can fine tune deployment details according to the subscriber population size and distribution using small angles to reach points farther away, or wider angles to distribute the available bandwidth among different customers.

WiMAX can also be used to establish connectivity between certain key points in the infrastructure and backhaul traffic from local access networks. Later in this chapter we describe, for example, how WiMAX can be used to backhaul environmental monitoring data and audio/video streams from remote areas. In a similar fashion, WiMAX can connect a remote village with the closest operator infrastructure connection point. Inside the village, residents can connect using a Local Wireless Access Network (LWAN), as illustrated in Figure 3.2. In this scenario the WiMAX SS is located in a central building, such as the town library, community center or an administrative building. The WiMAX SS acts either as a network router (layer 3 device) or a layer 2 bridge/switch and provides local access to a range of qualified WiFi connected devices. Users do not have to purchase WiMAX-enabled devices or cards: they access the network using inexpensive, often integrated, and in general widely available WiFi cards.

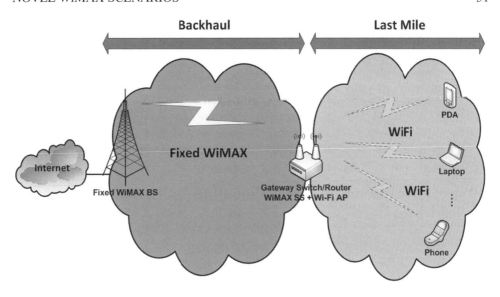

Figure 3.2 WiMAX backhaul for WiFi.

Of course, the WiMAX SS does not have to be collocated with the WiFi access point. Instead, it can be connected with a commodity Ethernet switch to a set of WiFi access points, which can provide access to public buildings such as the village school and the local police station, as well as to residents in a common area, such as the town square or a city park. This kind of deployment could be subsidized, initiated and managed by, for example, the local community. Alternatively, a commercial network operator could deploy the WiMAX SS and then use WLANs to reach potential customers, avoiding the installation of the entire cable infrastructure.

Another possibility is to actually install a Digital Subscriber Line Access Multiplexer (DSLAM) in the village and use the existing copper infrastructure for regular phone lines to provide DSL services to residents and small businesses alike. This is illustrated in Figure 3.3, where the extension of the network is achieved not by WLANs but by copper lines. A network operator can then offer services in a harmonized way with the rest of the offerings in the area or the country. The WiMAX backhaul will be transparent from the user's perspective, as a wired equivalent backhaul would be as well.

Although the nominal capacity of the WLAN is considerably higher than the nominal capacity of typical fixed WiMAX commercial off-the-shelf (COTS) equipment, in practice it is more than sufficient (Pentikousis et al., 2008b). First, fixed WiMAX operators can establish policies at the gateway router limiting the reservations for specific access points, thus making sure that customers receive the service level that they have signed up for. Second, fixed WiMAX has been shown in the lab to be more reliable and stable in performance than COTS WiFi access routers (Pentikousis et al., 2008a). The central scheduling performed by the WiMAX BS and IP-level policing are sufficient to enforce proper sharing of resources. Third, the WiMAX uplink and downlink are controlled independently and traffic in one direction does not affect traffic in the other.

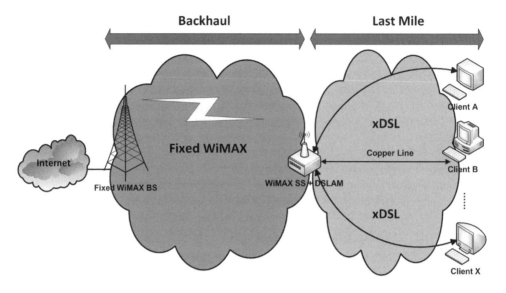

Figure 3.3 WiMAX backhaul for xDSL.

Finally, and beyond generic Internet data traffic backhauling, there have been proposals to use WiMAX to backhaul cellular data and voice traffic. The inherent QoS capabilities of WiMAX can meet the stringent requirements of cellular communication. Of course, the larger one-way delay is a concern, but for remote areas WiMAX can provide an acceptable price/performance trade-off (Pentikousis *et al.*, 2008a,b). Telecommunications operators have long been using proprietary microwave links to connect different offices or points of presence. Adopting a standard microwave solution such as WiMAX cannot be considered far-fetched. However, until now and to the best of our knowledge, WiMAX backhaul for cellular has yet to be deployed.

3.2.1.2 Urban Broadband Access

Although remote areas are very demanding to cover due to natural terrain challenges, urban zones are in fact one of the most difficult areas to provide truly cost-effective broadband connectivity. Mainly due to the large number of buildings, it is expensive and time consuming to deploy a wired broadband solution such as DSL or cable in these environments. Furthermore, cities with old or inadequate infrastructures are not encouraging network operators to install a wired solution. Historic, traditional and under preservation buildings also need network coverage, but cannot always be wired.

Fixed WiMAX is a very attractive solution to overcome the demanding challenges posed by urban areas. It is easy and very fast to install, without requiring major construction effort or structural interventions in buildings. Consequently, WiMAX proves to be significantly more cost effective when compared with wired alternatives and can deal with all types of physical environments, including NLOS conditions.

A typical urban scenario using fixed WiMAX as the access technology is shown in Figure 3.4. Potential subscribers, which may be interested in this type of deployment, include

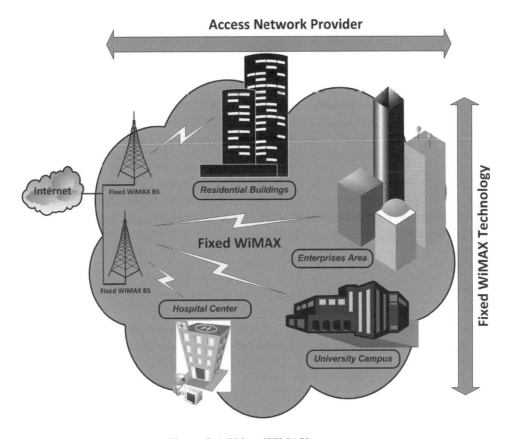

Figure 3.4 Urban WiMAX access.

residential customers, small and large enterprises, university campuses, public institutions and hospitals. Providing connectivity between the operator network and the customer is often straightforward. It simply requires the installation of a WiMAX SS in the roof of the target building(s).

To address the radio signal propagation challenges mainly posed by tall buildings, the scenario depicted in Figure 3.5 includes a WiMAX relay system, based on the IEEE 802.16j (IEEE, 2008b) draft standard. IEEE 802.16j will enhance WiMAX with the capability to provide Mobile Multi-hop Relay (MMR) by introducing Relay Stations (RSs) between the WiMAX BS and the receiving nodes. As a result, the BS cell coverage can be extended significantly. Furthermore, the IEEE 802.16j specification guarantees that the receiving entities will not notice the existence of WiMAX RSs in the path towards the WiMAX BS, that is, it is completely transparent to the terminal.

Based on the RS capabilities, MMR systems can operate in two distinct modes, namely, transparent or nontransparent. In transparent mode, the transparent RS, also referred to as simple RS, is only capable of processing and forwarding data traffic, whereas the signaling messages, such as the MAC management messages, must be sent towards the controlling BS

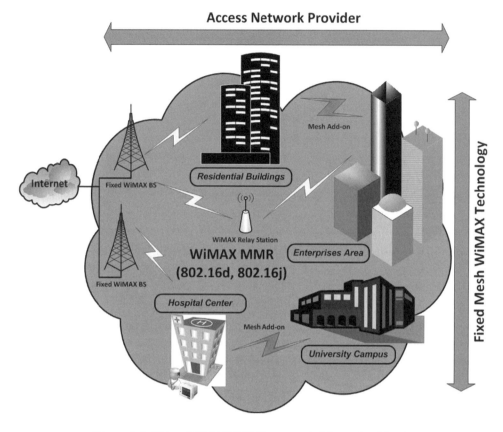

Figure 3.5 Urban WiMAX MMR access (with mesh add-on).

(or master BS). In nontransparent mode, the RS has the capacity to forward both signaling and data traffic towards the final nodes. Therefore, the nontransparent RSs are able to locally manage and control the WiMAX link, without synchronizing with the master BS.

A WiMAX mesh topology is also illustrated in Figure 3.5, allowing for direct communication between the WiMAX SSs installed on the roof of the buildings. Combining both mesh and MMR capabilities, fixed WiMAX is able to meet the challenges posed by urban landscapes. In this scenario, a WiMAX operator can provide broadband access, overcoming NLOS conditions and optimizing traffic routing by directly connecting different SSs.

3.2.2 Advanced Mobile WiMAX

Presently, wireless telecom operators are trying to optimize their third-generation wireless networks in order to deliver a new set of services to customers while capitalizing on the significant investment in their existing infrastructure. At the same time, operators are also seeking for next-generation access technologies, capable of integrating all-IP architectures, with QoS support and high throughputs, as well as seamless mobility mechanisms. Users of next-generation wireless networks will look for a combination of services that are currently

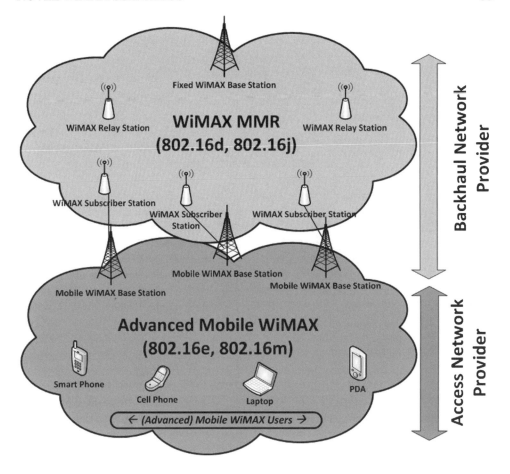

Figure 3.6 (Advanced) Mobile WiMAX access backhauled by WiMAX MMR.

offered in fixed environments, such as Voice over IP (VoIP), video streaming and data communications, with their mobile lifestyle, enabling more futuristic application scenarios.

WiMAX is built upon an all-IP framework, able to deliver triple-play services to mobile, nomadic and portable terminals. Bearing in mind these unique features, one of the most interesting scenarios for WiMAX is to provide ubiquitous connectivity to mobile users, as illustrated in the bottom of Figure 3.6. Users will be able to access broadband wireless Internet using different devices, such as a mobile phone, laptop or PDA.

Although IEEE 802.16e-2005 (IEEE, 2005), can deliver high end-user data rates, ITU IMT-Advanced, which will supersede ITU IMT-2000, states that the next-generation wireless technologies, to be deployed by 2015, must provide much greater data rates while allowing for high user mobility and delivering a wide range of services to users. More specifically, ITU IMT-Advanced dictates that data rates should exceed 100 Mbps in high mobility scenarios and 1 Gbps in low mobility scenarios. Recently, the IEEE 802.16 working group started the specification of a new draft standard titled Air Interface for Fixed and Mobile Broadband

Wireless Access Systems – Advanced Air Interface (IEEE, 2008a), which is an evolution of IEEE 802.16e-2005. Recall that the latter has already been accepted by ITU as one of the technologies for fourth-generation wireless communications (ITU, 2007).

Moreover, the WiMAX Forum, which currently certifies mobile WiMAX equipment based on IEEE 802.16e-2005, has already declared that, as soon as the IEEE 802.16m work is complete and the standard is finalized, it will be included on the WiMAX Forum roadmap. Another important consideration is that IEEE 802.16m will be backwards compatible with IEEE 802.16e-2005, enabling both standards to use the same WiMAX BS. The last mile of the scenario illustrated in Figure 3.6 addresses both Mobile WiMAX (based on IEEE 802.16e-2005) and Advanced Mobile WiMAX (based on IEEE 802.16m) access technologies, using the same WiMAX BS.

Nowadays, it is not possible for telecom operators to provide an efficient and cost-effective solution to connect the BSs to their core network infrastructure. Frequently, fiber or copper lines are necessary to establish a wired link between them. Owing to the high costs associated with wired solutions, it is not always economically feasible for operators to provide connectivity to certain remote areas, in which the number of broadband subscribers is very small and thus potential revenue is marginal, as discussed earlier in this chapter. As a result, a large number of areas are still not covered by mobile broadband wireless access. In Figure 3.6 we illustrate the use of the IEEE 802.16j draft standard, which allows multi-hop relay communication in the backhaul part of the network through WiMAX RSs.

Moreover, instead of using a simple WiMAX MMR topology, a very interesting solution is to integrate the mesh topology as well, as illustrated by the wireless communication links marked 'Mesh Add-on' in the middle of Figure 3.7. By exploiting mesh topologies, operators will be able to route traffic directly between the WiMAX SSs without having to send it towards the WiMAX BS or RS, thus optimizing the use of all available resources. Figure 3.7 also shows the last mile connectivity using the IEEE 802.16e and/or IEEE 802.16m standards, as well as the backhaul connectivity through the use of the WiMAX relay system.

Furthermore, Figure 3.7 depicts the achieved optimization through the use of both mesh and RSs in the backhaul network. Three possible data paths, represented by the dotted, dashed and solid lines for the communication between Client A and Client C are shown. The dashed arrow shows the communication between Client A and Client C using a transparent RS (RS#1) without the mesh optimization. In this case, data is sent by Client A in the uplink direction towards the mobile WiMAX BS#1 and delivered to the WiMAX SS#1. Thereafter, since the WiMAX BS is not reachable, the data is sent to the WiMAX RS#1. Since in this case the RS#1 is operating in transparent mode, it forwards the data upwards to the WiMAX BS; the latter will then send the packets in the downlink direction to Client C, through the WiMAX RS#1, SS#2 and BS#2. On the other hand, if RS#1 is operating in nontransparent mode, it would route the data towards the WiMAX SS#2, as shown by the dotted line, which forwards the packets to Client C. Finally, the solid arrow represents the optimization provided by the mesh topology. In this case, when the data packets reach the WiMAX SS#1, they are forwarded directly to WiMAX SS#2 without having to be sent towards WiMAX RS#1. This process has fewer hops between the source and the destination clients and therefore provides an optimized path for the communication.

Summarizing, enhancing the network with relay and mesh capabilities opens up several possibilities for routing traffic more efficiently by taking advantage of its locality and avoid-ing the traditional communication with the WiMAX BS. The operator has the opportunity

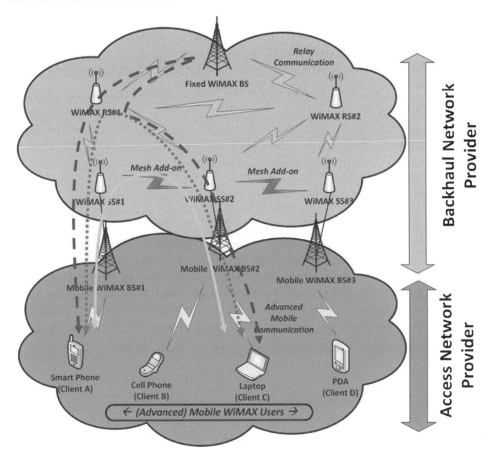

Figure 3.7 (Advanced) Mobile WiMAX access backhauled by WiMAX MMR (mesh add-on).

to use different paths to balance traffic, depending on user profiles, the type and priority of different flows and the link utilization. However, network operators must be cautious to ensure that the resulting bandwidth per user is sufficient for the running services, as bandwidth is shared between all of the WiMAX SSs and users.

Next, we build upon what was presented in this section and explain how WiMAX can also be used for telemedicine applications, on-site ambulance and accident monitoring, and environmental monitoring. All of these often require surveillance of remote and/or otherwise inaccessible locations. For example, we describe how WiMAX can be employed in emergency and high-mobility scenarios as well as volcano monitoring and fire prevention applications.

3.3 Telemedicine Applications

E-health is one of the areas where WiMAX technologies can substantially contribute to improve the daily activities and therefore enhance the quality of life of the citizens. Today a large number of medical activities are carried out with limited success, unnecessary costs and human difficulties because of the inability to exchange real-time information between different elements of the chain that are not at fixed locations. Some of the possible advanced medical services that can benefit from the integration of a wireless broadband access technology are as follows.

- Remote diagnosis: need to transmit urgent data in order to make an immediate basic diagnosis, for example, street accidents, people in special health conditions (pacemaker bearers, pregnant women, to name a few).

- Need to intervene on nontransportable patients (e.g. accidents) may require off-air transmission of critical data or images.

- Remote monitoring: today the elderly are monitored remotely when at home, but not when traveling or simply moving around town. This limits their ability to enjoy life and be integrated in the community around them.

- Remote follow-up: today patients travel to distant hospitals to be followed-up after therapies or surgical interventions.

3.3.1   Remote Patient Monitoring

During the last decade, the healthcare system has been giving much more attention to remote monitoring and assistance of patients. Remote patient monitoring poses difficult challenges in order to give a differentiated support to each patient, creating a personalized follow-up plan, which depends on the patient illness. Furthermore, it is very important that a bidirectional, secure, reliable and trustworthy relationship between the patient and the healthcare provider is established, allowing the patient to proactively trigger communication with the hospital and share important information or symptoms with the medical team. By the same token, hospital doctors must also be able to remotely control and monitor patient measurements, as well as send periodical reminders to the patient in order to take his medication.

Another important category comprises patients with chronic diseases that can be remotely monitored during the day, without having to change their daily routine. For instance, if the patient's professional life demands a constant need to travel abroad, they can easily maintain their work routine. Keeping chronic patients physically away from the healthcare center, not only improves the patients' quality of life, but is also cost effective for healthcare institutions, as it allows for operation with less medical personnel and requires a smaller number of hospital beds.

Despite the benefits provided by the telemedicine applications, until this moment no access technology was able to fulfill its challenging requirements, namely, wireless broadband availability, including QoS differentiation, as well as intrinsic mobility support. WiMAX, as a broadband wireless access technology for metropolitan environments, is a very promising candidate technology to overcome this gap and allows both healthcare institutions and patients to benefit from this application, as illustrated in Figure 3.8. Here, the mobile version of WiMAX is employed, allowing the patients to be monitored remotely by the healthcare institution.

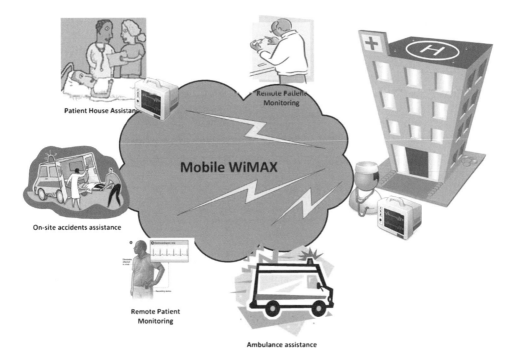

Figure 3.8 Remote patient monitoring.

3.3.2   On-site Medical Assistance

Another important scenario for telemedicine applications is to provide the remote medical staff with the capability to immediately establish a secure and reliable communication channel with the healthcare center. This scenario, using mobile WiMAX, is also illustrated in Figure 3.8 (Patient House Assistance).

Using mobile WiMAX, when a doctor assists a patient at home (e.g. senior citizen, pregnant women, patient with a chronic illness), they can easily establish a reliable communication channel with the hospital and retrieve important information about the patient, as well as send the portable ultrasound device results to the hospital. It also allows the doctor to exchange information and ask for a second opinion from their colleagues at the hospital. Furthermore, if necessary, a video-conference session can be established in real time using the doctor's laptop. If the doctor is visiting a pregnant woman, he can upload the ultrasound pictures of the baby to the hospital server and configure it to distribute the pictures to the father of the baby.

Another interesting scenario, also illustrated in Figure 3.8 (Ambulance Assistance), is when a doctor is on duty in an ambulance and is called to intervene on a car accident. The ambulance is equipped with a portable ultrasound device, connected to a WiMAX notebook. An injured man is found and to allow the fastest possible intervention, while travelling towards the hospital, the doctor collects important information about the patient's condition and sends it to their colleagues in the hospital. Based on this input, the medical staff can start to prepare equipment, as well as share important information about the patient with the ambulance doctor.

3.4   Environmental Monitoring

WiMAX technology also fits very well into a large number of environmental monitoring applications, mainly due to its wide coverage potential in remote areas but also as a result of its inherent support for broadband applications and QoS mechanisms. In this section, four environmental monitoring application areas will be discussed: monitoring of seismic and volcanic activity, fire prevention and other applications such as meteorological forecasting, pollution monitoring and law enforcement.

3.4.1   Seismic Activity

Europe has several regions with high seismic activity, requiring permanent monitoring in order to prevent calamity situations. The main affected regions are in the south, due to the fault zone between the Euro-Asian and African plates, and in the north, due to the fault zone between the Euro-Asian and North-America plates. This section describes typical seismic activity scenarios and presents the main challenges associated with volcano monitoring, which relies strongly on the observation of seismic activity patterns.

3.4.1.1   Seismic Activity Monitoring

Monitoring of seismic activity plays a critical role on the minimization of the impact of its effects on populations and facilities. The effect of seismic activities, or earthquakes, have a direct impact on the ground shaking and rupture, causing damage to service structures, such as dams and bridges, but can also be responsible for other nonexclusive natural disasters such as fires, tsunamis and floods. Therefore, the detection or even prediction of seismic activities is critical for planning preventive actions, as well as for all of the remedial actions to be carried out during and after such occurrences.

Nowadays, there are several sensor networks deployed in critical regions, dedicated to the collection of seismic signals. Usually, the information is stored in local storage devices and regularly sent to the acquisition center. However, the continuous and real-time transmission of data to the acquisition center would improve the reaction time in the event of a crisis. In this context, WiMAX technology introduces significant value, due to its broadband capabilities over long distances. Namely, by providing, in real-time, data from seismic sensors of different sources, many located in places farther than 30 km, the prediction of earthquakes may become more efficient and focused.

3.4.1.2   Volcano Monitoring

From the different events associated with seismic activity, volcano eruption is one of the most feared, especially in places with a long history of volcano activity, such as Iceland, Italy and Portugal. It is therefore critical an increased coverage to monitor the seismic activity associated with volcanic activities. This is even more important, given that seismic activity preceding volcanic eruptions, are only observed approximately one hour before the magma reaches the surface. With such a hard constraint, real-time transmission of data is vital for early warning systems and preparation for the oncoming risk.

In addition to relying on seismic activity observation, volcano monitoring is complemented by the use of gas monitoring equipment, which is located close to the peak of the volcano. In addition, video surveillance cameras are placed in mountains around the volcano in order to expose the location of the eruptive site and to follow the ascension of the volcanic column into the upper atmosphere.

In this scenario, wired technologies are naturally inappropriate, and most wireless technologies do not fulfil the connectivity at long distances with high bandwidth. Moreover, the access to the seismic and gas monitoring sensors, as well as to the video surveillance cameras can benefit from communication both with LOS and NLOS. Therefore, the use of WiMAX as a broadband wireless access technology becomes appealing and a strong candidate to fulfil the communication requirements of such demanding scenarios.

In addition to being an important contribution for the successful deployment of the application scenarios described above, the WiMAX technology also brings benefits to the field personnel who installs and maintains the monitoring equipment, as well as to the civil protection personnel and rescue teams in disaster situations. Namely, the wireless broadband access with mobility features allows for real-time communication between the personnel and the acquisition center, as well as with emergency services.

3.4.2 Fire Prevention

For Southern European countries, such as Portugal, Spain, Greece and southern France, forest fires represent a dramatic loss of life and property, as well as an increased risk of desertification and the higher emissions of greenhouse gases. The same applies, to some extent, to some areas of the United States, Australia and other countries. In order to address this situation, considerable public investment has been made in fire prevention (e.g. proactive land management and preventive clearing of ground vegetation), fire detection networks (surveillance posts, aerial patrols, car patrols, semi-automated fire detection mechanisms) and fire fighting resources (aerial resources, firefighting vehicles, firefighters and civil protection personnel, etc.). Nevertheless, due to climatic factors, increased desertification and land usage practices, these efforts remain largely unsuccessful.

3.4.2.1 Fire Detection

One of the key success factors in this area of environmental monitoring is early fire detection, since an unattended fire ignition can evolve, depending on weather conditions, into uncontrollable proportions in less than an hour. However, forest areas are typically scarcely populated and difficult to reach. In these areas, fire detection is traditionally performed by human spotters, on top of surveillance towers or, in some cases, small airplanes, searching for smoke signals and communicating their findings to centralized control centers. These methods are not cost-efficient: each tower requires the permanent presence of human watchers. Moreover, verbal radio communication can be ineffective as the control center loses precious minutes processing the spotter description and attempting to locate the potential fire precisely, before dispatching an airborne first response team. A small country such as

Portugal, for instance, needs to keep a network of more than 230 surveillance towers, just to get basic coverage of fire-sensitive territories.

Electronic surveillance mechanisms (for instance, video surveillance with automatic detection of smoke or heat sources) are not fully autonomous, since human confirmation of potential fires is always required. Nevertheless, they do improve the efficiency of surveillance networks.

- Personnel costs are greatly reduced. Even with plain and simple remote video surveillance, one centrally located operator can easily manage multiple surveillance towers.

- Fire location becomes faster and more precise, due to automated triangulation between adjacent surveillance towers and association with Geographic Information Systems (GIS).

- False alarms are easier to identify. Instead of relying in verbal descriptions from remote human spotters, the centralized operator can have simultaneous access to the video and data gathered at each surveillance tower that covers the potential fire location. The decision process becomes faster and more precise, leading to reduced costs with unnecessary dispatches of helicopter crews.

- Electronic surveillance can operate continuously, unlike traditional surveillance networks which, due to their high costs, are usually limited to a few months per year, despite the fact that recent climatic changes increase the frequency of violent off-season fires.

- Recorded data has great forensic value for criminal investigation and scientific research.

Video is the basis of electronic fire surveillance. Plain video surveillance already provides very positive results, and more sophisticated tools can be easily added for automated detection of smoke of heat sources. Other data, such as local meteorological data and digital compasses, play an important role during the detection and monitoring phases, but video is usually the key component (and also the most bandwidth demanding component).

One of the major shortcomings of electronic surveillance is communication between surveillance towers and control centers. Broadband access technologies such as DSL or 3G are not available in remote forests, and alternatives such as GPRS, GSM, TETRA or UHF radio-links, in addition to also presenting coverage limitations, rarely provide the bandwidth required for effective electronic monitoring.

WiMAX technology appears to fit well in this scenario. Given its potential range, bandwidth and adaptability to environmental conditions, WiMAX may provide connectivity to remote monitoring systems, capable of effectively providing early detection of fire in a more efficient and cost-effective manner.

3.4.2.2   Fire Monitoring and Firefight Coordination

Electronic surveillance is also critical during the firefighting phase. Senior coordinators, located at the operations control centers, can themselves follow the evolution of the fire live, often from multiple angles. Once again, this is a tremendous advantage over making strategic decisions based on verbal information received from remote watchers and firefighters.

At another level, WiMAX may also play an important role in the coordination of partners involved in the firefight, for instance allowing the instant exchange of data between firefighters and the control center (such as the status and GPS location of the firefighter vehicles, meteorological data, GIS data, video and images of the fire gathered at multiple locations, instructions from the control centre, etc.). This is not much different from classical warfare scenarios, where increased access to shared intelligence improves the global capacity of the military team.

It should be noted, however, that the requirements for electronic surveillance and firefight coordination are different. Electronic surveillance is based on a limited number of fixed network nodes (the surveillance towers), with very large distances between them but usually with LOS conditions, due to their placement in strategically high locations. Fixed WiMAX mesh networks, with PTP or PTMP links, provide a satisfactory answer in this case. Firefight coordination, on the other hand, requires support for mobile or nomadic terminals. Distances involved are potentially smaller, but the location of each network node is suboptimal (e.g. vehicles in valleys and ravines without LOS), imposing more demanding coverage problems. In this context mobile WiMAX is probably more adequate.

3.4.2.3   A WiMAX Testbed for Electronic Fire Surveillance

A small-scale demonstrator of the WiMAX technology for fire prevention is taking place in the surroundings of Coimbra, Portugal (Neves *et al.*, 2007), in the context of the WEIRD Project (Neves *et al.*, 2008). A WiMAX network was deployed, connecting the University of Coimbra (UC) with two surveillance towers, which are part of the nationwide surveillance network. The first surveillance point, located at 'Alto do Trevim, Serra da Lousã' (SL), is 22 km away from the UC, while the second surveillance point, located at 'Serra do Carvalho' (SC), is 19 km further away from SL.

Figure 3.9 shows the topology of the WiMAX network, which is based on two PTP fixed WiMAX links. According to the number and location of surveillance towers, the network could be based on PTP or PTMP links. Furthermore, since surveillance towers are usually located at strategically high locations and within sight of each other, it would be relatively easy to extend coverage by adding more WiMAX links, using multi-hop topologies.

The first surveillance tower has two video cameras, which provide 360° coverage with more than 20 km radius. The second surveillance tower provides similar coverage, with a single video camera. Collected data (periodic panoramic photos, digital compass data and meteorological information) is sent to a central application, where it is complemented with data from the GIS database and other surveillance towers and where prospective fires are further investigated (for instance, manually pointing and zooming the remote cameras to the area in question). Support for firefighting coordination activities, communication with firefighters and collection of data from mobile vehicles, will be added later, when mobile WiMAX equipment becomes more widely available.

Both WiMAX links use 14 MHz channels. Table 3.1 presents the WiMAX configuration of each link, as well as maximum link performance measured with Iperf (IPERF, 2005). Despite the use of 90° antennas for the WiMAX BSs, due to logistic reasons, and very poor LOS conditions for one of the links, available bandwidth was more than enough to support

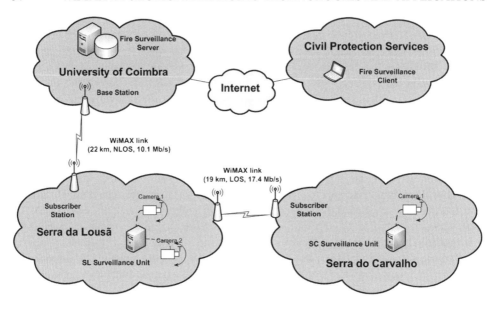

Figure 3.9 Fire prevention demonstrator of the WEIRD project.

Table 3.1 Fire Prevention Demonstrator of the WEIRD Project.

		Capacity		
Link	Antennas	Uplink (Mbps)	Downlink (Mbps)	Total (Mbps)
UC/SL: 22 km, 14 MHz, poor LOS	BS: 90°C, 23 dBm SS: 15°C, 16 dBm	4.5	5.6	10.1
SL/SC: 19 km, 14 MHz, good LOS	BS: 90°C, 23 dBm SS: 15°C, 16 dBm	7.8	9.6	17.4

the surveillance network. Live video streaming from each camera requires less than 2 Mbps, but even a much larger number of video cameras (more surveillance towers, infrared cameras, very high-resolution cameras, etc.) could be supported with simple optimization techniques such as reducing the frame rate of each camera in 'automatic scan' mode.

To the best of our knowledge, this is the first WiMAX-based fire surveillance network. It has been in experimental use since 2007, with considerable stability and availability (despite the harsh conditions at SL, located 1200 meters above sea level and frequently facing winds over 100 km/hour). During the summer of 2008 the system will be actively used by the Civil Protection Services for fire detection and fire monitoring, allowing a more extensive evaluation study. A more detailed description of the testbed can be found at Neves *et al.* (2007).

3.4.3 Other Applications

In addition to the already mentioned scenarios, WiMAX technology can also play an important role in other applications, such as pollution and generic environmental monitoring and video surveillance in wild areas.

3.4.3.1 Pollution and Generic Environmental Monitoring

The category of pollution monitoring and more generic environmental monitoring encompasses all sorts of telemetry targeted at specific environmental indicators. Examples include monitoring of pollutant levels in the air, soil or water; collection of local meteorological data; river and stream flow monitoring; monitoring of water reservoir levels, and so on.

From a communications point of view these applications are usually based on regular transmission of measured data to remote locations. Usually it is acceptable that some delay occurs between data collection and data transmission, and the size of data to transmit is relatively contained.

From a logistic perspective these applications may have two important requirements to take into account: location and energy consumption. The location of telemetry units is often inflexible (they need to be close to the monitored objects), which may impose coverage problems for a number of wireless transmission technologies. Quite often, these telemetry units are also far away from the energy grid, which means they have to rely on batteries or solar power. This imposes limitations both on the measurement equipment and on the communication equipment that might imply, for instance, nonpermanent connections and wireless technologies with lower power requirements. This is not the most favorable scenario for the WiMAX technology: other solutions, such as GPRS or even GSM/SMS might provide enough bandwidth for these telemetry applications, with lower power consumption (e.g. using nonpermanent connections).

However, in other cases, WiMAX still brings some potential advantages, especially when there is a need for considerable bandwidth or when GPRS or GSM coverage is simply not available. Communication costs also need to be taken into account, since the costs of GPRS-based telemetry can become quite high. Although the business model for commercial WiMAX use is still not clear, costs will probably be lower in these cases.

WiMAX networks can also interconnect 'sink' components of wireless sensor networks. Instead of directly communicating with a remote collector, each telemetry unit might use limited-range wireless technologies to communicate with a relatively close 'sink' (a common concept in wireless sensor networks, which also applies to generic telemetry networks). This 'sink' can then use WiMAX to relay monitoring data to the remote data collectors. Sinks have less energy restrictions, can have better WiMAX coverage and have higher bandwidth requirements. In this context, WiMAX is a very competitive technology.

3.4.3.2 Video Surveillance in Wild Areas

Another category of environmental monitoring relates with video surveillance in wild areas. Video surveillance, in these areas, can be used, for instance, to detect illegal activities (hunting, illegal logging, etc.) or to monitor wildlife activities (e.g. wild cattle or bird activity).

For the same reasons already discussed for fire surveillance, WiMAX fits quite well into this kind of applications, easily covering wild areas and providing the required bandwidth. In fact, once a video surveillance network is installed in a wild area, it is expected to be used simultaneously for a number of activities: fire prevention, detection of illegal activities and, sometimes, monitoring of wildlife.

Logistic problems are not much different from fire surveillance scenario, although the location of surveillance units can be less favorable (e.g. worst LOS conditions) and availability of energy sources can also be a problem. Nevertheless, energy sources are a problem associated with the remote surveillance unit, not specifically with WiMAX technology (no other technology can provide the required bandwidth with lower energy consumption).

3.5   Conclusions

It is unquestionable that WiMAX represents an important step on the evolution of communication scenarios. The great advantage of providing broadband communication to remote locations without wires enables the support of numerous different scenarios that were not envisioned before, both business and residential, in rural and urban areas.

This chapter has overviewed the most relevant scenarios, considering different standards, fixed, mobile and mesh, from simple scenarios, such as using fixed WiMAX to serve as backhaul to several buildings in an urban scenario, to the most complex scenarios, such as using mesh mode and RSs balancing the traffic and serving as backhaul to mobile WiMAX. In terms of applications in addition to normal communication between users, two main types were described, namely, medical and environmental monitoring applications. The broadband and reliability characteristics of the communications using WiMAX make it suitable to be used in emergency situations, such as on-site medical assistance, where it is required to maintain constant connectivity in usually remote locations. The distant communication required by environmental monitoring is also strongly benefiting from WiMAX, as it enables simultaneously reliability, QoS, broadband and remote access.

Acknowledgements

Part of this work was conducted within the framework of the IST 6th Framework Programme Integrated Project WEIRD (IST-034622), which was partially funded by the Commission of the European Union. Study sponsors had no role in study design, data collection and analysis, interpretation or writing the report. The views expressed do not necessarily represent the views of the authors' employers, the WEIRD project, or the Commission of the European Union. We thank our colleagues from all partners in WEIRD for fruitful discussions.

References

IEEE (2004) IEEE Standard for Local and Metropolitan Area Networks; Part 16: Air Interface for Fixed Broadband Wireless Access Systems, *IEEE Standard 802.16-2004*, IEEE 802.16 Working Group.
IEEE (2005) IEEE Standard for Local and Metropolitan Area Networks; Part 16: Air Interface for Fixed and Mobile Broadband Wireless Access Systems; Amendment 2: Physical and Medium Access

Control Layers for Combined Fixed and Mobile Operation in Licensed Bands, *IEEE Std. 802.16e-2005*, IEEE 802.16 Working Group.

IEEE (2007) IEEE Standard for Information Technology; Telecommunications and information exchange between systems; Local and Metropolitan Area Networks; Specific Requirements; Part 11: Wireless LAN Medium Access Control (MAC) and Physical Layer (PHY) Specifications, *IEEE Std. 802.11-2007*, IEEE 802.11 Working Group.

IEEE (2008a) Draft Amendment to IEEE Standard for Local and Metropolitan Area Networks; Part 16: Air Interface for Fixed and Mobile Broadband Wireless Access Systems – Advanced Air Interface, *IEEE P802.16m/D0.0*, IEEE 802.16 Working Group.

IEEE (2008b) Draft Amendment to IEEE Standard for Local and Metropolitan Area Networks; Part 16: Air Interface for Fixed and Mobile Broadband Wireless Access Systems – Multi-hop Relay Specification, *IEEE P802.16j/D5.0*, IEEE 802.16 Working Group.

IPERF (2005) *Iperf*, NLANR DAST, http://dast.nlanr.net/Projects/Iperf/.

ITU (2003a) *ITU Press Release*, ITU Group, http://www.itu.int/newsroom/press_releases/2003/30.html.

ITU (2003b) *ITU-R, Framework and Overall Objectives of the Future Development of IMT-2000 and Systems Beyond IMT-2000, M.1645*, ITU Group.

ITU (2007) *ITU Press Release*, ITU Group, http://www.itu.int/newsroom/press_releases/2007/30.html.

Neves, P., Nissila, T., Pereira, T., Ilkka, H., Monteiro, J., Pentikousis, K., Sargento, S. and Fontes, F. (2008) A vendor independent resource control framework for WiMAX. *Proceedings of the 13th IEEE Symposium on Computers and Communications (ISCC)*, Marrakech, Morocco.

Neves, P., Simões, P., Gomes, A., Mário, L., Sargento, S., Fontes, F., Monteiro, E. and Bohnert, T. (2007) WiMAX for emergency services: an empirical evaluation. *Proceedings of the 1st Broadband Wireless Access Workshop (BWA), Collocated with 1st Next Generation Mobile Applications, Services and Technologies (NGMAST)*, Cardiff, UK.

Pentikousis, K., Pinola, J., Piri, E. and Fitzek, F. (2008a) A measurement study of Speex VoIP and H.264/AVC Video over IEEE 802.16d and IEEE 802.11g. *Proceedings of the Third Workshop on MultiMedia Applications over Wireless Networks (MediaWiN)*, Marrakech, Morocco.

Pentikousis, K., Pinola, J., Piri, E. and Fitzek, F. (2008b) An experimental investigation of VoIP and video streaming over fixed WiMAX. *Proceedings of the Fourth International Workshop on Wireless Network Measurements (WiNMee)*, Berlin, Germany.

WiMAX Forum (2008) WiMAX End-to-End Network Systems Architecture Stage 2–3: Architecture Tenets, Reference Model and Reference Points, *Release 1, Version 1.2*, WiMAX Forum.

4

Pricing in WiMAX Networks

Ioannis Papapanagiotou, Jie Hui and Michael Devetsikiotis

4.1 Introduction

Pricing for multi-service and multi-topology Broadband Wireless Access (BWA) networks remains a major issue for the efficient design of WiMAX networks. With the proliferation of Voice over IP (VoIP) , entertainment video, triple play services and towards the evolution of the future Internet, an exorbitant demand for wireless broadband technologies has been created. However, little effort has been devoted to how users should be charged both from the economic interest of the provider but also from the way that a WiMAX network is designed.

In this chapter we discuss the use of economics and pricing in the design of a WiMAX network. Our initial focus shall be on whether the investment of building a WiMAX network is profitable or not, using terms such as Capital Expenditure (CAPEX), Operational Expenditure (OPEX), Net Present Value (NPV) and identifying who shall pay for the extra charges. Another major issue is the role of pricing in Quality of Service (QoS), congestion control and provisioning of a network (DaSilva, 2000). This provides insight into why prices for network services are not only a marketing and strategic decision but also an engineering concern. Moreover, the users' preferences need to be captured as well as their sensitivity when changes tend to happen in the form of a utility function, so as to compose the network utility maximization problem or that equilibrium that will satisfy all of the entities in a multi-topology scenario.

In the following we also study ways to build WiMAX networks from an economic perspective and describe the role of pricing in bandwidth allocation. We are interested in different topologies depicting most of the applications of WiMAX networks. The first topology shall be the one-hop topology in a singe cell environment when the Mobile Station (MS) is directly connected to the Base Station (BS). To capture the dynamics of such a model, a two-player game is built between the two functioning entities. In the second case the model

WiMAX Evolution: Emerging Technologies and Applications   Edited by Marcos D. Katz and Frank H.P. Fitzek
© 2009 John Wiley & Sons, Ltd

is extended to the multihop mesh mode case, where the MS is not in the range of the BS and uses other nodes, called Relay Stations (RSs), to route their packets towards it. The pricing model will encourage participation and cooperation among the nodes in order to accomplish the goal of accessing the information provided by the BS (called 'the Internet'), as in the case described in the latest effort of IEEE 802.16j in the Mobile Two-Hop Relay case.

Those two models are described in terms of unlimited capacity, which is a case when the Internet provider has dimensioned the network in such a way so that there is no congestion on the wireless channel and the user gains instant access without competition with the rest of the nodes after paying the price for it. Next we investigate the way to allocate resources in a congested channel according to the IEEE 802.16 and based on the price that the user intends to pay. The last application model includes a scenario where WiMAX and WiFi network providers are cooperating to provide the optimum pricing strategy for bandwidth sharing. Such models will generally be used in rural environments where short-range high-bandwidth wireless connections through WiFi are encouraged in terms of spectrum usage and low-cost deployment.

4.2   Economics in Network Engineering

4.2.1   Building a Business Model

As WiMAX is evolving through different standardization procedures, it is expected that WiMAX will be the technology that extends the current capabilities of broadband networks. Our intention in this chapter is to present the cost and charges for a broadband network from a traditional financial and economic perspective. According to Mason and Varian (1994) the costs for a network are divided as follows.

1. *Fixed cost* of providing a network infrastructure.

2. An incremental cost of connection to the network, which is usually paid by the user in the form of a *connection cost*.

3. *Cost of expanding* the network's capacity. Users who are willing to defer their transmission during congestion times should not pay for the expansion of network's capacity.

4. Incremental *cost of sending an extra packet*. This cost should be very low or zero without congestion, since the bandwidth of a broadband network is typically a shared resource.

5. *Social cost* defined as the extra delay incurred to other users by the transmission of a packet

The costs that the provider takes into account are usually divided into CAPEX and OPEX. CAPEX are the expenditures for investments in the purchase and installation of network equipment and we denote them by the variable C. There are cases that such initial investment is regarded as a sunk cost (and for that reason is disregarded in a charging scheme). CAPEX includes the BSs, spectrum, site preparation, site installation and backbone network equipment costs. One of the advantages of WiMAX in the battle with LTE is that it is an open standard developed by the IEEE 802.16 workgroup, which leads to less cost per device, and less CAPEX, since the *modus operandi* of 3GPP is to provide a closed standard for LTE.

The other category is OPEX ($O(y)$), which includes the operation, administration, management and provisioning costs (e.g. site leasing, equipment maintenance, network expansion and customer acquisition). However, such expenditures are difficult to predict since technology evolution plays a crucial role when calculating them (Pareek, 2006). We define y here as the discrete time usually in the scale of years. In order for a provider to cover the costs and make profit from the investment, a charging scheme is imposed on the users. It is thus straightforward that the profit for the provider can be defined as $P(y, p_y) = \text{Revenue}(y, p_y) - O(y)$. The standard economic measure for evaluating the value of investment at time y is called the Discounted Cash Flow (DCF), which takes into account the rate used to discount future cash flows to their present values (r):

$$\text{DCF} = \frac{P(y, p_y)}{(1+r)^y}.$$ (4.1)

To measure the profitability of the investment a widely used value is the NPV. It is used as a mapping of the value of the investment, to the present value (PV) and is given by

$$\text{NPV}(T) = \sum_{y=1}^{T} \text{DCF} - C,$$ (4.2)

where T is the total time of the project. According to the NPV method a provider shall invest in the project if it is positive. If it is zero it does not add any monetary value and other criteria shall be taken into account. Two other economic terms can be defined here. The Internal Rate of Return (IRR), which specifies the rate of return of the investment r, in which $\text{NPV}(T) = 0$. There is also the time to break even, which is the smallest time y from which the provider shall recoup the entire investment, $\text{NPV}(y) > 0$.

In Figure 4.1 we show the cost flow diagram which specifies both the economic and engineering considerations and results from the deployment of a WiMAX network (single point-to-point, mesh/relay mode and WiFi/WiMAX) as an extension to Gunasekaran and Harmantzis (2006). In the following we study the pricing schemes in order to define the appropriate p_{tm} and the possible charges that the provider can impose on the users.

4.2.2 Control and Pricing

According to Walrand and Varaiya (1996) the charges imposed on the user can be split into four categories:

1. *access charge*, which is the amount the user is required to pay for accessing the network;

2. *usage charges* are imposed to recover the variable costs of running the network (OPEX);

3. *congestion charges* reflect the additional cost when transmitting under congestion of the network (pricing schemes exist that differentiate charges according to congestion periods);

Cost flow diagram

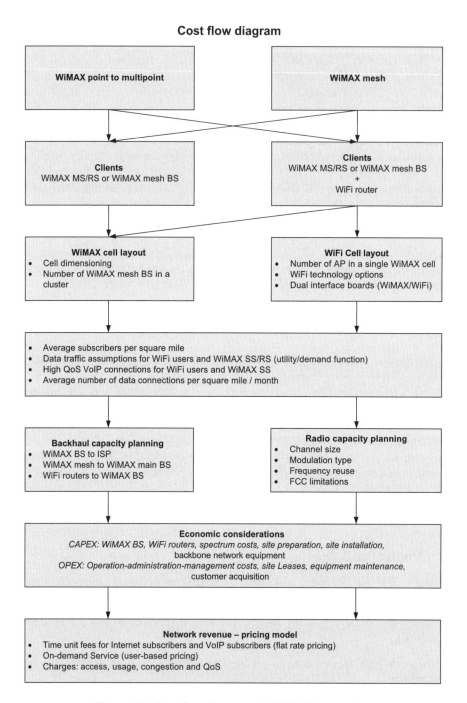

Figure 4.1 Cost flow diagram for WiMAX networks.

4. *QoS charge*, which the user has to pay to receive a certain QoS (e.g. an extra charge paid for a user to have VoIP access through a UGS traffic class).

Pricing schemes are those methods that define how the users of a network are billed and tend to be optimum when they succeed social welfare. However, complex pricing schemes are not always applicable since they must take into account a number of economic, social and technical factors, for example, will the prices prevent poor users from accessing a WiMAX network, how the charges should be accounted for, who is paying for extra cell development? Another property of a pricing scheme is that it is sometimes used by network engineers to control traffic and dimension the network (Courcoubetis and Weber, 2003). Setting higher prices for a product can reduce the demand, while decreasing the prices can lead to the opposite effect and so pricing can be regarded as a communication mechanism between the provider and the client. In the following we summarize the main features of pricing.

- *Congestion control.* A monetary charge could be imposed on the congestion signals such that to alleviate congested routes or reduce the transmission rate of a user. Thus, the demand function of a user is closely related to the price they are charged for transmitting a specific unit (e.g. packets, calls). The simplest pricing schemes do not differentiate congestion pricing in terms of the class of service (priority pricing) but treat each user equally, applying the same charge.

- *Admission control.* There are cases where users have access to the network according to the price paid, as described in their Service Level Agreement (SLA). Service is provided only to those who afford to pay the extra price, or even to those who wish to share their traffic descriptor and are rewarded for providing more information to the provider.

- *Overprovisioning.* Time of day pricing and dynamic pricing policies may influence the distribution of traffic over some time intervals, for example, providers can provide lower charges during night hours or weekends leading to equal share and better management of resources.

Another issue can be the billing, which requires the collection, maintenance and processing of network usage patters. In order to design the network efficiently, one must model both the user preferences and the provider objectives, both of which have an influence on the applied pricing scheme and are studied in the following sections.

4.3 Building the Pricing Schemes

4.3.1 Utility, Demand Functions and Optimization Objectives

Utility denotes the 'level of satisfaction' of a user or the performance of an application. In most of the networking cases it is a nondecreasing function of the amount of bandwidth, and it is a common practice to assume that the function is concave. If, in addition, we assume that the utility function $U(\cdot)$ is strictly concave, there is a unique maximizer that solves $\max\{U(BW) - BW * p_y\}$. The resulting strictly decreasing function is called the demand function.

The advantage of the utility function is that it can reflect the user's QoS requirements with the level of satisfaction and quantify the adaptability of an application. Ideally utility should be expressed as a function of actual QoS parameters, for example, delay or packet loss. However, in most cases it is impossible to predict such quality measures in advance and utility is expressed as a function of the resources allocated such as bandwidth or number of connections.

The traffic offered to a network is assumed to belong to three classes and can be mapped with different utility functions.

- *Class I: inelastic traffic (real-time nonadaptive).* This usually refers to applications that can have strict bandwidth requirements such as VoIP. If the allocated bandwidth is less than a B_{min}, users utility will drop to zero. Once the bandwidth threshold is met, more bandwidth allocation will not lead to performance enhancement (e.g. in G.711, $B_{min} = 64$ kbps). Such traffic is similar to the UGS traffic flow of WiMAX in which dedicated slots are allocated to those applications. The utility function is expressed as a step function.

- *Class II: partially elastic traffic (real-time adaptive).* This refers to applications which can adapt to the available bandwidth through adaptive coding. However, it requires the network to provide a minimum level of performance guarantees which can vary from some kilobits per second (as in online gaming) to megabits per second (as in streaming IPTV). The model is depicted as an S-type curve where B_{min} represents the bandwidth derived from the minimum encoding rate, whereas B_{max} is the intrinsic encoding rate. Such a traffic model is similar to rtPS traffic flow in WiMAX.

- *Class III: elastic (nonreal-time data).* Those applications belonging to this class are rather tolerant to delay. Sometimes it is preferred to buffer nonreal-time data at a network node (e.g. BS) and transmit them at a slower rate. It can function in a similar manner to the BE or even nrtPS traffic flows of WiMAX only by minor changes to some tunable parameters.

In Figure 4.2 we sketch the utility functions of the above classes, an overview of the utility functions can be found in Liu *et al.* (2007). The two main methods of pricing users are as follows.

4.3.2   Flat-rate Pricing

According to this pricing scheme the user is charged a fixed amount per time unit (e.g. month), irrespective of the usage. It is believed to be that pricing scheme that will initially dominate (WiMAX-TELECOM in Austria and Irish Broadband already use such a pricing scheme and US Sprint is looking into applying it). The reasoning for this is that flat pricing is simple and convenient: there is no need for measurements, complex billing and achieves social fairness since there is no distinction on the service level provided among poor and rich users. In the following it is proved that in some cases, such as when the network is designed in such a way that no congestion occurs, flat-rate pricing is the optimal strategy for the period that the client wishes to connect.

Flat-rate pricing tends to have many disadvantages compared with user-sensitive dynamic pricing methods. Demand for bandwidth in a WiMAX network is expected to grow in

Class I - Inelastic traffic   Class II - Partially Elastic traffic   Class III - Elastic

$$u(BW) = V \frac{sgn(BW - B\min) + 1}{2}$$ $$u(BW) = V \frac{1}{1 + (1/\varepsilon - 1)e^{\frac{-2\ln(1/\varepsilon - 1)}{B\max}}}$$ $$u(BW) = V \frac{\log(BW + 1)}{\log(B\max + 1)}$$

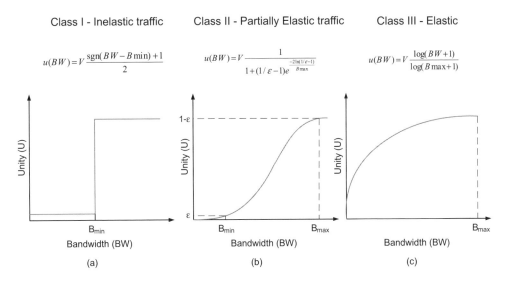

Figure 4.2 Utility functions of common applications in terms of bandwidth.

great paces and the only experience right now is from the cellular networks point of view. However, data traffic has very different nature and usage patterns than voice calls. Thus, network providers may find it hard to provide sufficient resources over all time periods, particularly when new generation applications tend to be correlated with social patterns (e.g. P2P downloading). Therefore, low load users (email, Web browsing) may be penalized with respect to high load users (multimedia application, streaming). In that sense flat rate pricing does not improve economic efficiency of a QoS-enabled network, since user preferences are not taken into account into the formalization of the pricing scheme (Edell and Varaiya, 1999).

4.3.3   User-based Pricing

While trying to solve the inefficiencies of flat-rate pricing, many pricing schemes were proposed based on the user preferences as expressed by their utility functions. Since many researchers have provided different versions of user-based pricing, our focus will be on those that can be amended in such a way as to include the specific characteristics of WiMAX. However, a general survey of most of them can be found in Falkner *et al.* (2000).

In *priority pricing* users are selecting one of the priority levels according to the QoS needed. Since the IEEE 802.16 standard has a connection-oriented nature, packets are classified into different service flows in which different pricing policies can be applied. This relation between the user's utility and the priority pricing raises the economic efficiency of the network, since the user satisfaction level is increased compared with flat-rate pricing. However, when the network is congested this pricing scheme increases the social cost for the low-priority users (the users in the BE traffic class have to contend with accessing in the uplink whereas UGS/VoIP traffic class are allocated regular intervals by the BS).

Another pricing scheme is the *smart-market pricing* initially envisioned by Mason and Varian (1994). The auctioneer places a specific usage cost based on the marginal congestion cost. Then, prior to transmitting a packet, the user either accepts the extra cost or if the user's bid does not exceed the current cutoff amount set by the auctioneer, they do not gain access to the network ('the Internet'). Although the applicability of Smart market pricing is difficult due to the significant technical changes required, in a WiMAX network the scheduler of the BS, which is fully aware of the congestion level of the network, may cooperate with the auctioneer to impose the correct amount on the users, thus making a great candidate that encourages both network and economic efficiency.

In *effective bandwidth pricing* the network sets prices to reflect the demand for effective bandwidth and publishes a number of pricing curves, differentiated according to the QoS service flow (Courcoubetis and Siris, 1999). Prices are then determined over a long time scale and users then accept those prices in their signed SLA. Such a pricing scheme can be very effective, but requires the traffic patterns of the user, which can be either measured in a test period or can be approximated through field test and user surveys.

In the following section we apply those pricing methods in different WiMAX topologies.

4.4   Pricing in Different WiMAX Topologies

WiMAX networks are to be deployed in metropolitan areas where users may either connect instantly to the BS, or through other WiMAX nodes (as in the case of RSs) and through WiFi APs that provide access to local areas. In this section we analyze all of these topologies in terms of how the provider is going to charge the users (optimum price allocation) for the service offered, and also based on how the network is designed. For these reasons mathematical tools such as game theory and convex optimization are used. The understanding of the following section requires basic knowledge of terms associated with those theories.

4.4.1   Point-to-point Unlimited Capacity

In this architecture, which is based on the work of Musacchio and Walrand (2006), it is assumed that each MS participates in a single cell scenario with one BS. Moreover, the MS is not able to have access in multiple contenting BSs leading to a pricing topology of a single seller single buyer. Time is divided into time slots, which can be in the scale of minutes or hours. In each time frame t the MS requests for a service and the BS replies with a p_t (e.g. the client requests access to the Internet for 10 minutes and the BS replies with the corresponding value, for example, when waiting in the airport). The MS then either accepts the price and the game continues for the next time slot or rejects the price and the game ends immediately. It is also assumed that the network is never congested, there is no QoS differentiation and thus the game of each MS with the BS is independent of the other MSs.

Three quantities are specified: τ is a discrete random variable representing the number of time slots the client intends to connect and T the discrete time slots the MS receives for the specified service; finally, U is a continuous random variable which represents the utility function of the user in a time slot.

The utility function of that model is specified by

$$F(T, \tau) = U \cdot \min(T, \tau). \tag{4.3}$$

The MS is aware of both τ and U, whereas the BS may only obtain their PDF (in case it had full knowledge, then $p_t = U$ and the MS would always connect).

- The MS connects or remains connected in slot t if and only if $t \leq \tau$ and $U > p_t$.

- The BS charges a nondecreasing price sequence p_t such that

$$p_t \in \arg \max_p p P(U > p). \tag{4.4}$$

Musacchio and Walrand (2006) proved that it is a Probability Bayesian Equilibrium (PBE) for the BS to choose a single maximizing value of $p P(U > p)$ and charge it for all time slots ($p_t = p^*$). Moreover, a maximum is achieved in $[0, \infty)$, because for $y(p) = \arg \max_p p P(U > p)$, we know that $y(0) = 0$ and $\lim_{x \to \infty} y(p) = 0$. In the end of the game the MS has a net payoff of $F(T, \tau) - \sum_{t=1}^{T} p_t$ and the BS a profit equal to $\sum_{t=1}^{T} p_t$. In order to find the revenue of the BS provider, as specified in Section 4.2.1, we need to sum the profit over all users connected N, plus the mean number of reconnections $E[R]$ multiplied by the users, $K = N + E[R]N$ (K can be easily measured by the provider either on the BS or on the billing system). Now the intended connection time of each user $k \in [1, \ldots, K]$ is a function $T(k)$. Since it is proved that the same price is charged over the entire duration of $T(k)$ the revenue is

$$\text{Revenue}(y, \ p_y) = \sum_{k=1}^{K} T(k) p_t, \tag{4.5}$$

where $y = \max(T(K))$, which is the duration of measurement.

4.4.2 Mesh Mode Operation

The multi-hop case, as shown in Figure 4.3, can be abstracted to an aggregate branch of the scheduling tree in order to involve all of the nodes from the BS to the last MS. We suppose that the MS is not in the range of the BS, thus all of the packets are relayed from the relaying nodes towards the BS. As described by Lam *et al.* (2007), the BS charges the one-hop RS with a price p_t^N, which in turn charges the next hop a price p_t^{N-1}, until we reach the MS where the price is specified as p_t^1. It is straightforward from the previous analysis that the payoff of the BS is a function of all of the prices from the MS to the BS. Equivalently for each RS its price is a function of the rest of the stations towards the client. The price received by MS after price p^i has been set by node i is marked up by all its downstream relays:

$$m^i(p^i) = p^{1*}(p^{2*}(\ldots (p^{i*}))) \quad \text{for all } i \in 2, \ldots, n. \tag{4.6}$$

Since the last node 1 does not have any other downstream relays $m^1(p^1) = p^1$. The payoffs are as follows.

- For the MS: $F(T, \tau) - \sum_{t=1}^{T} p_t^1$.

- For the RS: $\sum_{t=1}^{T} (p_t^i - p_t^i)$ for $i = 1, \ldots, N - 1$.

- For the BS: $\sum_{t=1}^{T} p_t^N$.

As proved by Lam *et al.* (2007) the optimal price can be found as a PBE with the following strategy profile:

- the MS connects or remains connected in slot t if and only if $t \leq \tau$ and $U > p_t$;

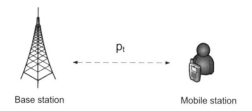

Single hop case

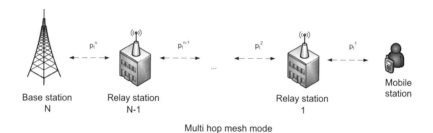

Multi hop mesh mode

Figure 4.3 Point-to-point and mesh mode operation of WiMAX networks.

- the RS picks up a price $p^{i*}(p^{i+1})$ that satisfies

$$p^{i*}(p^{i+1}) \in \arg \max_{p^i}[(p^i - p^{i+1})P(U \geq m^i(p^i))]; \qquad (4.7)$$

- the BS charges a nondecreasing price sequence

$$p_t^N \in \arg \max_{p^N}[p^N P(U \geq m^N(p^N))]. \qquad (4.8)$$

Using Equation (4.5) the Revenue(y, p_y) is found in this scenario.

4.4.3   Point-to-point Limited Capacity

The assumption of noncongestion may not always hold especially since overprovisioning wireless metropolitan area networks is rather difficult. So the WiMAX bandwidth allocation mechanism must be taken into account when congestion may happen due to limited capacity.

The WiMAX standard allocates specific minislots to each user i in the downstream. Let us assume that there are N active users ($i = 1, \ldots, N$), each of which receives a duration of $t_i \in [0, T)$ where T is the total duration of the downstream part of the frame. So $\sum_{i=1}^{N} t_i < T$.

4.4.3.1   Maximize Profit

Suppose that the user pays a price p_i per received packet and thus the problem is to find an optimum transmission time allocation vector $\mathbf{t}_i^*$ which maximizes the revenue for the BS.

Owing to the erroneous nature of the wireless environment the transmission rate in each slot is based on the channel conditions. Suppose that the channel conditions are summarized in a state h_i, which gives a rate of packets $r(h_i)$ (Marbach and Berry, 2002). The packets that the receiver is now able to accept, according to the allocated time, is $x_i = r(h_i)t_i$ (in low BER wired networks r is taken as constant, thus $x_i = $ Capacity). The maximization problem becomes

$$\text{maximize} \quad \sum_{i=1}^{N} x_i p_i$$

$$\text{subject to} \quad \sum_{i=1}^{N} \frac{x_i}{r(h_i)} \leq T$$

$$x_i \geq 0 \quad \text{for } i \in [1, N].$$

In the above problem we are dealing with $N + 1$ constraint functions. The Jacobian of the constrain functions is

$$\mathbf{Dh(x^*)} = \begin{pmatrix} \dfrac{1}{r(h_1)} & \dfrac{1}{r(h_2)} & \cdots & \dfrac{1}{r(h_N)} \\ -1 & 0 & \cdots & 0 \\ 0 & -1 & \cdots & 0 \\ 0 & 0 & \cdots & 0 \\ 0 & 0 & \cdots & -1 \end{pmatrix}.$$

Since its columns are linearly independent, it has a rank N, and the Nondegenerate Constraint Qualification (NDCQ) holds at any solution candidate (Simon and Blume, 1994). The unique maximizer is the one that satisfies that the Hessian matrix of the Lagrangian L with respect to $\mathbf{t}_i^*$ at $(\mathbf{t}_i^*, \mu^*)$, $D_x^2 L(\mathbf{t}_i^*, \mu^*)$ is negative definite. In order to maximize the revenue, the BS should allocate more resources t_i to users who pay more and tend to receive more packets. This scheme allows some flexibility, since the users may pay a low price but download a lot, but also users can connect with a higher price in case they are not willing to download huge files and congest the network. However, this revenue maximizing policy is socially unfair since rich users who download more can monopolize the transmission allocation time.

4.4.3.2 Maximize Social Welfare

In this case the network provider is not interested in achieving the maximum revenue but in achieving an optimum revenue while maximizing the utility of the users, otherwise expressed as social welfare. We suppose that each BS can support four traffic classes (UGS, rtPS, nrtPS and best effort), mapped into three utility functions $U_i^j(x_j)$, as mentioned in Section 4.3.1 where $j = [1, 2, 3]$. The downstream subframe is now divided into mapped intervals for each station, as shown Figure 4.4, and each of which is divided into regions of QoS $\mathbf{w} = [w_1, w_2, w_3]$, as described in IEEE 802.16 (IEEE, 2004). Now the utility

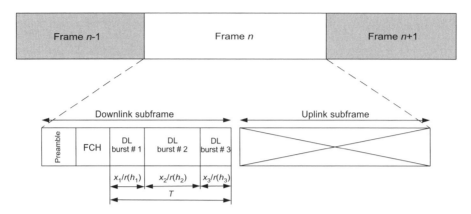

Figure 4.4   IEEE 802.16 frame structure explanation for point-to-point limited capacity model.

maximization problem is defined as

$$\text{maximize} \quad f(\mathbf{x}) = \sum_{i=1}^{N} \sum_{j=1}^{3} (w_j U_i^j (x_j))$$

$$\text{subject to} \quad \sum_{i=1}^{N} \frac{x_i}{r(h_i)} \leq T$$

$$\mathbf{w} \cdot e = 1$$

$$x_i \geq 0 \quad \text{for } i \in [1, N].$$

The above optimization problem involves inequality constraints and a set of non-negative constraints. Thus, the solution is obtained by using the Lagrange multiplier with Karush–Kuhn–Tucker (KKT) conditions. The Kuhn–Tucker Lagrangian is given by

$$\widetilde{L}(\mathbf{x}, \lambda) = f(\mathbf{x}) - \lambda \left[\sum_{i=1}^{N} \frac{x_i}{r(h_i)} - T \right] \tag{4.9}$$

with first-order conditions given by

$$\frac{\partial \widetilde{L}}{\partial x_i} \leq 0, \quad \frac{\partial \widetilde{L}}{\partial \lambda} \geq 0, \quad x_i \frac{\partial \widetilde{L}}{\partial x_i} = 0, \quad \lambda \frac{\partial \widetilde{L}}{\partial \lambda} = 0. \tag{4.10}$$

The above solution gives the Pareto optimum $\mathbf{x}_* = [x_1^*, \ldots, x_N^*]$, while satisfying $\lambda \geq 0$. By dividing with $r(h_i)$ we may find the optimum transmission time allocation vector that satisfies the social welfare in the point-to-point WiMAX network.

The advantage of a WiMAX network is that through the uplink bandwidth request mechanism, the scheduler is informed of the intentions of the MSs and the decision for allocating bandwidth during the downlink subframe can be a function of the price. Thus, the optimization solution can be solved from the scheduler in order to allocate the resources

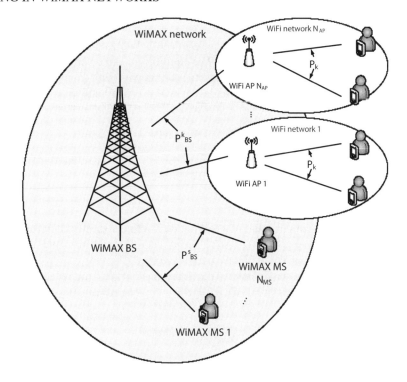

Figure 4.5 WiFi WiMAX cooperation network.

fairly among users, and increase their satisfaction leading to more profit for the provider. There are also other pricing methods that could be applied to find how the resources are distributed based on the bids per packet of the users. Economic theory suggest that efficiency is greater in such cases and waste of resources is reduced. However, such complex schemes may deter users from using the network services and slow the expansion of the network.

4.4.4 WiMAX/WiFi Architecture

In this model we suppose that the WiMAX BS and the WiFi APs are run by different service providers. WiMAX runs as a backbone to offer to Internet access to the APs in case they cannot be connected to an infrastructure network (see Figure 4.5). The presented pricing model is an extension of Niyato and Hossain (2007), and is based on adaptive bandwidth sharing. The WiMAX BS charges an adjustable price P_{BS}^k to each of the WiFi routers and P_{BS}^s to the serving MSs. The solution of the problem can be derived by a game-theoretic model based on Stackelberg equilibrium. We suppose that TDMA/TDD access mode is used based on single carrier modulation (WirelessMAN SC air interface). Each AP is working as a router with two interfaces, a WiMAX client interface and the regular WiFi interface.

4.4.4.1 The Model

The reason for this choice of game is because in the network there are two entities that deploy their own strategies to theirs clients. The equilibrium can be obtained by backwards induction, in which the follower (AP) chooses its profit maximizing quantity given its belief for the choice of the leader (BS). Then the leader will go ahead and maximize their profit already knowing how the follower (AP) will respond.

- *Players*: WiMAX BS, WiMAX SSs and WiFi APs.

- *Strategies*: WiMAX BS, the price charged to the WiFi AP and to the WiMAX SSs. For the WiFi AP, the price charged to the WiFi clients.

- *Payoffs*: The corresponding profit.

Assumption 1: The bandwidth demand of a WiFi node is a linear strictly decreasing function $D_j(P_k) = a_j - b_j P_k$ of user $j \in [1, \ldots, N_k]$, for $a_j, b_j \in \mathbb{R}_+$ and for which there exists a price $P_k^{\max} = a_j/b_j$ such that $D_j(P_k^{\max}) = 0$, for all $P_k \geq P_k^{\max}$.

The profit of the AP k can be defined as the difference between the revenue of the WiFi networks and the cost of offering a service $\pi_k = R_k - C_k$. The revenue R_k can be defined as

$$R_k = \sum_{j=1}^{N_k} P_k * b_j(P_k), \tag{4.11}$$

where $b_j(P_k)$ is the demand function of node j served by the AP k and is a function of the price P_k charged by AP k, which is assumed to be the same offered to all users. The cost function is given by

$$C_k = \sum_{j=1}^{N_k} P_{\mathrm{BS}}^k * b_j(P_k) + F_k = P_{\mathrm{BS}}^k \sum_{j=1}^{N_k} b_j(P_k) + F_k, \tag{4.12}$$

which is the sum over all clients N_k served by the AP, of the price set to the AP by the WiMAX BS P_{BS}^k plus the fixed cost (costs paid for installation, equipment, etc.). The first step of the problem is to define the optimal price P_k^* charged by the WiFi AP to the nodes, which is obtained by solving a straightforward Lagrangian constrained optimization problem

$$\text{maximize} \quad \pi_k = R_k - C_k$$

$$\text{subject to} \quad P_k^* \leq P_k^{\max}$$

$$P_k \geq 0.$$

The leaders payoff (BS) can be defined equivalently as the difference of the revenue obtained by the BS from the APs and the MSs minus the expenditures in a given time described as follows:

$$\pi_{\mathrm{BS}} = R_{\mathrm{MS}} + \sum_{k=1}^{N_{\mathrm{AP}}} R_k - F_{\mathrm{BS}} \tag{4.13}$$

the revenue obtained by the SSs is a function of their demand function $b_s(P_{\mathrm{MS}})$, whereas the revenue obtained by the AP is a function of the AP's price P_k^* found by the previous optimization. Thus,

$$\pi_{BS} = \sum_{s=1}^{N_{MS}} P_{MS} b_s (P_{MS}) + \sum_{k=1}^{N_{AP}} P_{BS}^k \sum_{j=1}^{N_k} b_j (P_k^*) - F_{BS}. \qquad (4.14)$$

Assumption 2: The bandwidth demand of a MS node is a linear strictly decreasing function $D_s(P_{MS}) = a_s - b_s P_{MS}$ of MS $s \in [1, \ldots, N_{MS}]$, for $a_s, b_s \in \mathbb{R}_+$ and for which there exists a price $P_{MS}^{max} = a_s/b_s$ such that $D_s(P_{MS}^{max}) = 0$, for all $P_{MS} \geq P_{MS}^{max}$.

The problem is again a Lagrangian constrained optimization, which is easily solved in order to find the optimal price vector $\mathbf{P}_{BS}^* = [P_{BS}^{k,*}, P_{BS}^{s,*}]$ set by the BS to each of the WiFi APs and each MS (since both share the same wireless channel). Having the optimal price vector which maximizes the profit of the BS, the $\pi_{BS}^* = P(t, p_t)$ and the NPV can be found.

4.5 Conclusion

To summarize, this chapter has dealt with the design of a WiMAX network in terms of economic theory, business modeling and engineering applications. Initially a business model was presented as a way to measure the profitability of deploying a WiMAX network taking into account profits and ways to charge the users in order to either maximize the revenue or the social welfare. Multiple network topology analyses were presented from which the provider may decide which depicts their application scenario.

From the pricing point of view we have showcased that although flat-rate pricing is simple and convenient it does not provide economic efficiency. Moreover, we have proved that when the network is designed in such a way that there is no congestion (unlimited capacity) in the wireless link between BS and MS, the Nash equilibrium is to provide an equal price for the connection time of the user (e.g. paying a standard amount of dollars for intended duration of connection). However, when the MSs compete for accessing the channel (limited capacity) the provider shall price the users a specific amount of dollars for each transmitted packet (user-based pricing). In the last model the pricing was used to allocate the resources fairly in a WiFi/WiMAX cooperated network. Such an analysis can be extended to other 4G wireless technology cooperating scenarios.

References

Courcoubetis, C. and Siris, V.A. (1999) Managing and pricing service level agreements for differentiated services. *Proceedings of the 7th International Workshop on Quality of Service*, pp. 165–173.

Courcoubetis, C. and Weber, R. (2003) *Pricing Communication Networks*. John Wiley & Sons Ltd, Chichester.

DaSilva, L.A. (2000) Pricing for QoS enabled networks: a survey. *IEEE Communication Surveys and Tutorials*, **3**(2), 2–8.

Edell, R. and Varaiya, P. (1999) Providing Internet access: what we learn from index. *IEEE Network*, **13**(5), 18–25.

Falkner, M., Devetsikiotis, M. and Lambadaris, I. (2000) An overview of pricing concepts for broadband IP networks. *IEEE Communications Survey*, **3**(2).

Gunasekaran, V. and Harmantzis, F. (2006) Financial assessment of city wide WiFi deployment. *Communications and Strategies*, **i63**.

IEEE (2004). Part 16: Air Interface for Fixed Broadband Wireless Access Systems. IEEE Standard 802.16, IEEE.

Lam, R.K., Chiu, D. and Lui, J. (2007) On the access pricing and network scaling issues of wireless mesh networks. *IEEE Transactions on Computers*, **56**(11), 1456–1469.

Liu C, Shi, L. and Liu, B. (2007) Utility based bandwidth allocation for triple play services. *Proceedings of the 4th European Conference on Universal Multiservice Networks ECUMN'07*, pp. 327–336.

Marbach, P. and Berry, R. (2002) Downlink resource allocation and pricing for wireless networks. *IEEE Infocom*, pp. 1470–1479.

Mason, J.K.M. and Varian, H.R. (1994) Pricing the Internet. *Proceedings of the International Conference of Telecommunication System Modeling*, Nashville TN, pp. 378–393.

Musacchio, J. and Walrand, J. (2006) WiFi access point pricing as a dynamic game. *IEEE/ACM Transactions on Networks*, **14**(2), 289–301.

Niyato, D. and Hossain, E. (2007) Integration of WiMAX and WiFi: optimal pricing for bandwidth sharing. *IEEE Communications Magazine*, **45**(5), 140–146.

Pareek, D. (2006) *The Business of WiMAX*. John Wiley & Sons Ltd, Chichester.

Simon, C.P. and Blume, L. (1994) *Mathematics for Economists* (1st edn). Norton, New York.

Walrand, J. and Varaiya, P. (1996) *High Performance Communication Networks* (1st edn). Morgan Kaufmann, San Francisco, CA.

Part IV

Advanced WiMAX Architectures

5

WiMAX Femtocells

Chris Smart, Clare Somerville and Doug Pulley

5.1   Introduction

5.1.1   A Brief History of Cell Sizes

It is over 60 years since it was first proposed that terrestrial radio coverage of indefinite extent could be achieved using multiple transmitters and continued reuse of a small number of radio channels (Brinkley, 1946). The simple presentation of the coverage area of each Base Station (BS) represented as a regular hexagon (a special case of a Voronoi polygon (Voronoi diagram, 2008)) with the base site at its center has since become the defining symbol of the cellular industry.

Traffic densities and coverage challenges in dense urban areas soon demanded more strategically placed BS sites, using antennas sited at or below the skyline. These more closely packed 'microcells' use buildings to tailor the coverage area, minimizing interference to local co-channel BSs. The 'picocell' rapidly joined its larger 'siblings' as the BS carefully targeted at the highest user densities in airports, railway stations, hotels and large enterprise buildings (Saunders and Aragón-Zavala, 2007).

5.1.2   Definition of a Femtocell

The term 'femtocell' is a relatively recent introduction and may be taken to imply a number of different functional and architectural alternatives. It is therefore important to establish exactly what is being referred to here as a femtocell in the WiMAX context.

The following is a list of key attributes that define a WiMAX femtocell with respect to a more traditional macrocell base station.

(A) Deployment scenario:

  • indoor;

WiMAX Evolution: Emerging Technologies and Applications   Edited by Marcos D. Katz and Frank H.P. Fitzek
© 2009 John Wiley & Sons, Ltd

- residential or Small Office/Home Office (SOHO);
- backhaul over subscriber's broadband connection (e.g. Asymmetric Digital Subscriber Line (ADSL));
- 'plug-and-play' installation by the user using self-configuration and self-optimization to minimize management overhead;
- typically 50–100 m cell radius;
- omnidirectional antenna (no sectorization).

(B) Capacity:

- typically constrained to <10 residential, <50 SOHO users;
- at least two connections per user;
- for a SOHO deployment maximum throughput is not affected, so typically 40 Mbps; for residential deployment, typically 10–20 Mbps.

(C) Closed Subscriber Group (CSG):

- defined user list rather than public access;
- not an extension to the macro network.

(D) Low cost:

- greatly reduced Bill Of Materials (BOM) compared with traditional BS ('consumer electronics' construction and pricing);
- installation and operating costs (backhaul, power, etc.) met by the end-user.

(E) Operator requirements and feature set:

- continuity of 'user-experience' so providing the same features as available in the macro network;
- potential for new femto-specific applications and services;
- no (or acceptably small) impact to existing macro network;
- compatible with all regular Mobile Stations (MSs) (no special modifications required for operations with femtocells;
- neighbor BSs should require no modification to support femtocell deployments.

5.2   Architecture of a WiMAX Femtocell

5.2.1   WiMAX Network Architectures for a Femtocell

The location of a femtocell within a WiMAX network is shown in Figure 5.1. A femtocell resides within an Access Service Network (ASN) probably dedicated to serving femtocells. Owing to the high number of femtocells serviced in a given area, aggregators are likely to be used between the femtocell and gateway. The gateway in the femtocell ASN is termed a 'femtocell gateway' to highlight that it may have additional functions beyond a traditional ASN gateway to support the unique features of a femtocell.

The femtocell ASN may communicate with other ASNs (femtocell or macrocell) or directly with the Connectivity Services Network (CSN).

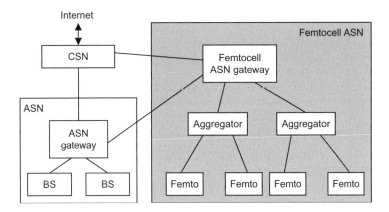

Figure 5.1 WiMAX network elements.

5.2.2   Femtocell Deployment Configurations

In a conventional network the radio resource is centrally managed; however, this will not be the case for femtocells which are, by definition, much more autonomous. Hence, to all intents and purposes, a femtocell will be seen by the macro network as a hostile interferer. Where possible, the femtocell 'layer' could be deployed on a separate Radiofrequency (RF) channel to the macrocell layer but this is not necessarily the case with all operators. Even in the dedicated RF channel case, interference between neighbor femtocells needs to be considered. To this end several deployment configurations have been proposed.

5.2.2.1   Closed, Dedicated Channel, Fixed Power Deployment

Closed in this sense means that the femtocell is available to a restricted set of subscribers (e.g. those who usually dwell in a residence) agreed between the femtocell owner and the operator.

The femtocell is deployed on a dedicated RF channel, that is, an RF channel that is not used within the macro layer. The worst-case dedicated RF channel deployment is the adjacent channel. The worst-case adjacent RF channel deployment is when the adjacent RF channel is owned by a different operator. Although the femtocell is deployed on a dedicated frequency with respect to the macro network, a co-channel interference scenario remains between femtocells: this is especially significant within a dense population of femtocells. It is possible that operators would use a dedicated RF channel for all types of small cell in general, hence femtocells may also experience/create interference from/to picocells and microcells.

A fixed maximum transmit power limit is defined for this deployment option, with a value of 5 dBm suggested in 3GPP. This limit must be defined such that the femtocell will cause an acceptably low level of interference in the worst case, that is, in a weak macrocell environment, and so may be too restrictive for general deployment.

5.2.2.2 Closed, Dedicated Channel, Adaptive Power Deployment

This option is similar to the previous configuration in that the femtocell is configured for a closed subscriber group and for a dedicated RF channel. The difference is that the maximum femtocell transmit power may be set as high as allowed by the standards or by the equipment capability, using a self-configuring algorithm. Higher power levels should only be used when appropriate for the deployed environment and when the resulting interference is acceptable.

5.2.2.3 Closed, Co-channel Deployment

The femtocell is deployed on the same RF channel frequency as the surrounding macro layer (co-channel) but has the ability to control its transmit power in order to minimize its impact on the macro layer by monitoring the macro layer interference. Isolation between the two systems is provided by the penetration losses of the building materials. This configuration becomes especially challenging when visitors to the home operate on the same RF channel but are power controlled by the macrocell and may cause interference (due to so-called 'near–far' effects).

5.2.2.4 Closed, Partial Co-channel Deployment

This option applies where more than one RF channel frequency is used on the macro layer but there is no frequency dedicated solely for the use of femtocells. The femtocell is required to select a RF channel frequency from the available set by determining one that is not deployed in the immediate vicinity. This helps provide an additional degree of isolation, relative to full co-channel deployments, to minimize interference and distribute the load between the femtocell and the macro layer.

   Provided that the correct choice of RF channel frequency can be automated, the RF performance of this configuration should be similar to that of dedicated channel deployment options.

5.2.2.5 Open Femtocell Deployment

In this configuration the femtocell is deployed as part of an open system, that is, shared amongst all of the subscribers attached to a single operator. This option may therefore be considered to be beyond the scope of femtocell deployments and instead fall into the category of more traditional pico- or microcells.

   Any of the previous four deployment configurations could be valid for an open deployment, but with the consideration that the femtocell is no longer a hostile interferer but should instead be considered an extension to the macro network.

5.3 Femtocell Fundamentals

In addition to the WiMAX air interface PHY (physical) and MAC (Medium Access Control), the femtocell includes the following key components.

(A) Femtocell management:

   • timing and synchronization;

- self-configuration of operational parameters.

(B) Remote/local configuration:

- Remote configuration options:
  - maintain a femtocell Management Information Base (MIB) using the Simple Network Management Protocol (SNMP);
  - maintain a femtocell data model using TR-069;
- Local configuration:
  - user configuration for CSG management.

(C) Network interface:

- authentication;
- backhaul security;
- handovers.

In the following sections we investigate these functional elements, focusing on those aspects that are unique to femtocells and providing suggestions for solutions to the special issues that arise.

5.3.1 Synchronization

WiMAX femtocells, like all BSs, are required to be synchronized with each other in time such that the preamble is broadcast from all BSs at the same instant. Specifically the requirement is for the time alignment to be accurate to within 1 μs. In addition, the transmit frequency from each BS must be accurate to within ±2 ppm of the RF channel frequency.

5.3.1.1 GPS

This is the traditional method for providing synchronization to a WiMAX BS. GPS is combined with a cheaper oscillator to provide accurate frequency synchronization. In addition, GPS provides a regular pulse which is used to provide time synchronization. GPS does not traditionally work well inside buildings but indoor solutions are now starting to become available driven to a large extent by the femtocell market.

5.3.1.2 NTP/PTP

The Network Time Protocol (NTP) and the IEEE1588 Precision Time Protocol (PTP) are both packet-based synchronization schemes that can extract timing information from the backbone. Their use would be combined with a cheap oscillator to provide an accurate frequency reference and also to generate the regular timing pulse required for time synchronization.

5.3.1.3 Other BSs

In areas where WiMAX coverage is already available, the femtocell can synchronize itself to a nearby macrocell. This requires the ability to receive Downlink (DL) signals from the macrocell, derive timing and frequency estimates from that cell and discipline a local oscillator to align with it. The DL Receive mode of operation to support this approach is discussed further in the following section.

5.3.2 Self-configuration

A key differentiator between a femtocell and a traditional macrocell is the manner in which it is installed. The femtocell must be a 'plug-and-play' device.

Part of the self-installation process involves the femtocell achieving accurate frequency and time synchronization, as defined above. Other information which needs to be self-configured includes:

- transmission frequency;

- maximum transmission power;

- preamble selection;

- neighbor cell list (WiMAX and other technologies, e.g. GSM).

A key enabler of femtocell self-configuration is the ability to operate in a DL Receive mode: in other words to operate, temporarily, as though it were a WiMAX MS. This allows the femtocell to detect any surrounding WiMAX macrocells, and possibly other femtocells, in order to integrate into the overall radio network without causing significant degradation to the performance of these neighbor cells.

5.3.2.1 Transmission Frequency

Selection of a RF channel frequency will determine to a large extent how the femtocell interferes with surrounding macrocells.

Single WiMAX frequency   In some cases operators will be constrained to the use of a single frequency, for both macrocell and femtocell networks, in which case the considerations of a 'co-channel' deployment as discussed earlier will come into play. In this case the femtocell can be statically provisioned by the operator to use the defined frequency without any self-configuration. However, the selection of Tx power, as discussed below, will become especially important for interference mitigation.

Single femtocell frequency   Another alternative is that the operator will identify a fixed RF channel frequency for femtocells but one that is distinct from that/those used by the macrocell network. This alternative will imply the 'dedicated channel' considerations defined earlier. As with the single frequency case, this option will not require self-configuration of the femtocell frequency: it can be statically defined by the operator. Further, in this dedicated channel case the selection of Tx power is less critical as far as macrocell interference is concerned and could potentially also be statically configured.

Self-configuration A third option is to introduce a pre-defined set of frequencies available for the femtocell network, possibly the same as or overlapping with the macrocell frequency set, and allow the femtocell to make its own selection in order to minimize interference with neighbor cells. This self-configuration procedure requires that the femtocell uses its DL Receive mode to scan for frequencies used in the surrounding cells and then selects the channel on which the lowest energy is present.

5.3.2.2 Maximum Transmission Power

Selection of the femtocell maximum transmission power, in conjunction with the RF channel frequency choice, will affect the level of interference that the femtocell generates towards neighbor cells.

Co-channel vs dedicated channel The choice of transmission power is especially critical in co-channel deployments. In this case the femtocell DL transmissions will appear as in-band interference to neighbor cells, with a direct impact on their coverage: the femtocells are effectively punching a hole in the geographical area served by the macrocells. In dedicated channel scenarios there is a little more flexibility in the selection of transmission power as the interference is now out of band, but this can still have an impact on neighbor cell coverage especially in terms of adjacent channel rejection performance. All DL transmissions will to a large extent be contained within the building where the femtocell is installed, but this cannot be assumed for all deployments; for example, the femtocell could be sited next to a window. Therefore, control of the transmission power is a key requirement.

Fixed maximum limit One option that is potentially attractive in dedicated channel deployments is to set the maximum transmission power of the femtocell at a pre-defined fixed limit. This level needs to be carefully selected to ensure that it is sufficiently low to cause no more than an acceptable level of interference in the majority of deployment scenarios. This limit can be statically defined by the operator and requires no self-configuration functionality in the femtocell. However, there will always be deployment cases where this pre-defined level is inappropriate, both in terms of minimizing interference and also providing effective coverage within the femtocell area.

Self-configuration A more flexible option, applicable in both co-channel and dedicated channel deployments, is for the femtocell to make measurements of the power levels received from neighbor cells and use these to determine its own transmission power. The general principle is that if the femtocell can hear a strong macrocell signal then so can any nearby MSs: therefore, it can afford to transmit at a higher level without generating too much interference. Clearly this self-configuration procedure requires that the femtocell is capable of detecting DL transmissions from surrounding cells: the DL Receive mode.

5.3.2.3 Preamble Selection

The DL preamble is the first OFDMA symbol transmitted in each frame. It spans all sub-channels, dividing them into three sets, each set mapping to a specific segment. One of 32 possible Pseudorandom Noise (PN) sequences is carried on the preamble for each segment, identifying the 'IDcell' parameter for the BS.

One possible use for the preamble would be to carry an IDcell parameter which identifies the BS as a femtocell, rather than a macrocell: this would allow femtocells to identify each other and avoid creating 'circular references' during self-configuration (for example, trying to synchronize to each other's frequency reference). This approach would require the operator to partition IDcell into ranges; some values valid for macrocells and others for femtocells. The macrocell would be provisioned as normal to use an IDcell from the macro range but pre-programming a single fixed value in the femtocell could create contentions when multiple femtocells are deployed in the same area. Therefore, it makes sense to configure the femtocell with a definition of the valid IDcell range it can use, and allow it to self-configure from this range. This would require detection of the IDcell value being used by any neighbor femtocells in order to select a locally unique value: the DL receive mode again.

Another option would be to have no distinction between femtocells and macrocells in their use of IDcell, and simply allow the femtocell to select from the entire range available based again on detection of neighbor usage. This would permit locally unique IDs to be deployed, but would not provide a mechanism for identifying other femtocells.

5.3.2.4   Advertising Neighbor Cells

WiMAX BSs advertise information about their neighbor cells in order to assist the MS handover procedure. For macrocells this information would be provisioned by the operator based on knowledge of the geographical deployment of the BSs, but for femtocells such provisioning is clearly not feasible.

A solution to this is for the femtocell to construct its own neighbor list. This list would be based directly on the results of scanning for surrounding macrocells using the DL Receive mode.

In order to avoid advertising neighbor femtocells as handover candidates, there would need to be a mechanism defined for distinguishing between femtocells and macrocells. One such mechanism would be to use subsets of the IDcell parameter carried on the DL preamble, as described above.

5.3.3   Remote Configuration

A macrocell typically includes a MIB, as defined by the IEEE (2007). This standardised MIB enables a network operator to remotely configure the BS. It also gives the network operator the ability to gather statistics from the BS.

The self-configuration of the femtocell makes many of these MIB functions redundant. However, some MIB features, such as statistics gathering, are still useful. One option is for a femtocell to implement a subset of this MIB; however, a standardized subset MIB for femtocells has not yet been defined.

An alternative would be to use TR-069 which has been developed to allow remote configuration of Customer Premises Equipment (CPE). An extension to provide a standardized management system for a WiMAX femtocell could be defined.

5.3.4 User Configuration

The femtocell is a plug-and-play device, which requires a minimum of user configuration. One element that does require input from the end-user is that of the allowed subscriber list.

To allow rejection of unauthorized users the femtocell should be configured with the identity of any permitted MS: this is the concept of a CSG. A good candidate identity for the CSG is the MAC address of the MS. The MAC address is unique to an MS and sent to the BS during network entry.

The management of the CSG could be either performed by the user, via a Web browser, or by the user contacting their network operator who then manages the CSG via remote management.

5.3.5 Backhaul Security

A network operator will have security concerns regarding the connection of femtocells to its network. The femtocell model, where an unknown entity performs installation and activation, is inherently less secure than a macrocell. The network operator will want to ensure that the femtocell was purchased from a legitimate source and that the control messages it transmits into the operator's network have not been tampered with. Of equal importance is that the femtocell user requires confidence that its data remains secure between the femtocell and ASN GW.

One option to ensure the validity of the femtocell is to provide the same authentication mechanism as used by the MS, namely the Extensible Authentication Protocol (EAP). This would allow the WiMAX network to authenticate a femtocell with the same network components and signalling protocols.

An alternative would be to use Internet Protocol Security (IPsec) for authentication. IPsec is also a strong candidate for ensuring both control messages, and user data remain private and unmodified across the backhaul connection.

5.3.6 Handovers

Handover in a WiMAX network is designed to guarantee quality of service providing a seamless multimedia experience to the user. This handover consists of several different stages each designed to make sure the MS selects the most appropriate BS and that the transfer from one BS to another is as fast as possible. In the first stage the MS surveys the signal quality of neighboring BSs. Based on the results, the MS may decide to investigate handover to a neighbor BS: part of this stage involves the current (serving) BS and potential (target) BS communicating over the backbone network. If the negotiation is successful the MS leaves its serving BS and establishes communication with the target BS. This network re-entry is fast and efficient due to the information exchanged over the backbone. Handover, between a femtocell and macrocell should provide the same seamless behavior.

5.3.6.1 Femto-to-Macrocell Handover

To achieve a seamless femto-to-macrocell handover there are several requirements that should be met. The MS must be aware of the neighboring BS, allowing it to locate potential target BS, and the femtocell should be able to communicate with neighboring BS via the backbone network. This section assumes that the femtocell ASN has a backbone connection to the macrocell ASN.

The femtocell self-configuration phase described earlier can provide the MS with a list of neighbor macrocells for potential handover. This list will allow the MS to perform the same controlled handover used in macrocell-to-macrocell handover.

If the femtocell was unable to detect any macrocells during self-configuration, then its neighbor cell-list will be empty. In this case the MS will need to perform an uncontrolled handover. In an uncontrolled handover a macrocell will not be expecting a MS: however, by using the backbone network it can retrieve information from the femtocell to aid network re-entry. Uncontrolled handover also occurs in macrocell-only networks.

5.3.6.2   Macro-to-Femtocell Handover

A seamless macro-to-femtocell handover is more challenging. The same requirements must be met as femto-to-macrocell handover, namely, the MS must be aware of the femtocell, and a backbone connection exists. For macro-to-femtocell handover the challenge is make the MS aware of the femtocell. Possible options are detailed in this section.

Advertise the femtocell in the macrocell neighbor cell-list   After the self-configuration phase, where the femtocell detects the surrounding macrocells, the network operator could use remote configuration to add the femtocell into the neighbor cell-list of these macrocells.

A disadvantage of this method is the potential number of femtocells which would need to be advertised by a macrocell. This could make the macrocell neighbor cell-list large. Also, since MS are unaware of the difference between femto and macrocells, there is a risk that MSs will encounter a degraded quality of service as they attempt to perform handover into femtocell which will not provide them with a service.

Network detects when MS is near femtocell   After the surrounding macrocells are determined by self-configuration, the network operator could use remote configuration to extract the neighbor cell-list from a femtocell. When the network detects a CSG MS is located in one of these macrocells it will request the MS handover into the femtocell. This method requires a new module in the network to associate MSs with femtocells, and femtocells with nearby macrocells.

When the network has detected the potential for macro-to-femtocell handover it has several options. It could instruct the MS to measure the signal quality of the femtocell and based on this information request the MS to perform handover. It could periodically suggest the MS should perform handover into the femtocell. The MS can then make the assessment on whether or not to perform handover. Finally, it could force the MS to perform handover into the femtocell. With this option, the network should be confident of the signal quality between the MS and femtocell.

MS detects when MS is near femtocell   A MS could build up a list of macrocells that are close to a femtocell it has permission to access. When present in one of these macrocells it would begin assessing the femtocell for handover. However, this would require changes to the MS. A specific aim of femtocell deployment is to not require MS changes; therefore, this method must be discounted.

5.4 Femtocell–Macrocell Interference

5.4.1 Interference Scenarios

In addition to coverage, the performance of a femtocell will be governed by the degree to which it is able to withstand interference. It is important to anticipate the level of interference these devices will be exposed to in order that they can be designed accordingly.

5.4.1.1 Downlink Interference

In some circumstances, an external, nearby macro user could experience some interference from a femtocell located in a nearby residence. As a result the macrocell transmitter will increase power to that user to overcome the interference. Where the macrocell user is on the edge of a cell it is likely the macrocell transmitter will be unable to increase the power sufficiently and the macrocell user may experience outage. However, it should be stressed that this type of interference only affects macro users near edge of coverage and near a house equipped with a femtocell. The effect may be exacerbated if the femtocell is located in an elevated position or is near a window which would offer less isolation to the macrocell.

On the plus side, the macrocell transmitter will have more power available to serve outdoor macrocell users since it no longer has to serve a significant portion of indoor users (now served by femtocells). More power is required to serve this type of user from outside the building due to the significant external wall losses.

To further counter interference to macro users, femtocells can be designed to 'listen' to the macrocell transmitter, determine the interference loading on the macrocell and limit their power accordingly. Femtocells can also be designed to have a degree of intelligence such that they only transmit just enough power to serve their users. Femtocells will only rarely operate at full power.

5.4.1.2 Uplink Interference

In some situations, a femtocell could interfere with a macrocell receiver, limiting its ability to receive distant users. As a result the macrocell will instruct distant users to increase their transmit power. Where distant users are at the edge of cell they may experience outage, resulting in diminished coverage at the cell edge. Again, the effect may be made worse if the femtocell user is at the edge of femtocell coverage and is therefore transmitting at a higher than average power level or if they are in an elevated position with a clear view to the macrocell itself.

On the other hand, with a population of femtocells across the network, the macrocell receiver will experience a reduced demand for resources thus leaving the macrocell with greater ability to overcome interference in isolated cases.

5.4.1.3 Interference Matrix

In the following interference matrix a macrocell is used to designate any outdoor wireless infrastructure and may actually be a macrocell, microcell or picocell. It is assumed that femtocells are deployed in an uncoordinated fashion.

Table 5.1 Matrix of interference cases.

Case	Aggressor	Victim
1	MS attached to femtocell	Macrocell UL
2	Femtocell DL	MS attached to macrocell
3	MS attached to macrocell	Femtocell UL
4	Macrocell DL	MS attached to femtocell
5	MS attached to femtocell	Neighbor femtocell UL
6	Femtocell DL	MS attached to neighbor femtocell

The combinations listed in Table 5.1 are too numerous to analyze them all here. Instead we take one case to look at in more detail, that of DL interference (Case 2).

5.4.2   Downlink Coverage Definitions

For the purposes of this analysis, the femtocell under consideration is assumed to be deployed in an area already covered by a WiMAX macrocell. The femtocell is further assumed to be operating with a CSG such that only a limited set of authorized MSs are allowed to gain network access via the femtocell; a 'macro MS' (i.e. one not on the CSG list) cannot attach to the femtocell.

A key feature of the femtocell is the ability to detect and measure characteristics of the surrounding WiMAX neighborhood: DL Receive mode. As a minimum for this analysis, the femtocell is assumed to be able to measure the Received Signal Strength Indicator (RSSI) of the WiMAX macrocell.

The following three definitions of DL femtocell coverage area are defined for the purposes of this analysis (see Figure 5.2).

- *Cell border*: The area bounded by the locus at which the femtocell preamble and macrocell preamble are received by an MS at equal power levels.

- *Femto quality*: The area within which the Signal-to-Noise Ratio (SNR) of the femtocell DL is sufficient to support a target Modulation and Coding Scheme (MCS), that is, a MS attached to the femtocell will have a guaranteed quality of service. The 'femto quality' coverage area will therefore vary with the MCS in use by the femto MS.

- *Macro deadzone*: The area within which the interference caused by the femtocell causes the SNR of the macrocell DL to be less than the sensitivity for each MCS, that is, a MS attached to the macrocell will fail to receive an adequate quality of service. The 'macro deadzone' area will therefore vary with the MCS in use by the macro MS.

To define the femto quality and macro deadzone coverage, typical SNR values are assumed for each combination of modulation and coding and are defined at the demodulator. In order to reference them to the antenna they so need to be adjusted to take into account the receiver noise figure and implementation margin using the following values:

$$\text{Receiver Noise Figure} = 8 \text{ dB},$$

$$\text{Implementation Margin} = 5 \text{ dB}.$$

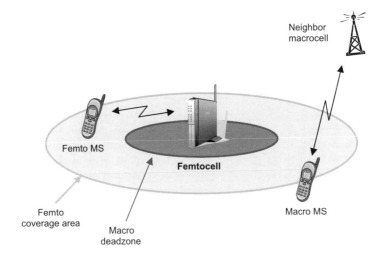

Femto MS

Femtocell

Femto
coverage area

Macro
deadzone

Macro MS

Figure 5.2 Femtocell coverage.

In the following calculations, the coverage area is first calculated in terms of the DL path loss in decibels; this is then converted into a range in meters using the ITU indoor propagation model. Therefore, the range is defined simply as an indoor distance and does not account for penetration of the external walls of the premises where the femtocell is installed.

5.4.3 Downlink Coverage Analysis

5.4.3.1 Cell Border

The parameters used in the cell border analysis are as follows:

P_{femto} = maximum transmit power from a femtocell;
RSSI = macrocell RSSI in the absence of the femtocell (i.e. measured in DL Receive mode);
N_{thermal} = thermal noise at the MS (in a 10 MHz bandwidth);
L_{border} = path loss femto to MS.

Recalling the definition for cell border coverage, 'femto preamble power at MS = macro preamble power at MS', it can be shown that

$$L_{\text{border}} = \frac{P_{\text{femto}}}{\text{RSSI} - N_{\text{thermal}}}. \tag{5.1}$$

Figure 5.3 shows the femtocell coverage resulting from Equation (5.1) for a range of P_{femto}, that is, the size of the area within which the femtocell transmissions are received at a higher level than that of the macrocell.

5.4.3.2 Femto Quality

The parameters used in the femto quality analysis are as follows:

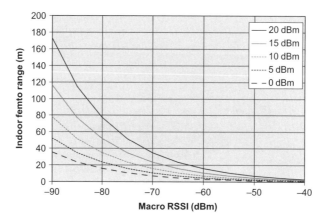

Figure 5.3 'Cell border' coverage.

P_{femto} = maximum transmit power from femtocell;

RSSI   = macrocell RSSI in the absence of the femtocell (i.e. measured in DL Receive mode);

N_{thermal} = thermal noise at the MS (in a 10 MHz bandwidth);

L_{quality} = path loss femto to MS;

ACIR   = Adjacent Channel Interference Ratio (39.8 dB for macro and femto on adjacent channels);

$\text{SNR}_{\text{femto}}$ = SNR target for MCS.

Recalling the definition for Femto Quality coverage, 'Femto SNR at MS = target SNR', it can be shown that:

$$L_{\text{quality}} = \frac{P_{\text{femto}}}{\text{SNR}_{\text{femto}} \cdot ((\text{RSSI} - N_{\text{thermal}})/\text{ACIR} + N_{\text{thermal}})}. \tag{5.2}$$

Figure 5.4 shows the femtocell coverage for a QPSK 1/2 MCS resulting from Equation (5.2) for a range of P_{femto}, assuming the femtocell is deployed on an adjacent channel to the macrocell.

From these results it can bee seen that adjacent channel femto coverage is independent of macro RSSI at low macro levels. Only when the macro RSSI is above about −70 dBm does it start to have an impact. Even then the femto coverage remains sufficient for a typical residential deployment up to the highest macro RSSI levels.

5.4.3.3   Macro Deadzone

The parameters used in the macro deadzone analysis are as follows:

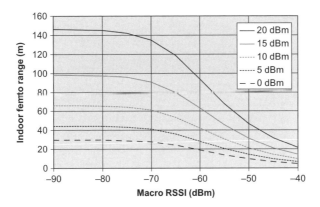

Figure 5.4 'Femto quality' coverage, adjacent channel, MCS = QPSK 1/2.

P_{femto} = maximum transmit power from femtocell;

RSSI = macrocell RSSI in the absence of the femtocell (i.e. measured in DL Receive mode);

N_{thermal} = thermal noise at the MS (in a 10 MHz bandwidth);

L_{deadzone} = path loss femto to MS;

ACIR = Adjacent Channel Interference Ratio (39.8 dB for macro and femto on adjacent channels);

SNR_{macro} = SNR target for macrocell MCS;

α = femtocell activity factor.

Recalling the definition for the macro deadzone boundary, 'macro SNR at MS = target SNR', it can be shown that

$$L_{\text{deadzone}} = \frac{\alpha \cdot P_{\text{femto}}}{\text{ACIR} \cdot ((\text{RSSI} - N_{\text{thermal}})/\text{SNR}_{\text{macro}} - N_{\text{thermal}})}. \tag{5.3}$$

Figure 5.5 shows the macro deadzone for QPSK 1/2 resulting from Equation (5.3) for a range of P_{femto}, assuming that the femtocell is deployed on an adjacent channel to the macrocell and has an activity factor $\alpha = 50\%$.

These results demonstrate that the adjacent channel macro deadzone around the femtocell can be kept below just a few meters by controlling P_{femto}.

5.4.4 Setting the Maximum Femtocell Transmit Power

The analysis above determined the size of the macro deadzone for a measured macro RSSI across a range of femto transmit powers (P_{femto}). This analysis can be inverted in order to determine the value of P_{femto} that creates a defined macro deadzone size.

So from a measured value of macro RSSI, the femtocell can determine what Tx power P_{femto} to configure for the required size of macro deadzone, that is, the femto can self-configure to introduce no more than a defined level of interference to the neighbor macrocell.

Figure 5.6 shows the effect of calculating P_{femto} in this way for a range of macro deadzone sizes; specifically 40, 50 and 60 dB, giving deadzone sizes of approximately 0.7, 1.5 and

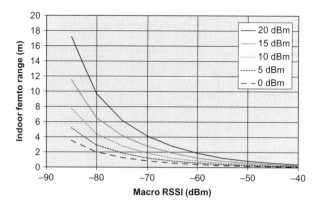

Figure 5.5 'Macro deadzone', adjacent channel, MCS = QPSK 1/2.

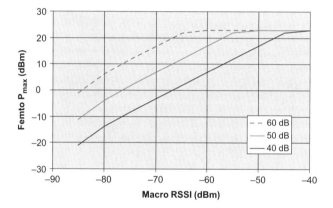

Figure 5.6 Femto P_{max} for defined 'macro deadzone'.

3.2 m, respectively. In addition, P_{femto} is limited to a maximum of 23 dBm to represent a typical upper power limit.

The value of P_{femto} calculated in this way can then be applied to the earlier femto quality analysis to determine the resulting coverage area within which the Femto MS will receive the required SNR.

Figure 5.7 shows the femto quality coverage areas resulting from the application of P_{femto}, calculated to create the specified macro deadzone sizes.

This demonstrates how a macro deadzone of less than a few meters is easily achieved for adjacent channel deployments with a Femto Tx Power in the range -20 to $+23$ dBm, and how the Femto Quality coverage resulting from such a self-configuration is sufficient to meet the requirements of typical residential deployments.

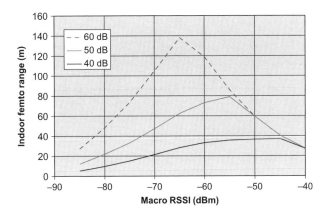

Figure 5.7 'Femto quality' coverage for defined 'macro deadzone', MCS = QPSK 1/2.

References

Brinkley, J.R. (1946) A method of increasing range of V.H.F. communication systems by multi-RF channel amplitude modulation. *Journal of the IEE*, **93**(3), 159–166.

IEEE (2007) Air Interface for Fixed Broadband Wireless Access Systems. Management Information Base Extensions. *P802.16i*, IEEE.

Saunders, S. and Aragón-Zavala, A. (2007) *Antennas and Propagation for Wireless Communication Systems* (2nd edn), John Wiley & Sons Ltd, Chichester.

Voronoi diagram, http://en.wikipedia.org/wiki/Voronoi_diagram (Accessed 2008).

6

Cooperative Principles in WiMAX

Qi Zhang, Frank H.P. Fitzek and Marcos D. Katz

6.1 Introduction

In this chapter the application of cooperative principles in WiMAX is advocated. Even though WiMAX is a new promising communication standard, which supports higher data rates than most existing infrastructure-based mobile or wireless communication systems already, cooperative principles will enrich and improve mobile communication in WiMAX systems even further. In this chapter we introduce the main ideas of cooperation and later give several dedicated examples where cooperation is applied and describe the benefits that can be achieved.

Cooperation describes an action of individuals collaborating towards a common or individual goal. Each of the individuals may have egoistic motivation to join the cooperation. With the exception of individuals driven by pure altruistic behavior, egoistic cooperation will take place as long as individuals gain by the collaborative activity. Cooperative communication has many different interpretations and various extensions. One of the main research areas on cooperative communication is the cooperative diversity exploiting the broadcast nature of the radio channel, as in Boyer *et al.* (2004), Gastpar *et al.* (2002), Gupta and Kumar (2003), Hunter and Nosratinia (2002), Laneman *et al.* (2004) and Sendonaris *et al.* (1998). In fact, the concept of the cooperative diversity originates from the relay wireless network which is not new and can be traced back to Cover and Gamal (1979) and van der Meulen (1971). The motivation for cooperative diversity is to exploit the spatial and temporal diversity to improve the reliability of communications, for example, in terms of outage probability, or symbol error probability, for a given transmission rate, etc. (Laneman, 2006). The following is a list with important related variations based on the relay concept.

WiMAX Evolution: Emerging Technologies and Applications   Edited by Marcos D. Katz and Frank H.P. Fitzek
© 2009 John Wiley & Sons, Ltd

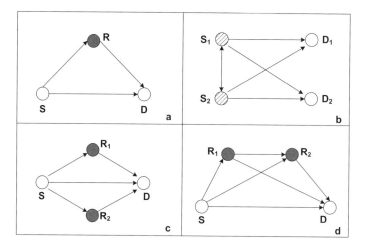

Figure 6.1 The related variations from a relay-based wireless broadcast channel (S: source; R: relay; D: destination).

- The relay channel (see Figure 6.1(a); Cover and Gamal (1979)).

- User cooperative diversity (see Figure 6.1(b); Sendonaris *et al.* (1998)).

- Cooperative coding (see Figure 6.1(b); Hunter and Nosratinia (2002)).

- Parallel relay channel (see Figure 6.1(c); Gastpar *et al.* (2002)).

- Multihop diversity (see Figure 6.1(d); Boyer *et al.* (2004), Gupta and Kumar (2003)).

Another research area of cooperative communications is the so-called Cellular-Controlled Peer-to-Peer (CCP2P) network architecture (Fitzek and Katz, 2006, 2007a). CCP2P is a dynamic approach to bridge cellular and peer-to-peer network architectures (Fitzek and Katz, 2007a). In CCP2P networks, a wireless device not only has a connection with an outside world through a cellular link but it can also communicate with the neighboring wireless devices in its proximity using short-range links. It should be noted that the cellular link is not limited to the radio link in traditional cellular networks but it can be more generically understood as the main access link to the service. Generally speaking, in CCP2P the cellular network has a infrastructure-based network topology with a relatively large coverage. Moreover, the cellular approach mostly uses the licensed spectrum and is characterized by a relatively high Energy-per-Bit Ratio (EpBR). In contrast to the cellular network, the short-range nomadic network has relatively smaller coverage, typically without a central fixed infrastructure, works in the license-free spectrum and exhibits a much lower EpBR. One of the motivations for CCP2P is to exploit the short-range link to improve the on-going communication performance on the cellular link, in terms of energy efficiency, delay and throughput, etc.

Cooperation can also be distinguished between network cooperation and user cooperation. Network cooperation is much easier to realize as each network entity, such as the relay station (RS), is under control of the network operator. The decision whether to cooperate or not is taken by the network operator and not by individuals. Network cooperation is characterized by altruistic behavior of the participating entities. Representatives cases of network cooperation are the parallel relay channel scenario (see Figure 6.1(c)) and the multihop diversity scenario (see Figure 6.1(d)) when the relays are installed by the network operator. User cooperation on the other side is realized by individuals with their own pay-off schemes and no overlay decision maker is present. Unlike network cooperation where the collaborative interactions are embedded in the system and completely transparent to the user, the user behind a wireless device has a key role in the user cooperation approaches. Altruistic behavior is not the main driving force in user cooperation, it is egoist behavior. Even though egoistic behavior is dominating, cooperation will be established as long as all participating individuals receive some benefits from it. Therefore, user cooperation is the most interesting and challenging field. The user cooperative diversity scenario and cooperative coding scenario (see Figure 6.1(b)) are examples of user cooperation in cooperative diversity area. Moreover, the cooperation scenarios in CCP2P network architectures are clear examples of user cooperation.

In the following we discuss why cooperation is needed for wireless networks. Current cellular communication systems are grouped into different types of generations depending on their basic technology that they support. With the introduction of GSM, the second generation (2G) was launched. The main novelty was the change from analog mobile systems to digital mobile systems. In recent years, following the worldwide success of 2G, the third generation (3G) was introduced. 3G changed the access scheme from Time Division Multiplex Access (TDMA) towards wideband Code Division Multiplex Access (CDMA), although it did not introduce novel services. 3G was targeting mainly Internet access with high data rates. This evolution is continued now with new technologies such as WiMAX and Long Term Evolution (LTE), sometimes referred to as 3.75G. However, all of these new technologies will not provide a solution to the limitation of the coverage-rate trade-off defined by the Shannon law. This means that to achieve reliable transmission the signal-to-noise ratio of each received bit should be above a certain threshold. At a given transmission power, the higher data rate results in less energy at each bit. Therefore, the increased data rate reduces the coverage range. Exploiting cooperative diversity, it can extend the coverage range with enhanced data rate (Pabst *et al.*, 2004, Sadek *et al.*, 2006, Sendonaris *et al.*, 1998, 2003a,b) and reduce the outage probability (Laneman *et al.*, 2001).

Furthermore, the new technologies such as WiMAX and LTE will be unable to overcome the problem of the power and energy consumption on the mobile device (MD). As MDs are battery driven, the energy that can be pulled out of it is limited. The quicker the energy is consumed, the shorter the operational or stand-by time of the MD. Users are quite well aware on the available service time of their MD. In fact, it has been recognized as one of the major selling criteria. With the introduction of 3G MDs, the standby time was nearly cut in half compared with those available in 2G. This had to do with the larger energy consumption of 3G chipsets and the 3G network situation at that time. 3G chipsets are real 'power hogs' as Steve Jobs mentioned recently in one of his iPhone presentations. The more complex the air interface of a MD becomes, the more energy will be consumed. To support a variety of services with higher data rate and better Quality of Service (QoS), the trend is to make the air interface even more complex in the future, introducing technologies such as Multiple Input Multiple Output (MIMO) for the MD. From approximately half to two-thirds of the overall power requirements of current mobile devices corresponds to the communication functions (e.g. baseband processing, Radiofrequency (RF) chains, and others). Clearly, there is a large margin for improvement. Furthermore, there seems to be a clear tendency for higher data rates at the cost of higher energy requirements. The increased complexity can be supported by Moore's law. Unfortunately developments in battery technology cannot keep up, and battery capacity only doubles roughly every decade only. From the users' perspective, there will be better services supported by higher data rates but the operational time will decrease. This tendency has already been recognized by the mobile manufactures and even standardization bodies and major efforts are being undertaken to reduce the overall battery consumption. In addition to the energy consumption, the power consumption is also relevant as it is directly linked to the heating problem of a mobile device. The more energy consumed in a given time period, the more the device will be heated. As the heating is produced at given hot spots, the heat needs to be distributed over the whole device as quickly as possible. So far mobile phones are cooled in a passive way, but if the tendency for greater energy consumption and smaller form factors continues, active cooling may be the only option. Active cooling does not only disturb the user by producing annoying noise, it also consumes additional energy. Fitzek and Katz (2006, 2007a) described a technique exploiting user cooperation in a CCP2P network architecture as a powerful way out of this problematic situation.

Introducing cooperation among users in the CCP2P network architecture means basically breaking the very old and fundamental paradigm of cellular communications. Very quickly after the early demonstrations by Marconi, mobile communication systems were dominated by a centralized architecture, as given in Figure 6.2. As can be appreciated, each MD is connected to an overlay BS (or access point (AP)) and each device is equipped with a set of capabilities that it can make use of. However, with the clear trend of more and more wireless devices populating the world and surrounding each other[1], the centralized architecture should be enriched by the communication among MDs. By allowing communication links among MDs in proximity to each other, MDs can form a cooperative cluster. Since the MD is composed of several entities or functionalities grouped into user interfaces (camera, keyboard, sensors, etc.), communication interfaces (cellular and short-range) and a number of built-in resources as given in Figure 6.3), this cooperative cluster can be referred to as a *wireless grid* as the devices are able to share their essential resources such as battery, CPU, wireless links, etc. By accumulating those resources into a virtual entity, the cooperative cluster may use those parts more efficiently than any stand-alone device could ever do.

As given in Figure 6.4 cooperative clusters are formed by multiple MDs within proximity using short-range technology. It is obvious that all capabilities (or resources) of the MDs can be used in different novel ways, exploiting cooperative principles. For instance, resources can be shared by some or all of the participating nodes, or they can be moved to a particular node, if needed. A simple case in point, assuming that each MD has a certain data rate R and J MDs form one cooperative cluster, then the cluster has a virtual data rate of $R \cdot J$. The virtual data rate can be used by all MDs, a subset of MDs or by a single MD. Nevertheless, the virtual data rate can be used by the cooperative cluster in a more clever way than the stand-alone device. In the worst case the resource allocation will give each device the data rate R, which is the same as the stand-alone device. In all other cases the resources can be shared in a better way and this is not only true for the data rate but also for all other capabilities.

---

[1] According to the visions of the Wireless World Research Forum (WWRF), by year 2017 there will be about 1000 wireless devices per person on the globe.

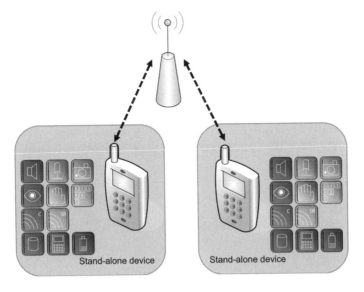

Figure 6.2 Stand-alone mobile devices in a cellular environment.

Figure 6.3 Available entities on a MD grouped into user interfaces, communication interfaces and built-in resources.

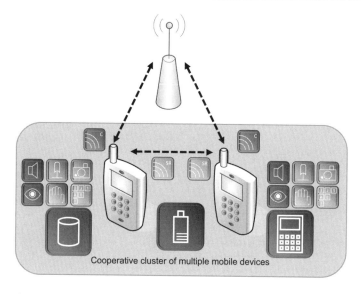

Figure 6.4 Cooperative MDs forming a wireless grid in a cellular environment.

Figure 6.5 depicts two approaches to deliver information to (from) a wireless device (AP or Base Station (BS)). Figure 6.5(a) shows the conventional case where the downlink information is delivered directly to the target node (node j) from the AP or BS, or vice versa in the uplink. This is the approach cellular systems have always used, although recently repeaters have been used between the source and destination in order to increase the data rate, particularly at the cell edges. The direct purely centralized delivery was originally conceived for voice-centric applications, although it is also used today for transferring data. Figure 6.5(b) illustrates an expected development in future wireless communications, namely the information is delivered to (from) the target device through a wireless grid composed of other nodes in close proximity. One could devise different cooperation systems between the cellular and short-range networks, depending on the goals (i.e. maximizing energy efficiency at the wireless devices, minimizing average transmitted power at the AP/BS, maximizing the throughput of the network, and others). The wireless grid concept in general assumes that nodes have at least two air interfaces, a trend already well represented in the current generation of MDs. Indeed, commercially available MDs today support 2G, 3G, WiMAX, Bluetooth and WLAN (IEEE 802.11) on the same device. One of the most attractive characteristics of wireless grids is that they allow a better utilization of resources, not only radio resources, but also other physical resources onboard of nodes, as discussed already. Cellular and short-range networks can be seen as highly complementary to each other, one exploiting a centralized architecture and the other typically (but not always) a distributed *ad hoc* architecture. Once again, it is highlighted that transmitting bits over short-range links is much more energy efficient than transmitting through cellular links, and typically the spectrum in the former case is unlicensed. Combining these two topologies and taking into account some fundamental differences will result in very attractive solutions, as discussed by Fitzek and Katz (2006, 2007a).

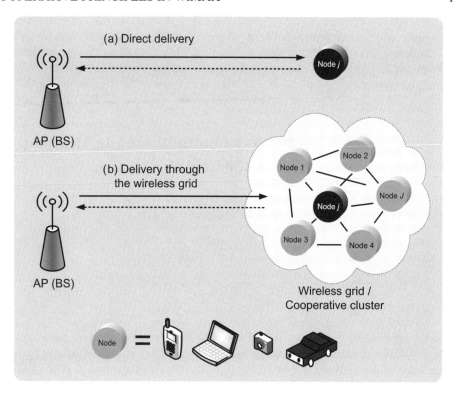

Figure 6.5 Delivery of information to a target node (or AP/BS): (a) conventional approach; (b) through a wireless grid.

It has been recognized that the gains of cooperative clustering based on the cellular-controlled peer-to-peer network architecture impact the whole value chain, from the user to network and service providers to equipment manufactures Fitzek and Katz (2006, 2007a). We list some of the benefits.

- Better resource allocation for CPU (Brodlos *et al.*, 2005) or spectrum (Kristensen and Fitzek, 2007).

- Improved energy consumption (Albiero *et al.*, 2007, 2008, Perrucci *et al.*, 2008).

- Increased robustness through diversity (Albiero *et al.*, 2008, Madsen *et al.*, 2007, Zhang and Fitzek, 2007).

- Less cost per service in a cooperative cluster (Fitzek and Katz, 2007b).

- Availability of new services (not supported by the stand-alone device) (Fitzek and Katz, 2007b).

The interested reader is referred to Fitzek and Katz (2006, 2007a) for deeper discussions on the benefits of the cooperative clusters over the stand-alone device. In addition to those benefits directly linked to the user, the mobile manufacture and network operator will also benefit from cooperative principles. Even though network operators are often scared by the term peer-to-peer (and cooperation involves mobile peer-to-peer), cooperation among users will enable network providers with new business segments without changing their pre-installed cellular networks at all. In the following some representative examples for the cross over of cooperative principles and WiMAX are presented and discussed.

6.2   Cooperative Diversity Schemes in Mobile Multihop Relay Based WiMAX (802.16j)

WiMAX offers high data rates over a relatively large coverage. However, the current deployment of WiMAX suffers from the following issues:

- coverage limitations or low Signal-to-Interference Noise Ratio (SINR) at cell edge caused by significant signal attenuation at high spectrum;

- poor signal reception due to shadowing or even coverage holes;

- limited spectrum.

Deploying a dense network with WiMAX BSs is a possible solution but is not cost efficient. A more efficient solution to extend coverage and enhance throughput is to deploy low-cost RSs in the network. A new task group 802.16j was established in 2006 to support the Mobile Multihop Relay (MMR) operation in WiMAX. The basic idea is that the WiMAX MDs in unfavorable locations can communicate with BS through the intermediate RS at high data rates. The RS can be fixed, nomadic or mobile. The main targeted usage scenarios of MMR WiMAX is as follows (802.16j Task Group, 2006):

- fixed infrastructure;

- in-building coverage;

- temporary coverage;

- coverage on mobile vehicle usage.

As mentioned in Section 6.1, the relay-based wireless network is not new and can be traced back to Cover and Gamal (1979) and van der Meulen (1971). The performance of a wireless relay network has been thoroughly studied from an information-theoretic point-of-view (Cover and Gamal, 1979, Host-Madsen, 2006, Kramer *et al.*, 2005, Laneman *et al.*, 2004). The communication between BS and MD through one intermediate RS is the simplest cooperative scenario. Many of the latest research efforts have been focused on cooperative diversity schemes (Kramer *et al.*, 2005, Laneman, 2002) and resource allocation (Can *et al.*, 2007b, Hammerstrom *et al.*, 2004, Li and Liu, 2006, Lin *et al.*, 2005, Onat *et al.*, 2007) in wireless relay networks. The main scheduling issues in cooperative relay-based wireless networks are: (i) which entity (or entities) make the cooperation decision; (ii) whether to use relay or not; and (iii) how to relay the data (different cooperative diversity schemes). In user cooperative diversity it is quite a challenge to choose the decision-maker entity, either BS, relay or MD. However, in MMR WiMAX, the decision is made by the BS, therefore, we focus on the latter two issues.

Before explaining how to make a cooperation decision, a short description of the basic relay transmission system is given. We assume a two-hop point-to-multipoint (PMP) MMR WiMAX system[2]. The RSs are established by the operator, therefore, MMR WiMAX represents a typical network cooperation as defined in Section 6.1. MMR WiMAX is characterized by altruistic behavior of the RSs, namely, there is no energy consumption constraint and no cooperation incentive issues on the RSs. In the relay-based wireless network, the transmission has two phases, namely from source to relay ($S \rightarrow R$) and from relay to final destination ($R \rightarrow D$).

For those MDs that are beyond the BS coverage, the only option for the MDs is to communicate through the RSs. However, for those MDs that can communicate directly with BSs with different data rates, it becomes a question of whether it is worth breaking the direct communication into two phases. In other words, is it worth using a relay or does cooperation pay off? The key to the question is to evaluate the cooperation gain which can be defined as throughput gain, that is, the throughput with relay to the original throughput ratio. Here we do not consider the detailed link adaptation on each subchannels. Generally speaking, the cooperation gain of a single user can be expressed as

$$G_{u,\text{coop}} = \eta(\gamma_{u,k}) - \eta(\gamma_{u,\text{SD}}). \tag{6.1}$$

Here, $\eta(\gamma_{u,k})$ is the achievable throughput of user u by employing a RS under the cooperative diversity scheme k and $\eta(\gamma_{u,\text{SD}})$ is the throughput of user u by direct communication with a BS.

Different cooperative diversity schemes exploiting spatial and temporal diversity are given by Can *et al.* (2008). They are extensions based on the parallel relay channel scenario mentioned in Section 6.1 (see Figure 6.1(c)). The schemes can be summarized as follows. The illustration of different cooperative relay schemes is shown in Figure 6.6.

- *Cooperative transmit diversity I* (given in Figure 6.6(a)). The MD and RS(s) listen to the transmission of BS at the first phase. Then, in the second phase, the BS and RS(s) simultaneously transmit to the MD (Can *et al.*, 2008). BS and RS(s) use Space Time Block Codes (STBC) to transmit the redundant data streams. The receiver uses Maximum Ratio Combining (MRC) techniques to combine the multiple received data streams. This scheme requires BS and RS(s) to use the same Modulation and Coding Scheme (MCS) at two phases, therefore, the two phases have equal duration. The drawback is that the good channel between RS(s) and BS has not been fully exploited because the same MCS is required for both phases.

- *Cooperative transmit diversity II* (given in Figure 6.6(b)). Different from the cooperative transmit diversity I, in this scheme only RS(s) listen to BS in the first phase. The advantage of this scheme is that MCS in the two phases can be chosen independently. Therefore, high-level MCS can be used in the first phase.

- *Cooperative receive diversity* (given in Figure 6.6(c)). The MD and RS(s) receive data from BS in the first phase. In the second phase, only the RS(s) repeat the transmission

---

[2]Theoretically, there can be many hops between the BS and a MD.

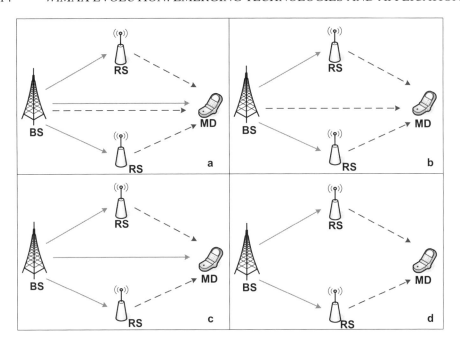

Figure 6.6   Illustration of different cooperative relay schemes (solid line: the first phase; dashed line: the second phase).

with the same MCS. This scheme has no potential to outperform the cooperative transmit diversity scheme.

- *Cooperative selective diversity.* In this scheme, the BS dynamically chooses conventional relaying (given in Figure 6.6(d)) or direct transmission (Can *et al.*, 2008). In the conventional relaying the RS(s) receive data from the BS then forward to the MD. The received data at the MD only depends on the data on the RS(s) to the MD links. The advantage of this scheme is that the MCS choice is flexible in the two phases.

- *Adaptive cooperative diversity scheme.* This scheme is to choose the best scheme among the aforementioned four schemes considering the cooperation gain and complexity. The increasing complexity is ordered as follows: direct transmission, conventional relaying, cooperative transmit diversity II and cooperative diversity I (Can *et al.*, 2007a).

Under the different cooperative diversity schemes, the corresponding Signal-to-Noise Ratio (SNR) at RS(s) and MD included in the Channel State Information (CSI) will be fedback to the BS. Then, according to the available MCS in WiMAX, as given in Table 6.1, the BS can calculate the achievable throughput by different cooperative relay schemes. Thus, BS can evaluate the cooperation gain to make the right cooperation decision. The detailed throughput derivation can refer to Can *et al.* (2007b). An example of the achievable throughput gain by introducing relays is illustrated as Figure 3 in Can *et al.*

Table 6.1 Type of MCS according to the received SNR from Table 266 in 802.16e Task Group (2006).

Modulation	Coding rate	Received SNR (dB)	Useful bits/Symbol
BPSK	1/2	3	96
QPSK	1/2	6	192
	3/4	8.5	288
16QAM	1/2	11.5	384
	3/4	15	576
64QAM	2/3	19	768
	3/4	21	864

(2008). It presents the overall average throughput per channel use as a function of the total number of relays in the cell.

6.3 Cooperative Schemes for Multicast Broadcast Services in WiMAX

Multicast communication has been identified as an effective way to disseminate information to a potential group of receivers sharing the same service interest (Yamamoto, 2004). Furthermore, as said by Phil McConnell, CEO of Data Connection, 'Multicast services are becoming very important and potentially very significant revenue drivers for service providers'. Indeed, along with widespread deployment of wireless networks, the fast-improving capabilities of MDs, and an increasingly sophisticated mobile work force worldwide, content and service providers are increasingly interested in supporting multicast communications over wireless networks (Varshney, 2002). In the conventional cellular network, 3GPP has standardized Multimedia Broadcast Multicast Service (MBMS) to efficiently support multicast/broadcast services. As a good competitor of 3G and a promising substitute of Digital Subscriber Line (DSL), efficiently supporting multicast services becomes one of the critical issues in WiMAX success.

Generally speaking, multicast services can be categorized into multicast multimedia services and reliable multicast services. The former has lower requirements for packet loss/error but is delay- and jitter-sensitive such as IPTV, video conferences, distant education, etc. Therefore, there is no acknowledgement in multimedia multicast services. The latter is more delay tolerant but is very sensitive to the packet error/loss. The reliability is often achieved by the acknowledgement and retransmission. Reliable multicast services include software distribution, data distribution and replication, mailing list delivery, Atwood (2004) and many new m-commerce services such as mobile auctions, etc. (Varshney, 2002). From the viewpoint of the network operator, the goal is to provide decent QoS to both multicast services at the meantime to keep good throughput of the network. To reach the goal, the two proposed cooperative schemes for multimedia multicast services and reliable multicast services are introduced and discussed in the following sections.

6.3.1   Cooperative Transmission for Multimedia Multicast Services

As for multimedia multicast services in WiMAX, the challenge is in the radio access part, that is, the transmission from BS to the end users. The motivation of introducing the cooperative schemes is that each member in a multicast group has a different channel condition at a time instant. An important issue is how to balance the trade-off between the user outage probability and the network throughput in the wireless multicast services. Conventionally, the BS, not only in the traditional cellular network but in WiMAX, chooses the data rate according to the channel condition of the worst user. By this approach a user who suffers from deep fading or shadowing will deteriorate the network throughput performance. A cooperation scheme has been proposed as one feasible solution to improve the network throughput by Hou *et al.* (2008).

The basic idea of cooperative transmission for multimedia multicast service is quite similar as the so-called cooperative diversity selective scheme in the MMR WiMAX (see Figure 6.6(d)). Namely, the BSs transmit multicast data with high MCS in the first phase then the users with good channel condition receive the data, decode and forward to those who are not able to successfully decode the received data in the first phase. (In this chapter the users who can successfully decode the data in the first phase are called group 1, denoted as G_1. The remaining users are called group 2, denoted as G_2.) The main differences between the relaying in this scenario and the relaying in MMR WiMAX are as follows. First, the RSs here are not built by operators but some MDs with good channel conditions. The capabilities of those MDs have certain limitations such as energy consumption constraints and antenna height. Second, it is a big challenge to select a proper transmission rate at the BS for both transmission phases. The reasons lie in that the BS needs to figure out how many members of the multicast group should be grouped in G_1 and the geographic relation between the users in G_1 and G_2. Third, in this scenario, some MDs work not only as end user but as a RS. Since these double-role MDs are individual entities, they need incentive to cooperate. To some extent, the network operator cannot force the individuals to forward data together. However, the network operator can use some kind of rewarding-based mechanism encouraging the MDs to cooperate. Therefore, this application scenario essentially belongs to network cooperation although it looks like user cooperation.

The feasible transmission rate selection algorithm is discussed in the following. To select the transmission rate at the BS at the first phase, R_1, there are two options.

- It selects the average of the data rate which can be supported by individual member in the multicast group.

- It selects the data rate which can be supported by certain percentage, such as, 50%, of the multicast group members, which is used by Hou *et al.* (2008).

We denote the SNR at member j as γ_j; the data rate that can be supported by member j, R_0^j, is

$$R_0^j = F_{\text{MCS}}(\gamma_j), \tag{6.2}$$

where F_{MCS} is the MCS discrete mapping function according to Table 6.1. So the set of the data rate that can be supported by each member in the multicast group, G_R, is

$$G_R = \{R_0^j, j \in G\}, \tag{6.3}$$

where, G is the set of all of the members in the multicast group.

In the conventional multicast, the data rate R_0 is the minimum value of G_R, that is, $R_0 = \min\{G_R\}$, if the minimum SNR at the members is no less than 3 dB (see Table 6.1). Otherwise, R_0 is the minimum data rate which can be supported in WiMAX. In the latter case the users that cannot support the minimum available data rate will be out of service.

If using the aforementioned mean data rate based selection algorithm, it is clear that the selected transmission rate at the BS at the first phase equals to the average of G_R, that is, $R_1 = E(G_R)$. Correspondingly, the selected transmission rate at the BS in the first phase of the second algorithm, R_1, can be expressed as

$$P(G_R \geq R_1) \geq \zeta \cap P(G_R \leq R_1) \geq 1 - \zeta, \tag{6.4}$$

where ζ is the percentage of the members that can support the data rate R_1.

To decide the data rate at the second phase, R_2, it considers the combined received signal from all of the cooperation partners. Therefore, it depends on how the cooperative transmission is done and how many cooperative partners are involved in the second phase.

In the cooperative transmission scheme, each MD in G_2 can measure the link quality to know the received SNR from its cooperation partners. Assuming there are M and N members that belong to G_1 and G_2, respectively. The received SNR at MD_n on the link between MD_m and MD_n is denoted by $\gamma_{m,n}$. The data rate can be supported by MD_n at second phase, R_2^n. Here R_2^n depends on the SNR of the combined received signal $\tilde{\gamma}_n$ according to the MCS mapping (Table 6.1). Hence, R_2^n can be expressed as

$$R_2^n = F_{\text{MCS}}(\tilde{\gamma}_n)$$

$$= F_{\text{MCS}}\left(\sum_{m \in G_1} \gamma_{m,n}\right), \quad n \in G_2. \tag{6.5}$$

The data rate can be received by the MDs in G_2 in the second phase, R_2, can be expressed by

$$R_2 = \begin{cases} \min\{R_2^n, n \in G_2\} & \min\{\tilde{\gamma}_n, n \in G_2\} \geq 3, \\ \xi & \min\{\tilde{\gamma}_n, n \in G_2\} < 3. \end{cases} \tag{6.6}$$

where ξ is the minimum available data rate in WiMAX and $\tilde{\gamma}_n$ is the SNR of the received combined signal at a member n in G_2.

For the multicast multimedia service, a utility function is defined as the product of the number of the supported multicast group members and the throughput of each member. Thus, the utility function of the conventional multicast services U_{con} is

$$U_{\text{con}} = (1 - \sigma)R_0|G|, \tag{6.7}$$

where R_0 is the data rate in the conventional multicast services, $|G|$ is the number of the members in the multicast group and σ is the probability of the multicast members that cannot support R_0. Usually those are the members that cannot support the minimum available transmission rate defined in Table 6.1. Correspondingly, the utility function of the cooperation scheme U_{coop} is

$$U_{\text{coop}} = R_1|G_1| + (1 - \sigma')R_2|G_2|, \tag{6.8}$$

where R_1 and R_2 are the transmission rate of the first and second phase, respectively, $|G_1|$ and $|G_2|$ are the number of multicast members that belong to $G1$ and $G2$, respectively, and σ' is the probability of the members in G_2 that cannot support R_2. Those are often the members that cannot support the minimum available data rate defined in Table 6.1.

To make the cooperation decision, the BS has to evaluate the cooperation gain. The cooperation gain is defined as the gain of the utility function by the cooperation scheme to that of the original. Hence, the cooperation gain can be expressed as

$$
G_{\text{coop}} = \frac{U_{\text{coop}} - U_{\text{con}}}{U_{\text{con}}}
$$

$$
= \frac{(R_1 G_1 + (1 - \sigma') R_2 G_2) - ((1 - \sigma) R_0 G)}{(1 - \sigma) R_0 G}. \tag{6.9}
$$

Based on the cooperation transmission scheme and the two transmission rate selection algorithms for the first phase, we ran a simulation to show the cooperation gain that can be achieved by the cooperative transmission scheme. In the simulation the channel between the BS and MD is modeled by the IEEE 802.16 (SUI) model (Anderson, 2003). The model constant is based on Terrain Type C. The BS and MD height are 30 m and 2 m, respectively. The signal attenuation on the link between MDs is modeled by the path loss with exponent equal to 3 (Anderson, 2003). The two transmission rate selection algorithms for the first phase are both simulated. We refer to the first algorithm as the mean data rate based selection since it chooses the mean data rate as R_1. The second algorithm is referred to as the median data rate based selection since it chooses the data rate R_1 which 50% of MDs can support. The cooperation gain in the simulation is shown in Figure 6.7 which illustrates the great potential of the cooperation scheme. From Figure 6.7 we can also see that the cooperation gain increases with the increasing number of the multicast group members. The reason is that the more users join the cooperation transmission, the more cooperative diversity can be obtained. The relative cooperation gain relation between the two algorithms is not absolute, because it depends on the distribution of the MDs. Under the simulation condition that the MDs are uniformly distributed in a geographic area, the median SNR-based selection algorithm outperforms the mean SNR-based selection algorithm.

6.3.2   Cooperative Retransmission Scheme for Reliable Multicast Services Using Network Coding

For reliable multicast services in the wireless network, one of the key challenges is error/loss recovery. The reasons are as follows. First of all, the traditional error/loss recovery schemes such as Forward Error Correction (FEC), Automatic Repeat Request (ARQ) and Hybrid ARQ (HARQ) are not efficient in the wireless multicast services due to implosion and exposure issues. Implosion is a result of the multiple Negative Acknowledgments (NACKs) from many receivers. It might swamp the sender and the network, even other receivers (Zhang *et al.*, 2007). Exposure occurs when the retransmitted packets are delivered to those receivers who did not lose the packets (Zhang *et al.*, 2007). Both implosion and exposure are fatal impediments for multicasting in wireless networks (Zhang *et al.*, 2007). Second, the heterogeneity of the wireless channel conditions results in heterogenous error/loss.

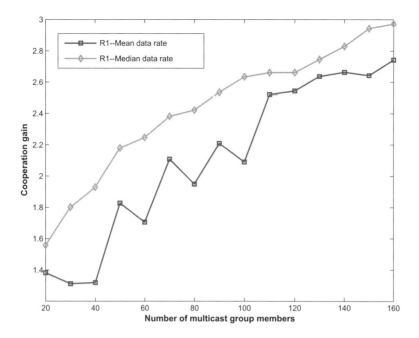

Figure 6.7   The cooperation gain comparison of the two transmission rate selection algorithms for the first phase. (Mean data rate algorithm: the average of the supportable data rates; Median data rate algorithm: the data rate which can be supported by 50% of all of the MDs.)

The cry baby issue[3] will bring about the exposure issue and deteriorate the overall network performance. Third, because of wireless channel time correlation characteristics (Sgardoni *et al.*, 2007), when a radio link suffers from the instantaneous bad channel condition, it cannot effectively help itself out by requesting retransmission. To solve these challenging issues of error/loss recovery in the wireless reliable multicast services, the cooperative retransmission scheme was proposed by Zhang *et al.* (2007). The goal is twofold: to reduce the throughput waste on sending NACK and retransmission, and to save the energy consumption of the MD, especially to reduce the energy waste on receiving the duplicate packets.

The proposed cooperative retransmission scheme is based on the assumption that multiple MDs are in the proximity of each other and they can group into a cooperative cluster. Since the MDs in the cluster have the heterogeneous error/loss characteristics (namely different MDs missed the different packets), the error/loss can be recovered within the cluster without or with very little BS involvement. The cooperative retransmission scheme is based on the CCP2P network architecture (Fitzek and Katz, 2006, 2007a), as introduced in Section 6.1, that is, the MD can communicate not only with BS through the cellular link but also with the neighboring MDs through the short-range link. The short-range links among the MDs in the cluster usually have more favorable channel conditions.

---

[3]The frequent retransmission request from one user results in that the BS has to make a retransmission for this single user. An analogy is a crying baby and the mother assuming that all of her babies are crying.

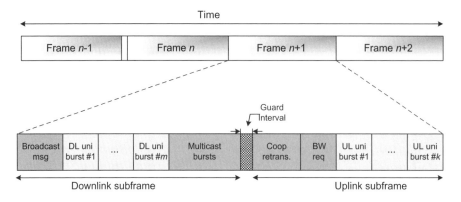

Figure 6.8 Frame structure in the cooperative retransmission scheme (Zhang *et al.*, 2007).

Zhang *et al.* (2007) designed a frame structure as shown in Figure 6.8 on the cellular link with TDD mode which fits very well with the WiMAX frame structure. In Zhang *et al.* (2007), the short-range link has the same frequency as the cellular link, therefore, the BS reserves time slots for the cooperative retransmission. More generically, the short-range communication can use different time, frequency, codes or even different air-interfaces as long as it is orthogonal to the cellular link. In case the short-range communication does not use the WiMAX frequency, the BS can allocate the cooperation retransmission slots to other services. Anyway, the designed frame structure works in the WiMAX network.

The basic idea of the cooperative retransmission scheme of Zhang *et al.* (2007) is as follows. The MDs in a cluster are grouped by a logical token ring topology. The cooperative retransmission is composed of two procedures: counting the missed packets within the cluster by marking the Lost Packet Matrix (LPM), as well as the local retransmission. In the first procedure the LPM is passed through all of the MDs in the cluster until it returns to the first MDs in the token ring (see Figure 6.9, for more details refer to Zhang *et al.* (2007)). In the local retransmission procedure, a couple of MDs called primary MDs share the retransmission task. In case the primary MDs are not able to completely recover the error/lost packets, the other MDs called auxiliary MDs can help to complete the recovery. The worst case is that the recovery cannot be completed within the cluster, then the cluster can request the BS to complete the recovery. Since the wireless channels between the MDs and the BS are independent, the probability that multiple mobile devices lose the same packet is very low. Therefore, requesting a retransmission from BS has a very low probability. The illustration of the local retransmission procedure is shown in Figure 6.10 (see Zhang *et al.* (2007) for additional details).

Although the great potential of the cooperative local retransmission scheme has been illustrated by Zhang *et al.* (2007), we can further improve the retransmission procedure by encoding the retransmitted packets with network coding. The reason lies in the coding mechanism of network coding and the error/loss heterogeneity. By network coding, a MD can make a linear combination of the packets that it receives to generate new packets. When other MDs receive the new coded packets, they can use their known packets to decode the missed packets which are encoded in the retransmitted packets. The advantage of network

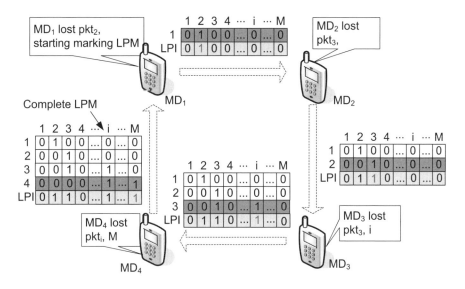

Figure 6.9 The first marking LPM procedure (Zhang *et al.*, 2007).

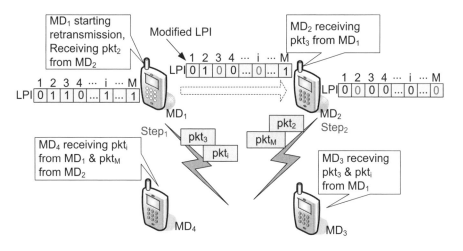

Figure 6.10 The second marking LPM procedure (Zhang *et al.*, 2007).

coding is that although different MDs miss the different packets, it is possible for them to use the same coded packets to recover the different missed packets. The detailed encoding and decoding algorithm of network coding can refer to Fragouli *et al.* (2008). With network coding the number of local retransmission packets can be significantly reduced; in particular, when the missed packets at each MD are highly heterogeneous. In other words, it is possible to attain higher cooperation gain by network coding. A simple case in point is that assuming there are N MDs in a cluster and the cooperation retransmission is done every M packets

($M > N$). Assuming that each MD misses one packet (but each misses a different ones), then the cluster can complete the recovery by $1 + 1 = 2$ coded retransmission packets instead of retransmitting N packets.

The theoretical analysis is given in the following. The set of the missed packets of a cluster can be expressed by

$$L = L_1 \cup L_2 \cup \cdots \cup L_n, \tag{6.10}$$

where $L_1, L_2, \ldots, L_n$ are the sets of the missed packets of the mobile device $1, 2, \ldots, n$, respectively.

Therefore, the number of the retransmission packets N_{lr} is

$$N_{lr} = |L|, \tag{6.11}$$

where $|L|$ is the cardinality of set L, that is, the number of the total lost packets in the cluster. With network coding, the number of the retransmission packets is N_{nc}, which is given by

$$N_{nc} = \begin{cases} \max\{|L_1|, |L_2|, \ldots, |L_n|\} + \min\{|L_1|, |L_2|, \ldots, |L_n|\} & \Lambda = \phi, \\ \max\{|L_1|, |L_2|, \ldots, |L_n|\} + \min\{|L_1|, |L_2|, \ldots, |L_n|\} - |\Lambda| & \Lambda \neq \phi, \end{cases} \tag{6.12}$$

where $|L_1|, |L_2|, \ldots, |L_n|$ represent the cardinality of set $L_1, L_2, \ldots, L_n$. Here Λ is the intersection of set $L_1, L_2, \ldots, L_n$, that is, $\Lambda = L_1 \cap L_2 \cap \cdots \cap L_n$. When $\Lambda = \phi$, it means that all of the missed packets can be recovered within the cluster. In case $\Lambda \neq \phi$ the cluster needs to request $|\Lambda|$ packets from the BS. Note that the expression of N_{nc} here does not consider the error/loss of the retransmitted packets.

A simulation is conducted to show the potential of network coding in cooperative retransmission. The assumptions are given as follows. The cooperative retransmission is done every M sequent packets with $M = 20$ or 40. There are 5 to 100 MDs in a cluster. The Packet Loss Rate (PLR) is 1% or 5%. The comparison of cooperative retransmission with/without network coding is shown in Figure 6.11. It should be clear by now that with network coding, the number of retransmitted packets is much lower than that of the conventional cooperative retransmission. For instance, when the number of packets in one packet flow is 20 and the PLR is 1%, under the condition that there are 100 MDs in the multicast group the number of retransmitted packets with/without network coding is 2 and 13, respectively. With the same number of MDs in one multicast group, when the number of packets in one packet flow is increased to 40 and the PLR is increased to 5%; the number of retransmitted packets without network coding increases to 40. However, the number of retransmissions increases to 6 with network coding. This indicates that with the increased PLR or the number of packets in one packet flow, the variation of the number of the retransmitted packets is much smaller with network coding. Furthermore, it should be noted that with network coding the number of the retransmitted packets is quite stable with the varying size of the cluster.

The advantages of applying network coding for cooperative retransmission are threefold:

- network coding shortens the retransmission time and it improves the overall network throughput performance;

- network coding reduces the times of transmission and reception within the cluster, therefore, it has the potential to save more energy on the MDs;

- with the characteristics of the stable retransmission times, it is easier for the BS to reserve certain number of time slots for the cooperative retransmission beforehand.

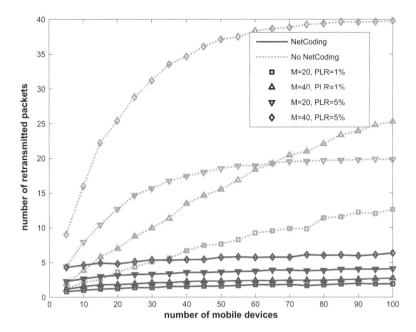

Figure 6.11 Number of retransmitted packets within a cluster by cooperative retransmission with/without network coding.

6.4 Network Coding Implementation in the Commercial WiMAX Mobile Device

Current MDs are equipped with multiple air interfaces supporting both cellular and short-range networking, for example, 3G data access, IEEE802.11 WLAN and Bluetooth, which makes them highly suitable for cooperative wireless networking. For example, Nokia has launched a WiMAX enabled Internet tablet that supports Bluetooth and IEEE 802.11 WLAN in addition to the WiMAX technology. The N810 Internet tablet is depicted in Figure 6.12.

Using the N810 device, researchers of Aalborg University have shown the possibility of implementing cooperative principles. Assuming that each N810 device has WiMAX connectivity, the devices will download partial information from the overlay WiMAX cell and distribute (i.e. exchange) that information via IEEE802.11 WLAN until all devices have the full information. By doing so, the cooperative cluster will obtain the information faster and with less energy in comparison with stand-alone devices. In this chapter we do not treat energy savings in detail. However, by reducing the number of retransmissions in the cluster, it is clear that energy saving can be obtained.

In one of the projects at Aalborg University, the exchange of the partial information over IEEE802.11 was boosted by network coding concepts. In Figure 6.13 the initial packet distribution is given for four MDs. Light boxes represent packets that are already available on the MD and dark boxes represent the missing packets. The losses created for each MD are made in such a way that it is easy for the reader to follow the explanation. The full information

Figure 6.12 N810 WiMAX edition (courtesy of Nokia).

is already available in the cluster, but not at each device. Therefore, the MDs will start to convey the missing packets to each other. In case of unicast, one packet per transmission can be repaired. In case multiple devices are missing the same packet, broadcast may result in less transmissions. In the best case, broadcast could convey $J - 1$ packets in one transmission, where J is the number of MDs in one cooperative cluster. Such a favorable situation cannot be found very often and less packets are repaired by one broadcast transmission. Nevertheless this example gives some insights of the potential of state-of-the-art transmission schemes.

Network coding is not dependent on favorable packet loss distribution. As long as all $J - 1$ devices are missing at least one packet, network coding can transport $J - 1$ valuable information packet per transmission. For example, in the case of four MDs where device A has packet set {1, 2, 4}, B has packet set {2, 3, 4} and C has {1, 3, 4}, MD D, which has all four packets already, will send only one coded packet to all neighboring devices to repair all three different losses. In this case using network coding, the coded packet will be a result of the linear combination of packets 1, 2 and 3 and some header information about which packets are encoded and the coefficients of the linear combination. The receiving devices will then decode the coded packet with all available packets to retrieve the missing packet. For instance, MD A will decode the coded packet with the already existing packets 1 and 2 obtaining packet number 3, which was missing. In this simple example the coding gain, defined as how many useful packets are conveyed by one transmission, is three. The final packet distribution after performing network coding is given in Figure 6.14. Now all packets are available for each MD. All packets labeled with a '2' indicate a packet that was rebuilt by the linear combination of the received coded packet and two further packets.

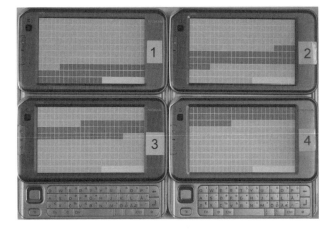

Figure 6.13 Initial packet distribution over four MDs.

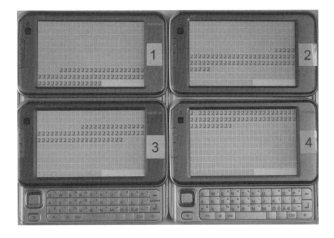

Figure 6.14 Final packet distribution with coding gain per packet.

6.5 Conclusion

In this chapter we have discussed the underlying concepts behind cooperative principles for WiMAX communication systems. Even though WiMAX can support high data rates already, cooperation within the network or between users has the potential to improve key aspects of WiMAX including throughput, data rate and energy efficiency. Loosely speaking, cooperation opens up a new dimension for WiMAX systems. Indeed in addition to enhancing WiMAX performance, cooperation will bring up new concepts for advanced services for the mobile users. In this chapter some examples on cooperative principles were given.

Acknowledgements

We would like to thank Nokia for providing technical support as well as mobile phones to carry out the implementation. Special thanks to Mika Kuulusa, Gerard Bosch, Harri Pennanen, Jarmo Tikka, Nina Tammelin, Helena Hattinen and Per Moeller from Nokia. Furthermore we would like to thank the student group 08gr654 (Morten Tychsen, Peter Østergaard, Jakob Sloth Nielsen and Karsten Fyhn Nielsen) for carrying out the implementation of network coding on N810 devices. This work was partially financed by the X3MP project granted by the Danish Ministry of Science, Technology and Innovation.

References

802.16e Task Group (2006) IEEE Standard for Local and Metropolitan Area Networks Part 16: Air Interface for Fixed and Mobile Broadband Wireless Access Systems Amendment 2: Physical and Medium Access Control Layers for Combined Fixed and Mobile Operation in Licensed Bands and Corrigendum 1. *IEEE Standard 802.16e-2005 and IEEE Standard 802.16-2004/Corrigendum 1-2005 (Amendment and Corrigendum to IEEE Standard 802.16-2004)*, pp. 1–822.

802.16j Task Group I (2006) Harmonized contribution on 802.16j (mobile multihop relay) usage models. *Technical Report*, IEEE 802.16 Broadband Wireless Access Working Group.

Albiero, F., Fitzek, F.H. and Katz, M. (2007) Cooperative power saving strategies in wireless networks: an agent-based model. *Proceedings of the IEEE International Symposium on Wireless Communication Systems (ISWCS 2007)*, vol. 1, Trondheim, Norway, pp. 287–291.

Albiero, F., Katz, M. and Fitzek, F.H. (2008) Energy efficient cooperative techniques for multimedia services over future wireless networks. *Proceedings of the IEEE International Conference on Communications (ICC 2008)*, vol. 1, Beijing, China, pp. 2006–2011.

Anderson, H.R. (2003) *Fixed Broadband Wireless System Design* (1st edn). John Wiley & Sons Ltd, Chichester.

Atwood, J. (2004) A classification of reliable multicast protocols. *IEEE Network*, **18**, 24–34.

Boyer, J., Falconer, D. and Yanikomeroglu, H. (2004) Multihop diversity in wireless relaying channels. *IEEE Transactions on Communications*, **52**(10), 1820–1830.

Brodlos, A., Fitzek, F. and Koch, P. (2005) Energy aware computing in cooperative wireless networks. *Proceedings of Cooperative Networks, WirelessCom 2005*, vol. 1, Maui, HI, pp. 16–21.

Can, B., Portalski, M., Simon, H., Lebreton, D., Frattasi, S. and Suraweera, H. (2007a) Implementation issues for OFDM-based multihop cellular networks. *IEEE Communications Magazine*, **45**(9), 74–81.

Can, B., Yanikomeroglu, H., Onat, F., De Carvalho, E. and Yomo, H. (2008) Efficient cooperative diversity schemes and radio resource allocation for IEEE 802.16j. *Proceedings of the 2008 IEEE Wireless Communications and Networking Conference*, pp. 36–41.

Can, B., Yomo, H. and Carvalho, E.D. (2007b) Link adaptation and selection method for OFDM based wireless relay networks. *Journal of Communications and Networks (Special Issue on MIMO-OFDM and its Applications)*, **9**(2), 118–127.

Cover, T.M. and Gamal, A.A.E. (1979) Capacity theorems for the relay channel. *IEEE Transactions on Information Theory*, **25**(5), 572–584.

Fitzek, F.H. and Katz, M.D. (eds) (2006) *Cooperation in Wireless Networks: Principle and Applications*. Springer.

Fitzek, F.H. and Katz, M.D. (eds) (2007a) *Cognitive Wireless Networks: Concepts, Methodologies and Visions*. Springer, Berlin.

Fitzek, F. and Katz, M. (2007b) *Cognitive Wireless Networks – Cellular Controlled Peer to Peer Communication*. Springer, Berlin, pp. 31–59.

Fragouli, C., Boudec, J.Y.L. and Widmer, J. (2008) Network coding: an instant primer, http://algo.epfl.ch/ christin/primer.ps.

Gastpar, M., Kramer, G. and Gupta, P. (2002) The multiple-relay channel: coding and antenna-clustering capacity. *Proceedings of the IEEE International Symposium on Information Theory*, p. 136.

Gupta, P. and Kumar, P. (2003) Towards an information theory of large networks: an achievable rate region. *IEEE Transactions on Information Theory*, **49**(8), 1877–1894.

Hammerstrom, I., Kuhn, M. and Wittneben, A. (2004) Channel adaptive scheduling for cooperative relay networks. *Proceedings of IEEE 60th Vehicular Technology Conference (VTC2004-Fall)*, vol. 4, pp. 2784–2788.

Host-Madsen, A. (2006) Capacity bounds for cooperative diversity. *IEEE Transactions on Information Theory*, **52**(4), 1522–1544.

Hou, F., Cai, L.X., She, J., Ho, P.H., Shen, X. and Zhang, J. (2008) Cooperative multicast scheduling scheme for IPTV service over IEEE 802.16 networks. *Proceedings of the 2008 IEEE International Conference on Communications*, pp. 2566–2570.

Hunter, T.E. and Nosratinia, A. (2002) Cooperation diversity through coding. *Proceedings of the IEEE International Symposium on Information Theory*, p. 220.

Kramer, G., Gastpar, M. and Gupta, P. (2005) Cooperative strategies and capacity theorems for relay networks. *IEEE Transactions on Information Theory*, **51**(9), 3037–3063.

Kristensen, J. and Fitzek, F. (2007) *Cognitive Wireless Networks – Cellular Controlled P2P Communication Using Software Defined Radio*. Springer, Berlin, pp. 435–455.

Laneman, J., Tse, D. and Wornell, G. (2004) Cooperative diversity in wireless networks: efficient protocols and outage behavior. *IEEE Transactions on Information Theory*, **50**(12), 3062–3080.

Laneman, J., Wornell, G. and Tse, D. (2001) An efficient protocol for realizing cooperative diversity in wireless networks. *Proceedings of the IEEE International Symposium on Information Theory*, p. 294.

Laneman, J.N. (2002) Cooperative diversity in wireless networks: algorithms and architectures. *PhD thesis*, Massachusetts Institute of Technology, Cambridge, MA.

Laneman, J.N. (2006) *Cooperation in Wireless Networks: Principle and Applications*. Springer, Berlin, pp. 163–187.

Li, G. and Liu, H. (2006) Resource allocation for OFDMA relay networks with fairness constraints. *IEEE Journal on Selected Areas in Communications*, **24**(11), 2061–2069.

Lin, Z., Erkip, E. and Ghosh, M. (2005) Adaptive modulation for coded cooperative systems. *Proceedings of the IEEE 6th Workshop on Signal Processing Advances in Wireless Communications*, pp. 615–619.

Madsen, T., Zhang, Q. and Fitzek, F. (2007) *Cognitive Wireless Networks – IP Header Compression for Cellular-Controlled P2P Networks*. Springer, Berlin, pp. 499–511.

Onat, F., Adinoyi, A., Fan, Y., Yanikomeroglu, H. and Thompson, J. (2007) Optimum threshold for snr-based selective digital relaying schemes in cooperative wireless networks. *Proceedings of the 2007 IEEE Wireless Communications and Networking Conference*, pp. 969–974.

Pabst, R., Walke, B., Schultz, D., Herhold, P., Yanikomeroglu, H., Mukherjee, S., Viswanathan, H., Lott, M., Zirwas, W., Dohler, M., Aghvami, H., Falconer, D. and Fettweis, G. (2004) Relay-based deployment concepts for wireless and mobile broadband radio. *IEEE Communications Magazine*, **42**(9), 80–89.

Perrucci, G., Fitzek, F. and Petersen, M. (2008) *Heterogeneous Wireless Access Networks: Architectures and Protocols – Energy Saving Aspects for Mobile Device Exploiting Heterogeneous Wireless Networks*. Springer, Berlin (accepted).

Sadek, A., Han, Z. and Liu, K. (2006) A distributed relay-assignment algorithm for cooperative communications in wireless networks. *Proceedings of the 2006 IEEE International Conference on Communications*, vol. 4, pp. 1592–1597.

Sendonaris, A., Erkip, E. and Aazhang, B. (1998) Increasing uplink capacity via user cooperation diversity. *Proceedings of the IEEE International Symposium on Information Theory*, p. 156.

Sendonaris, A., Erkip, E. and Aazhang, B. (2003a) User cooperation diversity. Part I. System description. *IEEE Transactions on Communications*, **51**(11), 1927–1938.

Sendonaris, A., Erkip, E. and Aazhang, B. (2003b) User cooperation diversity. Part II. Implementation aspects and performance analysis. *IEEE Transactions on Communications*, **51**(11), 1939–1948.

Sgardoni, V., Ferre, P., Doufexi, A., Nix, A. and Bull, D. (2007) Frame delay and loss analysis for video transmission over time-correlated 802.11a/g channels. *Proceedings of the 2007 IEEE 18th International Symposium on Personal, Indoor and Mobile Radio Communications*, pp. 1–5.

van der Meulen, E. (1971) Three-terminal communication channels. *Advances in Applied Probability*, **3**(1), 120–154.

Varshney, U. (2002) Multicast over wireless networks. *Communications of the ACM*, **45**(12), 31–37.

Yamamoto, M. (2004) Multicast communication in next-generation Internet. *Proceedings of the 10th IFAC/IFORS/IMACS/IFIP Symposium on Large Scale Systems: Theory and Applications*, pp. 639–645.

Zhang, Q. and Fitzek, F. (2007) *Cognitive Wireless Networks – Cooperative Retransmission for Reliable Wireless Multicast Services*. Springer, Berlin, pp. 485–498.

Zhang, Q., Fitzek, F.H. and Iversen, V.B. (2007) Design and performance evaluation of cooperative retransmission scheme for reliable multicast services in cellular controlled P2P networks. *Proceedings of the 18th Annual IEEE International Symposium on Personal, Indoor and Mobile Radio Communications (PIMRC07)*.

7

The Role of WiMAX Technology in Distributed Wide Area Monitoring Applications

Francesco Chiti, Romano Fantacci, Leonardo Maccari, Dania Marabissi and Daniele Tarchi

7.1 Monitoring with the WSN Paradigm

Wireless sensor networks (WSNs) represent an inherently disruptive approach specifically designed to detect events or phenomena, collect and process related data, and transmit sensed information to final users in a distributed way (Akyildiz *et al.*, 2001).

Although WSNs exhibit several features common to wireless *ad hoc* or mesh networks, such as self-organizing capabilities or short-range broadcast communication with a multihop routing, they present additional constraints in terms of limitations in energy, transmit power, memory and computing power. Further, the operative conditions usually require cooperative efforts of sensor nodes in the presence of frequently changing topology due to fading and node failures or node mobility (Ilyas and Mahgoub, 2005).

A typical WSN is comprised of the following basic components:

- A set of distributed or localized sensors;

- An interconnecting network usually, but not always, wireless-based;

- A central point of information clustering;

- Computing resources at the central point (or beyond) to handle data correlation, event trending, status querying and data mining.

These elements allow a system administrator to observe and react to events and phenomena in a specified environment. The administrator can be a civil, governmental, commercial or industrial entity. Typical application fields can span from agriculture (AgroSense, 2007–2010, GoodFood, 2004–2006) and environmental monitoring (DustBot, 2007–2009), civil engineering, disaster management (InSyEme, 2007–2010), military applications up to health monitoring and surgery (Chiti and Fantacci, 2006).

Generally speaking WSN systems can be classified into two basic categories:

- Mesh-based systems with multihop radio connectivity between WSNs, utilizing dynamic routing in both the wireless and wired portions of the network.

- Point-to-point or multipoint-to-point (*star-based*) systems generally with single-hop radio connectivity to WSNs, utilizing static routing over the wireless network.

The latter scheme is composed of networks in which the end devices (i.e. the sensors) are one radio hop away *forwarding node*, for example, a wireless router connected to the terrestrial network via either a wired or a point-to-point wireless link. On the other hand, the former approach allows end devices (sensors) to be more than one radio hop away from a routing or forwarding node. The forwarding node is a wireless router that supports dynamic routing, while wireless routers are often connected over wired links covering a wider area.

Presently WSNs have been largely focused on dense, small-scale homogeneous deployments to monitor a specific physical phenomenon. Nevertheless, the integration of multiple heterogeneous sensor networks operating in different environments could provides the ability to monitor diverse physical phenomena at a global scale, as addressed in WP122 (2008). In addition, such remote integration will make the infrastructure able to query and fuse data across multiple, possibly overlapping, sensor networks in different domains. Moreover, new types of sensor networks based on *mobile sensor platforms* are becoming available, for example, vehicles in the urban grid or firefighters in a disaster recovery operation equipped with a variety of sensors (Ilyas and Mahgoub, 2005, InSyEme, 2007–2010) (location, video, chemical, radiation, acoustic, etc). The vehicle grid then becomes a sensor network that can be remotely accessed from the Internet to monitor vehicle traffic congestion and to prevent accidents, chemical spills and possible terrorist attacks. Likewise, on-site operators as firefighters might be equipped with several wearable devices such as cameras or sensors, allowing the commander to be aware of the conditions in the field and to direct the operations to maximize the use of the forces, while preserving the life of his responders.

As a consequence, one of the most interesting applications of an integrated WSN is the ability to create a *macroscope* to take a look at a picture of the monitored environment wider than the areas monitored by a single WSN (WP122, 2008). There have been several attempts, for instance, the WSN deployed on redwood trees, a wildlife monitoring site on Great Duck Island, tracking zebras in their natural habitat and monitoring volcanic eruptions. All of these systems have been deployed in remote locations with limited access: some areas might be accessible only once in several months, straining the lifetime of sensors with limited battery power. Many are subject to harsh elements of nature that cause rapid device and sensor malfunction. Network links to back-end monitoring and collection systems may be intermittent due to weather or other problems, while in-network data storage is limited, leading to important observations being missed.

In addition to this, there is an increasing interest in real-time connecting heterogeneous devices with application to building/commercial automation (security, lighting control, access control) or industrial control (asset management, process control, environmental, energy management). These applications usually addressed as *invisible computing* involve different technologies such as (WPANs: Wireless Personal Area Networks in which IEEE 802.15.4/ZigBee standard plays a crucial role), Wireless Local Area Networks (WLANs; mainly IEEE 802.11a/b/g/h, etc. standards) and metropolitan transport (for which IEEE 802.15.3/WiMAX standards are ideally suited).

For the aforementioned applications, the sensor networks cannot operate in a stand-alone manner; there must be a way to monitor an entity to gain access to the data produced by the WSN. By connecting the sensor network to an existing network infrastructure such as the global Internet, a local-area network, or a private Intranet, remote access to the sensor network can be achieved. Given that the TCP/IP protocol suite has become the *de-facto* networking standard, not only for the global Internet but also for local-area networks, it is of particular interest to look at methods for interconnecting sensor networks and IP core networks. Sensor networks often are intended to run specialized communication protocols, for example IEEE 802.15.4 or Zigbee, therefore an all-IP-network will not be viable, due to the fundamental differences in the architecture of IP-based networks and sensor networks. It is envisaged that the integration of sensor networks with the Internet will need gateways in most cases. A proxy server at the core network edge is able to communicate both with the sensors in the sensor network and hosts on the TCP/IP network, and is thereby able to either relay the information gathered by the sensors, or to act as a front-end for the sensor network. It is also envisaged that sensing devices will be equipped with interfaces to wireless access networks such as 2/3G and WLAN enabling total *ubiquitous connectivity*.

7.2   Overall System Architecture

As discussed previously, a wide area WSN could be achieved by integrating specialized and even heterogeneous subnetworks through a reliable transport backbone. For many kinds of applications a wireless connection represents a flexible and cost-effective solution. In particular, the Worldwide Interoperability for Microwave Access (WiMAX), provides wireless broadband services on the scale of the Metropolitan Area Network (MAN). WiMAX brings a standards-based technology to a sector that otherwise depends on proprietary solutions: the standardized approach ensures interoperability between WiMAX equipment from vendors worldwide reducing costs and making the technology more accessible. This technology can provide fast and cheap broadband access in areas that lack infrastructure (fiberoptics or copper wire) such as rural areas, unwired countries and disaster recovery scenes where the wired networks have broken down (WiMAX can be used as backup links for broken wired links). Very often WSNs are used in these areas to monitor the environment, to prevent natural disasters or to aid rescue operations. WiMAX can also provide the last mile coverage in urban areas where the monitoring and control of anthropic processes taking place, including buildings, streets, factories and storehouses. WiMAX attributes open the technology to a wide variety of applications: with its wide coverage range and high transmission rate, WiMAX can serve as a backbone for integrating specialized sensors subnetworks and for connecting the WSN to the data processing center. Alternatively, users can connect mobile devices such as laptops and handsets directly to WiMAX base stations; WiMAX is able to support vehicular speeds of up to $125 \, \text{km hour}^{-1}$ providing ubiquitous

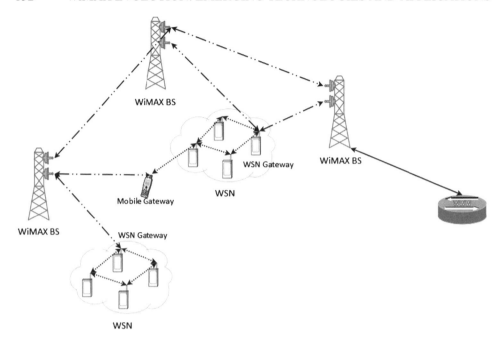

Figure 7.1 Envisioned system architecture providing interoperability among nonoverlapping WSNs through mesh-based WiMAX backhaul, as well as supporting mobile data mule.

mobile services. In this way the information coming from the WSNs can be distributed to the users: a user may be using a wireless videophone, a laptop or a PDA to access data while also dynamically interacting with the remote server by means of additional content uploading regarding a monitored area. For example, a fire team could download the internal map of a building to support or enhance the localization information coming from the WSN.

The IEEE 802.16 standard was designed mainly for point-to-multipoint topologies, in which a base station distributes traffic to many subscriber stations but it also supports a *mesh mode*, where subscriber stations can communicate directly with one another. The latter mode allows the Line-of-Sight (LOS) requirement to be related and the deployment costs for high-frequency bands to be eased by using subscriber stations to relay traffic to one another. In addition the use of multihop relay stations can extend the coverage area and improve throughput at a feasible economical level. The mesh mode can be a feasible solutions to connect distributed WSNs that must be connected together or to the same data processing center.

A possible architecture enabling the integration of heterogeneous WSNs, is depicted in Figure 7.1.

In addition to the previous advantages the choice of WiMAX technology is based on its Physical (PHY) and Medium Access Control (MAC) features that make a reliable, flexible and secure wireless system. The IEEE 802.16 MAC layer supports Quality of Service (QoS) for stations through adaptive allocation of the uplink and downlink traffic. This is very important to manage different data flows provided by different WSNs. In addition the security sublayer provides functionalities such as authentication, secure key exchange and encryption assuring the WSNs data is not corrupted. The MAC of 802.16 supports different transport technologies such as Internet Protocol version 4 (IPv4), IPv6, Ethernet, and Asynchronous Transfer Mode (ATM) making the integration of heterogeneous networks easier.

Finally, PHY layer is characterized by a high level of flexibility in the allocated spectrum. the work frequency can be chosen to improve performance and interterence and the bandwidth can be varied depending on the requirements (i.e. IEEE 802.16e uses the Scalable Orthogonal Frequency Division Multiple Access (S-OFDMA) scheme). The adaptive features at the PHY allow trade-offs between robustness and capacity and the robustness to the adverse propagation conditions permits the use of Non-Line-of-Sight (NLOS) communications, differently from alternative technologies currently available for fixed broadband wireless supporting only LOS coverage.

7.3   Efficient Access Management Schemes

As stated in the previous section, WiMAX represents a promising and reliable solution to provide a transport backbone among sensor subnetworks. As a matter of fact, it allows the establishment of effective communications handling different data flows with specific QoS requirements in terms of priority level, throughput, delay and jitter as well. The dynamic and flexible resource allocation scheme WiMAX is able to support typical application features concerning, for instance, the monitoring of synchronous processes, dispatching warnings and alarms and the reliability of data exchange.

The IEEE 802.16 family of standards (IEEE, 2005, 2004), supported by the WiMAX commercial consortium, defines the PHY and MAC layers specifications for a Broadband Wireless Access (BWA) communication protocol. Both MAC and PHY are designed to have a flexible access scheme and an adaptive resource management. This aspect is very important in the proposed architecture to manage a high number of distributed users including sensors subnetworks also under mobility conditions.

As for the PHY layer, among several alternatives, the IEEE 802.16 standard proposes the use of Orthogonal Frequency Division Multiplexing (OFDM) for mitigating frequency-dependent distortion across the channel band and simplifying the equalization in a multipath fading environment (van Nee and Prasad, 2000). The basic OFDM principle is parallelization: by dividing the available bandwidth into several smaller bands that are called subcarriers, the transmitted signal over each subcarrier may experience flat fading. Moreover, Orthogonal Frequency Division Multiple Access (OFDMA) is used to provide a flexible multiuser access scheme: disjunctive sets of subcarriers and OFDM symbols are allocated to different users.

To have more flexible and efficient OFDM/OFDMA systems, adaptive OFDM schemes are adopted to maximize the system capacity and maintain the desired system performance (Bohge *et al.*, 2007, Keller and Hanzo, 2000b). In particular, in an OFDM-based wireless system, the inherent multi-carrier nature of OFDM allows the use of link adaptation techniques according to the behavior of the narrow-band channels: the bit-error probability of different OFDM subcarriers, transmitted in time-dispersive channels, depends on the frequency-domain channel transfer function (Bohge *et al.*, 2007, Keller and Hanzo, 2000b).

Transmission techniques which do not adapt the transmission parameters to the fading channel require a fixed link margin or coding to maintain acceptable performance under deep fade conditions. Thus, these systems are effectively designed for the worst-case channel conditions, resulting in an insufficient utilization of the available channel bandwidth. Conversely, if the channel fade level is known at the transmitter, Shannon capacity is achieved by matching transmission parameters to time-varying channel: the signal transmitted to and by a particular station can be modified to take into account the signal quality variation.

Usually, wireless systems adopt power control as the preferred method for link adaptation. In a system with power control, the power of the transmitted signal is tuned in order to maintain the quality of the received signal at each individual subcarrier. Therefore, the transmit power will typically be low when a user is close to the base station (BS) and it will increase with the distance from the BS.

Power control is based on the water filling theorem: given a certain power budget, more transmit power is applied to frequencies experiencing lower attenuation. Thus, given the transfer function, the optimal power distribution is similar to inverting the transfer function and pouring a liquid (i.e. power) into the shape.

Although the use of just power control can improve the system performance in terms of the Bit Error Rate (BER), the total channel capacity is not used efficiently at any transmission time, if the modulation scheme is fixed. To overcome this drawback, Adaptive Modulation and Coding (AMC) or subcarrier allocation should be considered. In a system with AMC, the power of the transmitted signal is held constant but the modulation and coding orders are changed to match the current received signal quality. Users close to the BS are typically assigned higher-order modulations and higher code rates but the modulation order and/or the code rate usually decrease when their distance from the BS increases (Keller and Hanzo, 2000a).

Furthermore, in a multiuser OFDMA wireless network where the given system resources are shared by several terminals an adaptive subcarrier allocation strategy can significantly increases the system capacity by exploiting the *multiuser diversity*: the channel characteristics for different users are almost mutually independent; more attenuated subcarriers for a user may not result in a deep fade for other users. Subcarrier allocation strategies dynamically assign subcarriers with the best frequency response to the users.

Subcarrier allocation strategies can follow different criteria, such as having a fair data rate distribution among users or to maximize the overall network throughput. A possible subcarrier allocation strategy has been proposed by Rhee and Cioffi (2000) with the aim of obtaining almost equal data rates for all users. With this strategy more resources are allocated to users with bad channel conditions or far away from the BS. As a consequence, the capacity of users with good channel conditions are not fully exploited. Absolute fairness may lead to low bandwidth efficiency. However, throughput maximization is sometimes unfair for those users with bad channel conditions.

Adaptive subcarrier allocation techniques have been addressed, also jointly with other resource-allocation strategies, by Kim *et al.* (2005), Kulkarni *et al.* (2005), Wong *et al.* (1999) and Ermolova and Makarevitch (2007). The adaptive subcarriers and bits assignment scheme presented by Kulkarni *et al.* (2005) has the aim of minimizing the total transmitted power over the entire network while satisfying the data rate requirement of each link. Ermolova and Makarevitch (2007) considered a low-complexity suboptimal power and subcarrier allocation for OFDMA systems, proposing a heuristic noniterative method as an extension of the ordered subcarrier selection algorithm for a single user case to OFDMA systems.

The traffic types and services made by the devices composing a WMAN are strongly specialized and they have to be scheduled respecting transmission time and used bandwidth constraints. In order to provide the compliance of service parameters and QoS, a traffic management model is needed: IEEE 802.16 divides all services in four different groups, distinguished by traffic parameters, bandwidth request and resource-allocation techniques. However, WiMAX does not specify any *uplink* or *downlink* scheduling algorithm.

Recently the scheduling issue of multimedia traffic in wireless network has become a hot topic for the research community. Song and Li (2005) proposed a utility-based function for resource allocation and scheduling for downlink traffic in an OFDM-based communication system by exploiting wireless channel status jointly with packet queue information. Cai *et al.* (2005) proposed a downlink resource management technique for OFDM wireless communication systems, considering different traffic types and by exploiting subcarriers and power allocation. Liu *et al.* (2001) discussed the principles of opportunistic scheduling in resource-sharing wireless communication by focusing on the time varying conditions of the physical channel. Wong *et al.* (1999) devoted particular attention to the resource allocation in OFDMA systems. In particular, a suitable adaptive subcarrier, bit and power allocation algorithm is proposed for the case of a frequency selective wireless channel. Likewise, Ergen *et al.* (2003) proposed a fair scheduling technique that exploits subcarriers and bit allocation for an OFDMA wireless system.

Cicconetti *et al.* (2007), Lee *et al.* (2005) and Niyato and Hossain (2006) investigated the resource management for the case of IEEE 802.16 systems. Cicconetti *et al.* (2007) presented a performance evaluation for the different scheduling services offered in the IEEE 802.16 standard, by focusing on a Frequency Division Duplexing (FDD) system. Niyato and Hossain (2006) performed a queue analysis for IEEE 802.16 networks by considering real-time services and their impact on the highest priority traffic. Finally, Lee *et al.* (2005) investigated the VoIP service in IEEE 802.16 networks by focusing on the uplink and a mobile environment.

7.3.1 System Model and Problem Formulation

The system under consideration, as specified by the IEEE 802.16e standard, exploits the OFDMA among users for allocating the resources.

On the other hand, each user terminal is supposed to have different requirements of bandwidth and bit rate due to the type of traffic to be sent out and the QoS constraints especially in terms of priority (e.g. video-monitoring sensor networks or alarm delivery). Finally, we assume that the BS and user terminal transmits on each subcarrier with the same power, that is fixed and independent from the total available power and number of allocated subcarriers (Rhee and Cioffi, 2000).

For optimizing the access scheme two main aspects have to be considered: the adaptation of the modulation and coding and the optimization of the scheduling. Both aspects depends on the channel behavior as well as on the QoS requirements of each device in the coverage area.

7.3.1.1 AMC techniques

AMC denotes the possibility of choosing the most suitable modulation and channel coding scheme according to the propagation conditions of the radio link (channel state) known at the transmitting end.

For our purposes we have considered that the channel quality degradation is mainly due to the path loss and multipath fading.

There are mainly two types of AMC technique: maximum throughput AMC, in which the Modulation and Coding Scheme (MCS) is selected to achieve the best overall throughput without any constraint on the data reliability (i.e. bit error probability), and minimum bit error probability AMC, in which, conversely, the main goal is to meet specific data reliability constraints and, hence, the MCS is selected accordingly.

7.3.1.2 Scheduling Strategy

The aim of a scheduling strategy is to perform an optimal allocation of the network resources among the users in order to maximize the overall network throughput and meet the user QoS constraints in terms of minimum bit rate needed, R_{min}^k, and $\bar{P}_{ber}$. From above, it follows that the objective is to search for the subcarriers allocation matrix $\mathbf{M}$ for which we have:

$$\max_{\mathbf{M}} \sum_{k=0}^{K-1} \sum_{n \in \mathbf{M}_k} r_k(n) \quad \text{such as} \quad \begin{cases} \sum_{n=0}^{N-1} r_k(n) \geq R_{min}^k, & \text{for all } k, \\ \sum_{k=0}^{K-1} \delta\left[r_k(n)\right] \leq 1, & \text{for all } n, \\ P_{ber} \leq \bar{P}_{ber}, \end{cases} \tag{7.1}$$

where $r_k(n)$ is the bit rate achieved by user k on subcarrier n, $\delta[\cdot]$ is the Kronecker delta function, $\mathbf{M}_k$ is the allocation matrix for user k, K is the total number of users and N the total number of subcarriers. Note that $\mathbf{M}$ can be considered as a time-frequency grid with the x-axis and y-axis formed, respectively, by the number of OFDM symbols contained in a frame and all of the subcarriers.

7.4 Secure Communications Approaches

Security services are essential for any modern network, whether they are general purpose networks, such as a network for access delivery or specific purpose-driven networks. In general, security services such as *authentication*, *confidentiality*, *access control*, *integrity* and *availability*, as defined by Stallings (2006) have to be guaranteed in most real-life scenarios. In the context of monitoring of wide areas, the security of the overall system depends on the security features of each component and from their interaction. In this section we briefly review the security features of the building blocks of the monitoring network.

The most delicate component of the system is surely the WSN, for the hardware limitations and for the specific requirements it presents. In general, security in WSN is an underestimated feature, meaning that the stringent hardware requirements force the designers to give more attention to other details. Still, the application that a WSN can be targeted to are often critical and some security services should be guaranteed. We give three practical example scenarios.

- In a surveillance network an intruder must be unable to alter the flow of information toward the gateway, or pollute the gateway with fake messages. We can imagine an attacker that wants to access a restricted area using a laptop to flood the nodes, or to hijack routing protocols in order to prevent the gateway from receiving alarms.

- In a WSN for home automation it is imperative to guarantee some form of access control in order to avoid an attacker from taking control of the environment.

- In a body network the parameters that are monitored are private and should not be disclosed. If the parameters are processed automatically to control health equipment authentication and access control is, again, fundamental.

The hardware used for WSN is unable, in most cases, to perform the computation necessary to use public/private key schemes which is a great limitation since most of the modern security protocols are based on such algorithms. This makes the WSN a stand-alone network that must be interfaced with the rest of the system, but cannot be easily integrated.

Another important issue is the degree of distribution that a WSN implements. In general, security schemes are easier to implement when there is a strict hierarchy between the components of a network. If the WSN presents a star topology, with a single gateway always reachable by all of the nodes, the security associations between the nodes and the gateway can be easily pre-loaded. Otherwise if the WSN is a multi-hop mesh network with an unpredictable (and even varying) topology the security association between a couple of nodes that share the same link must be automatically negotiated using some pre-shared credentials, which introduces scalability problems.

Lastly, WSN are generally unattended, nodes can be stolen, their keys can be compromised and even the software can be reprogrammed. Large-area WSNs should resist the presence of a certain number of compromised nodes.

A completely different situation can be found in the transport WiMAX network. First of all, most of the time WiMAX networks use a centralized model which helps the creation of a security hierarchy, then WiMAX standards mandate that the equipment must be capable of performing computations needed for public/private key cryptography; lastly, robust security protocols can be used. Nevertheless some severe vulnerabilities have been found in IEEE 802.16d, in part fixed in the later IEEE 802.16e.

The security scheme used in WiMAX is distinct for the so-called Point-to-Multipoint (PMP) mode, in which a BS serves various clients, and the *mesh* mode in which a certain number of peers form a distributed flat network. In the first case the authentication of a IEEE 802.16d network is based on an original protocol designed for WiMAX which makes use of RSA certificates. The standard mandates that each network client should be in possession of a factory installed certificate that bounds the MAC address of the device to an RSA key, and this key is used to perform authentication to the BS. With a packet exchange defined in the standard, two fresh symmetric keys are generated during authentication, the so-called *AK* and *TEK*, where the first is used as a proof to perform periodical re-authentications the second is used for encrypting and authenticating data. The introduction of mandatory RSA certificates and hardware capable of performing public key cryptography is a new feature that distinguishes WiMAX from previous standards. With such feature it should be possible to prevent an attacker from stealing the MAC address of another client in order to access the network. In a WiFi network, for example, MAC-based access lists are widely used but they are much more insecure then certificate-based authorization. Nevertheless, the overall security scheme of IEEE 802.16d has been proven to be insecure (Maccari *et al.*, 2007). Briefly, some of the defects that it presents are as follows.

- Authentication is always mono-directional. As happened in IEEE 802.11 networks with the Wired Equivalent Privacy (WEP) security standard the BS never authenticates itself with the subscriber stations. This gives to an attacker the possibility of creating a *rogue* network in which the clients might authenticate.

- There is no message authentication code in the frames, even after that the authentication has been accomplished and keys have been derived. This exposes the standard to reply attacks.

- The use of the Data Encryption Standard (DES) encryption algorithm (Cipher Block Chaining (CBC) mode) is unsafe if not associated with a message authentication code. The attacker might be able to interfere with the decryption of frames into the client nodes.

- Some sensitive information is sent from and to the BS without authentication, so that an attacker could inject false data. For instance, the frames that the client stations use to request the activation of new QoS profiles are not authenticated, so that an attacker can send spoof requests pretending to be any other client.

- There is no means of performing certificate management on the BS. Certificate and access lists are hard-coded into the BS and no reference has been made to the use of Authentication, Authorization and Accounting (AAA) protocols such as RADIUS (Remote Authentication Dial-In User Service). This makes the security features much harder to use.

- Authentication is based only on the device certificate, there is no user authentication.

The experiences conducted with WiFi networks highlight that when the price of the devices lowers, the network security is much more stressed because any commercial device can be used by an attacker. At present, WiMAX devices are still high-end devices, but when they reach a higher diffusion legacy, 802.16d devices will be much more difficult to defend.

In the 802.16e revision the 802.16 working group have addressed some of these problems. First of all the legacy authentication scheme, PKMv1, has been substituted by a modular and more modern PKMv2, based on EAP (Extensible Authentication Protocol) and RADIUS. Such a change is of great importance because it introduces a modular approach to authentication, that is not performed on the BS but is relayed to a separate authentication server. This allows a fully centralized user management that starts with a bi-directional authentication and continues with authorization (assignment of user profiles and capabilities based on the pair *user–device*) and accounting (profiling of user activities). EAP allows any kind of authentication to be performed, based on certificates, passwords or other credentials. Encryption has been upgraded to more robust algorithms and MAC has been added to authenticated frames. Still, some management frames have been left unauthenticated, which exposes IEEE 802.16e networks to the problems described before.

To understand the importance of this upgrade, note that the WiMAX Forum, which is able to give WiMAX certifications in the stage two version 1.0.0 specification, do not allow the certification of PKMv1 devices.

The introduction of a centralized authentication server recalls the model used for IEEE 802.11i, and detailed in IEEE 802.1X. It eases the management of a wide area network but also introduces great delays in the authentication operations, since every time a node has to be re-authenticated it needs to complete the authentication not with its own BS (with which it is connected with a wireless link) but with a server that can be several hops away, introducing a delay that can be even several seconds long.

A more complex situation is present for mesh networks. In a mesh WiMAX network as defined in the standard the authentication of nodes is performed exactly as in the PMP mode, but the derived AK key is the same for every node in the network. It is not clear how this key should be refreshed, which introduces more security problems.

In a mixed WSN–WiMAX scenario, the security of the overall system depends on the security of each component of the network, but also from the interfaces chosen to make them interact. In particular, the information should be secured along the entire chain of transmission, and the sources of information should be certified from bottom to top. This implies the creation of opportune Interworking Functions (IWFs) that match the security features of each single component respecting user attributes.

As an example, imagine a multi-hop WSN where each node has a security association with all of its closest one-hop neighbors. Data is collected from each single node, transmitted over a wireless link to a neighbor and conveyed over a multi-hop path to the gateway. Each link is secured by a key negotiated with an algorithm as in Fantacci *et al.* (2008) or Chan *et al.* (2003) and intermediate nodes can perform data fusion before the frames reach the gateway. The gateway has a WiMAX link that can end directly in the BS, or it can be part of a mesh network of BSs connected to Internet. Information can be gathered also by mobile data sinks, that walk across the WSN and collect the measures directly from the nodes of the WSN.

Given this generic scenario, let us analyze a possible organization of the network security scheme.

- Information is sensed by a node, and transported over a wireless link that is secured by a shared key (data is ciphered and authenticated). If the link is direct to the gateway, the gateway will map the data to a single node, if there is a multi-hop path to the gateway and each intermediate node can perform data fusion, there is no strict association between a single node and the information that reach the gateway.

- From the gateway, the link to the BS is secured by a WiMAX wireless connection that is authenticated with a RSA certificate or any other EAP method. The gateway will possibly aggregate and elaborate the sensed data, so that the data will be seen by the backbone as coming from the gateway itself.

- On the WiMAX backbone data will be moved using the mesh configuration, so authentication will be based on a single AK key.

- The WiMAX network will end with an IP gateway, connected to the Internet: from this gateway to a control center a Virtual Private Network (VPN) can be used to secure traffic.

Basically in this scheme each level of the network is masquerading as a lower layer, as represented in Figure 7.2. This configuration is easy to deploy because masquerading avoids the problem of defining low-level IWFs between the different protocol stacks.

Now let us imagine that the control center receives an alert that is later shown to be false, so that there is suspicion that any of the link of the chain has been compromised. The

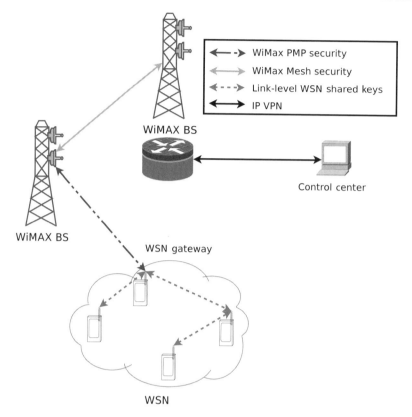

Figure 7.2 Security architecture based on link-layer protocols.

compromised link could be the single sensor node that generated the false information, but it could be also any node on the path to the gateway that performed data fusion. Alternatively it could be coming from a compromised gateway, or it could even have been injected over a compromised WiMAX link.

Another difficult issue to resolve is key revocation, when one of the mobile sinks is stolen. Mobile sinks draw data directly from the sensor nodes, so they need a key shared with each of them. It is not an easy task to reprogram every node in order to invalidate that key. We see that a layered approach has disadvantages under a security and management point of view. Now let us imagine a completely opposite scenario.

- Each sensor node is in possession of a shared key with a unique authentication server, for the whole network. Each time a node wants to create a link with one of its neighbors, it will communicate with the authentication server and ask for a fresh shared key. Communications with the authentication server must pass through the sensor technology, WiMAX links and Internet with proper encapsulation. In this way the WSN mesh is formed.

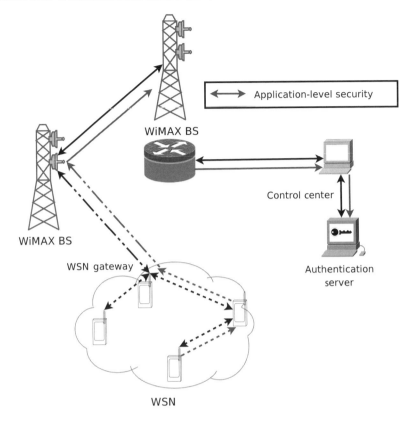

Figure 7.3 Security architecture based on end-to-end protocols.

- When the node wants to send information it will sent a frame ciphered and authenticated with its key, that will be encapsulated in the WiMAX link.

- The control center receives data that are authenticated directly by the single nodes.

We see that in such a configuration as depicted in Figure 7.3, there is an end-to-end authentication, so that a misbehaving sensor node can be recognized in the control center. Moreover, since links are dynamically created with the help of an authentication server, a stolen node can be easily excluded by the network revoking its keys and credentials from the server. On the other side such a configuration introduces new difficulties in managing the whole network. Since data is authenticated and ciphered, no fusion can be made along the path, and information can be lost on the way (for instance, the gateway in the field might know information useful for data fusion, such as the position of the sensors, which will not be available in the control center). Then, for each link that has to be created, a multi-hop, multi-technology handshake must be fulfilled. Lastly, specific encapsulation must be defined for every network bridge.

Acknowledgements

This work is partially supported by MIUR-FIRB Integrated System for Emergency (InSyEme) project under the grant RBIP063BPH and by the Italian National Project Wireless multiplatfOrm mimo active access netwoRks for QoS-demanding muLtimedia Delivery (WORLD), under grant number 2007R989S.

References

AgroSense (2007–2010) Wireless sensor networks and remote sensing – foundation of a modern agricultural infrastructure in the region, http://www.agrosense.org/.

Akyildiz, I., Su, W., Sankarasubramanian, Y. and Cayirci, E. (2001) Wireless sensor networks: a survey. *Computer Networks*, **38**, 393–422.

Bohge, M., Gross, J., Wolisz, A. and Mayer, M. (2007) Dynamic resource allocation in OFDM systems: An overview of cross-layer optimization principles and techniques. *IEEE Network*, **21**(1), 53–59.

Cai, J., Shen, X. and Mark, J.W. (2005) Downlink resource management for packet transmission in OFDM wireless communication systems. *IEEE Transactions on Wireless Communications*, **4**(4), 1688–1703.

Chan, H., Perrig, A. and Song, D. (2003) Random key predistribution schemes for sensor networks. *Proceedings of 2003 Symposium on Security and Privacy*, pp. 197–213.

Chiti, F. and Fantacci, R. (2006) Wireless sensor network paradigm: overview on communication protocols design and application to practical scenarios. *EURASIP Newsletter* **17**(4), 6–27.

Cicconetti, C., Erta, A., Lenzini, L. and Mingozzi, E. (2007) Performance evaluation of the IEEE 802.16 MAC for QoS support. *IEEE Transactions on Mobile Computing*, **6**(1), 26–38.

DustBot (2007–2009) Networked and cooperating robots for urban hygiene, http://www.dustbot.org/.

Ergen, M., Coleri, S. and Varaiya, P. (2003) QoS aware adaptive resource allocation techniques for fair scheduling in OFDMA based broadband wireless access systems. *IEEE Transactions on Broadcasting*, **49**(4), 362–370.

Ermolova, N.Y. and Makarevitch, B. (2007) Low complexity adaptive power and subcarrier allocation for OFDMA. *IEEE Transactions on Wireless Communications*, **6**(2), 433–437.

Fantacci, R., Chiti, F. and Maccari, L. (2008) Fast distributed bi-directional authentication for wireless sensor networks. *Journal on Security and Communication Networks*, **1**(1), 17–24.

GoodFood (2004–2006), http://www.goodfood-project.org/.

Ilyas, M. and Mahgoub, I. (2005) *Handbook of Sensor Networks: Compact Wireless and Wired Sensing Systems*. CRC Press, Boca Raton, FL.

InSyEme FIMFR (2007–2010) Integrated System for Emergency, http://www.unifi.it/insyeme/.

Keller, T. and Hanzo, L. (2000a) Adaptive modulation techniques for duplex OFDM transmission. *IEEE Transactions on Vehicular Technology*, **49**(5), 1893–1906.

Keller, T. and Hanzo, L. (2000b) Adaptive multicarrier modulation: a convenient framework for time-frequency processing in wireless communications. *Proceedings of the IEEE*, **88**(5), 611–640.

Kim, K., Han, Y. and Kim, S.L. (2005) Joint subcarrier and power allocation in uplink OFDMA systems. *IEEE Communications Letters*, **9**(6), 526–528.

Kulkarni, G., Adlakha, S. and Srivastava, M. (2005) Subcarrier allocation and bit loading algorithms for OFDMA-based wireless networks. *IEEE Transactions on Mobile Computing*, **4**(6), 652–662.

Lee, H., Kwon, T. and Cho, D.H. (2005) An enhanced uplink scheduling algorithm based on voice activity for VoIP services in IEEE 802.16d/e system. *IEEE Communications Letters*, **9**(8), 691–694.

Liu, X., Chong, E.K.P. and Shroff, N.B. (2001) Opportunistic transmission scheduling with resource-sharing constraints in wireless networks. *IEEE Journal on Selected Areas in Communications*, **19**(10), 2053–2064.

Maccari, L., Paoli, M. and Fantacci, R. (2007) Security analysis of IEEE 802.16. *Proceedings of the 2007 IEEE International Conference on Communications.*

Niyato, D. and Hossain, E. (2006) Queue-aware uplink bandwidth allocation and rate control for polling service in IEEE 802.16 broadband wireless networks. *IEEE Transactions on Mobile Computing*, **5**(6), 668–679.

Rhee, W. and Cioffi, J.M. (2000) Increase in capacity of multiuser OFDM system using dynamic subchannel allocation. *Proceedings of IEEE VTC 2000-Spring*, vol. 2, Tokyo, Japan, pp. 1085–1089.

Song, G. and Li, Y.G. (2005) Utility-based resource allocation and scheduling in OFDM-based wireless broadband networks. *IEEE Communications Magazine*, **43**(12), 127–134.

Stallings, W. (2006) *Cryptography and Network Security*. Prentice Hall, Englewood Cliffs, NJ.

IEEE (2005) Amendment to IEEE Standard for Local and Metropolitan Area Networks – Part 16: Air Interface for Fixed Broadband Wireless Access Systems – Physical and Medium Access Control Layers for Combined Fixed and Mobile Operation in Licensed Bands. *IEEE Standard 802.16e-2005.*

IEEE (2004) IEEE Standard for Local and Metropolitan Area Networks – Part 16: Air Interface for Fixed Broadband Wireless Access Systems. *IEEE Standard 802.16-2004.*

van Nee, R. and Prasad, R. (2000) *OFDM for Wireless Multimedia Communications*. Artech House.

Wong, C.Y., Cheng, R.S., Letaief, K.B. and Murch, R.D. (1999) Multiuser OFDM with adaptive subcarrier, bit, and power allocation. *IEEE Journal on Selected Areas in Communications*, **17**(10), 1747–1758.

WP122 (2008) Report on experimental capabilities of the integrated test bed. *Technical Report FP6-IST-4-027738-NoE, 'CRUISE'.*

8

WiMAX Mesh Architectures and Network Coding

Parag S. Mogre, Matthias Hollick, Christian Schwingenschloegl, Andreas Ziller and Ralf Steinmetz

8.1 Introduction

The seminal work of Ahlswede *et al.* (2000) introduced the notion of Network Coding (NC) as a means to achieve bandwidth savings for multicast data transmissions. The authors demonstrated that suboptimal results in terms of required bandwidth are achieved in general, if the information to be multicast is considered as a fluid which is to be routed or replicated on a set of outgoing links at each node relaying the information in the network, that is, the transportation network capacity is not equal to the information network capacity. The concept of network coding was later also extended to unicast information transmission. Recently, the application of network coding in wireless networks (Wireless Network Coding (WNC)) is being investigated intensively. In particular, the deployment of WNC in Wireless Mesh Networks (WMNs) to achieve bandwidth savings and throughput gain is very promising. We demonstrate this functionality with the help of Figure 8.1. In a WMN, nodes typically send data to destinations via multi-hop routes. Here, a number of nodes relay the data packets between the source and destination. Readers can find more background information about WMNs and a survey of respective research challenges in Akyildiz *et al.* (2005).

Consider a simple linear WMN topology as shown in Figure 8.1. Assume that node N_1 and node N_2 transmit data to each other, which is relayed by the node R_1. Figure 8.1(a) shows the behavior in WMNs without application of WNC. Here, N_1 transmits data to the next hop R_1 in slot (or transmission number) 1. The data received is relayed by R_1 in slot 2 to node N_2. Similarly, data transmitted by N_2 addressed to node N_1 is transmitted and relayed in slots 3 and 4, respectively. Figure 8.1(b) shows how the same data can be transferred to the

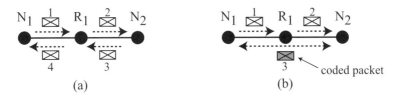

(a)   (b)

Figure 8.1 Sample topology showing a simple WNC constellation: (a) transmission schedule using traditional packet forwarding; (b) transmission schedule using network coding.

destinations using a simple form of network coding. The node R_1 receives the packets to be relayed in slots 1 and 2. The node R_1 can then code the received packets together using the XOR function and transmit this XOR-coded packet in slot number 3. If the nodes N_1 and N_2 preserve local copies of the packets they transmitted, they can XOR the received coded packet with the packet they transmitted before to recover the data addressed to them. Comparing Figures 8.1(a) and (b) we see that in this simple setup one transmission opportunity or slot for data transmissions can be saved using network coding. This illustrating example clearly shows that WNC is a promising way to improve the throughput of wireless networks.

The work of Katti *et al.* (2006) was one of the first to depart from mainly theoretical investigations and to deploy WNC in real networks using standard off-the-shelf protocol stacks. This has been followed by other work also looking at practical deployments of WNC. However, the majority of research investigating WNC (both practical deployments as well as theoretical investigations) assumes the use of generic IEEE 802.11 or similar Medium Access Control (MAC) layers. Recent standardization developments show a trend towards highly sophisticated mechanisms at the MAC layer for supporting stringent Quality of Service (QoS) requirements of multimedia and real-time traffic expected in future WMNs; the IEEE 802.16 standard (see IEEE (2004)) and the upcoming IEEE 802.11s standard being examples.

In this chapter we choose to study the IEEE 802.16 standard as a prototype for MAC layers providing radically different medium access mechanisms when compared with the generic IEEE 802.11 MAC. The fundamental difference between the contemporary IEEE 802.11 and the IEEE 802.16 standard arises due to the reservation-based medium access supported by the latter. In this work we investigate network coding within the context of WMNs built using the IEEE 802.16 MeSH mode (also referred to as MeSH throughout this document). We analyze the issues involved in deploying COPE-like (see Katti *et al.* (2006)) basic network coding solutions in WMNs using the IEEE 802.16 MeSH mode. The fundamentally different medium access mechanisms of the IEEE 802.16 and 802.11 standards make the direct adoption of network coding solutions designed and developed within the scope of 802.11 inefficient, if not impossible. In this work, we first analytically model the bandwidth reservation mechanism in the IEEE 802.16 MeSH mode, thus motivating the need for investigating deployment issues for network coding from a novel perspective. We break away from the myopic IEEE 802.11-only view of many WMNs. We instead present extensions to the current IEEE 802.16 MeSH mode specifications to enable efficient support for practically deploying network coding in IEEE 802.16-based WMNs. Finally, we present simple yet meaningful metrics for quantifying the gain obtained by deploying network coding in the latter WMNs.

This book chapter is structured as follows. In Section 8.2 we introduce the reservation schemes supported by the IEEE 802.16 MeSH mode and provide some background information on the MeSH mode. In Section 8.3 we present an analytical model for the MeSH mode's bandwidth reservation scheme and derive design principles for WNC deployment. In Section 8.4 we present extensions to the MeSH mode specifications which enable efficient deployment of WNC. Section 8.5 discusses relevant related work, and Section 8.6 draws conclusions for the work presented in this chapter and also gives pointers for further research in this context.

8.2   Background on the IEEE 802.16 MeSH Mode

The IEEE 802.16 MeSH mode (see IEEE (2004)) specifies the MAC and the Physical (PHY) layers to enable the deployment of WMNs. In particular, it specifies the framework for medium access and bandwidth reservation. The algorithms for bandwidth reservation are, however, not defined and left open for optimization by individual vendors. The MeSH mode uses Time Division Multiple Access/Time Division Duplex (TDMA/TDD) to arbitrate access to the wireless medium, where the time axis is divided into frames. Each frame is composed of both a control subframe and a data subframe. The data subframe is further divided into minislots (or slots) carrying a data payload, while MAC layer messages meant for network setup and bandwidth reservation are transmitted in the control subframe. Contention-free access to the wireless medium in the control subframe can be both centrally regulated by a Mesh Base Station (MBS), which may also provide access to external networks such as the Internet or provider networks, or managed by the individual Subscriber Stations (SSs) in a distributed fashion. In the latter case, the SSs manage the access to the medium directly among each other using the distributed mesh election algorithm specified by the standard (see IEEE (2004), Mogre *et al.* (2006) and Cao *et al.* (2005)).

Reservation of bandwidth for transmission of data messages in the data subframe can be both centrally managed by the MBS, that is, centralized scheduling, or a contention-free transmission schedule can be negotiated by the nodes individually without involving the MBS, that is, distributed scheduling. Centralized scheduling is limited to scheduling transmissions on a scheduling tree specified and rooted at the MBS. Distributed scheduling is more flexible and can be used to schedule transmissions on all of the links, including those in the scheduling tree in the WMN. Using distributed scheduling, a SS negotiates its transmission schedule via a three-way handshake with the neighboring node to receive the transmission (see Figure 8.2(a)). Given the limitations of centralized scheduling, without loss of generality, we assume that only distributed scheduling is used for the rest of this chapter.

Nodes in the mesh network use a three-way handshake to request and reserve a range of minislots for a contiguous range of frames (e.g. reservation $Resv(e, 2\text{--}3, 102\text{--}105)$ is used to denote that minislots numbered 2 to 3 are reserved for transmission on link with identifier e for the frames numbered 102 to 105). The number of minislots reserved is termed the demand level, denoted as $\Delta(MS)$, and the number of frames for which the reservation is valid as demand persistence, denoted as $Per_{\Delta F}$, where ΔF is the number of frames for which the reservation is valid. Where as per the standard's specification $\Delta F \in \{1, 2, 4, 8, 32, 128, \infty\}$. We may thus have reservations with demand levels $1 \ldots$ maximum number of minislots; and with demand $Per_1, Per_2, Per_4, \ldots, Per_\infty$. Only slots reserved with persistence Per_∞

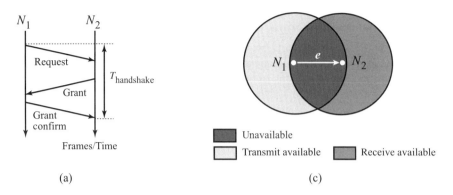

(a)   (c)

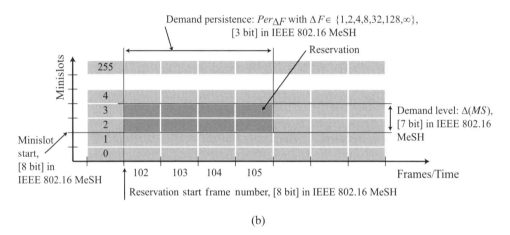

(b)

Figure 8.2   Basic elements of distributed scheduling: (a) three-way protocol handshake; (b) scope of validity of a minislot reservation using distributed scheduling; (c) minislot status for a transmission from N_1 to N_2.

can be freed when no longer required via a cancel three-way handshake. The latter special case of reservation of slots with persistence Per_∞ is what we call a persistent reservation. Figure 8.2(b) illustrates minislots reserved using distributed scheduling. To compute conflict-free schedules, each node needs to maintain the states of all minislots in each frame.

Depending on the activities which may additionally be scheduled in a slot, the slot has one of the following states: *available* (*av*: transmission or reception of data may be scheduled), *transmit available* (*tav*: only transmission of data may be scheduled), *receive available* (*rav*: only reception of data may be scheduled), *unavailable* (*uav*: neither transmission or reception of data may be scheduled). Consider edge $e = (N_1, N_2) \in E$ in Figure 8.2(c), with E representing the set of edges in the WMN. Figure 8.2(c) shows how nodes in the network will update their slot states when a transmission is scheduled on edge e, provided that all of the nodes were in state *av* at the beginning of the handshake. Neighbors of the receiver (N_2) overhear the grant and update the state for the granted slots to reflect that they may not transmit in the granted slots. Neighbors of the transmitter (N_1) overhear the grant confirm message and update their local slot states to reflect that they cannot receive any other transmission without interference in the confirmed slots. This handshake process is similar to the Request to Send (RTS)/Clear to Send (CTS) mechanism used by 802.11-based nodes. A transmission may be scheduled on an edge $e = (N_1, N_2)$ in a given slot m and frame f if and only if $s_m^f(N_1) \in \{av, tav\}$ and $s_m^f(N_2) \in \{av, rav\}$, where $s_m^f(N)$ denotes the state of slot m in frame f at node N. Additional details about the McSII mode and the data structures and control messages used can be found in IEEE (2004). To make the material more accessible to readers unfamiliar with the MeSH mode, we provide a detailed overview of the MeSH mode specification in Mogre *et al.* (2006).

8.3 Design Principles for Network Coding in the IEEE 802.16 MeSH Mode

Having outlined the function of the MeSH mode's distributed scheduling in the previous section, we now discuss the pitfalls in implementing COPE-like (see Katti *et al.* (2006)) practical network coding solutions in the MeSH mode. A core principle of COPE's packet coding algorithm is to not delay the transmission of packets just for the sake of enabling coding of packets. This is especially important for the case of delay-sensitive applications and multimedia traffic, which is expected to be the core beneficiary of the sophisticated QoS features offered by the MeSH mode. COPE can code and transmit packets as soon as a set of matching codable packets are available at the transmitting node. This is not the case for the MeSH mode due to its reservation-based nature. To understand the former issue, we look at the reservation of slots in the MeSH mode using distributed scheduling in detail. Transmissions in the MeSH mode are scheduled in a contention-free manner using explicit reservation of slots for individual links before transmission of data on those links.

We next formulate our model. Consider the parameters outlined in Table 8.1. Assume that the parameters hold for a given frame. Let K be the number of neighbors (identified individually by their index k) which should receive the coded packet. This subset of neighbors is selected by looking at the next hops of the packets available for coding similar to COPE. As we cannot transmit data to neighbors without reserving bandwidth for the transmission, we first need to reserve sufficient bandwidth for the multicast transmission[1]. Let us assume that we use enhanced handshake procedures to allow us to reserve multicast bandwidth, and let us consider that we need to reserve d slots for the transmission in a given frame having the parameters as shown in Table 8.1. Now let S_T and S_k denote the set of slots suitable for scheduling at the transmitter and at receiver k, respectively. For the transmitter to be able to successfully negotiate and reserve the same d slots for the multicast transmission to the K receivers, we require that $|(S_T \cap S_k)| \geq d$ for all k. For the given model parameters, using counting theory, we derive the probability that a common set of d slots for the transmission is available as given by

$$P_{\text{succ}}^K = \frac{C_T \prod_k \sum_{j=d}^{\min(T, R_k)} \binom{T}{j}\binom{N-T}{R_k-j}}{C_T (\prod_k C_{R_k})}. \tag{8.1}$$

---

[1] Here, we face a severe pitfall: the IEEE 802.16 MeSH mode does not natively support mechanisms to reserve multicast bandwidth. See Section 8.4 for our solution to introduce multicast reservations in the MeSH mode of IEEE 802.16.

Table 8.1 Parameters for modeling the bandwidth reservation mechanism of the IEEE 802.16 MeSH mode.

Parameter	Interpretation of parameter
N	Number of slots for distributed scheduling in a frame
d	Number of slots to be reserved (demand)
T	Number of slots suitable for scheduling transmission at the sender (status *av* or *tav*)
K	Number of receivers to which the transmission is to be scheduled
k	$1, \ldots, K$, index for intended receivers
R_k	Number of slots suitable for reception at receiver k (status *av* or *rav*)
C_T, C_{R_k}	Combinations $\binom{N}{T}$, $\binom{N}{R_k}$, respectively

Figure 8.3 shows plots of the success probability (P^K_{succ}) given by Equation (8.1) for a handshake with $K = 1$ neighbors for a demand (slots to be reserved) of 1, 5, 10 and 20 slots, respectively. The total number of slots per frame is 100. The x-axis shows the number of slots suitable for transmission on the transmitter side. The y-axis shows the number of slots suitable for reception at the receiver(s). The plot shows the case where all of the receiving neighbors are assumed to have the same number of slots available for reception. Comparing Figures 8.3(a) and (b), (c) and (d), we note that a higher number of slots need to be available for transmission at the sender, and a higher number of slots need to be available at the receiver(s) with an increase in the number of slots to be reserved, to successfully reserve the required slots in a given frame with a high probability. In short, the probability of successfully reserving d common slots for transmission to a fixed number of receivers in a given frame decreases with increasing d, given that the number of receivers, the number of available slots at the transmitter and the receiver(s) remain unchanged.

Figure 8.4 shows contour plots for P^K_{succ} showing the minimum number of slots suitable at the transmitter and the receivers beyond which P^K_{succ} exceeds the values shown in the graph for reserving d minislots in a given frame. Comparing Figures 8.4(a) and (b) or Figures 8.4(c) and (d) we see that for the same demand d, the number of suitable slots needed at the transmitter and receiver(s) for successfully reserving (with a certain probability of success) the required number of slots increases with the number of receivers (K) involved in the handshake. Analysis of the figures and Equation (8.1) reveals that P^K_{succ} decreases drastically as soon as the transmitter or one of the receivers has a low number of slots suitable for the intended communication in the given frame. Further, with increasing d, a large number of slots needs to be free for the intended communication, to be able to successfully negotiate and reserve slots with a high probability. For the same demand d and the given number of free slots at both the transmitter and receivers, P^K_{succ} decreases with an increase in the number of intended receivers K. In practice, not all receivers share the same number of available slots; a single receiver having a low number of slots suitable for reception results in P^K_{succ} to be very low. We can conclude that on-demand reservation of slots for network coding transmissions cannot be achieved with high success, which means that, unlike COPE, we need to set up the reservation for the multicast network coding transmission prior to the arrival of a set of packets which can be coded.

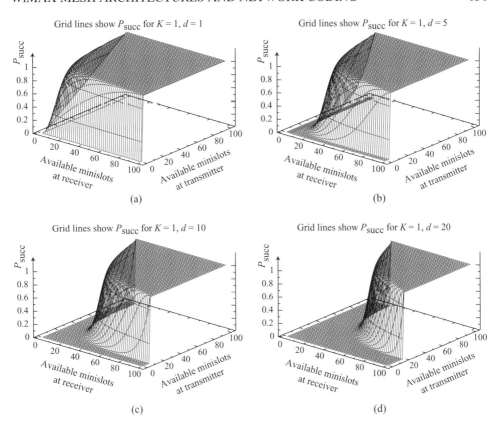

Figure 8.3 Plots showing the probability P_{succ}^K of successfully reserving d slots in a given frame for simultaneous reception at K neighboring nodes: (a) $K = 1, d = 1$; (b) $K = 1, d = 5$; (c) $K = 1, d = 10$; (d) $K = 1, d = 20$.

An important aspect of the IEEE 802.16 MeSH mode is the way the reservations are carried out. Nodes perform the three-way handshake shown in Figure 8.2(a) to reserve bandwidth for links to individual neighbors; distributed scheduling messages (MSH-DSCH) containing Request, Grant and Grant-confirmation are exchanged to reserve a set of slots for the required transmission. In the above analysis we have computed the value for P_{succ}^K considering the entire range of available slots at the transmitter and all of the receivers. However, due to message size restrictions, with the bandwidth request in a MSH-DSCH message, the transmitter can only advertise a subset of the slots suitable for transmission to the receivers. This effectively reduces the value of P_{succ}^K by reducing the number of slots available at the transmitter for negotiating the reservation. However, a more important problem with multicast reservation is that each node maintains its own independent state for all of the minislots. Thus, the individual receivers for the multicast transmission do not possess a common view about which slots are suitable at the other receiver(s) and may, hence, issue grants for different slot ranges. Such disjoint grants require multiple transmissions

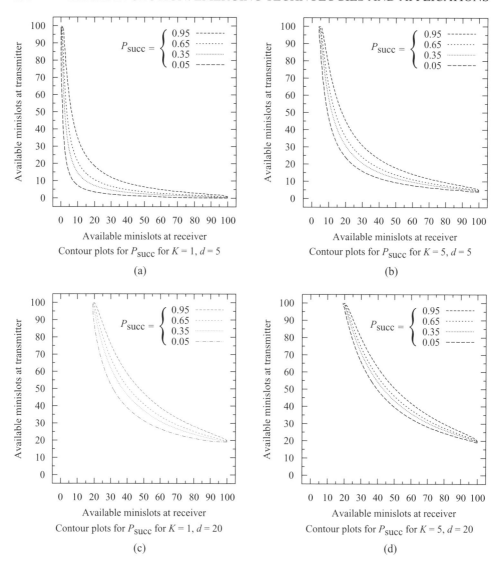

Figure 8.4 Contour plots showing the probability of successfully reserving d slots in a given frame for simultaneous reception at K neighboring nodes: (a) $K = 1$, $d = 5$; (b) $K = 5$, $d = 5$; (c) $K = 1$, $d = 20$; (d) $K = 5$, $d = 20$.

(one to each neighboring node in the worst case), thereby defeating the goal that coded transmissions should be simultaneously received by multiple neighbors.

Another critical aspect that needs to be considered by network coding solutions designed for 802.16's MeSH mode is the three-way handshake overhead. Each node may transmit the control messages (MSH-DSCH) for the three-way handshake only in transmission opportunities belonging to the control subframe, which have been won by the node using the mesh election algorithm specified by the standard. Cao *et al.* (2005) provide an analytical model for mesh election and analyze the three-way handshake delay. Let the mean three-way handshake duration between transmitter t and receiver k be H_k^t. The standard's scheduling constraints require that slots granted by the receiver may be used for transmission only after the three-way handshake is complete. Hence, it is only meaningful to grant slots in frames occurring after the completion of the three-way handshake. For a multicast handshake as required for network coding it implies that the nodes should start searching for the required d slots in frames after a duration $T_H^K = \max_k (H_k^t)$. Let P_i be the probability of successfully being able to reserve the required slots in frame i for the intended communication (i.e. given a set of transmitter, receivers and d required slots and N total slots in the frame). The mean number of frames that need to be considered starting from a given start frame to reach the first frame in which the demand can be satisfied is given by

$$F_{\text{mean}} = \sum n = 1^\infty n\, P_{(sf+n)} \prod_{j=1}^{n-1} (1 - P_{(sf+j)}). \qquad (8.2)$$

Here, sf is the number of the frame after completion of the multicast handshake. Hence, if the duration of a frame is F_D, the mean waiting time before the reserved frame for the multicast transmission starting from the start of the multicast bandwidth reservation handshake is given by

$$T_{\text{mean}} = T_H^K + F_{\text{mean}} F_D. \qquad (8.3)$$

From the above analytical model we can obtain the following design criteria for network coding solutions for the MeSH mode of IEEE 802.16.

- *Principle 1*. On-demand reservation of multicast slots for WNC, that is, reserving slots after a set of packets for coding is present, is not feasible without prohibitive overhead; hence, reservation of multicast slots should ideally be performed *a priori*.

- *Principle 2*. The higher the number of neighbors in the multicast reception set, the more difficult it is to obtain an agreement on a common set of slots for reception, especially in presence of background traffic and different number of available slots at the involved parties. The success probability of such a reservation in a given frame further diminishes with an increase in the demanded slots. Hence, the size of the receiver set should be kept as small as possible.

- *Principle 3*. The three-way handshake delay combined with the overhead of reserving the required multicast slots mean that the number of such three-way handshakes required should be kept to a minimum. If possible, the handshake should optimize the probability of obtaining a successful multicast reservation.

8.4 Enabling WNC for the IEEE 802.16 MeSH Mode

In this section we present our solution to enable practical deployment of network coding in the IEEE 802.16's MeSH protocol stack. The presented solution is based on the design principles derived in Section 8.3. For the purpose of the current discussion, without loss of generality, we restrict the size of the set of receivers for each multicast transmission to two

(*Principle 2*), that is, a node reserves bandwidth for simultaneous transmission of coded packets to at most two of its neighbors. Further, our solution only uses persistent (*Per$_\infty$*) reservations, that is, the reservation remains valid till the reserving node explicitly cancels the reservation (see Section 8.2), for the multicast network coding transmissions. This means that a node reserves a common set of slots for an infinite number of frames for transmission to two neighbors which shall receive the coded packets. This design choice of using only persistent reservations has *Principle 3* as a background and reduces the number of three-way handshakes for multicast bandwidth reservation. Still a valid design needs to address *Principle 1*. However, to enable *a priori* reservation of bandwidth for network coding, one needs to be able to associate a figure of merit with the use of network coding at a given node. Towards this end, we next present an analytical model, which enables the quantification of the bandwidth savings obtained in TDMA-based WMNs such as IEEE 802.16's MeSH mode.

8.4.1 Modeling the Coding Gain

Definition 8.1 (Degree of freedom of a slot) We define the degree of freedom or scheduling freedom of a slot as the number of types of activities, which may be scheduled in a particular slot given its current status. The degree of freedom of a slot is given by the function $\lambda(s)$ where s is the slot status. We define $\lambda(s)$ as follows:

$$\lambda(s) = \begin{cases} 0 & \text{for } s \in \{uav\}, \\ 1 & \text{for } s \in \{tav, rav\}, \\ 2 & \text{for } s \in \{av\}. \end{cases} \tag{8.4}$$

The values for $\lambda(s)$ reflect the scheduling possibilities a node has in a given slot. In slots with status av the node can either schedule a transmission or reception of data, that is, it has two possibilities, hence $\lambda(av) = 2$. It follows that $\lambda(rav) = \lambda(tav) = 1$ (only one degree of freedom left at the node) and $\lambda(uav) = 0$ (node possesses no degree of freedom). From the above we can define the degree of freedom of the entire WMN for a given range of frames as the summation of $\lambda(s)$ for all s at all the nodes in the network. This total degree of freedom reflects the capability to set up additional transmissions in the WMN.

We use the above measure as an aid to decide when to deploy network coding in the network. Consider Figure 8.5; three nodes (N_1, N_2, R_1) and two links (e_1, e_2) form the network to be analyzed. The circles depict the reception ranges for transmissions by nodes at the center of the circle. Here $NB(X)$ represents the set of neighboring nodes of node X. We can now define the cost of a transmission on a link L by $\mu(L, n)$ as the loss in degree of freedom of the network for the range of frames for which the n slots are reserved for transmissions on link L. For example, assume that n slots are reserved for transmission on link e_1 in Figure 8.5, and also assume that all of the nodes have the slots with status av before the transmission is scheduled. Then the cost of the transmission: $\mu(e_1, n) = n(|NB(N_1)| + |NB(R_1)|)$ (as a change to status rav or tav from av corresponds to a cost of one degree of freedom per slot per node, and a change to status uav corresponds to a cost of two degrees of freedom per slot per node). Similarly $\mu(e_2, n) = n(|NB(N_2)| + |NB(R_1)|)$. Thus, the total costs for the two transmissions gives $C_{\text{forwarding}} = \mu(e_1, n) + \mu(e_2, n)$, where, to simplify the computations, the cost of a set of transmissions is defined as the sum of the cost of the individual transmissions. Let us now look at replacing the above two transmissions via

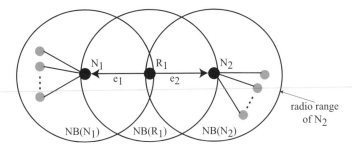

Figure 8.5 Relay constellation for analyzing the network coding gain using the IEEE 802.16 MeSH mode's distributed scheduling.

a multicast transmission on links e_1 and e_2 simultaneously using network coding. Assume that we intend to code data in both directions transmitted within n slots; due to the additional coding overhead $n + \epsilon$ slots are needed to be reserved for the multicast transmission. Thus, the cost for the multicast coded transmission is

$$C_{coding} = \mu(e_1, e_2, n + \epsilon)$$

$$= (n + \epsilon)(|NB(N_1)| + |NB(N_2)| + |NB(R_1)| - |NB(N_1) \cap NB(N_2)|).$$

Now, the gain in the scheduling degrees of freedom in the WMN equals $C_{forwarding} - C_{coding}$.

The nodes in the IEEE 802.16 WMN should choose to deploy network coding and persistently reserve multicast bandwidth only if the gain obtained is positive. The appropriate choice of n, that is, how many slots need to be reserved, remains. For this purpose we use the running average of the required bandwidth (in slots per frame) for the data cross flows (e.g. in Figure 8.5 the cross flows at node R_1 are the packets from N_1 to be forwarded to N_2 and vice versa).

8.4.2 Network Coding Framework

Figure 8.6 shows the logical building blocks of the MAC Common Part Sublayer (CPS) we propose for supporting WNC in IEEE 802.16's MeSH mode. The MAC CPS contains the core functionality for MAC within the IEEE 802.16 MeSH mode specifications. Packets arriving at the MAC layer from the network or higher layers are classified by a service-specific packet classifier which is located in the Convergence Sublayer (CS) of the MAC layer. The packet classifier enables classification of packets according to different scheduling services applicable to the packets. Transmissions/receptions at the PHY layer occur either in the control subframes or in the data subframes, as shown in Figure 8.6. The MAC management module is responsible for handling/processing the default protocol management messages of IEEE 802.16's MeSH mode at the MAC layer. Management messages defined for the purpose of supporting WNC in the MeSH mode are processed by the network coding management module. Regular unicast data transmissions are regulated by the unicast data management module, which transmits queued data for each outgoing link in slots reserved for transmission on the respective links. The network coding data management module is responsible for the multicast transmission of coded data packets using packets from the fragment pool. In addition, the

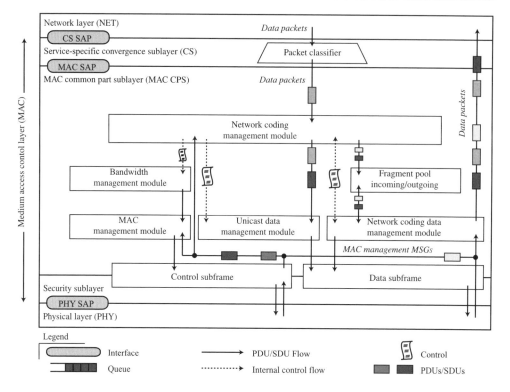

Figure 8.6 Block diagram showing the logical components of our framework extending IEEE 802.16's MeSH mode to support WNC.

network coding data management module is responsible for decoding received coded packets before these can be further processed either at the MAC or higher layers.

8.4.3   Reservation Strategies

The IEEE 802.16 MeSH mode lacks mechanisms to enable reservation of bandwidth for multicast transmissions which is needed for enabling WNC. Hence, we introduce additional management messages for reserving bandwidth for multicast transmissions[2]. Towards this end, we extend the three-way handshake used for reserving bandwidth for a single outgoing link to be able to reserve slots for simultaneous transmission on multiple outgoing links. In particular, we propose two different strategies for reserving slots for the multicast network coding transmissions.

We now consider the policies for reserving slots for network coding, again referring to Figure 8.5 as an example. Once the cross-flows N_1 to be forwarded to N_2 and vice versa are detected and are stable at node R_1, the node can compute the gain obtained for deploying network coding for the flows in question and replace the two transmissions on links e_1

---

[2]In this context multicast transmission implies packets transmitted by a node on the wireless medium which are intended to be received simultaneously by multiple direct neighbors. This is not to be confused with transmission of multicast data to a set of nodes in the network, where this set may consist of non-neighboring nodes.

and e_2 with a single multicast transmission. Here, let $s(e_1)$ and $s(e_2)$ denote the sets of n slots reserved for the unicast transmissions on links e_1 and e_2, respectively. Node R_1 may now select a set of suitable slots still free and use these to reserve $n + \epsilon$ slots for the multicast coded transmission to nodes N_1 and N_2. However, with increasing traffic (either the cross-flows, or other unrelated background traffic) the number of slots additionally available is reduced, implying the decrease of the probability of successfully reserving multicast bandwidth, as shown in Section 8.3. We introduce our novel slot allocation strategy termed the *replacement strategy* to counteract this decrease in slots: the core idea is to consider the reuse of the slots already reserved for transmission to the nodes N_1 and N_2 in addition to the additionally available slots at the transmitter to negotiate and reserve a common set of slots for transmission to neighbors N_1 and N_2. In addition to the available slots at the transmitter, the sets $s(e_1)$ and $s(e_2)$ are therefore also sent with the request for multicast bandwidth by node R_1. Nodes N_1 and N_2 may then use these slot ranges for the grant if reception is allowed by the current network schedule in these slots. The additional range of slots available for the grants increases the probability of successful reservation of the multicast bandwidth. Here, we see that N_1 is guaranteed to be able to receive in slots $s(e_1)$ and N_2 is guaranteed to be able to receive in slots $s(e_2)$. Thus, in the ideal case, the multicast handshake is now reduced to the case of a unicast handshake, thereby further increasing the probability of successful reservation (*Principles 2 and 3*).

Figure 8.7 illustrates our advanced two-phase handshake mechanism for reserving slots for the network coding transmissions using the same base topology as in Figure 8.5. We refer to Figure 8.7(a), (b), and (c) for the following discussion. Each subfigure shows the network topology augmented by the status of the reservation (illustrated using the particular slot numbers of the reserved slots; one-sided dotted arrows indicate unicast reservations, two-sided dotted arrows indicate multicast reservations). The topology shown in Figure 8.7(a) depicts the reservation state prior to the network coding handshake. Figure 8.7(b) and (c) show the reservation state of the new allocation strategy and the replacement strategy after the handshake, respectively. Unicast slots can correspondingly be freed for the given example after a successful multicast reservation has been established (please note, however, that we do not show the protocol interactions to actually free these slots). The message sequence chart in Figure 8.7(a) shows the initialization of the handshake process, which is common to both handshake variants. Figures 8.7(b) and (c) show the subsequent message sequence of the new allocations strategy and the replacement strategy, respectively.

Let us consider the reservation state as shown by the topology in Figure 8.7(a). Here, node R_1 may deploy network coding for relaying the packets between nodes N_1 and N_2. We employ our two-phase handshake mechanism for reserving slots for coded transmissions: the two phases are the initial handshake shown in Figure 8.7(a) followed by either the handshake to reserve as yet unused slots in Figure 8.7(b) or the handshake to repurpose existing unicast reservations in Figure 8.7(c), depending on our strategy used for reserving slots for network coding. The two-phase network coding handshake should be followed by normal three-way handshakes to free any superfluous slots reserved for unicast transmissions. Here, we should recall that we will only reserve slots for network coding for flows which are stable and, hence, will usually have persistent reservations (Per_∞). Similarly, for the current discussion we assume that slots are reserved for network coding with persistence[3] Per_∞.

---

[3]It is also possible to reserve slots for NC with persistences less than Per_∞; here the two-phase handshake presented can be optimized and adapted slightly for efficiently reserving non-Per_∞ slots.

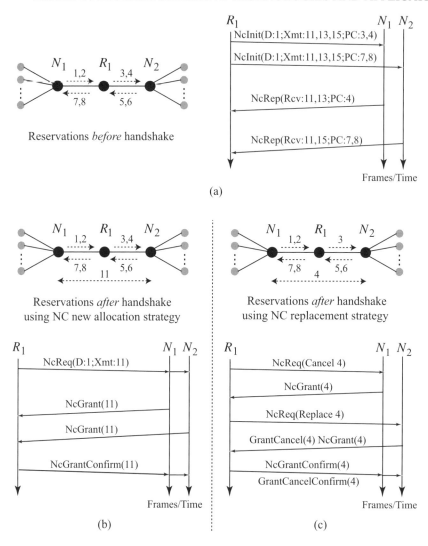

Figure 8.7   Example of our two-phase handshake variants for multicast bandwidth reservations supporting network coding in IEEE 802.16 MeSH mode: (a) NC initialization hand shake; (b) NC new allocation strategy; (c) NC replacement strategy.

8.4.4   Implementation Issues

We define new message types in addition to the existing protocol messages in IEEE 802.16's MeSH mode to carry the information related to network coding multicast reservations. The messages NcInit (Network Coding Initialization), NcRep (Network Coding Reply), and NcReq (Network Coding Request) are preferably transmitted in the data subframe in slots reserved for transmission to the addressed node, thus minimizing the latency of the handshake. The messages NcGrant (Network Coding Grant) and NcGrantConfirm

(Network Coding Grant Confirmation) are transmitted in the schedule control subframe. The message `NcInit` is used to initiate the process of reservation of slots to multiple neighboring nodes for the transmission of coded data, it contains the following fields.

- `D`: the value for `D` specifies the number of slots to be reserved for the transmission of coded data to multiple neighbors.

- `Xmt`: this field specifies the slots which are suitable and available for transmission of data at the node initiating the network coding handshake.

- `PC`: pseudo cancel, which specifies the set of slots which the addressed node should also check for their suitability for receiving data transmissions considering that the node initiating the handshake will free these slots reserved by it for transmissions to some other node.

The nodes addressed by the `NcInit` message reply using the `NcRep` message. The `NcRep` message has the following fields.

- `Rcv`: this field specifies the set of slots which are suitable at the node for receiving data transmissions.

- `PC`: this field specifies the set of slots which would be suitable for reception at the node if the transmitter of the initiating `NcInit` message would free these currently reserved slots.

After this initial handshake (Figure 8.7(a)), the initiator of the handshake knows which slots are suitable for transmitting the coded packets to the intended receivers. The intended receivers on the other hand know which slots should not be used in the near future for concurrent grants, as the relay would be initiating the next phase of the NC reservation process using these indicated slots. The next phase of the NC reservation may use either the new allocation strategy (Figure 8.7(b)) or the replacement strategy (Figure 8.7(c)). We next discuss the remaining handshake messages followed by a brief outline of both the reservation strategies.

The `NcReq` message is the NC counterpart for the normal request message used in the three-way handshake (Figure 8.2(a)) for distributed scheduling. `NcReq` can specify the number of slots to be reserved with the slots to be used for transmission as shown in the new allocation strategy. `NcReq` has the following fields.

- `D`: the value for `D` specifies the number of slots to be reserved for the transmission of coded data to multiple neighbors (similar to `NcInit`).

- `Xmt`: this field specifies the particular slots selected for the multicast network coding reservation (a subset of the `Xmt`-slots given in `NcInit` and `Rcv`-slots given in `NcRep`).

- `Cancel`: this field indicates that the given slots shall be cancelled and repurposed for a novel multicast reservation.

- `Replace`: this field specifies the slots for which the novel multicast reservation is to be issued.

For the replacement strategy, as discussed previously, the relay reuses some slots already reserved by itself for unicast transmission to one of the intended recipients of the coded data to schedule new coded transmissions to multiple recipients. Hence, when using the replacement strategy, based on which neighboring node is being addressed, the `NcReq` message is used with differing intentions. The relay sends a `NcReq` message with a `Cancel` indication, notifying the neighbor that it is cancelling the slots specified by the `Cancel` field (which had been reserved previously for a unicast transmission to some other neighbor) and that the node being addressed should reserve these slots for the multicast NC transmission. The neighbor then replies using a `NcGrant` message granting the slot for the NC transmission (the semantic of the `NcGrant` message is similar to the `Grant` message in IEEE 802.16). The relay uses the `NcReq` message with a `Replace` indication to address a node to which it has reserved the slots specified in the `Replace` field. This tells the neighbors that these slots which had been reserved previously for the unicast transmission from the relay to itself will be used for the transmission of coded (multicast) data by the relay to itself. The addressed neighbor then responds by simultaneously cancelling the unicast reservation and granting the same slots for the NC transmission. A `NcGrantConfirm` confirms the novel multicast reservation to all neighbors (similar to the `GrantConfirm` in IEEE 802.16). Readers interested in the exact implementation details and extensions to the MeSH mode (the control message formats and extensions to the MeSH mode specifications) can find them in Kropff (2006).

8.5   Related Work

The seminal work of Ahlswede *et al.* (2000) introduced network coding and demonstrated that bandwidth savings are possible when network coding is deployed. This was followed by literature which further investigated the benefits that can be obtained theoretically by applying network coding (see, e.g., Li *et al.* (2003) and Sagduyu and Ephremides (2005)). The work of Katti *et al.* (2006) is one of the first that considered the deployment of network coding in a realistic setting. Katti *et al.* (2006) present their COPE architecture, which uses opportunistic network coding to combine multiple packets from different sources before forwarding. The authors show that gains obtainable via opportunistic network coding can overhaul the gains in the absence of opportunistic listening. They deployed their architecture in a mesh network which uses the IEEE 802.11a MAC layer. In a MAC based on IEEE 802.11 (see IEEE (1999)) a basic access scheme as well as an RTS/CTS scheme are specified to enable access to the medium for unicast data transmissions. The RTS/CTS scheme ensures that when a node transmits unicast data, all nodes in the direct neighborhood of both the sender as well as the receiver do not transmit any data simultaneously. Thus, when a node transmits any data or acknowledgements following the successful RTS/CTS handshake, all of its neighbors remain silent themselves and will be able to receive the unicast data/acknowledgement transmission as long as none of their neighbors transmits simultaneously. This provides a conducive environment for opportunistic listening. However, as shown in Section 8.2, the neighbors of the node transmitting data in a slot may schedule simultaneous transmissions making opportunistic listening difficult if not impossible. Further, the IEEE 802.16 standard introduces a security and privacy sublayer in the MAC layer which encrypts data on a per link basis before transmission making opportunistic listening impossible. Another key aspect of the IEEE 802.16 MAC is the need for *a priori* reservation of minislots before transmission of data can take place, the implications of which have been presented in Section 8.3. The key question which the work of Katti *et al.* (2006) addresses is to determine which of the pending outgoing packets should be combined together before transmission, using information obtained via opportunistic listening as well as heuristics based on usage and availability of the Expected Transmission Count (ETX) routing metric (see De Couto *et al.* (2003)). In contrast, we do not rely on the availability of any particular routing metric or routing algorithm. The key questions which we addressed in this book chapter are as follows.

- When does network coding help in terms of throughput/bandwidth savings considering advanced MAC layers?

- How can we dynamically manage the multicast reservations for network coding without leading to conflict with other existing data transmission schedules in reservation-based MAC layers?

- How do we design practical solutions to deploy WNC in IEEE 802.16's MeSH mode?

8.6   Conclusions and Outlook

We discussed the applicability of WNC for WMNs based on the IEEE 802.16 MeSH mode with particular emphasis on MAC layer issues. First, we presented an analytical model for the distributed bandwidth reservation process of the IEEE 802.16 MeSH mode. The analysis was used to derive design principles for implementing and deploying efficient network coding solutions for the MeSH mode. We next designed solutions for deploying network coding in the MeSH mode using the derived design principles as a roadmap. Furthermore, advanced strategies for reserving slots for transmission of coded data were presented. The presented solutions have been initially discussed and investigated by means of simulations in Mogre *et al.* (2008); the results provide a proof of concept for the presented solutions and give pointers for future investigation.

However, MAC layer mechanisms alone are not sufficient for obtaining the maximum possible gain via deployment of network coding in IEEE 802.16's MeSH mode. Our work presented in Mogre *et al.* (2007) presents a first step towards effectively deploying network coding in IEEE 802.16's MeSH mode, where routing, scheduling and network coding are optimized simultaneously. In the future, we will look for further improvements to the latter work, and perform an evaluation of more advanced multicast bandwidth reservation strategies. Solutions which are suitable for coding real-time data and other delay sensitive data are of special interest, as these form the major class of traffic which benefits from the use of advanced bandwidth reservation mechanisms provided by the MeSH mode.

In summary, any solutions for network coding to be deployed in MAC layers supporting bandwidth reservation need to be able to work seamlessly with the specified reservation schemes. Furthermore, in most cases a one-to-one mapping of network coding solutions designed and developed within the scope of the IEEE 802.11 standard will not work optimally in advanced MAC layers. Owing to this fact, a lot of interesting and challenging aspects persist for further research in deploying network coding efficiently in next-generation WMNs.

Acknowledgements

We thank Matthias Kropff and Nico d'Heureuse for stimulating discussion on the subject of Wireless Network Coding for the MeSH mode of IEEE 802.16 as well as for supporting the task of implementing a simulation environment. Moreover, regards go to Rainer Sauerwein for critical discussion on the achievable performance gains under realistic usage scenarios, and Andreas Reinhardt for improving the readability of this manuscript. The research carried out has been supported by Siemens Corporate Technology.

References

Ahlswede, R., Cai, N., Li, S.R. and Yeung, R.W. (2000) Network information flow. *IEEE Transactions on Information Theory,* **46**(4), 1204–1216.

Akyildiz, I.F., Wang, X. and Wang, W. (2005) Wireless mesh networks: a survey. *Computer Networks*, **47**(4), 445–487.

Cao, M., Ma, W., Zhang, Q., Wang, X. and Zhu, W. (2005) Modelling and performance analysis of the distributed scheduler in IEEE 802.16 mesh mode. *Proceedings of the 6th ACM International Symposium on Mobile Ad Hoc Networking and Computing (MobiHoc 2005)*, pp. 78–89.

De Couto, D., Aguayo, D., Bicket, J. and Morris, R. (2003) High-throughput path metric for multi-hop wireless routing. *Proceedings of the 9th Annual International Conference on Mobile Computing and Networking (MobiCom 2003)*, pp. 134–146.

IEEE 1999, IEEE Standard for Information Technology, Telecommunications and Information Exchange between Systems, Local and Metropolitan Area Networks, Specific Requirements, Part 11: Wireless LAN Medium Access Control (MAC) and Physical Layer (PHY) Specifications, *IEEE Standard 802.11-1999.*

IEEE 2004, IEEE Standard for Local and Metropolitan Area Networks, Part 16: Air Interface for Fixed Broadband Wireless Access Systems, *IEEE Standard 802.16-2004.*

Katti, S., Rahul, H., Hu, W., Katabi, D., Medard, M. and Crowcroft, J. (2006) XORs in the air: practical wireless network coding. *ACM SIGCOMM Computer Communication Review*, **36**(4), 243–254.

Kropff, M. (2006) Network coding for bandwidth management in IEEE 802.16 mesh networks. *Master's Thesis* TU Darmstadt.

Li, S.R., Yeung, R.W. and Cai, N. (2003) Linear network coding. *IEEE Transactions on Information Theory*, **49**(2), 371–381.

Mogre, P.S., d'Heureuse, N., Hollick, M. and Steinmetz, R. (2007) A case for joint near-optimal scheduling and routing in TDMA-based wireless mesh networks: a cross-layer approach with network coding support (poster abstract and poster). *Proceedings of the IEEE International Conference on Mobile Adhoc and Sensor Systems (MASS 2007).*

Mogre, P.S., Hollick, M. and Steinmetz, R. (2006) The IEEE 802.16-2004 MeSH mode explained, *Technical Report, KOM-TR-2006-08.* KOM, TU Darmstadt, Germany, ftp://ftp.kom.tu-darmstadt.de/pub/TR/KOM-TR-2006-08.pdf.

Mogre, P.S., Hollick, M., Kropff, M., Steinmetz, R. and Schwingenschloegl, C. (2008) A note on practical deployment issues for network coding in the IEEE MeSH mode. *Proceedings of the IEEE International Workshop on Wireless Network Coding (WiNC 2008).*

Sagduyu, Y.E. and Ephremides, A. (2005) Joint scheduling and wireless network coding. *Proceedings of the First Workshop on Network Coding, Theory, and Applications (NetCod 2005).*

9

ASN-GW High Availability through Cooperative Networking in Mobile WiMAX Deployments

Alexander Bachmutsky

9.1 Introduction

High Availability (HA) is one of the most important features on the operator's requirements list, but it is left out of scope of most standardization bodies. The reason is simple: HA is usually considered as implementation specific and has to be covered somehow by every network element internally.

Before diving into the HA topic, it is necessary to clarify a few definitions. The first is *reliability*, which means correct functionality as per specification. The second is *availability*, which can be defined as a period of time when the system is reliable; the network element is usually considered as *highly available* if it performs its tasks reliably for 99.999% of the time, the measure sometimes referred to as 'five nines' or as a *carrier-grade* implementation; it is translated into at most 5.26 minutes per year of unplanned down time. The third related term is *redundancy*, which assumes having one or more extra functional elements capable of taking over failed identical functional elements; two main types of redundancy are *Active–Standby*, where the standby functional block does not perform any service until the active element fails, and *Active–Active*, where all functional blocks are performing their work at all times, and one or more blocks can increase their work dynamically by taking over tasks from the failed function. One-to-one (1:1) active–standby redundancy means that for every single active component there is a corresponding dedicated standby one. N:1 active–standby redundancy means that a single standby component protects N active components. N:M active–standby or active–active redundancy means that M standby or active functions protect

N active components. An additional level of the active–standby redundancy is it being either *cold*, when the standby starts only after the active failure, or *warm*, when the standby is started and always ready, but does not have all active states and has to recreate these states in some way, or *hot*, when the standby is started, ready to take over at any point of time with all active states always in sync. The fourth definition to remember is the *resiliency* that can be viewed as capability to maintain the reliability even in the case of failure. Resiliency can be improved through, for example, redundancy, but it is not the only way. Resiliency of a processor can be enhanced with an automatic power degradation when temperature rises above certain level; resiliency of an operating system can become better with a particular architecture or implementation (take carrier-grade Linux as an example); applications can also be designed to be resilient to denial-of-service attacks (for example, rate limit Transmission Control Protocol (TCP) SYN or ICMP packets, detect port scanning attempts) or failure of other applications (for example, Linux memory protection between processes). In most cases the resiliency directly affects the availability.

There is one confusion (coming frequently from operators) related to the reliability. As per the definition above, reliability means correctness. In general, it does not include performance and scalability dimensions. On the other side, operators do not want to call the network reliable if it can serve only a fraction of all intended subscribers. Some operators even define that the reliable function includes 100% scalability and 100% performance. However, the same operators are absolutely fine with products that have an extra redundant component for every function (for example, in a bladed system each control card, line card or processing card has its own standby redundant peer); in many cases the same box without redundancy can practically double the scalability and performance. That just means that these operators agree to reduce the capacity by half as a starting point, and after that require 100% reliability. Therefore, this paper adopts the definition that the function performs reliably when it serves correctly with at least 50% of the planned capacity for a nonredundant configuration and 100% of the planned capacity in a fully redundant configuration.

Network element availability is frequently used instead of the network availability or vice versa. The truth is that operators do not really care about a specific network element or even the entire network availability; they care about the service availability: capability to serve subscribers most of the time. However, service availability is not equal to the network element availability. For example, if we have the network with two routers, and one router can perform the job for the second router in the case of failure, we can obtain practically 100% of the service availability while every separate router availability is only 50%, assuming that both routers never fail at the same time (a so-called single point of failure in the network).

Another misconception is in calculating the network availability based on Mean Time Between Failures (MTBF). For example, we can compare a network element that fails on average once a year, but comes back into service in 10 minutes, with a network element that fails every month on average but comes back to service in 10 seconds. MTBF for the first network will be much higher, but availability for the second network will be higher. Regarding the service availability in our example, it is even more complex calculation. On the one hand a network element that boots longer can usually (not always) serve more subscribers, meaning that failure affects more sessions. On the other hand, a network element failure might require subscribers to renew their sessions bringing out-of-service time much beyond just a network element coming back into service; taking that into account might bring the service availability down for our faster booting and smaller network element.

It is also necessary to remember that availability is not a static number, it is a curve with time as one of the function parameters. For a new product the probability of Hardware (HW) failure is usually lower, but the probability of Software (SW) failure is much higher. When time passes, HW failure probability grows, but SW becomes more mature with bug fixes, so SW failure probability drops. We need to take into account both HW and SW when calculating projected network element availability.

At the end, availability calculations are often meaningless and provide just failure probability calculations. The only real way to measure the availability is to calculate retroactively the network element and/or service availability and derive from that whether it *was* highly available or not.

9.2 Classic HA Implementation

In many systems high availability is achieved through redundancy. Frequently, it is a bladed system with 1:1 active–standby redundancy for the control card (sometimes called also system card), 1:1 active–standby redundancy for line cards and a variety of redundancy schemes for processing/application cards: usually 1:1 active–standby, sometimes 1:1 active–active, in rare cases other implementations. In some systems the hardware even provides either dedicated switch fabric for redundancy management or dedicated backplane connection between adjacent slots for easier 1:1 protection. All that exists in addition to redundant power supplies, at least one redundant cooling fan, redundant disks, redundant ports and more.

The diagram of Nortel ASN-GW in Figure 9.1 is an example of a classic HA design. Classic design has its advantages and disadvantages. One of the biggest pros for such design is (excuse the pun) the availability of HA components. First of all, many vendors reuse previous products by adding new functionality achieving faster time-to-market and higher reliability because of integrated previously productized SW and HW components. Also, one can find some commercial HW and SW parts even when building the solution from scratch: redundant power supplies, redundant switch fabrics, open standards defined by the Service Availability Forum (SAF) and correspondingly many middleware products ready to be integrated into the system. All of those components create good justification for many Telecommunications Equipment Manufacturers (TEMs) to choose that path.

The classic implementation also has its cons. The biggest is a very significant additional complexity. It requires much more code for middleware, additional Application Programming Interfaces (APIs), complex state synchronization between active and standby, more debugging, more maintenance, more negative and corner use cases during failover. All of the above reduces SW reliability and MTBF defeating the original redundancy purpose, and definitely increases not only HW but also SW and operational costs. Another disadvantage is the failover time in cold and warm redundancy: it takes time to load SW and/or recreate all static and dynamic states, and as a result it increases the time between the failure and readiness to serve. In a hot standby mode the failover time is shorter, but the performance suffers because of the requirement for constant state synchronization even when there is no failure at all: both active and standby work continuously really hard to maintain everything in sync only to protect against a potential crash one or few times a year.

ASG Rear View

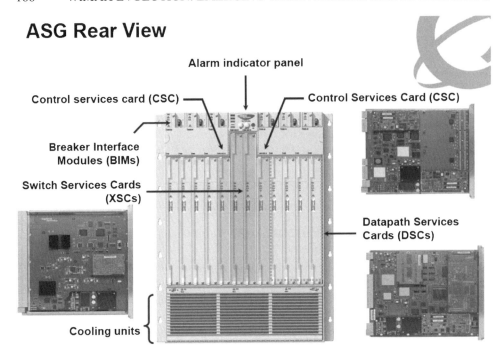

Figure 9.1 Nortel ASN-GW. (Source:http://www.nortelforum.org/files/7684/WiMAX.pdf.)

To understand the performance issue, let us assume that active function has to update its state (database entry, sequence number, etc.). If it updates the state before updating the standby, the system can crash right between these updates, and standby will not have the correct information to take over. If the requirement is to ensure 100% hot standby functionality, the common rule is that standby should 'know' at any point of time at least the same if not more compared with active (it knows more if the state was updated on standby, but active had crashed before updating its own state). This creates so-called check-points between active and standby: active sends a state to standby, receives an acknowledgement that it was received and/or processed, and then updates its internal state. Imagine performing that for TCP session where every packet in both directions creates or updates at least one state. Many years ago startup Amber Networks (later acquired by Nokia) did just that, and succeeded to achieve *only* 20% performance degradation becoming a base for stateful redundant Border Gateway Protocol (BGP) routing protocol implementation. It was a very complex project: the entire code was rewritten three times to cover all corner cases. One example is TCP timers and timer differences between active and standby: the timer on active can expire a little before the timer on standby, active takes some action, crashes, switchover occurs, standby becomes active and then 'the same' timer expires on the new active and another action is taken. Also, when standby boots (either after its own failure or after failover when the old active becomes a new standby) we have to update it in-service with all existing states from the active: by the time the last entry of a table is synchronized, the first might be modified again and again needs a synchronization.

Stateful TCP redundancy might be viewed as an extreme example, but WiMAX Access Services Network Gateway (ASN-GW) does not have smaller challenges, especially taking into account its scalability. First of all, there are protocols similar to TCP in some aspects: IP Security (IPsec) can protect R4 and R6 control channel communication between network elements, Generic Routing Encapsulation (GRE) with sequence numbers is used for the data integrity (lossless handover) mechanism. There are also transaction states and IDs being modified with received or sent packets, hundreds or even thousands of subscribers are entering or leaving the ASN-GW every second, Quality of Service (QoS) states with multiple QoS layers (per flow, per subscriber, per port, etc.) are changing with every user packet, millions of timers are being updated every second, and so on.

The solution is usually to give up the requirement for 100% hot standby redundancy and not to replicate all states all of the time, but such a decision can cause the active sessions to be dropped for many mobiles when the active failure occurs. Performance issues might also bring some trade-offs in a way of additional algorithm complexity or protection much below 100%. For example, the checkpoint can be implemented without an acknowledgement; in most cases the message will arrive at standby without problems, but sometimes that state update message will be lost, and the corresponding state will not be recovered after a failure. The worst part here is that message losses will happen right before the active crash because of some problems already happening on active for some time before the real failure is detected. In addition, there could be dependency between states, and the loss of one message can cause a different system view on standby. A classic example of this is table management functions with two sequential operations for the same element: delete the current operation, add another. If the first message (delete) has been lost, the addition of the second operation might fail, or duplicate elements in the table can be created. In the case of the second message loss standby will not have that element in the table at all; it can be service affecting for either specific subscriber or the entire system.

To summarize, a very complex solution is created, and 100% protection against failure is not achieved.

9.3 Network-based Resiliency Solutions for Routing

As mentioned above stateful TCP and BGP redundancy is a very good solution, but it is complex to implement. There is, however, a different way to go: network-based resiliency that is based on cooperation from other network elements. The concept is simple: if everything is OK, the network element can do everything on its own, but if a problem occurs, it will ask for a help from its neighbors. In the case of routing it is actually pretty much built-in as a part of routing architecture, because neighbors usually have all advertised routes. When the failed router comes back, it just asks its neighbors to send their entire table, and after that it can start serving traffic normally. Special protocols and concepts were created for that (Virtual Router Redundancy Protocol (VRRP), Hot Standby Router Protocol (HSRP), nonstop forwarding, nonstop routing; Cisco Systems is naturally one of their biggest developers and promoters), but the implementation is definitely simpler compared with hot standby redundancy.

The diagram in Figure 9.2 has been borrowed from the Cisco Web site and describes the VRRP concept: two or more routers share the same virtual IP address; there is one master (selected by the protocol) that usually processes the traffic; when the master fails, the slave/secondary router takes over.

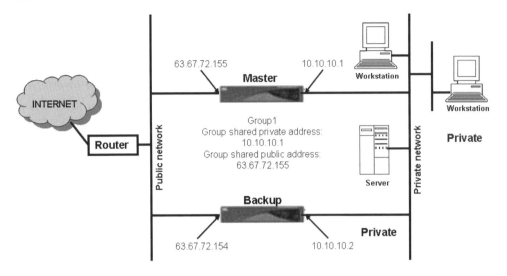

Figure 9.2 VRRP concept.

A more complex use case would be when boxes in the diagram above are not just routers, but firewalls that create constantly dynamic states to allow or block specific traffic. These states always have to be shared with standby/backup to ensure reliable behavior after the failure. If the number of workstations served by a single firewall in the diagram is close to the number of mobile subscribers served by a single ASN-GW, the use cases are similar.

9.4   WiMAX Network Elements R4/R6 Health Management

The WiMAX network already has some basic components for efficient network cooperation: the same Base Station (BS) can be served by multiple ASN-GWs, all ASN-GWs are interconnected in a full mesh logical connectivity. The missing pieces are protocols and the division of roles.

There is one huge advantage of WiMAX mobile network compared with above diagram for firewalls: mobile subscribers can also cooperate (or being forced to cooperate) with the network unlike workstations in a stationary deployment.

Let us start from a basic mobile WiMAX deployment diagram, but adding the capability of the BS to be served by multiple (in our example, two) ASN-GWs as shown in Figure 9.3.

The conceptual difference with active–standby or VRRP redundancy is that active–active configuration will be used. For simplicity reasons the description initially assumes a case of 1:1 active–active protection with two ASN-GWs that can be located in the same Central or Regional Office or placed in geographically distinct locations to protect against location-specific failures – power interruption, natural disaster, and similar.

The same ASN-GW can play a role of anchor Data Path Function (DPF) for one MS, serving DPF for another, anchor Paging Controller (PC) for the third mobile subscriber, and

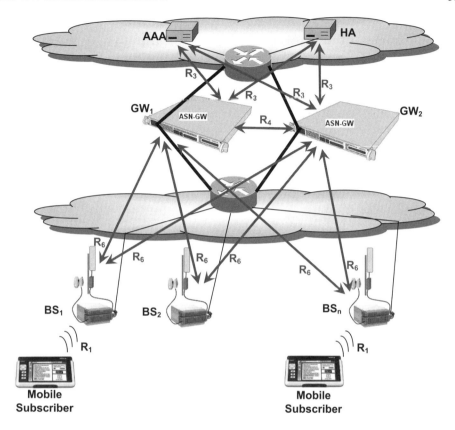

Figure 9.3 WiMAX BS connected to multiple ASN-GWs.

relay PC for the fourth. Since each physical ASN-GW will have its own protecting peer in a one-to-one scenario, the network diagram would look as shown in Figure 9.4.

We have actually introduced a new term in the above diagram, redundancy domain, which can be defined as a set of ASN-GWs protecting each other and all BSs served by these ASN-GWs.

The first element to introduce is the ASN-GW health management functionality by means of 'keepalive' messages *R4-Status-Req* and *R4-Status-Rsp* between ASN-GWs and correspondingly *R6-Status-Req* and *R6-Status-Rsp* between ASN-GWs and BSs to indicate their health change. The main part of these messages will be the *Status* that includes all required information about the network element wellness. To minimize the amount of required messages, it is suggested to include a capability to piggyback any other message by adding *Status* there too. The basic rule is that *Status* information has to be sent and received at least once during a preconfigured period of time, meaning that dedicated *R4-Status-Req* is used when there is no other message to be piggybacked, and *R4-Status-Rsp* always has to be used as a response to it (no response piggybacking using other messages is allowed because of WiMAX transaction management restrictions). The timer value should be a few (usually three) times smaller than a preconfigured status timeout to accommodate

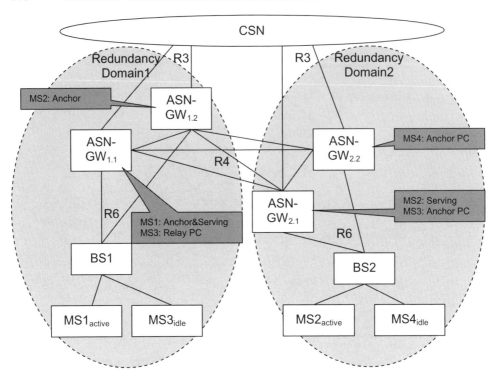

Figure 9.4 WiMAX network diagram with redundant ASN-GWs.

for occasional message loss. If a *Status* informational element is not received on-time for any reason, the corresponding peer can be considered being in the *FAILED* state. Alternatively, instead of the immediate failure decision, some implementations might send explicit *R4-Status-Req* to inquire about the peer health. In any case, a total time of not knowing about a failed peer defines a *failure detection time*.

In general, it is possible to run keepalive messages also over R6, but with potentially thousands of BSs served by a single ASN-GW and a usual requirement for a short failure detection time causing multiple keepalive messages per second for every peer, the preference would be to notify BSs only in the case of failure, and such a notification can be sent either by the failed network element or its peer with the inclusion of the failed ASN-GW identification. It is obvious that ASN-GW will report its own healthy state, but it might be counterintuitive to mention that failed ASN-GW can send any notification. One of use cases considered here is an operator-initiated ASN-GW graceful restart, when it is possible to notify all neighbors before actually going down. In addition to periodic *Status* information exchanges over R4, it also has to be sent on both R4 and R6 when the health status changes or a network element has been added (configured or learned) or restarted. For example, BS after its restart should send *R6-Status-Req* to all connected ASN-GWs stating the reason for this message (valid reasons are *PERIODIC*, *STATUS-CHANGE* and *RESTART*).

The WiMAX standard is built as a functional model; therefore, not every ASN-GW might have all functions implemented. To cover such implementation, *Status* information can

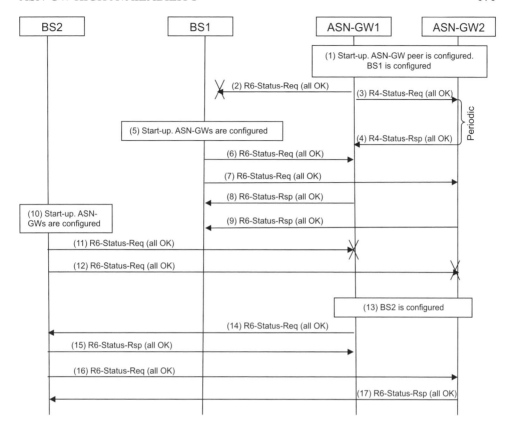

Figure 9.5 WiMAX network start-up use case.

include health conditions for separate functions, for example, by including the *Authenticator Status* information element for its authenticator health state. Depending on implementation, ASN-GW can decide to send separate messages per function; it can be useful if different functions are located on different blades or other physical units in a single logical ASN-GW. After the implementation of this part of the protocol, ASN-GWs and BSs will be aware of the health of all functions for all of the neighbors.

It is important to emphasize that for status exchange to work between ASN-GWs there has to be a very reliable network serving R4 interfaces, including redundant routers/switches and redundant physical paths, to avoid misidentification of a network failure to be the ASN-GW failure when *Status* does not arrive on-time.

Status information can also be used for other purposes (load exchange, capabilities exchange, etc.), but it is out of the scope of the ASN-GW HA topic.

Figure 9.5 shows status messages exchanged during system start-up or new network element deployment (some configuration events are also shown). The diagram is pretty much self-explanatory, but two comments are important:

- in step (2) the message did not reach BS1, because it was still powered down;

- in steps (11) and (12) the message is dropped by ASN-GWs because they were not configured to serve BS2; alternate behavior is to respond with error to minimize retransmissions.

9.5   R6 Load Balancing

While R6 load balancing can be used for different purposes in the network, the consideration here is only for the ASN-GW HA scenario.

As was already mentioned above, the proposed ASN-GW protection scheme is based on active–active configuration, meaning that all ASN-GWs serve mobile subscribers while being in the *HEALTHY* state. It brings the requirement to somehow divide traffic between ASN-GWs. Again, for simplicity of explanation, the example is given for the network with two ASN-GWs and one-to-one protection.

The recommended scheme is to load balance the traffic per mobile subscriber. When mobile subscriber enters a particular BS or prepares Handover (HO) to a particular BS, that new serving or target BS selects one of the connected *HEALTHY* ASN-GWs to become anchor/serving/target ASN-GW for the mobile subscriber. To help the network optimization during predictive HOs, serving BS should include current serving ASN-GW (or ASN-GW functions) as a part of the mobile subscriber context transfer. It allows the target BS to minimize R4 HOs if it has R6 connection to the same ASN-GW. When the mobile subscriber context does not exist in any directly (through R6) connected ASN-GW, the BS will choose one of *HEALTHY* ASN-GWs based on some internal algorithm: round robin or weighted round robin based on number of served subscribers, traffic load or other parameters. Specifics of the algorithm can be implementation-specific or even deployment-specific; for example, it can be different when both the BS and ASN-GW are from the same vendor and the load balancing scheme can be more efficient.

As a result of the proposed load balancing mechanism, some mobile subscribers in our example network will be served by GW1 and, some by GW2.

A message diagram for R6 load balancing would look as shown in Figure 9.6.

9.6   ASN-GW Failure and Recovery

ASN-GW failure is detected either based on the *Status* information with a *FAILED* state indication or lack of any status for some preconfigured period of time.

The failed ASN-GW played potentially different roles for different active mobile subscribers, serving, anchor or target, and the serving or target BS knows these roles. In some cases the failed ASN-GW did not play any role for any mobile subscriber served by the BS; in such case the failure notification can be just ignored by this BS.

Target BS in the simplest implementation can just ignore target ASN-GW failure; the worst-case scenario is that predictive HO will become unpredictive if that BS is selected by the MS; smarter implementation would be to reassign new *HEALTHY* target ASN-GW and establish data paths to the anchor ASN-GW as needed.

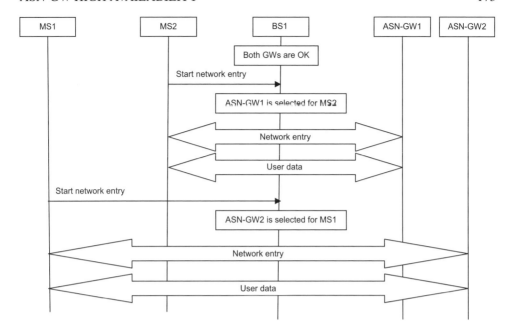

Figure 9.6 R6 load balancing.

The serving BS detecting the failure of the serving non-anchor ASN-GW can similarly reassign new *HEALTHY* Serving ASN-GW and establish all data paths to the anchor through this new serving ASN-GW. Very important BS functionality to handle this scenario is *the BS capability to be connected to at least two ASN-GWs*. The mobile subscriber is not even involved in the use case in Figure 9.7 (periodic R4-Status-Req/Rsp is shown as a Keepalive).

The most complex scenario is when the anchor ASN-GW with active mobile subscriber state fails (see Figure 9.8). There are multiple possible ways to handle such a situation; two of them are as follows.

(a) *Force the MS to fully re-enter the network while maintaining all its active sessions.* The reset command cannot be used, because the mobile subscriber will drop all sessions. The best available scheme is the network-initiated HO, but in usual cases of ASN-GW failure there is no need to change the BS; therefore, the mobile subscriber has to be ready to re-enter the network at the same BS as suggested by the network. This is one additional example when network cooperation (including terminals) really helps. The network entry can be treated as a regular HO similar to the situation when the mobile subscriber crosses the authentication domain and there is no way to retrieve the mobile subscriber context: the mobile subscriber will be reauthenticated, it will perform Mobile IP (MIP) registration (through the Dynamic Host Configuration Protocol (DHCP) and Proxy MIP or Client MIP) and continue normal functionality.

The network reentry procedure can take some time and therefore cause some time-sensitive sessions to fail, but it is not much different from the mobility across authentication domains. Load on Authentication, Authorization and Accounting (AAA) servers

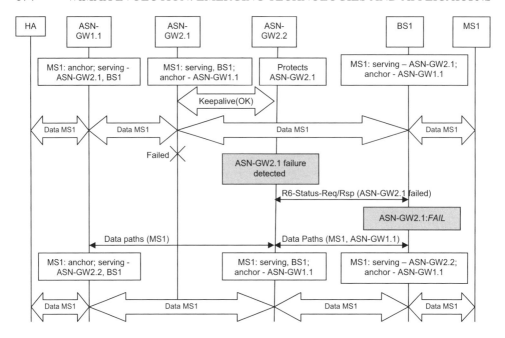

Figure 9.7 Use case of the serving non-anchor ASN-GW failure.

and home agents has to be considered, because many terminals will perform network entry simultaneously. It would be possible to share the security information between ASN-GWs, but current specification does not allow that. First of all, it is impossible to request such information from AAA server because of the requirement of the AAA protocol to forget such information as soon as it is sent to the authenticator. Second, there is no way to transfer the information over R4 interface because of security concerns. We do not share that view restricting the information transfer even between ASN-GWs located in the same regional office. Generally speaking, if both ASN-GWs have a direct or indirect trust relationship with the same AAA server, and there is also a trust relationship between these ASN-GWs, there should be no limitations on the sharing of security information. It will help not only for redundancy purposes, but also in the HO processing with authentication domain change or network-initiated R3 HOs.

A BS has all required information about active subscribers, but in many implementations the BS does not keep any state for idle subscribers. In this mechanism we have to know at least the list of idle mobile subscribers that were served by failed anchor paging controller. This list can be created and updated on the protecting ASN-GW, and the easiest way to achieve this is to use the existing transaction *Anchor_PC_Ind/Ack* from anchor PC to protecting PC.

The proposed information flow is shown in Figure 9.9 for both active and idle modes when the failed ASN-GW played both the anchor and serving roles.

If protecting ASN-GW has the list of idle subscribers served by the failed anchor PC, it can wake-up these mobile subscribers gradually (to avoid high load spike for the

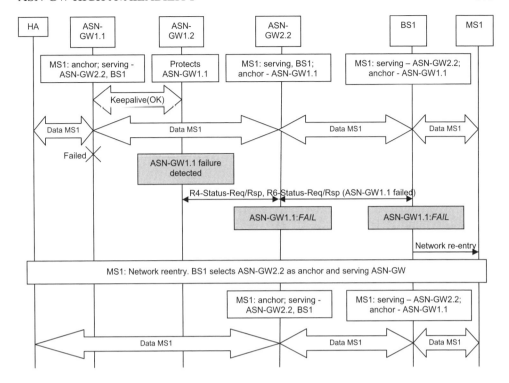

Figure 9.8 Use case of the non-serving anchor ASN-GW failure.

air interfaces and the network) forcing them to re-enter the network and reauthenticate themselves through new ASN-GW. If the list does not exist or a particular mobile subscriber is not on the list for any reason, there is no immediate impact on mobile subscribers, but they will not be paged if any incoming data arrives for them; if any MS not on the protection list moves in the idle mode after the anchor PC failure, it will be forced into the unsecured location update procedure and full network re-entry; if such a mobile subscriber wakes up on its own, it will also be forced into the reauthentication.

(b) *Predictive CSN anchored mobility (R3 relocation).* The idea behind it is similar to the predictive Access Service Network (ASN) anchored mobility, but preparations are made by the network, not the mobile subscriber, and it can be applied for both active and idle terminals. The procedure can be used not only for HA, but also to speed up regular R3 relocations.

The proposal is to add messages for a prerelocation between ASN-GWs, these messages can use the same information elements as used for regular R3 relocation. The most important part is the mobile subscriber context replication, there is no need to preestablish any R4/R6/R3 connections. As it was already mentioned above, there is a need to allow security context exchange between a pair of ASN-GWs over R4 reference point.

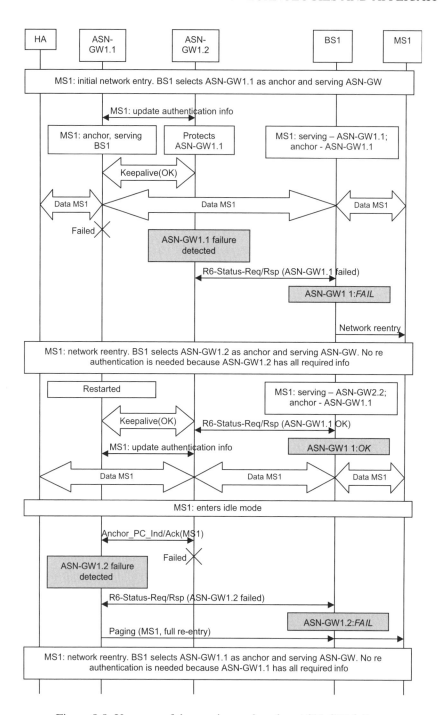

Figure 9.9 Use case of the serving and anchor ASN-GW failure.

When the failure occurs, the protecting ASN-GW can finish the R3 relocation (for active subscribers reestablish all R4 and R6 data paths, perform PMIP registration or force the mobile subscriber to perform new Client Mobile IP (CMIP) registration by means of waking the mobile subscriber up if needed and initiating foreign agent advertisement) without the need to request any info from the now failed anchor ASN-GW.

While the scheme can help in saving on any reauthentication (major load and relocation latency bottleneck), there is still a need to perform MIP reregistration in the case of MIP deployments. For the Simple IP case the procedure after failure is similar to R4 HO without preestablished data paths, so it is really simpler for the described use case.

The protecting ASN-GW can take over some IP addresses of the failed ASN-GW. For example, it can start using and accepting the failed foreign agent IP address(es). It can advertise these addresses as its own through routing protocols and gratuitous Address Resolution Protocol (ARP). The mechanism can potentially help to avoid the need for a MIP reregistration. The limitation of such 'virtual' addresses is that it can be used only in a 1:1 or N:1 redundancy; it will not work in the N:N redundancy discussed below.

It is also possible to preestablish R4 data paths to serving non-anchor ASN-GW, because we do not expect the mobile subscriber to move frequently through multiple ASN-GWs while being constantly in the active session. This functionality would save some additional time after failure, but will require cleanup to be triggered when these data paths are no longer needed.

In a hopefully rare scenario when a highly reliable network responsible for R4 connectivity fails, all ASN-GWs will assume that their peers are not functioning properly and will notify BSs about that event. It would be the BS responsibility to understand that it is just a transport failure, because it will receive contradicting reports that every ASN-GW is OK, but some or all peers are failed. One of the potential behaviors would be for BSs to not do anything when anchor and serving ASN-GWs are collocated; otherwise, forcing either HO or even full network re-entry.

This is a byproduct benefit that we are getting from our health management procedures: with the network cooperation we can detect and recover from both network elements and some transport network failures.

ASN-GW recovery is treated as a state change event, and the notified BSs will start using it for load balancing. Depending on the load, protecting ASN-GW can trigger R3 relocation of some terminals to the recovered ASN-GW.

9.7 N:N Redundancy

One serious limitation of 1:1 active–active redundancy is very similar to 1:1 active–standby redundancy: scalability and performance. If the requirement is to ensure 100% protection at any point of time, every ASN-GW can be loaded not more than 50%, which effectively doubles the number of ASN-GWs in the network. However, the operator under normal conditions does not want to load its network and network elements above certain level (let us say, 70%). In active–standby configuration the active function is loaded 70% in such a case, the standby function is not loaded, thus bringing the average load to 35%. In active–active configuration we can load both ASN-GWs to 50% and still fully survive a single failure, meaning that it still has advantages over the active–standby case.

Sometimes active–standby resiliency is implemented as N:1 redundancy: a single standby function protects N active components. Any active failure will trigger the standby to take over. The problem is that it is relatively simple for cold and warm redundancy, but becomes complex in hot redundancy, because the standby function has to receive and maintain states from all active functions. In some situations standby works much harder than any individual active causing stability issues on the standby itself. While standby failure is not that severe, it can affect the performance of the active function because of the required in-service update of all states at once (so-called bulk update) after the standby recovery. Also, failure of standby means that there is a window when the system is not protected at all, and a longer standby recovery time means longer period without protection, an increased probability of the unrecoverable active failure and a lower availability as a result of that.

There is also N:M active–standby redundancy when M standby functions protect N active components. The scheme brings a higher availability compared with N:1, but increased complexity limits such implementation. Also, average scalability and performance in N:M is lower than in N:1 because of the greater number of standby functions.

That is where active–active redundancy really shines. Let us analyze the diagram of WiMAX ASN connectivity in Figure 9.10.

All BSs are divided into clusters C1 to C20 for more efficient HOs; all BSs in the same cluster are connected to the same ASN-GWs; a cluster can be as small as a single BS or include arbitrary number of BSs. Clusters from C11 to C20 will be connected in the same way as the clusters from C1 to C10 correspondingly; it is made just to illustrate that multiple clusters can be connected to the same set of ASN-GWs. From ASN-GW point of view the connectivity is as follows:

ASN-GW1 is connected to clusters C1, C2, C3, C4, C11, C12, C13, C14;

ASN-GW2 is connected to clusters C1, C5, C6, C7, C11, C15, C16, C17;

ASN-GW3 is connected to clusters C2, C5, C8, C9, C12, C15, C18, C19;

ASN-GW4 is connected to clusters C3, C6, C8, C10, C13, C16, C18, C20;

ASN-GW5 is connected to clusters C4, C7, C9, C10, C14, C17, C19, C20.

For every BS this connectivity is not different from 1:1 active–active connectivity described previously, its role in handling health management and ASN-GW failure and recovery is exactly the same. ASN-GW implementation is also the same except the need to keep full mesh health exchange between all ASN-GWs. The distinction is in the distributed protection. For example, when ASN-GW1 fails, the first subset of mobile subscribers served by BSs in clusters C1 and C11 will be moved to ASN-GW2; the second subset of mobile subscribers served by BSs in clusters C2 and C12 will be moved to ASN-GW3; the third subset of mobile subscribers served by BSs in clusters C3 and C13 will be moved to ASN-GW4; finally, the fourth subset of mobile subscribers served by BSs in clusters C4 and C14 will be moved to ASN-GW5. The result is that each ASN-GW is protected by all other ASN-GWs.

With just a smarter configuration and without any implementation change the proposed diagram increases the scalability and performance of each ASN-GW in our example up to

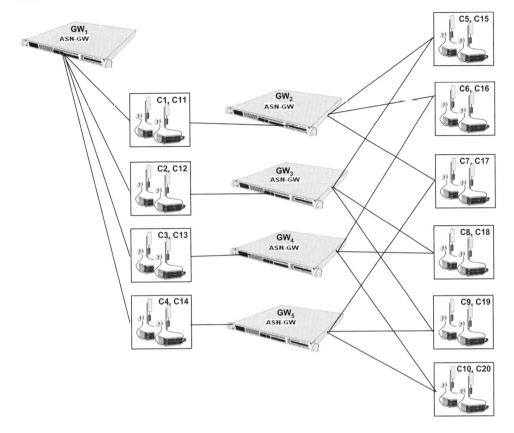

Figure 9.10 $N:N$ ASN-GW redundancy.

80%, because in the case of failure the load is divided between four other ASN-GWs (20% each; after that all four ASN-GWs will be loaded 100%), and the network will continue to serve all existing subscribers until the failed ASN-GW recovery. These numbers assume ideal load balancing: in the real scenario there is some inefficiency, so it is recommended even in the above example with five ASN-GWs not to exceed 70%–75% of maximum load. It is still a very significant average load compared with 35% in 1:1 active–standby or 50% in 1:1 active–active configurations, and this average will grow with more ASN-GWs; for example, a set of 10 ASN-GWs allows up to 90% load with full protection against a single ASN-GW failure at any point of time.

This special configuration proposal unleashes the real power of $N:N$ active–active redundancy to achieve the goal of ASN-GW HA.

If there is any concern regarding the increased configuration complexity, we would recommend automating it. For example, for a given number of N ASN-GWs and BS cluster X the recursive algorithm presented in Figure 9.11 can find ASN-GW indexes Z and Y that each BS in the cluster X has to be connected to.

Such an algorithm can become part of the provisioning solution for a WiMAX network.

Red (x)

$\{$ IF $\left(x \leq \dfrac{N*(N-1)}{2}\right)$

THEN $\{IF\ (x \leq (N-1))$ *THEN* *RETURN* $(Y=1;\ Z=x+1)$;
        ELSE *FOR* $(2 \leq i \leq (N-1);\ i++)$

$$\{IF\left[x \leq (N-1) + \sum_{j=2}^{i}(N-j)\right]$$

    IF $(i=2)$ *THEN* *RETURN* $(Y=i,\ X=i+x-(N-1))$;

$$ELSE\quad RETURN\left(Y=i;\ Z=i+x-(N-1)-\sum_{j=2}^{i-1}(N-j)\right);$$

    ELSE ;
    $\}$

  $\}$

ELSE $\left\{x=x-\dfrac{N*(N-1)}{2}\right.$;

    RETURN (Re d(x)); /* *RESTART THE ALGORITHM WITH THIS NEW X VALUE* */
   $\}$
$\}$

Figure 9.11 Algorithm for ASN-GW selection.

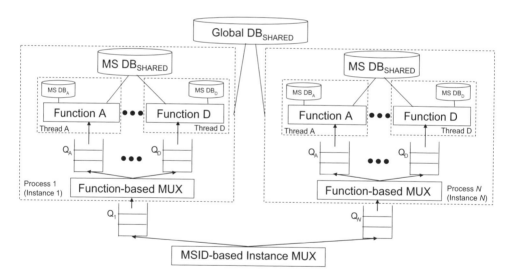

Figure 9.12 Concept of multi-instance ASN-GW.

9.8 Multi-instance ASN-GW

Many ASN-GW implementations use the concept of multi-instance application. Instance can be with HW boundary (blade in the chassis, CPU on the blade, etc.) or SW boundary (thread or process). The diagram in Figure 9.12 presents SW based multi-instance.

In both SW and HW cases there is usually an internal or external load balancing mechanism between these instances, and the example of the most obvious algorithm is based on mobile subscriber or flow identification (Mobile Subscriber ID (MSID), Network Access Identifier (NAI), GRE key, etc.) when every instance becomes responsible for a subset of all subscribers or flows served by the ASN-GW.

Multi-instance support is usually well shielded from the network. HA requirements can change that, because the failure cannot occur for the entire ASN-GW, but only for a specific instance. Of course, it can be assumed that the whole network element is unhealthy if at least one instance is unhealthy, but it is not a very good way to improve the availability.

One way is to expose the instance handling into the health management protocol by adding one more layer of hierarchy into the *Status* information element, *Instance Status*, that includes the instance ID and a function-specific status. An instance failure is treated in exactly the same way as an ASN-GW failure as explained previously, but BSs should have the mapping between instances and mobile subscribers. It can be achieved by the addition of *Instance Identification* in, for example, data path establishment messages or, alternatively, by the instance ID encoding into the data path ID (GRE key).

The health state of a multi-instance ASN-GW can be now represented by three states: *HEALTHY* (all instances are HEALTHY), *UNHEALTHY* (at least one instance is failed) and *FAILED* (more than half of ASN-GW instances are failed).

Since every instance handles only a subset of all served subscribers, and the most probable failure will occur for a single instance as opposed to the entire ASN-GW, we can significantly minimize the system impact by the health monitoring per instance.

9.9 The Proposal Summary

The condensed version of the new functionalities to implement the proposed ASN-GW HA scheme is as follows.

(a) Operations and Management (O&M): configure protecting ASN-GW peer(s); report current self or peer status to the management layer upon change, or periodically, or on request.

(b) ASN-GW: Periodically exchange keepalive messages between protecting peers with a timeout for a peer failure detection.

(c) ASN-GW and BS: R4/R6 redundancy status transaction to notify about the self or peer status change and the status after the network element restart.

(d) ASN-GW: transfer security context over R4 to the protecting peer after successful mobile subscriber (re)authentication.

(e) ASN-GW: Notify the protecting peer about mobile subscriber entering/exiting the idle state; keep the information about idle mobile subscribers in the protecting ASN-GW.

(f) ASN-GW: gradually wake up relevant idle mobile subscribers after taking over the failed ASN-GW peer.

(g) BS: maintain the health state for every connected ASN-GW; assign new MSs only to healthy ASN-GW(s).

(h) BS: force HO for terminals affected by anchor ASN-GW failure (in an alternate implementation it can be R3 HO forced by an ASN-GW peer); establish new R6 data path for mobile subscribers with only serving ASN-GW failure.

As is easy to see, there is not much needed during the normal data and control plane processing; the 'real' work only starts when the failure really occurs, as it should be in the efficient HA scheme. Previously it was possible only using cold or warm redundancy. A new proposal based on the network cooperation principles allows the same capability for a hot redundancy with virtually no performance and scalability degradation compared with the system without HA.

9.10   Conclusions

The proposed active–active network-assisted HA mechanism achieves the required carrier-grade reliability by distributing the functionality between multiple network elements, ASN-GWs and BSs, with much simpler implementation for each of them.

This chapter has concentrated on ASN-GW HA. When the concept of network-assisted resiliency is accepted, the scheme can be extended to HA for other network elements: AAA servers, home agents and others. This will enable more efficient deployments of WiMAX network with significantly reduced capital expenditure (CAPEX) as a result of simpler implementation and higher utilization of the network components.

Part V

WiMAX Extensions

10

Robust Header Compression for WiMAX Femto Cells

Frank H.P. Fitzek, Gerrit Schulte, Esa Piri, Jarno Pinola, Marcos D. Katz, Jyrki Huusko, Kostas Pentikousis and Patrick Seeling

10.1   Introduction

Even though WiMAX technology offers high data rates, the overall capacity over the air is shared among several WiMAX users. This is especially true for the WiMAX support of femto cells. In this scenario very small wireless cells, covered by short-range technologies such as WLAN or Bluetooth, are wirelessly connected to the Internet backbone by WiMAX links. The number of accumulated users will be high and any means to gain additional capacity are more than welcomed. An effective approach to increase capacity is to utilize header compression. In the wireless world, Robust Header Compression (ROHC), as standardized in RFC3095 (Bormann *et al.*, 2001), is a widely applied header compression scheme. ROHC is a method to reduce the overhead of the packet header information down to 10% or less. This method is especially effective if the payload is relatively small in contrast to the header itself. In this chapter we will motivate the use of ROHC in WiMAX, as proposed in the WiMAX documents (Fitzek *et al.*, 2004), and demonstrate the potential gain of using ROHC for WiMAX femto cells by implementation and measurements. The results presented here are the outcome of a collaborative work of acticom (acticom GmbH , 2008b) and VTT implementing acticom's ROHC protocol stack in a commercial WiMAX system. As shown throughout this chapter, ROHC demonstrates outstanding performance over the standard system.

This chapter is organized as follows. First, ROHC is introduced. Then the basic concept and motivation for applying ROHC in WiMAX systems is given. The Technical Research

WiMAX Evolution: Emerging Technologies and Applications   Edited by Marcos D. Katz and Frank H.P. Fitzek
© 2009 John Wiley & Sons, Ltd

Centre of Finland (VTT) and the German company acticom joined forces to integrate ROHC as defined by RFC3095 into fixed and mobile WiMAX systems. Measurement results using a novel WiMAX testbed with integrated ROHC technology are presented and discussed before we conclude. Measurements show that using ROHC for a given WiMAX scenario will more than double system capacity.

10.2   ROHC in a Nutshell

ROHC supports the efficient use of scarce wireless resources for multimedia data with real-time constraints. Reducing the overhead of Internet Protocol (IP)-based headers in packet-oriented communication networks significantly improves performance, especially for multimedia data services, such as audio or video transmissions. Thus, by employing ROHC, 2.5G and 3G network operators will be able to increase the amount of resources available to their customers, resulting in an improved return on investment. Robust header compression was standardized by the Internet Engineering Task Force (IETF) in RFC3095 and will be an integral part of releases 4 and 5 of the Third Generation Phone Partnership (3GPP) Universal Mobile Telecommunications System (UMTS) specification as well as standards for WiMAX networks. The efficient transport of multimedia data over packet-oriented (wireless) data networks has gained much interest in recent years. To meet the requirements in terms of delay, jitter and latency for interactive communication, for example, Voice over IP (VoIP), sampling and packetization delays must be minimized. When using the Real-time Transport Protocol (RTP) and RTP/User Datagram Protocol (UDP)/IP headers for encapsulating voice samples, the ratio between the size of the RTP/UDP/IP headers and the payload size is typically 2:1 (or 3:1 for IPv6). Thus there is a significant overhead and waste of resources. ROHC uses a connection-oriented approach to remove packet inter- and intra-dependencies and thus reduces the header payload significantly. In a multimedia scenario with real-time voice services and IPv6, the overall bandwidth can be reduced by a factor of four. The error characteristics on wireless links differ dramatically from wired links. ROHC was designed to operate in error-prone environments by providing error detection and correction mechanisms combined with robustness for IP-based data streams.

The basic motivation for IP-based header compression is based on the fact that packet header information has significant redundancy. The combined headers for a real-time multimedia stream using IPv4 includes the 20-byte IPv4 header, the 8-byte UDP header and the 12-byte RTP header. The headers for IPv6 total 60 bytes. Redundancy exists among the different headers (IP, UDP and RTP), but in particular between consecutive packets belonging to the same IP flow. The header fields can be separated into nonchanging and changing. The nonchanging group consists of static, static-known and inferred header fields and is more or less easy to compress. A large portion of the header fields are static or static-known and therefore can be compressed easily or will not be sent at all after the first successful transmission between sender and receiver. Other header fields are referred to as inferred. These fields can be inferred from other header fields and are also easy to compress. The changing group consists out of not-classified *a priori*, rarely-changing, static or semi-static changing, and alternating changing header fields. These types of header fields are more difficult to compress and it depends on the header compression scheme how the compression is implemented. The potential savings for voice and audio services are presented

Table 10.1 Potential capacity savings for header compression schemes.

CODEC	Mean bit rate (kbps)	IPv4 savings (%)	IPv6 savings (%)
LPC	5.6	74	81
GSM	13.2	55	65
G.711	60.0	21	29

in Table 10.1. For voice services, we present the Linear Predictive Coding (LPC) with 5.6 kbps, a GSM codec with 13.2 kbps, and a codec following the ITU-T standard G.711 with 60.0 kbps. Assuming a packet generation rate of 50 Hz (one sample every 20 ms), the results shown in Table 10.1 can be achieved.

These results show that the application of header compression can significantly increase the system capacity of a network operator. For LPC and IPv6, a system with header compression can accommodate five times more users than a system without. All header compression schemes before ROHC were unable to handle error-prone and long-delay sensitive links such as those typically found in the wireless environment.

ROHC in its original specification, as in RFC3095, is a compression scheme with profiles for three protocol suites: RTP/UDP/IP, UDP/IP and Encapsulating Security Payload (ESP) Protocol/IP. In case any other protocol suite is used, ROHC will not perform compression at all (uncompressed profile), but there are other profiles to support more protocol suites (IP only, Transmission Control Protocol (TCP)/IP). As illustrated in Figure 10.1, ROHC is located in the standard protocol stack between the IP-based network layer and the link layer. To offer the ability to run over different types of links, ROHC operates in one of three modes: unidirectional, bidirectional optimistic and bidirectional reliable. Similarly to the states described below, ROHC must start at the lowest mode (unidirectional) but then it can transit upwards if the link is bidirectional. In contrast to the states, in fact, modes are not related to the compression level, but they determine which actions ROHC must perform in every state and in state transitions (and according to the link characteristics in terms of feedback and context updates). Mode transitions can be initiated by the decompressor. The decompressor can insert a mode transition request in a feedback packet indicating the desired mode. ROHC uses a flow-oriented approach to compress packets. Each flow is mapped to a context at the compressor and the decompressor and is identified by a context identifier (CID). A context is a set of state variables and contains (among other variables) the static and dynamic header fields that define a flow.

The ROHC compressor and decompressor can each be regarded as a state machine with three states. Compressor and decompressor start at the lowest state which is defined as 'no context established', that is compressor and decompressor have no agreement on compressing or decompressing a certain flow. Thus, the compressor needs to send a ROHC packet containing all of the flow and packet information (static and dynamic) to establish the context. This packet is the largest ROHC packet that the compressor can send. In the second state, the static part of a context is regarded as established between compressor and decompressor while the dynamic part is not. In this state, the compressor sends slightly larger ROHC header than when it is in the third state, where the static as well as the dynamic part of a context are established. Fallbacks to lower states occur when the compressor detects a

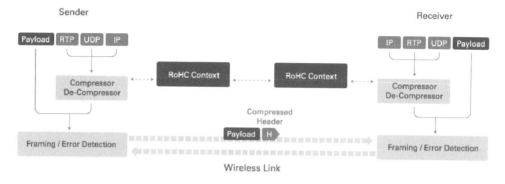

Figure 10.1 ROHC method. Reproduced with permission from © acticom, 2008.

change in the dynamic or static part of a flow, or when the decompressor detects an error in the dynamic or static part of a context. The compressor strives to operate as long as possible in the highest state under the constraint of being confident that the decompressor has enough and updated information to decompress the headers correctly. Otherwise, the compressor must transit to a lower state to prevent context damage and to avoid context error propagation. Compression of the static part of headers for a flow is trivial since they only need to be transmitted at context establishment and then remain constant. More sophisticated algorithms are needed for compression of the dynamic part. ROHC basically uses two algorithms to compress and decompress dynamic header fields: self-describing variable length values and Windowed Least Significant Bits (W-LSB) encoding. The first reduces the number of bits needed to transmit a field upon the actual value of that field (low values need fewer bits). The latter algorithm, after parametrizing to the dynamic change characteristic of the header field to be compressed, yields the minimum bits needed to be transferred to reconstruct the new value from an old value that the decompressor received previously. In particular, the W-LSB compression algorithm in combination with an elaborated scheme to protect sensible data in ROHC compressed headers contribute to the robustness of ROHC.

Using acticom's ROHC implementation it can be shown that the application of ROHC in a wide range of applications can bring benefits in:

- increased capacity when ROHC is used for multimedia services;

- improved robustness of the ROHC frames versus uncompressed; and

- decreased delay and jitter of ROHC compressed packets.

10.3   Scenario Under Investigation

Throughout this chapter we investigate ROHC for WiMAX femto cells. Femto cells are small hotspots typically located in office or home environments that are connected to a number of mobile devices on one side and to a wireless backbone on the other side. In Figure 10.2, we illustrate one WiMAX Base Station (BS) which is connected to multiple femto cells and the Access Service Network (ASN) for Internet connection. Each femto cell is covered by one

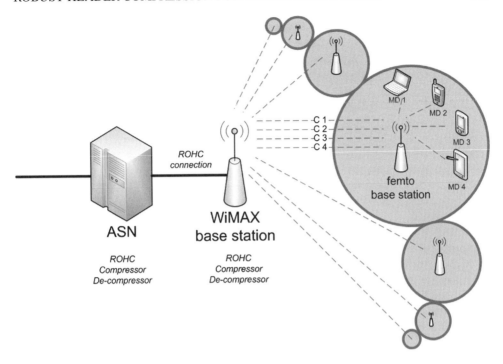

Figure 10.2   Scenario under investigation: multiple femto cells are connected with one WiMAX BS to gain access to the Internet via the ASN.

femto cell BS. WiMAX technology is used between the WiMAX BS and the femto BS. Any other or multiple other technologies can be used within the femto cell. One example could be that a femto cell is covering a home environment with Bluetooth and IEEE802.11 Wireless Local Area Network (WLAN) such that all possible devices (mobile phone, PDA, laptops, etc.) can be connected. All of these different technologies are aggregated and conveyed over WiMAX to the wireless Internet provider. One WiMAX BS is able to support multiple femto cells. As we consider WiMAX 16d as the main carrier, we assume that the femto BS will be stationary. For an in-detail treatment of femto cells in WiMAX, readers are referred to Chapter 5 of this book.

ROHC is used to reduce the IP overhead on the connection between the ASN and the femto BS. In turn, the compression/decompression of the IP headers takes place at the ASN and the femto BS. The individual link from the femto BS to the individual mobile device is not considered to employ ROHC. It is possible to use ROHC even within the femto cell, but this chapter focuses on the aggregated WiMAX link only. Interested readers are referred to acticom GmbH (2008a), where ROHC in combination with IEEE802.11 WLAN has been demonstrated on Nokia's Internet tablet N810.

In the case of downlink traffic, the ASN will perform the ROHC compression of all streams for a certain femto cell and aggregate those packets into one larger container packet. The aggregated packet will be sent to the femto BS, where the packet is unpacked and each ROHC packet is decompressed. The uncompressed packets are then forwarded by the femto

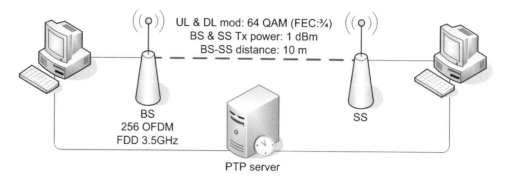

Figure 10.3 ROHC/WiMAX testbed setup.

BS to the dedicated mobile device over any given technology. In the case of uplink traffic, the procedure is reversed. The envisioned scenario is especially interesting for VoIP traffic support in office spaces or public places. A large number of end devices using VoIP services are accumulated within the femto cell. Multiple femto cells can support even a larger area, as illustrated in Figure 10.2.

10.4   WiMAX and ROHC Measurement Setup

The experimental facility employed in the empirical femto cell measurements is part of the VTT Converging Networks Laboratory. The schematic of the fixed WiMAX testbed is illustrated in Figure 10.3 and comprises one Airspan MicroMAX-SoC BS, one Airspan ProST Subscriber Station (SS), two GNU/Linux PCs symmetrically connected on the BS and SS sides and the Precision Time Protocol (PTP) synchronization server. The PCs act as traffic sources/sinks and are connected directly to the BS and the SS. The fixed WiMAX testbed operates in the 3.5 GHz frequency band with 3.5 MHz bandwidth using Frequency Division Duplex (FDD) to separate the uplink and downlink. Maximum goodput levels achieved with this configuration and a Maximum Transmission Unit (MTU) size of 1500 bytes for UDP are 9.4 and 5.5 Mbps for downlink and uplink, respectively, as measured in Pentikousis *et al.* (2008b). The testbed is deployed in an indoor laboratory with very short link span (BS–SS = 10 m). For this reason, the transmission power is set to 1.0 dBm. The short line-of-sight distance between BS and SS and relatively static indoor conditions keep the signal level relatively static and modulation constantly in 64 Quadrature Amplitude Modulation (QAM) (Forward Error Correction (FEC) = 3/4) for the uplink and the downlink. The Medium Access Control (MAC) scheduling is based on a best-effort scheme.

The clocks at both end-hosts are synchronized within an order of tens of microseconds using PTPd (Corell, 2008), which is an open-source software-only implementation of IEEE 1588 PTP (Correll *et al.*, 2005). The accuracy of PTPd when used over Ethernet is similar to the accuracy of GPS synchronized clocks running on Windows PCs, see Pentikousis *et al.* (2008a). Similar to the Network Time Protocol (NTP), PTP is based on synchronization messages sent over the network. In order to avoid interference between PTP traffic and

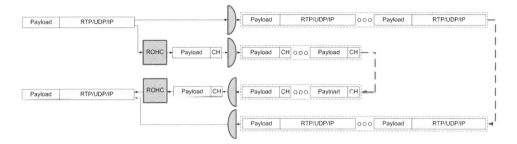

Figure 10.4 Aggregation of compressed packets. CH, Compressed Header.

measurement traffic, the end-hosts are connected to the PTP server via different Ethernet interface than the measurement interfaces.

Multiple parallel femto cell data streams are synthetically generated by employing a traffic generator implemented in Perl using additional modules (Barr, 2008, Humfrey, 2008, Wegscheid *et al.*, 2008). Linux is not natively a real-time kernel, which is the reason for the use of adaptive sleep intervals between packet injections into the traffic generator in order to improve the traffic accuracy. Via this operation, the achieved cumulative interval deviation for positive direction is of the order of 0.5–3.5%, which is sufficient for our experiments.

In the fixed WiMAX testbed, we have used acticom's ROHC module (version 3) that compresses UDP/IPv4 headers to 6 bytes and RTP/UDP/IPv4 to 9 bytes in most of the packets of a flow generated by the traffic generator introduced above. acticom's ROHC produces small compressed header information. Extra information was added to the headers to use Ethernet connections that were present in VTT's testbed. Small packets in Ethernet are prone to padding and since the length of a compressed packet is known, we appended this information to the packet itself. This approach destroys some of the compression gain. Nevertheless, we expected the gain to be large enough to be able to use some bytes for additional information. In the measurements, we observed compressed header sizes of 6 bytes for headers introduced by RTP/UDP/IPv4. As a payload we used 20 bytes and a packet sending interval of 20 ms which produces a payload bit rate of 8.0 kbps. These values do not try to replicate any VoIP CODEC specific values, but emulate the characteristics of common narrowband VoIP codecs. However, since the packet aggregator for the femto cell scenario is currently under development, we quantified the performance of this femto cell scenario with and without ROHC compression by emulating the ROHC module and the packet aggregator. The emulation is carried out by changing the payload size of the generated UDP/IPv4 packet in the traffic generator according to the aggregation level and ROHC compression efficiency. Figure 10.4 illustrates the emulated system for the transmitting and receiving sides. Packets arrive at the aggregator via the ROHC module or directly, bypassing the ROHC operations, depending on the measurement scheme. The aggregator contains buffer space for packets waiting for bundling operation.

Table 10.2 lists the packet sizes in bytes for different femto cell sizes (aggregation levels) with the noncompressed and compressed schemes used in the measurements. The potential capacity gains achievable by using ROHC are significant, namely more than 50%. Note that we did not measure ROHC with a femto cell size of one because the WiMAX equipment is

Table 10.2 Packet sizes for different femto cell sizes with and without ROHC.

Femto cell users	Without ROHC (bytes)	With ROHC (bytes)
1	60	—
2	120	52
4	240	104
8	480	208
16	960	416

connected to the end-hosts via Ethernet, and the minimum Ethernet frame size is 64 bytes, including 14 bytes of Ethernet header and 4 bytes of frame check sequence. Thus, the benefits of ROHC would have vanished because of padding in the injected packets over the WiMAX link for this specific WiMAX setup. Future measurements and implementations will not suffer from this drawback and real ROHC packets will be transmitted.

10.5   WiMAX and ROHC Measurements Results

For the WiMAX femto cell measurements, packet streams with and without ROHC were tested over the fixed WiMAX link with several different aggregation levels and with different packet sizes, as presented in Table 10.2. The performances of the uplink and downlink were evaluated separately. Each measurement campaign lasted 60 seconds and was repeated three times. The average value of those three repetition runs was calculated and factored into the results.

10.5.1   ROHC on WiMAX Downlink

In Figure 10.5, the measured packet loss rates for the WiMAX downlink are presented. As can be seen, significant performance increases can be achieved employing simple packet aggregation/multiplexing when small packets are forming the majority of the traffic transmitted over the WiMAX link. This observation corresponds with the findings noted in Pentikousis *et al.* (2008b), which also concludes that packet aggregation performed at network layer has significant impact on the total sustained VoIP flow amount in a fixed WiMAX link.

Without aggregating small packets into larger ones or compressing the headers with ROHC, a packet loss rate of 5% is reached already with 255 simultaneous data streams. By aggregating two small packets into a single larger packet, the number of simultaneous data streams increases to 348 before the packet loss rate exceeds 5%. When the 4, 8 and 16 packets are aggregated into one larger packet, the number increases to 385, 408 and 425 simultaneous data streams, respectively. When ROHC is introduced in addition to the packet aggregation, the performance gain is considerably larger. The 348 simultaneous data streams achieved without ROHC for two packets aggregated into one, increases to 510. If each packet transmitted over the WiMAX link with ROHC contains four femto cell packets, the number of simultaneous data streams increases from 385 to 782 before the packet loss rate for the link exceeds 5%. Furthermore, the use of ROHC increases the number of sustained simultaneous

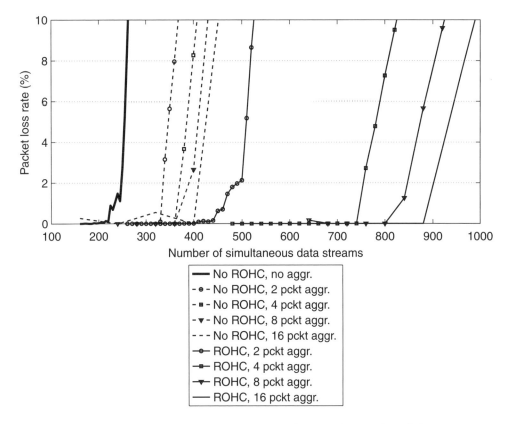

Figure 10.5 Measured packet loss rates for the WiMAX downlink.

data streams from 408 to 875 and from 425 to 935 for 8 and 16 aggregated femto cell packets, respectively.

When converted into femto cell data goodput values, the capacity increase due to aggregation and ROHC becomes even more intuitive. As the case without aggregation and ROHC yields a goodput of 2.04 Mbps, the value increases respectively to 2.78, 3.08, 3.26 and 3.40 Mbps when 2, 4, 8 and 16 packets are aggregated into one larger packet. When ROHC is employed, an even more impressive gain is achieved, as the femto cell data goodput increases to 4.08, 6.26, 7.00 and 7.48 Mbps when 2, 4, 8 and 16 packets are aggregated together, respectively. Clearing other words, using ROHC for the given scenario is even more than doubling the WiMAX cell capacity (even though the full compression of acticom's ROHC is not even enabled). Table 10.3 summarizes the measured femto cell data goodput values and the capacity gain achieved with the use of ROHC.

For all of the values presented above, the packet loss rate of the data streams exceeded 5% before the one-way delay of the WiMAX link crossed the 150 ms limit, which is often defined as the threshold for adequate quality for VoIP conversations. Only after the packet loss rate made the VoIP conversation quality unacceptable, did the one-way delay increase clearly above the 150 ms mark.

Table 10.3 Measured femto cell data goodput values and ROHC gain achieved with ROHC in the WiMAX downlink.

Number of femto cell users	Without ROHC (Mbps)	With ROHC (Mbps)	ROHC gain (%)
1	2.04	—	—
2	2.78	4.08	46.8
4	3.08	6.26	103.2
8	3.26	7.00	114.7
16	3.40	7.48	120.0

Table 10.4 Measured femto cell data goodput values and capacity gain achieved with ROHC in the WiMAX uplink.

Number of femto cell users	Without ROHC (Mbps)	With ROHC (Mbps)	ROHC gain (%)
1	0.88	—	—
2	1.68	1.73	3.0
4	1.82	3.48	91.2
8	1.92	4.17	113.1
16	1.88	4.41	134.6

10.5.2 ROHC on WiMAX Uplink

Similar to the preceding investigation of the ROHC gain for the WiMAX downlink, the ROHC performance for the uplink is evaluated in the following. Figure 10.6 illustrates the measured packet loss values for the WiMAX uplink. Similar behavior to that observed for the downlink can be seen, as the packet aggregation leads to large performance gains and employing ROHC in addition improves these values significantly.

The WiMAX uplink sustains 110 simultaneous data streams from the femto cells before the packet loss rate increases above 5%. Packet aggregation improves this value to 210, 228, 240 and 235 when 2, 4, 8 and 16 packets are aggregated into one packet, respectively. With ROHC, these numbers increase further to 216, 435, 521 and 551 by aggregating 2, 4, 8 and 16 packets, respectively.

If the goodput values resulting from the number of simultaneous data streams are observed for the WiMAX uplink, it can be seen that only 880 kbps femto cell data goodput is achieved if no aggregation or ROHC is used. By aggregating 2, 4, 8 and 16 packets together, the goodput value increases to 1.68, 1.82, 1.92 and 1.88 Mbps, respectively. If ROHC is used in addition, these values respectively increase to 1.73, 3.48, 4.17 and 4.41 Mbps for the aggregation of 2, 4, 8 and 16 packets. Table 10.4 summarizes the measured goodput values and presents the capacity gain achieved by using ROHC. The small difference between the plain aggregation and aggregation with ROHC for two femto cell packets aggregated into one larger packet is caused by the impairments in the used WiMAX equipment.

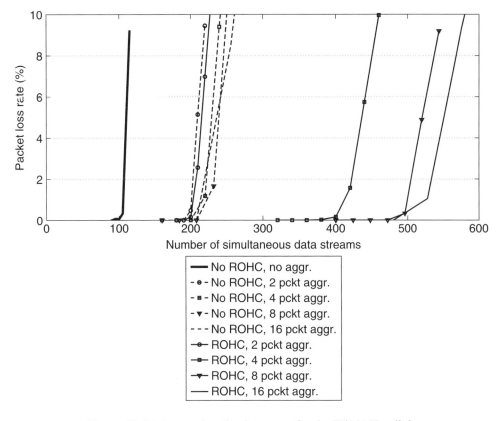

Figure 10.6 Measured packet loss rates for the WiMAX uplink.

Similarly, as explained in the context of the downlink measurement results, in the uplink the one-way delay stays under 150 ms when the packet loss rate is within acceptable limits. Only after the packet loss rate by itself would result in unacceptable VoIP conversation quality, did the one-way delay increase considerably.

10.5.3 ROHC Capacity Gain

Figure 10.7 illustrates the !ROHC capacity gain versus different aggregation levels for uplink and downlink. In the downlink, the goodput capacity gain achieved with ROHC increases already by 50% for a femto cell size of two. With higher aggregation levels the capacity effectively more than doubles. In the uplink, we measured only a 3% capacity gain for an aggregation level of two when using ROHC. For small aggregation levels, the effectively experienced gain in the uplink seems to be relatively small for the WiMAX hardware used in this setup. The underlying reason for this behavior is assumed to be the MAC scheduling efficiency and its limitations. When packet sizes grow according to larger aggregation levels, we started to observe significant gains with ROHC also in the uplink. For a femto cell size of 4 the gain is already more than 90%, whereas a femto cell size of 16 produces the largest

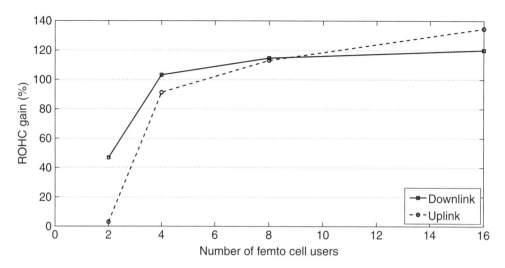

Figure 10.7 ROHC capacity gain for uplink and downlink.

gain of 134.6%. However, nearly the full ROHC gain can be achieved with four aggregated traffic flows. In some scenarios, one user can already have multiple traffic flows. Thus, we can expect aggregation levels larger than four.

10.6   Conclusion

This chapter has advocated the use of ROHC in WiMAX systems. The examples shown in this chapter focused on femto cells that are supported by overlay WiMAX networks. Owing to the impairments in the used WiMAX equipment, a combination of real measurements and emulated traffic was employed. In the future, the fixed WiMAX system will be replaced by a mobile WiMAX system, which is considerably newer and will even favor ROHC for single-stream traffic. Nevertheless, the measurement results show a clear capacity gain using ROHC over systems that operate with uncompressed headers. For the downlink the measured capacity gain was 120% for 16 aggregated femto cell users. For the uplink, the capacity gain even increased to 134%. In our future work, we will replace the fixed WiMAX testbed with a mobile WiMAX testbed. In that testbed ROHC is investigated for single user traffic (no aggregation). This new system, supporting smaller packet sizes, will favor acticom's ROHC implementation even more.

As ROHC has shown its potential in WiMAX systems for the outlined scenario, the future work by VTT and acticom will focus on the integration of ROHC into mobile WiMAX. In the mobile scenario, services such as:

- web browsing;

- gaming;

- voice services; and

- multimedia applications;

will be used to investigate the ROHC gain. For all of these services a large gain in capacity is expected using ROHC.

References

acticom GmbH (2008a). acticom shows RoHC enabled Nokia N810 Internet Tablet, http://acticom.de/en/news/2008-05-26/ (accessed 15 July 2008).

acticom GmbH (2008b). RoHC enabled 802.16 WiMAX network, http://acticom.de/en/news/2008-01-31/ (accessed 15 July 2008).

Barr, G. (2008) IO::Socket::INET – Object interface for AF_INET domain sockets, CPAN. http://search.cpan.org/~gbarr/IO-1.2301/IO/Socket.pm (accessed 15 July 2008).

Bormann, C. *et al.* (2001) RObust Header Compression (ROHC): Framework and four profiles: RTP, UDP, ESP, and uncompressed. Request for Comments *3095* IETF.

Correll, K., Barendt, N. and Branicky, M. (2005) Design considerations for software only implementations of the IEEE 1588 Precision Time Protocol. *Proceedings of the Conference on IEEE 1588.*

Corell, K. (2008) PTP daemon (PTPd), http://ptpd.sourceforge.net/ (accessed 15 July 2008).

Fitzek, F.-P., *et al.* (2004) Header compression schemes for wireless Internet access. *Wireless Internet.* CRC Press, Boca Ratan, FL.

Humfrey, N. J. (2008) Net::RTP – Send and receive RTP packets (RFC3550), CPAN. http://search.cpan.org/~njh/Net-RTP-0.09/lib/Net/RTP.pm (accessed 15 July 2008).

International Electrotechnical Commission (2004) Precision Clock Synchronization Protocol for Networked Measurement and Control Systems, *IEC 61588:2004(E), IEEE 1588-2002(E),* September 2004.

Pentikousis, K., *et al.* (2008a) An experimental investigation of VoIP and video streaming over fixed WiMAX. *Proceedings of the 4th International Workshop on Wireless Network Measurements (WiNMee),* March 2008.

Pentikousis, K., *et al.* (2008b) Empirical evaluation of VoIP aggregation over a fixed WiMAX testbed. *Proceedings of the 4th International Conference on Testbeds and Research Infrastructures for the Development of Networks and Communities (TRIDENTCOM),* March 2008.

Piri, E., *et al.* (2008) ROHC and aggregated VoIP over fixed WiMAX: an empirical evaluation. *Proceedings of the IEEE Symposium on Computers and Communications (ISSC),* May 2008.

Wegscheid, D., *et al.* (2008) Time::HiRes – High resolution alarm, sleep, gettimeofday, interval timers, CPAN. http://search.cpan.org/~jhi/Time-HiRes-1.9715/HiRes.pm (accessed 15 July 2008).

11

A WiMAX Cross-layer Framework for Next Generation Networks

Pedro Neves, Susana Sargento, Ricardo Matos,
Giada Landi, Kostas Pentikousis, Marília Curado and
Francisco Fontes

11.1 Introduction

Ubiquitous Internet access is one of the main challenges for the telecommunications industry. The number of users accessing the Internet is growing at a very fast pace. At the same time, the average customer uses more than one device to connect to the Internet, and downloads and uploads digital media of an unprecedented magnitude. The network access paradigm of 'always connected, anytime, anywhere' is a central requirement for Next Generation Networks (NGNs).

Such a requirement places a tall order to operators that ought to find ways to provide broadband connectivity to their subscribers independently of their location and access device. Furthermore, the popularity of high-bandwidth services (including those arising from social networking sites) and other demanding multimedia applications is expected to continue to increase. IEEE 802.16 (IEEE, 2004, 2005b) and the WiMAX Forum network architecture extensions (WiMAXForum, 2008) provide an attractive solution for this type of NGN environments. WiMAX is a Point-to-Multipoint (PTMP) technology, providing high throughputs, and it is oriented for Wireless Metropolitan Area Networks (WMANs). The built-in Quality of Service (QoS) functionalities through the use of unidirectional connections

WiMAX Evolution: Emerging Technologies and Applications   Edited by Marcos D. Katz and Frank H.P. Fitzek
© 2009 John Wiley & Sons, Ltd

and service flows between Base Stations (BSs) and Subscriber Stations (SSs) are also an important feature provided by this wireless technology.

Another aspect of NGN is the seamless integration of heterogeneous network technologies. Future network architectures will provide seamless QoS support, mobility and security, among other features, which are crucial for the success of the future networks. Taking into account the convergence scenario envisioned in the telecommunications area, it is essential that different access technologies, wired and wireless, are able to work together, allowing mobile users to Handover (HO) between them seamlessly. In this sense, in order to integrate WiMAX technology (WiMAXForum, 2008), in next generation environments, one needs to support a cross-layer framework that enables seamless communication between WiMAX and other access technologies, as well as with the QoS, security and mobility management protocols.

IEEE has been working on a standard which enables Media Independent Handovers (MIHs) (IEEE, 2008). The IEEE 802.21 standard (IEEE, 2008) is expected to be ratified during 2008 and its main objective is to assist in optimizing mobility processes through a set of services and interfaces that can be used in a standard way for different wired and wireless technologies and higher-layer mobility management protocols. MIHs and radio detail abstraction mechanisms have been explored in the research literature for some time and with IEEE 802.21 finalized it is expected that these concepts will play a dominant role in the integration of different technologies in future networks.

In this chapter we present a cross-layer framework that seamlessly integrates WiMAX in heterogeneous NGNs. We concentrate on QoS and mobility aspects and explain how Next Steps in Signaling (NSIS) (Hancock *et al.*, 2005) and IEEE 802.21 can be taken advantage of and integrated with the specifics of WiMAX technology in terms of service flow QoS provisioning and mobility management messages. This framework allows for standards-based end-to-end QoS support and seamless mobility management. The foundation of this framework is the architecture designed in the European WEIRD project (Guainella *et al.*, 2007). A prototype implementation of this framework has already been evaluated in the project's testbeds.

The chapter is organized as follows. Section 11.2 provides an overview of the WiMAX architecture and Section 11.3 introduces our proposed cross-layer framework for the integration of WiMAX networks in next generation networks. Section 11.4 presents a case study based on the WEIRD architecture; results from this study are given in Section 11.5. Finally, Section 11.6 concludes this chapter outlining future work items.

11.2   IEEE 802.16 Reference Model

The IEEE 802.16 standard (IEEE, 2004, 2005b) reference model comprises the data, control and management planes. The data plane protocol stack is illustrated in Figure 11.1. It includes the Physical (PHY) layer and the Medium Access Control (MAC) layer. Multiple physical layers are supported, operating in the 2–66 GHz frequency spectrum, and support single and multi-carrier air interfaces, depending on the particular operational environment.

The MAC layer is connection-oriented and provides QoS assurances through the use of service flows and scheduling services. A connection is defined as a unidirectional mapping between the BS and the SS/Mobile Station (MS) MAC layers for transporting a service flow's

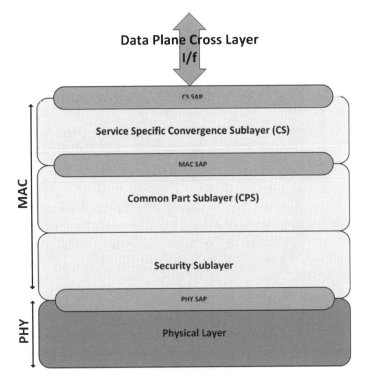

Data Plane Cross Layer I/f

CS SAP

Service Specific Convergence Sublayer (CS)

MAC SAP

Common Part Sublayer (CPS)

Security Sublayer

PHY SAP

Physical Layer

MAC

PHY

Figure 11.1 IEEE 802.16 data plane.

traffic. Each connection is identified by a unique Connection Identifier (CID). The MAC layer comprises three internal sublayers, namely, the Service Specific Convergence Sublayer (CS), the Common Part Sublayer (CPS) and the Security sublayer.

The CS resides on top of the MAC CPS and provides a set of convergence sublayers that map the upper layer packets into the 802.16 MAC Protocol Data Units (PDUs). CS is responsible for accepting higher layer MAC Service Data Units (SDUs) handed down through the CS Service Access Point (SAP), classifying them to the appropriate CID, and delivering the classified packets to the MAC CPS through the MAC SAP. The classifier is based on a set of packet matching criteria, which are applied to each packet. The pattern-matching criteria are based on protocol-specific fields, such as IP and MAC layer addresses, a classifier priority, and a reference to a particular CID. Each connection has a specific Service Flow (SF) associated with it that provides the necessary QoS requirements for every packet in the flow. If no pattern matches for a given packet, then a default action, depending on the equipment configuration, must be taken. For example, the packet can be discarded, sent through a default connection, or a new connection can be established for it, if enough resources are available. Note that the downlink and uplink classifiers, applied by the BS and SS, respectively, may be different. Figure 11.2 illustrates the downlink classification process.

The IEEE 802.16-2004 standard (IEEE, 2004) defines two general CSs for mapping services to and from the 802.16 MAC connections: the packet-convergence sublayer and the

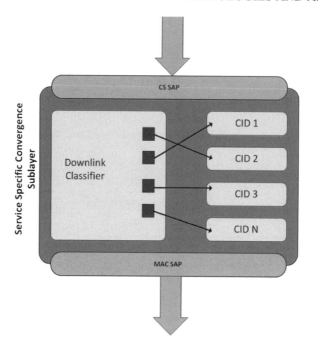

Figure 11.2 IEEE 802.16 downlink classification process.

Asychronous Transfer Mode (ATM) convergence sublayer. The packet-convergence sublayer is defined to support packet-based protocols, such as IPv4 (ISIUSC, 1981), IPv6 (Deering and Hinden, 1998), 802.1Q (IEEE, 2006) and 802.3 (IEEE, 2005a).

CPS is the intermediate MAC sublayer that links the CS and Security sublayers. CPS receives packets arriving from the CS through the MAC SAP and is responsible for a set of functions related to system access, such as bandwidth allocation, connection establishment and management, addressing, construction and transmission of the MAC PDUs, scheduling services management and contention resolution. Finally, the Security sublayer provides authentication, secure key exchange and encryption.

In an effort to provide a clear integration path of IEEE 802.16 in an IP-based network, the IEEE 802.16 working group specified the IEEE 802.16g standard (IEEE, 2007), an amendment to IEEE 802.16-2004 (IEEE, 2004). Recall that IEEE 802.16-2004 and IEEE 802.16e-2005 (IEEE, 2005b) define data plane functionalities. IEEE 802.16g defines the control and management plane functionalities, enabling interoperability with higher layers as well as efficient management of the network resources, QoS and mobility.

IEEE 802.16g specifies the Network Control and Management System (NCMS) entity, an abstraction layer between the IEEE 802.16 MAC and PHY layers and the higher layers of the protocol stack, for control and management purposes. NCMS, as illustrated in Figure 11.3, allows the IEEE 802.16 PHY and MAC layers to be independent of the network architecture. NCMS resides at both BS and MS entities. For control and management purposes, NCMS is the most important entity for cross-layering with the higher layers, whereas CS is responsible for data plane cross-layering with the higher layers.

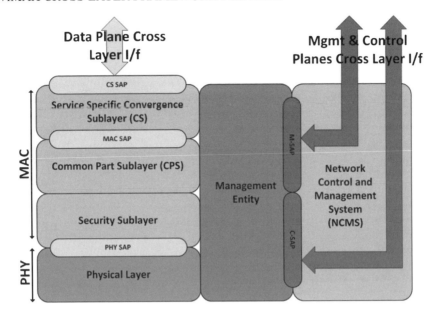

Figure 11.3 IEEE 802.16 reference model (data, control and management planes).

Furthermore, the IEEE 802.16g (IEEE, 2007) standard defines the Control SAP (C-SAP) and the Management SAP (M-SAP), which expose the control and management plane functions to the higher layers, respectively. The M-SAP is used for less time-sensitive management plane functionalities related with system configuration, monitoring statistics, notifications, triggers and multimode interface management. On the other hand, the C-SAP is used for time-sensitive control plane functionalities such as HOs, Mobility Management, Radio Resource Management, MIH Function Services and Service Flow (SF) Management (SFM).

Figure 11.3 highlights the different paths for the control and data planes. For control and management plane functionalities, the higher layers communicate with the 802.16 system through the NCMS, using the M-SAP and the C-SAP. With respect to data plane functions, the communication with the upper layers is established through the CS and CS SAPs.

To sum up, NCMS provides a range of different services for cross-layering with the higher layers. For example, it provides Authentication, Authorization and Accounting (AAA), Security, QoS (SFM), Multicast and Broadcast Management, Mobility Management as well as MIH Function Services, among others.

11.3   Cross-layer Design for WiMAX Networks

11.3.1   Cross-layer Mechanisms for QoS Support

IEEE 802.16 intrinsically supports QoS by using a connection oriented approach, based on SF and scheduling services. In the 802.16-2004 standard (IEEE, 2004), a SF is defined as a MAC transport service that provides unidirectional packet transport either for uplink or

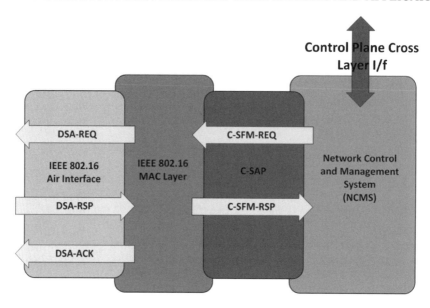

Figure 11.4 QoS (SFM service) in IEEE 802.16.

downlink. All packets traversing the MAC interface are associated with a SF, identified by a CID with some predefined treatment, and assigned resources for the duration of the connection. Several types of connections may be established in the 802.16 system, in particular, management, broadcast, multicast and transport connections, with associated QoS parameters.

Five scheduling services, associated with each connection during the system setup, are defined to meet different QoS needs: (i) unsolicited grant service, which supports real-time service flows that generate fixed size data packets on a periodic basis, such as Voice over IP (VoIP); (ii) extended real-time polling service, which supports real-time service flows that generate fixed size data packets on a periodic basis, but the resources allocation are dynamic; (iii) Real-Time Polling Service (rtPS) that supports real-time service flows with variable sized data packets on a periodic basis, such as video; (iv) Non-Real-Time Polling Service, which supports non-real-time service flows that require variable size data grants on a regular basis, such as high-bandwidth File Transfer Protocol (FTP); and (v) Best Effort, which provides efficient service to best-effort traffic, without throughput or delay guarantees.

SF activations, modifications or deletions can be initiated from both the BS and/or the SS/MS, with a three-way handshake. For SF creation, three MAC management messages are used. First, a Dynamic Service Addition Request (DSA-REQ) message is sent by the BS to the SS to request the allocation of a new SF, including the QoS parameters. Then, a Response (DSA-RSP) message is sent by the SS/MS as a response, indicating whether the requested QoS parameters are supported. Finally an Acknowledgment (DSA-ACK) is sent by the BS to the SS to confirm and acknowledge the reception of the DSA-RSP. The MAC management messages used for the SF establishment process are illustrated in Figure 11.4.

When dynamic SF modification is required, a similar set of messages is used: Dynamic Service Change Request (DSC-REQ), Dynamic Service Change Response (DSC-RSP) and Dynamic Service Change Acknowledgment (DSC-ACK). Finally, the group of messages used to delete service flows are Dynamic Service Deletion Request (DSD-REQ), Dynamic Service Deletion Response (DSD-RSP) and Dynamic Service Deletion Acknowledgment (DSD-ACK).

In order to trigger QoS procedures in a IEEE 802.16 system, NCMS provides the SFM service. The SFM service is composed of a set of primitives for supporting QoS management between the BS and SS/MS. The interaction between the NCMS and the 802.16 QoS system is done through the C-SAP.

The SFM service is based on a two-way handshake, shown on the right-hand side of Figure 11.4, based on following two primitives. First, a Control Service Flow Management Request (C-SFM-REQ) primitive is sent by the NCMS to the 802.16 MAC system to start a SFM procedure, such as SF creation. Effectively a C-SFM-REQ triggers the DSA-REQ MAC message, described earlier, or a modification (triggers a DSC-REQ MAC message), or deletion (triggers a DSC-DEL MAC message). Second, a Control Service Flow Management Response (C-SFM-RSP) primitive is sent back as a response to the requested SFM procedure (based on a DSA-RSP, DSC-RSP or DSD-RSP MAC messages, respectively).

A SF reservation process in the IEEE 802.16 system is shown in Figure 11.4, illustrating both the SFM primitives (C-SFM-REQ and C-SFM-RSP) in the C-SAP interface, as well the corresponding MAC management messages (DSA-REQ, DSA-RSP and DSA-ACK) in the IEEE 802.16 air interface.

Two main approaches have been proposed to achieve QoS support at the IP level, namely, the integrated services (Braden *et al.*, 1994) and the differentiated services (Blake *et al.*, 1998). While the former has shown scalability problems, the later is used in networks to provide qualified applications the service level that is adequate for their requirements. In order to provide different levels of service, the differentiated services framework includes a set of Per Hop Behaviors (PHBs), namely, Expedited Forwarding (EF) (Jacobson *et al.*, 1999), Assured Forwarding (AF) (Heinanen *et al.*, 1999) and best effort. The AF PHB is further divided into four subclasses with different requirements for buffer space and bandwidth.

End-to-end QoS support needs to resort to a signaling protocol in order to convey the resource reservation requests along the network. The NSIS framework (Hancock *et al.*, 2005) has been conceived in order to support network signaling, in general, with QoS signaling as its first application. The NSIS framework comprises two layers, namely, the NSIS Transport Layer Protocol (NTLP) and the NSIS Signaling Layer Protocol (NSLP). The NTLP layer, also known as General Internet Signaling Transport (GIST) (Schulzrinne and Hancock, 2008) is responsible for the transport of the signaling messages sent by NSLPs. The NSLP layer is specific to each application. The first NSLP defined, named QoS-NSLP, was designed to provide resource reservation signaling support (Manner *et al.*, 2008).

The QoS parameters specific for the QoS model of each context are defined on the QoS Specification (QSPEC) object defined in the NSIS framework (Ash *et al.*, 2006). Depending on the context, the QoS parameters may vary. In order to establish an end-to-end resource reservation which crosses different domains and technologies, there is a need to map between the QoS parameters associated with each context. This function of QSPEC mapping may be performed horizontally and vertically. In the first case, the mapping is done between the QoS models of different domains. In the second case, also known as cross-layer, the mapping is performed between two different layers of the protocol stack, for instance, between the IP layer and the network access technology dependent layer.

By using the NSIS framework (Hancock *et al.*, 2005) for end-to-end network signaling, and specifically, for resource reservation, we have to translate the IP QoS information conveyed by QoS-NSLP into the specific QoS parameters of the IEEE 802.16 technology. Although the IEEE 802.16g NCMS provides the C-SAP interface to manage IEEE 802.16 QoS, it does not define how to convert the generic QoS parameters from the upper layer entities and protocols to the IEEE 802.16-specific QoS parameters and scheduling services. Therefore, we define a new entity responsible for abstracting and translating the generic QoS parameters to IEEE 802.16 specific QoS parameters. Section 11.4 presents a QoS architecture that includes this abstracting entity, which we call the resource controller.

11.3.2   Cross-layer Mechanisms for Seamless Mobility Optimization

IEEE 802.16e-2005 standard (IEEE, 2005b) defines a framework for supporting mobility. Three HO methods are supported in IEEE 802.16e-2005, only one of which is mandatory; the remaining two are optional. The mandatory HO method is called the Hard Handover (HHO), or otherwise known as 'break-before-make HO' and it is the only HO type required to be implemented by mobile WiMAX. HHO implies that there may be an abrupt transfer of connection from one BS to another. The two optional HO methods defined in IEEE 802.16e-2005, which both fall under the category of 'make-before-break' or soft HOs, are Fast Base Station Switching (FBSS) and Macro Diversity Handover (MDHO). In FBSS and MDHO, the MS maintains a valid connection simultaneously with more than one BS and the connection with the target BS starts before service disconnection with the previous serving BS occurs.

HOs can be either initiated by the MS, Mobile Initiated Handover (MIHO), or by the BS, Network Initiated Handover (NIHO). Four MAC management messages are defined in the 802.16e-2005 standard (IEEE, 2005b) for integrating mobility support, including both MIHO and NIHO: Mobility Mobile Station/Base Station Handover Request (MOB-MSHO-REQ and/or MOB-BSHO-REQ), Mobility Base Station Handover Response (MOB-BSHO-RSP) and Mobility Handover Indication (MOB-HO-IND).

For a MIHO, the MS starts by sending a MOB-MSHO-REQ message to the serving BS with a list of the possible target BSs. The serving BS contacts the target BSs over the backbone network to check whether they have enough resources to support the requested QoS parameters for the HO. Note that the backbone communication between the serving and the target BSs is left undefined by IEEE 802.16. After receiving the response from the target BSs, the serving BS summarizes the results obtained from the target BSs and informs the MS using the MOB-BSHO-RSP MAC management message. The MOB-BSHO-RSP message includes a recommended list of the target BSs that can effectively support the MS handover. Finally, the MS selects the target BS and notifies the serving BS using the MOB-HO-IND message. Figure 11.5 illustrates the MIHO scenario, including the MAC management messages.

With respect to the NIHO scenario, the serving BS informs the MS that it is going to be switched to another BS using the MOB-BSHO-REQ message. Thereafter, if the MS is able to make the HO to one of the recommended BSs, it replies with the MOB-HO-IND message, indicating its commitment to the handover.

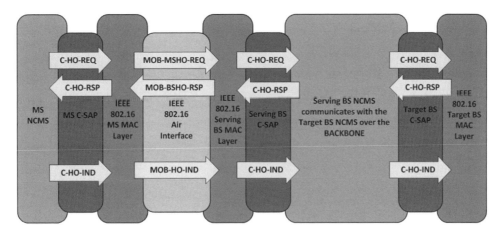

Figure 11.5 MM service in IEEE 802.16 without backbone communication.

For triggering the HO messages in the IEEE 802.16 system, the NCMS provides the Mobility Management (MM) service. The MM service provides a set of control primitives for the HO process, through the C-SAP, that support the HO procedures between the BS and the MS. The MM Service HO control primitives are based on a three-way handshake, based on the following primitives. First, a Control Handover Request (C-HO-REQ) primitive is used by the IEEE 802.16 system or NCMS to start a MIHO (triggering a MOB-MSHO-REQ MAC message) or a NIHO (triggering a MOB-BSHO-REQ MAC message). Then, a Control Handover Response (C-HO-RSP) primitive is used by the IEEE 802.16 system or by the NCMS to respond to the HO request (triggering the MOB-BSHO-RSP MAC message). Third, the Control Handover Indication (C-HO-IND) primitive is used to explicitly notify that HO execution (triggered by the MOB-HO-IND), cancellation or completion.

A MIHO process is illustrated in Figure 11.5, showing both the MM primitives (C-HO-REQ, C-HO-RSP and C-HO-IND) in the C-SAP interfaces and the mobility MAC management messages (MOB-MSHO-REQ, MOB-BSHO-RSP and MOB-HO-IND) through the IEEE 802.16 air interface.

As discussed earlier, IEEE 802.16e-2005 (IEEE, 2005b) specifies the communication between the BS and the MS using the MAC management messages shown in Figure 11.5. On the other hand, IEEE 802.16g (IEEE, 2007) specifies the control primitives to manage the HOs, using the NCMS MM primitives (also shown in Figure 11.5). Nevertheless, neither 802.16e nor 802.16g specify the communication between the serving and the target BSs in the backbone network. This communication link is very important to transfer the context information between the serving and the target BSs. To fill this gap, the WiMAX Forum has specified a network protocol that establishes communication between the serving and the target BSs. Three messages have been defined: (i) a Handover Request (HO-REQ) message is sent by the serving BS to inform the target BSs that the MS is requesting a HO (includes the list of QoS parameters required by the MS service flows); (ii) a Handover Response (HO-RSP) message is sent by the target BSs back to the serving BS announcing whether the required QoS parameters are available; and (iii) a Handover Confirmation (HO-CNF)

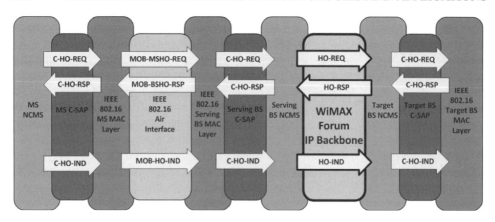

Figure 11.6 MM service in IEEE 802.16 including WiMAX Forum backbone communication.

message is sent by the serving BS to alert the chosen target BS for the MS handover. This three-way handshake is illustrated in Figure 11.6.

In order to optimize mobility procedures, it is important to integrate cross-layer mechanisms and provide the higher layer MM protocols with the required link layer information. For example, in order to trigger the HO process, MM protocols should obtain information about the current link states and available networks in order to improve their decision-making process. The IEEE 802.21 draft standard (IEEE, 2008), also known as MIH, defines an abstract framework that optimizes and improves horizontal and vertical handovers by providing information about the link layer technologies to the higher layers.

IEEE 802.21 (IEEE, 2008) introduces a new entity called the MIH Function (MIHF), which hides the different link layer technology specificities from the higher layer entities. A uniform interface, called Media Independent Handover Link Service Access Point (MIH-LINK-SAP), is defined to establish the communication between the link layers and the MIHF. The higher layer entities, also known as MIH Users (MIHUs) have access to the link layers information through a standardized uniform interface called the Media Independent Handover Service Access Point (MIH-SAP). Several higher layer entities (MIHUs) can take advantage of the MIH framework, including, for example, MM protocols, such as Mobile IP (MIP) (Johnson *et al.*, 2004, Koodli, 2005, Perkins, 1996), and Session Initiation Protocol (SIP) (Rosenberg *et al.*, 2002), as well as the Mobility and QoS Decision Algorithms.

Figure 11.7 illustrates the 802.21 MIH platform including the MIHU, MIHF and link layer entities, as well as the MIH-LINK-SAP and MIH-SAP standardized uniform interfaces.

In order to detect, prepare and execute the HOs, the MIH platform provides three services: Media Independent Event Service (MIES), Media Independent Command Service (MICS) and Media Independent Information Service (MIIS).

MIES provides event reporting such as dynamic changes in link conditions, link status and link quality. The events may be either local or remote and they may originate from MAC, PHY or MIH Function either at the MS or at the network Point of Attachment (PoA). Multiple higher layer entities may be interested in these events at the same time, so these events may need to be sent to multiple destinations.

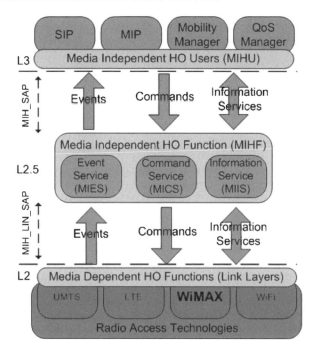

Figure 11.7 IEEE 802.21 MIH framework.

IEEE 802.21 defines two types of events: link events and MIH events. Link events originate from the lower layers and then are propagated by the MIHF to registered MIHUs. Events that propagate from the MIHF to the MIHUs are called MIH events. MIH events may be local or remote. Local MIH events may be propagated across different layers within the local protocol stack, whereas the remote MIH events traverse across the network to a peer MIHU.

MICS enables MIHUs to control the physical, data link and logical link layer. The higher layers may utilize MICS command services to determine the status of links and/or control a multi-mode terminal. Furthermore, MICS may also enable MIHUs to facilitate optional HO policies. The commands are classified in two categories: MIH commands and link commands. MIH commands originate from the higher layer(s) addressing the MIHF, and they may be local or remote. Local MIH commands are sent by higher layers to the MIHF in the local protocol stack. Remote MIH commands are sent by higher layers to a MIHF in the peer stack. Link commands originate from the MIHF and sent to the link layers. These commands mainly control the behavior of the link layer entities.

Finally, MIIS provides a framework by which a MIHF located in the MS or in the network side may discover and obtain network information within a geographical area to facilitate handovers. The objective is to acquire a global view of all of the heterogeneous networks in the area in order facilitate seamless HOs when roaming across these networks.

Next, we discuss how we employ the IEEE 802.21 framework in a novel WiMAX cross-layer design solution.

11.4 WEIRD: A Practical Case of WiMAX Cross-layer Design

The WiMAX Extension to Isolated Research Data networks (WEIRD) project aimed at enhancing the WiMAX technology through the seamless integration of WiMAX-based access networks into end-to-end IP architectures, which typically include heterogeneous domains with various network technologies. The prototypes of the WEIRD system have been deployed in four different testbeds located in Finland, Italy, Portugal and Romania connected through the European research network GÉANT2 and the National Research and Educational Networks (NRENs). Through the WEIRD system, isolated and remote areas can be reachable by exploiting the WiMAX technology, with support for complex and heterogeneous application scenarios, characterized by strict requirements in terms of QoS, security and mobility.

The WEIRD approach is based on cross-layer mechanisms involving the application, control and transport layers and the development of convergence layers that enable full interaction among the different planes. This solution allows the interoperability with different underlying technologies (IEEE 802.16d/e) at the transport plane and, at the same time, support for a large set of applications characterized by different QoS requirements and signaling capabilities that can exploit and take advantage of the QoS and mobility features assured by the WiMAX technology. Figure 11.8 provides an illustrative overview of the WEIRD architecture, highlighting the strong interactions between the lower layers that characterize the network technologies and the higher layers of the control and application planes.

The WiMAX Forum is currently extending the IEEE 802.16 architecture, by defining the Network Reference Model (NRM) (WiMAXForum, 2008). NRM is a logical representation of a WiMAX network and its main goal is to guarantee interoperability between distinct WiMAX vendor offerings. The WEIRD control plane architecture, illustrated in Figure 11.9, is fully compliant with the WiMAX Forum guidelines, and in particular with NRM. It is composed of a set of standardized interfaces, also known as WiMAX Reference Points, and by three functional entities: the Connectivity Service Network (CSN) which contains the core network entities, such as the DHCP, DNS, AAA and SIP servers, and establish connectivity with the IP backbone; the Access Service Network (ASN) comprising several WiMAX BSs connected to multiple ASN Gateways (ASN-GWs), which are responsible for establishing connectivity with the CSN; and the MS, terminal equipment responsible for establishing radio connectivity with the WiMAX BS.

The WEIRD control plane mechanisms provide features for both long/medium-term resource management and short-term resource control in the WiMAX access segments through the interaction with the WiMAX transport plane, based on a common technology-independent module and specific hardware-dependent drivers for device configuration. The full integration with the end-to-end IP architecture is assured through the seamless interaction between the WiMAX access and connectivity network and the core network, characterized by different underlying technologies. Communications between the WiMAX control plane modules, in charge of the ASN and CSN configuration, and the control plane of the external core network is based on the multi-domain NSIS signaling protocol, providing a comprehensive solution for end-to-end QoS and mobility.

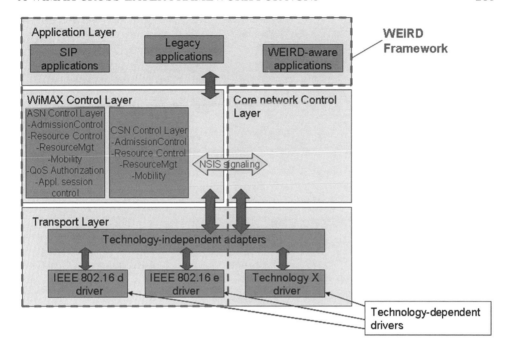

Figure 11.8   WEIRD architecture overview: application, control and transport layers interactions.

The integration of the application and the control plane allows the dynamic configuration of WiMAX network resources according to the actual requirements of the active services. The resource allocation on the rest of the end-to-end path is delegated to the control plane modules located in the external networks. The interface between the WEIRD framework and the heterogeneous external domains is based on NSIS signaling, in particular on the QoS NSLP protocol. This approach provides the required guarantees for a coherent QoS signaling along various networks exporting different transport technologies and resource control mechanisms, characterized by specific QoS metrics.

The WEIRD system supports a wide set of applications ranging from legacy applications without signaling capabilities to WEIRD-aware applications that can be developed from scratch or existing applications that can be updated in order to directly interact with the WEIRD control plane and SIP applications with their application-level signaling. The complete integration between the control plane infrastructure, strictly related to the WiMAX network architecture, and the application plane allows the dynamic reconfiguration of the wireless link through the creation and modification of SFs that fit the profile of the network traffic generated by the applications. In fact, the WEIRD control plane is able to interact with different application signaling schemes and to translate various types of application traffic descriptions into the QoS metric used in the 802.16 domain, based on the main SF parameters (scheduling class, QoS attributes and classifiers). This flexibility allows the WEIRD framework to adapt itself to a variety of application scenarios, characterized by specific signaling procedures.

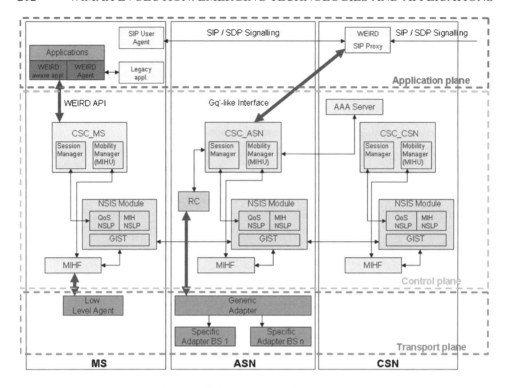

Figure 11.9 WEIRD architecture.

L2 and L3 MM is assured at the control plane level, through the interaction with the WiMAX lower layers using the IEEE 802.21 protocol. The PHY layer information is retrieved from the WiMAX devices and distributed to all relevant WEIRD control plane modules through the exchange of MIH messages supported by the MIH NSLP protocol (Cordeiro *et al.*, 2008). The low-level information is used to manage the HOs following a Make Before Break (MBB) approach, so that the WiMAX resources can be preallocated in the wireless segment between the MS and the target BS. The related session contexts are updated at the control plane level, so that the HO procedure is transparent for the applications.

11.4.1   WEIRD Architecture

At the control plane, each segment is managed by a module called Connectivity Service Controller (CSC). CSCs represent the main coordination points of the WEIRD infrastructure and control all procedures concerning the application sessions and the HOs in their related segment. The communication between the three existing instances of CSC (CSC-MS, CSC-ASN and CSC-CSN) is based on NSIS signaling.

11.4.1.1   WEIRD Architecture and QoS-related Procedures

As shown in Figure 11.9, the interaction between the application plane and the control plane follows two different approaches according to the application type. For legacy and

WEIRD-aware applications the WEIRD Application Programming Interface (API) allows the CSC-MS to retrieve all information about the current services, such as traffic type, required bandwidth, maximum supported latency and jitter, classifiers and authorization data. In this case, CSC-MS is the main coordinator and the initiator of the end-to-end QoS NSIS signaling. CSC-MS translates the application traffic description into an initiator QSPEC based on the WiMAX QoS model and initiates the NSIS signaling towards the core network. The QoS NSLP messages are intercepted by each NSIS node along the end-to-end path, where the QSPEC is processed by the related Resource Management Function (RMF) and the corresponding resources are allocated in the specific segment, according to the QoS attributes included in the QSPEC. In particular, CSC-ASN, located at the ASN-GW, has the role of the RMF for the WiMAX access network, while CSC-CSN acts as RMF for the CSN.

On the other hand, the QoS signaling for SIP applications is based on the network-initiated approach, following the IP Multimedia Subsystem (IMS) model (3GPP, 2008). An enhanced SIP proxy located at the CSN intercepts the incoming SIP messages from the SIP user agents and interacts with the CSC-ASN. This interface is based on the Gq' specification (Calhoun and Loughney, 2003): the description of the SIP session is retrieved from the SDP messages and all of the information concerning media types, bandwidth and classifiers is mapped in a diameter (Calhoun and Loughney, 2003) Authentication Authorization Request (AAR) message and sent to the CSC-ASN. In this case the CSC-ASN has a dual role: it is responsible for the resource allocation in the WiMAX segment and acts as NSIS initiator for the end-to-end QoS signaling through the core network, creating the initiator QSPEC.

The tight coordination between the SIP signaling at the application layer, and the resource reservation and the QoS NSIS signaling at the control plane allows the WEIRD system to support both QoS-enabled and QoS-assured models for SIP applications. In the former case, the actual bandwidth allocation along the end-to-end path does not have an impact on the procedures for the call setup, but only on the quality of the audio/video for that call. The signaling at the application layer and at the control layer are two distinct, parallel processes: the NSIS signaling is only triggered by SIP signaling, and then they can both proceed without interfering with each other. On the other hand, in the QoS-assured model the successful resource allocation along the full path is a precondition for the establishment of the SIP session and the two procedures must be coordinated at the ASN.

While QoS signaling can be managed following the host-initiated or the network-initiated model, the resource allocation for the WiMAX segment is always handled through the network-initiated approach and it is coordinated at the ASN by the CSC-ASN and the Resource Controller (RC). This choice allows the WEIRD system to adopt the procedures for the BS-initiated SFM, defined as mandatory in the IEEE 802.16 specifications and widely supported by different vendor hardware.

At the ASN, resource allocation must be authorized through diameter message exchanges between the CSC-ASN and the AAA server located in the CSN, following the specifications of the diameter QoS application. This procedure includes two different phases: first the user is authenticated through the user credentials conveyed by the NSIS signaling and thereafter the resource utilization is authorized, according to the user profile and the QSPEC specifications.

The actual resource reservation in the wireless segment is handled by the RC (detailed in Neves *et al.* (2008) and Sousa *et al.* (2008)). RC manages all resources in the Radio Access Network (RAN). The WiMAX SF creation and activation can be performed following the one-phase or the two-phase activation model, as defined by the IEEE 802.16 specifications.

Each SF is firstly created with the provisioned status allowing efficient resource utilization. When required by new user service sessions, the related SFs can be immediately activated (one-phase activation model) or first admitted and then activated (two-phase activation model). The WEIRD system adopts the latter model for SIP applications: during the procedure for the session setup the SFs are only admitted, and subsequently activated in order to carry the media traffic.

The device configuration for the SF management is handled at the transport layer by the adapter (Nissila *et al.*, 2007) through the Simple Network Management Protocol (SNMP) protocol (Case and Fedor, 1990), using the Management Information Base (MIB) (Case *et al.*, 2002), defined in the IEEE 802.16f specification (IEEE, 2005c). The adapter includes a single module, called the Generic Adapter (GA), that handles the common interface with the RC and one or more Specific Adapters (SAs) to control each specific WiMAX system, as illustrated in Figure 11.9. This solution provides the flexibility required to support a variety of vendor equipment, without any significant impact on the higher layers. New equipment can be added to the WEIRD system through the development of a single hardware-dependent driver supporting a subset of the API exported by the GA.

11.4.1.2   WEIRD Architecture and Mobility-related Procedures

The WEIRD cross-layer mechanisms can be used to provide seamless mobility by processing the lower-layer information retrieved from the MS and BS, as well as by dynamically reconfiguring the resources related to the current application sessions hosted by the MSs involved in the handovers. Lower-layer data about the links status are retrieved on the MS side by a module called a Low Level Agent (LLA). LLA translates link status events to MIH event messages, such as link up, link going down and link down. These messages are handed over to the MIHF modules located at the MS, ASN and CSN using the MIH NSLP protocol and sent to the MM entities residing on each CSC. The various Mobility Managers have the role of MIHUs: they elaborate the information about the imminent HOs in order to coordinate all mobility actions through the reconfiguration of resources on the network segments involved in the HO, the update of the session contexts and the management of the transport layer procedures through MIH command messages.

WEIRD MM is based on the MBB approach for both micro and macro mobility. The resource reservation along the new path is a precondition for the HO, so that when the MS moves towards the target BS it can immediately use the set of SFs required by its active sessions. This approach aims at providing seamless mobility while maintaining the QoS as perceived by the user during the HO. This mechanism requires multiple interactions between the lower layers, the MIHF modules, the mobility managers at different WiMAX entities and Mobile IP.

It should be noted that both the Mobility Managers and the MIP protocol act as MIHUs Users, but at different levels. The former are the main coordinators of the mobility procedures, handling at the same time the user application status, the links status (through the received MIH event messages) and the resource reservations. Mobility managers are the only entities that take active decisions about the HOs. MIP receives only some MIH commands messages originated by the mobility managers and enforces the HO.

WEIRD supports both ASN-anchored (micro-mobility) and CSN-anchored mobility (macro-mobility). In the former case the MS moves between two BSs controlled by the same ASN, while in the latter case serving and target BSs are located on different ASNs and managed by different CSC-ASNs. The procedures for micro-mobility require only the reconfiguration of the wireless and ASN segments. They are managed directly by the CSC-MS (MM) for legacy applications and by the CSC-ASN (MM) for SIP applications. For macro-mobility the reconfiguration involves also the CSN segment and the application context must be transferred from the serving ASN-GW to the target one. In this case, the HO procedures are controlled by the CSC-CSN (MM), with the active cooperation of both the serving and the target CSC-ASN (MM).

In the next section we present a set of measurements related with cross-layer mechanisms collected on a WiMAX testbed.

11.5 WEIRD Framework Performance Evaluation

In this section we present a set of measurements from our testbed performance evaluation of the WiMAX cross-layer framework described in Section 11.4. We start by evaluating the time necessary to establish an end-to-end SF reservation over a fixed WiMAX network. Then, we present a set of tests that evaluate the QoS performance of the implemented system using VoIP and IPTV services.

11.5.1 Cross-layer Signaling Measurements

In order to evaluate the efficiency of the implemented cross-layer mechanisms, it is important to integrate the WiMAX system in a NGN environment, which is able to support real-time services. Therefore, in addition to evaluating the performance of the standalone WiMAX cross-layer mechanisms, it is also important to evaluate the global performance of the mechanisms that interact with the WiMAX cross-layer entities. Hence, this section presents an evaluation of the Layer 3 QoS (L3QoS) framework, composed of the CSCs and the NSIS framework, as well as an assessment of the total end-to-end time required to perform a reservation over the WiMAX system.

11.5.1.1 Implemented Demonstrator and Tests Methodology

The demonstrator implemented to validate and evaluate the WiMAX cross-layering framework is illustrated in Figure 11.10. The testbed is composed of three main parts: the ASN, the WiMAX RAN and the Customer Premises Equipment (CPE). The ASN is generally composed of several gateways (ASN-GWs), which establish connectivity with the core network. Moreover, ASN performs relay functions to the core network in order to establish IP connectivity and Authentication, Authorization and Accounting (AAA) mechanisms. A Correspondent Node (CN) is connected to one of the ASN-GWs for the communication with the WiMAX terminals. The testbed WiMAX RAN includes one BS, which provides radio connectivity to the WiMAX SSs in a PTMP topology.

The testbed WiMAX BS operates at 3.5 GHz, with a 3.5 MHz channel using a 64 QAM modulation scheme with 3/4 Forward Error Correction (FEC) (see Table 11.1 for more details). The WiMAX BS is connected to the ASN via the ASN-GW. On the host side, a WiMAX Terminal (WT) is connected to each WiMAX SS.

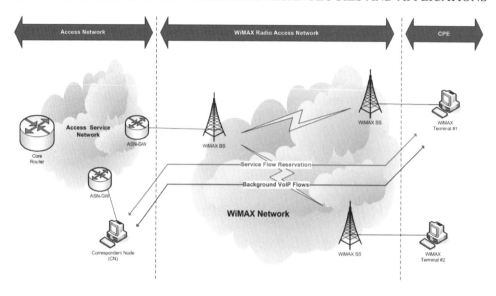

Figure 11.10 Implemented demonstrator for cross-layer signaling measurements.

Table 11.1 WiMAX testbed parameters.

Base station	Redline Communications RedMAX AN-100U
Subscriber stations	Redline Communications SUO
PHY	256 OFDM TDD
Frequency band	3.48 GHz
Channel bandwidth	3.5 MHz
Downlink Modulation	BPSK, QPSK, 16 QAM and 64 QAM
Uplink Modulation	BPSK, QPSK, 16 QAM and 64 QAM
MAC scheduling	Best effort, rtPS

The main focus of these tests is the evaluation of the approach described with respect to its capability to efficiently manage the WiMAX network and to integrate it with a NGN IP architecture. Each module in the chain has been evaluated, as well as the overall path towards the WiMAX system. We next present the L3QoS framework (CSCs, NSIS) results, in addition to the WiMAX cross-layer signaling results (RC, GA, SA and the WiMAX BS). The total end-to-end time necessary to establish a QoS reservation is also provided.

The processing times were measured with and without background traffic. For the background traffic, a variable number of background VoIP flows (50, 100, 150, 200) was used, as illustrated in Figure 11.10. The VoIP traffic was generated based on the ITU-T G.723.1 codec (ITU, 1996).

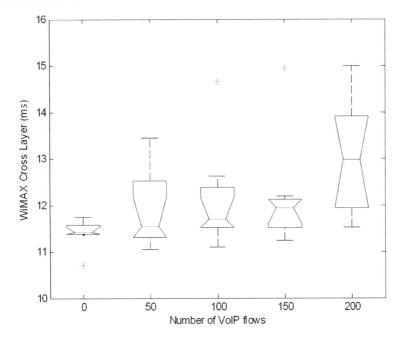

Figure 11.11 WiMAX cross-layer processing times versus number of background VoIP flows.

11.5.1.2 Evaluation Results

Figure 11.11 presents the processing time that the WiMAX cross-layer modules, namely CSC, RC, GA, SA and WiMAX BS, need to establish a QoS reservation. After the NSIS reservation message arrives at the ASN-GW, the CSC-ASN will enforce the QoS reservation in the WiMAX segment. The vertical axis represents the cumulative average processing time of the WiMAX cross-layer modules in milliseconds (ms), whereas the horizontal axis represents the number of background VoIP flows that are traversing the WiMAX link. The measurements are presented in 'box-whisker-plots', or simply boxplots. The box in each figure contains the middle 50% of the measured values. The line in the middle represents the median; the top and bottom of the box correspond to Q3 and Q1, respectively. Values outside the whisker lines, shown as crosses, are considered outliers.

Figure 11.11 shows that the WiMAX cross-layer processing time depends on the number of VoIP flows that are traversing the WiMAX link. Basically, this time includes the modules processing time (RC, GA and SA), the SNMP management messages exchange with the WiMAX BS, and the Dynamic Service Addition (DSA) MAC management messages, which represent the major time-consuming process in the chain; for further details, see Neves *et al.* (2008).

Without background VoIP flows, the average time in the cross-layer modules is very small (approximately 11.5 ms) and the results of the tests are similar. When there are 50 and 100 VoIP background flows, the average time is approximately the same, on median, but we observe higher variability. In the tests with 150 and 200 VoIP flows, the median measured

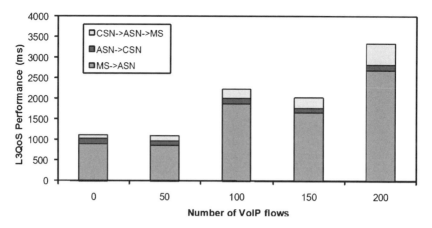

Figure 11.12 The L3 QoS performance time versus the number of background VoIP flows.

time increases slightly, to 12 and 13 ms, respectively. Note that the processing time is always less than 15 ms in all tested configurations.

In short, an increasing number of VoIP background flows slightly rises the time spent in establishing and activating a QoS reservation in the WiMAX link. This is expected, as the QoS signaling traffic is not differentiated from the background VoIP traffic and the request processing time increases due to larger number of entries in the hash tables and the interaction with the SNMP MIB tables in the WiMAX BS. Nevertheless, the overall processing time does not introduce a significant overhead, which is well suited for real-time applications and fast mobility environments, ensuring a fast resource reallocation, ranging from 11 ms without background VoIP flows up to 13 ms for 200 VoIP background flows. Therefore, the testbed results indicate that the impact of the WiMAX cross-layer system in establishing a QoS reservation is rather small and within acceptable bounds.

Figure 11.12 apportions the processing delays to each of the individual L3QoS modules. Each stack column is split in three parts, each one corresponding to a specific segment of the L3QoS communication in the end-to-end path, namely, between the MS and the ASN (bottom portion), the ASN and the CSN (middle portion) and finally between the CSN and the MS (top portion). For each column segment, the vertical axis represents the cumulative average time (in milliseconds) to successfully perform L3QoS request processing and communication between the different entities, whereas the horizontal axis represents the number of background VoIP flows that are traversing the WiMAX link, when we attempt the QoS reservation.

Without background traffic, the L3QoS performance between the MS and the ASN (including the WiMAX segment) takes an average time of 950 ms. When 50 background VoIP flows are injected into the testbed, the average processing time remains the same. When 100 and 150 background VoIP flows are introduced, the average time increases to almost 2 s and it reaches almost 3 s when 200 VoIP flows are transported over the WiMAX testbed. Analyzing the results between the MS and the ASN, we can conclude that, due to the increase of VoIP traffic, the WiMAX link saturates and therefore the L3QoS processing and communication time increases significantly.

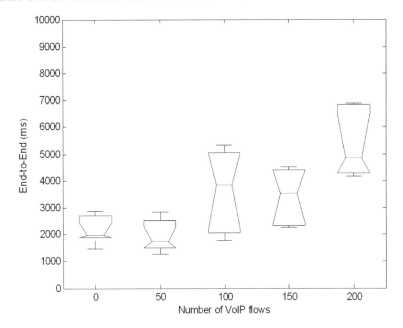

Figure 11.13 End-to-end processing time versus the number of background VoIP flows.

With respect to the L3QoS behavior between the CSN and the MS, the processing time also increases when the VoIP background traffic in the WiMAX channel increases. The L3QoS communication between ASN and CSN remains constant across the different experiment configurations and is only a small fraction of the total time (approximately 100 ms).

Finally, Figure 11.13 presents the time needed to establish an end-to-end QoS reservation. The vertical axis represents the total time needed to perform the QoS reservation; the horizontal axis depicts the number of background VoIP flows traversing the WiMAX link.

We can see that the end-to-end time for reservation establishment depends on the number of VoIP flows that are traversing the WiMAX link. The median end-to-end time is approximately 2 s without background VoIP flows and 5 s for 200 VoIP flows. We note a peak at approximately 7 s when 200 background flows are introduced. These results are mainly due to the L3QoS performance. Another significant component of the end-to-end processing delay is the time consumed by the diameter protocol communication between the CSC-ASN and the AAA, which is approximately 1 s, independently of the presence of any VoIP background flows.

11.5.2   QoS Evaluation

It is widely anticipated that the next generation wireless networks will handle an exponential growth of audio/visual (A/V) content. In order to evaluate the QoS performance over WiMAX, it is important to test the WiMAX system with real-time services, such as VoIP

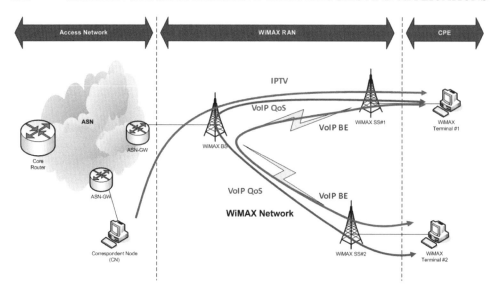

Figure 11.14 Implemented demonstrator to evaluate QoS performance.

and IPTV. Hence, this section presents an evaluation of the WiMAX performance using VoIP and video streaming services.

11.5.2.1   Implemented Demonstrator and Tests Methodology

The demonstrator implemented to validate and evaluate the WiMAX QoS performance is illustrated in Figure 11.14. We follow an approach similar to that presented by Pentikousis *et al.* (2008a,b,c) and Martufi *et al.* (2008) and employ multiple competing traffic sources over a PTMP WiMAX topology and measure the capacity of the WiMAX link to handle a multitude of VoIP flows between the SSs, while simultaneously delivering a variable number of IPTV streams. As depicted in Figure 11.14, we emulated an IPTV service, running between the CN (connected to the WiMAX BS) and WT1 (connected to SS1), in parallel with QoS and best-effort VoIP conversations, both running between WT1 and WT2. By gradually increasing the number of IPTV streams, we determined the saturation point of the WiMAX downlink channel. We repeated each run 10 times, with a fixed duration of 60 s, and measured the application throughput and packet loss.

To study the behavior of the WiMAX system using different service classes, we have used different service classes for each service, as described in Table 11.2. The rtPS service class was employed for both VoIP QoS and IPTV traffic, giving lower priority to the IPTV traffic. For VoIP QoS traffic, four SFs were created, two per SS (one for uplink and one for downlink), whereas for IPTV traffic, a downlink SF on SS1 domain has been created. The VoIP BE traffic between SS1 and SS2 is emulated in a similar way to the VoIP QoS traffic, but using the BE service class, in order to differentiate both VoIP services.

Before proceeding with the evaluation, we have measured the maximum throughput that can be attained in the fixed WiMAX testbed. We saturated the WiMAX link and measured

Table 11.2 Services involved in the QoS evaluation.

Service	Service class	SF	Direction
VoIP QoS (1)	rtPS	SS1 → BS; BS → SS2	WT1 → WT2
VoIP QoS (2)	rtPS	SS2 → BS; BS → SS1	WT2 → WT1
VoIP BE (1)	BE	SS1 → BS, BS → SS2	WT1 → WT2
VoIP BE (2)	BE	SS2 → BS; BS → SS1	WT2 → WT1
IPTV	rtPS	BS → SS1	CN → WT2

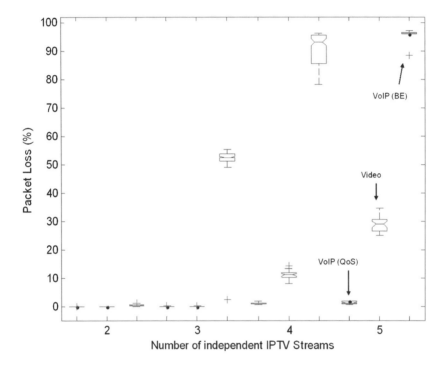

Figure 11.15 Measured packet loss versus the number of IPTV streams.

the maximum application-level throughput, also called goodput, in the downlink and uplink separately. For the uplink, the average maximum measured goodput was 4.75 Mbps and for the downlink it was 5.75 Mbps.

11.5.2.2 Evaluation Results

We start by presenting in Figure 11.15 the measured packet loss for VoIP (with and without QoS) and for IPTV. The WiMAX downlink between the BS and the SS1 can handle three simultaneous IPTV streams in parallel with the VoIP QoS traffic with negligible packet loss. For VoIP QoS traffic, the packet loss values are always closer to zero, even when increasing the number of IPTV streams, as VoIP QoS is the highest-priority traffic. With three

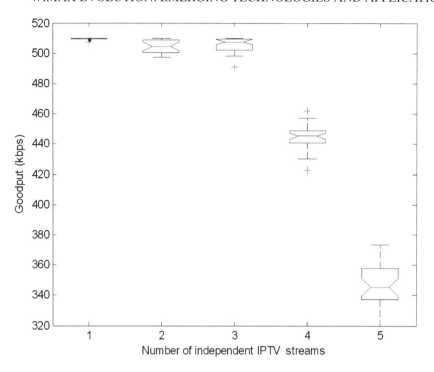

Figure 11.16 Measured IPTV goodput versus the number of IPTV streams.

simultaneous IPTV streams on the WiMAX channel, the packet losses do not exceed 5%. However, when we have more than three IPTV streams flowing in the WiMAX link, the video packet loss rapidly increases, exceeding 10%. The VoIP BE traffic presents a high level of packet loss, since it is the lowest priority traffic. With five IPTV streams, packet loss is almost one 100%.

The application throughput (goodput) results are depicted in Figure 11.16 and 11.17. When we injected more than three simultaneous IPTV streams, the WiMAX link becomes saturated and the goodput for the IPTV traffic decreases rapidly. The video median goodput decreases, and the variability in goodput in different runs increases considerably.

Since the VoIP QoS traffic has higher priority than the IPTV and VoIP BE traffic, it will be able to use the amount of bandwidth needed, independently of the number of IPTV streams that are traversing the WiMAX link. When the WiMAX link starts to saturate, there is no bandwidth available for the VoIP BE traffic due its lower class of service.

11.6   Summary

The integration of WiMAX technology in next generation environments requires a cross-layer platform that enables seamless communication between WiMAX and other access technologies, with full support for integrated QoS, security and MM protocols. This chapter presented an overview of the current issues in designing a cross-layer framework for

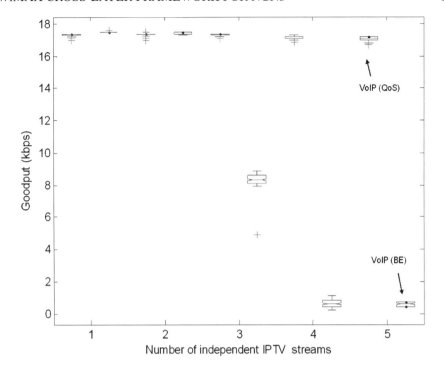

Figure 11.17 Measured VoIP goodput versus the number of IPTV streams.

supporting WiMAX operation in NGNs. Specifically, we discussed the issues of QoS and mobility support, and the integration of media independent protocols and frameworks, using NSIS for QoS and IEEE 802.21 for mobility, respectively, and provided specific details and mechanisms relevant to the WiMAX technology.

As an example and case study of cross-layer integration, we presented the architecture designed and developed in the European WEIRD project, and the performance results obtained, in terms of signaling complexity and QoS. With respect to the signaling results, the cross-layer processing times, as measured in our WiMAX testbed with a prototype implementation, are on median in the range of 12 ms and always less than 15 ms, even for 200 VoIP background flows are simultaneously injected in the testbed. On the other hand, the end-to-end reservation processing times are significant, due to the NSIS framework and the CSCs and diameter processing. Future work is still necessary in this domain, and is high on our research and development agenda. In terms of the level of QoS achieved, the proposed architecture is able to support the differentiated services requirements. The experimental results from our prototype implementation indicate that the proposed cross-layer architecture can be efficient and compliant with real-time services and next generation environments.

Acknowledgements

Part of this work was conducted within the framework of the IST 6th Framework Programme Integrated Project WEIRD (IST-034622), which was partially funded by the Commission of the European Union. Study sponsors had no role in study design, data collection and analysis, interpretation or writing the report. The views expressed do not necessarily represent the views of the authors' employers, the WEIRD project, or the Commission of the European Union. We thank our colleagues from all partners in WEIRD project for fruitful discussions.

References

3GPP (2008) Technical Specification Group Service and System Aspects; IP Multimedia Subsystem; Stage 2, *TS23.228 - Release 8*, 3GPP.

Ash, G., Bader, A., Kappler, C. and Oran, D. (2006) QoS NSLP QSPEC Template, *IETF Internet-Draft*, IETF.

Blake, S., Black, D., Carlson, M., Davies, E., Wang, Z. and Weiss, W. (1998) An architecture for differentiated services, *IETF Request for Comments 2475*, IETF.

Braden, R., Clark, D. and Shenker, S. (1994) Integrated services in the Internet architecture: an overview, *IETF Request for Comments 1633,* IETF.

Calhoun, P. and Loughney, J. (2003) Diameter base protocol, *IETF Request for Comments 3588*, IETF.

Case, J. and Fedor, M. (1990) A Simple Network Management Protocol (SNMP), *IETF Request for Comments 1157*, IETF.

Case, J., McCloghrie, K. and Rose, M. (2002) Management Information Base (MIB) for the Simple Management Protocol (SNMP), *IETF Request for Comments 3418*, IETF.

Cordeiro, L., Curado, M., Neves, P., Sargento, S., Landi, G. and Fu, X. (2008) Media Independent Handover Network Signaling Layer Protocol (MIH NSLP), *IETF Internet-Draft*, IETF.

Deering, S. and Hinden, R. (1998) Internet Protocol - Version 6 (IPv6) Specification, *IETF Request for Comments 2460*, IETF.

Guainella, E., Borcoci, E., Katz, M., Neves, P., Curado, M., Andreotti, F. and Angori, E. (2007) WiMAX technology support for applications in environmental monitoring, fire prevention and telemedicine. *Proceedings of the IEEE Mobile WiMAX Symposium*, Florida.

Hancock, R., Karagiannis, G., Loughney, J. and den Bosch, S.V. (2005) Next Step in Signalling (NSIS): Framework, *IETF Request for Comments 4080* IETF.

Heinanen, J., Baker, F., Weiss, W. and Wroclawski, J. (1999) Assured Forwarding PHB Group, *IETF Request for Comments 2597*, IETF.

IEEE (2004) IEEE Standard for Local and Metropolitan Area Networks; Part 16: Air Interface for Fixed Broadband Wireless Access Systems, *IEEE Standard 802.16-2004*, IEEE 802.16 Working Group.

IEEE (2005a) IEEE Standard for Information Technology - Telecommunications and Information Exchange between Systems - Local and Metropolitan Area Networks - Specific Requirements Part 3: Carrier Sense Multiple Access with Collision Detection (CSMA/CD) Access Method and Physical Layer Specifications, *IEEE Standard 802.3-2005*, IEEE 802.3 Working Group.

IEEE (2005b) IEEE Standard for Local and Metropolitan Area Networks; Part 16: Air Interface for Fixed and Mobile Broadband Wireless Access Systems; Amendment 2: Physical and Medium Access Control Layers for Combined Fixed and Mobile Operation in Licensed Bands, *IEEE Standard 802.16e-2005*, IEEE 802.16 Working Group.

IEEE (2005c) IEEE Standard for Local and Metropolitan Area Networks; Part 16: Air Interface for Fixed Broadband Wireless Access Systems - Amendment 1: Management Information Base, *IEEE Standard 802.16f-2005*, IEEE 802.16 Working Group.

IEEE (2006) IEEE Standard for Local and Metropolitan Area Networks. Virtual Bridged Local Area Networks, *IEEE Standard 802.1Q-2005,* IEEE 802.1 Working Group.

IEEE (2007) IEEE Standard for Local and Metropolitan Area Networks; Part 16: Air Interface for Fixed Broadband Wireless Access Systems; Amendment 3: Management Plane Procedures and Services, *IEEE Standard 802.16g-2007*, IEEE 802.16 Working Group.

IEEE (2008) IEEE Draft Standard for Local and Metropolitan Area Networks: Media Independent Handover Services, *IEEE P802.21/D10.0*, IEEE 802.21 Working Group.

ISIUSC (1981) Internet Protocol, *IETF Request for Comments 791*, IETF.

ITU (1996) ITU-T, Dual rate speech coder for multimedia communications transmitting at 5.3 and 6.3 Kbps, *ITU-T Recommendation G.723.1* ITU Group.

Jacobson, V., Nichols, K. and Poduri, K. (1999) An expedited forwarding PHB, *IETF Request for Comments 2598* IETF.

Johnson, D., Perkins, C. and Arkko, J. (2004) Mobility Support in IPv6, *IETF Request for Comments 3775* IETF.

Koodli R. (2005) Fast Handovers for Mobile IPv6, *IETF Request for Comments 4068* IETF.

Manner, J., Karagiannis, G. and McDonald, A. (2008) NSLP for Quality-of-Service Signalling, *IETF Internet-Draft*, IETF.

Martufi, G., Katz, M., Neves, P., Curado, M., Castrucci, M., Simoes, P., Piri, E. and Pentikousis, K. (2008) Extending WiMAX to new scenarios: key results on system architecture and test-beds of the WEIRD project. *Proceedings of EUMOB*, Oulu, Finland.

Neves, P., Nissila, T., Pereira, T., Ilkka, H., Monteiro, J., Pentikousis, K., Sargento, S. and Fontes F. (2008) A vendor independent resource control framework for WiMAX. *Proceedings of the 13th IEEE Symposium on Computers and Communications (ISCC)*, Marrakech, Morocco.

Nissila, T., Huusko, J., Harjula, I. and Katz, M. (2007) Adapter implementation between WiMAX specific layers and network/application layers. *Proceedings of the First Broadband Wireless Access Workshop (BWA), Collocated with 1st Next Generation Mobile Applications, Services and Technologies (NGMAST)*, Cardiff, UK.

Pentikousis, K., Pinola, J., Piri, E. and Fitzek, F. (2008a) A measurement study of Speex VoIP and H.264/AVC Video over IEEE 802.16d and IEEE 802.11g. *Proceedings of the Third Workshop on MultiMedia Applications over Wireless Networks (MediaWiN)*, Marrakech, Morocco.

Pentikousis, K., Pinola, J., Piri, E. and Fitzek, F. (2008b) An experimental investigation of VoIP and video streaming over fixed WiMAX. *Proceedings of the Fourth International Workshop on Wireless Network Measurements (WiNMee)*, Berlin, Germany.

Pentikousis, K., Pinola, J., Piri, E., Fitzek, F., Nissilä, T. and Harjula, I. (2008c) Empirical evaluation of VoIP aggregation over a fixed WiMAX testbed. *Proceedings of the Fourth International Conference on Testbeds and Research Infrastructures for the Development of Networks and Communities (TRIDENTCOM)*, Innsbruck, Austria.

Perkins, C. (1996) IP Mobility Support, *IETF Request for Comments 2002,* IETF.

Rosenberg, J., Schulzrinne, H. and Camarillo, G. (2002) Session Initiation Protocol (SIP), *IETF Request for Comments 3261,* IETF.

Schulzrinne, H. and Hancock, R. (2008) GIST: General Internet Signalling Transport, *IETF Internet-Draft*, IETF.

Sousa, B., Neves, P., Curado, M., Sargento, S., Fontes, F. and Boavida, F. (2008) The cost of using IEEE 802.16 dynamic channel configuration. *Proceedings of the IEEE International Conference on Communications (ICC)*, Beijing, China.

WiMAX Forum (2008) WiMAX End-to-End Network Systems Architecture Stage 2–3: Architecture Tenets, Reference Model and Reference Points, *Release 1, Version 1.2*, WiMAX Forum.

12

Speech Quality Aware Resource Control for Fixed and Mobile WiMAX

Thomas Michael Bohnert, Dirk Staehle and Edmundo Monteiro

12.1   Introduction

Hardly any subjects in telecommunications caused such controversy as the Quality of Service (QoS) issue in the Internet. The persistent advance in switching capacity over bandwidth demand by real-time services is dividing those calling for Internet QoS from their counterparts. The Internet2 QoS Working Group, for instance, concluded after years-long large-scale QoS deployment that conceptual issues plus a lack of demand, that is, missing real-time applications, severely inhibits the evolution of the Internet rather than fostering it (Teitelbaum and Shalunov, 2002). They even went as far as calling for banning end-to-end Internet QoS by regulation in order to guarantee a free and therefore powerful Internet based on its original principles. Meanwhile this conviction turned into a concept known as *net neutrality* and became subject of a fierce power struggle between telecommunication and service providers (Bohnert *et al.*, 2007).

While this conclusion might possibly hold for wireline communications it is a very different story for wireless communications and WiMAX is a perfect example of this. With the roll-out of mobile WiMAX the long-awaited IP convergence comes true and means that traditional real-time services, such as voice and video, will have to be delivered over the Internet infrastructure. However, subscribers will not change their expectations along with exchanging access technology and voice is therefore a very intuitive example. Mobile voice subscribers are meanwhile used to cellular-phone-like service quality and will expect a similar experience. The success of WiMAX will thus ultimately depend on the success of operators in providing competitive QoS levels.

This awareness has settled right at the beginning of IEEE 802.16 standardization and a designated feature of mobile WiMAX, based on IEEE 802.16-2005, commonly called IEEE 802.16e (IEEE, 2005) is its inherent QoS support which is largely based on the concepts of ATM's comprehensive QoS body (Eklund *et al.*, 2006). So WiMAX defines a set of services categorized and parameterized by their respective target applications. One of these services is voice and assumes standard Voice over IP (VoIP) technology at the application layer. The WiMAX QoS model in turn defines semantics adopted by VoIP to control packet delay, loss and jitter. Explicit implementation instructions, however, are not part of the IEEE 802.16 standards and the choice of which particular resource control algorithms to implement is entirely left to manufactures.

In the context of VoIP these two decisions, using the QoS model of the Asynchronous Transfer Mode (ATM) as the bottom line without implementation instructions, render themselves as a considerable challenge: the identification and implementation of efficient resource control components capable of achieving VoIP QoS support in WiMAX systems. By way of simulation we present two such resource control algorithms, Admission Control (AC) and scheduling, where they are both based on dynamic quality estimations derived from instantaneous local measurements. In order to do so we first discuss quality assessment as such in Section 12.2 before we introduce speech quality assessment as the only appropriate means for this purpose in Section 12.3. In Section 12.4 we then present our approach for dynamic and on-demand quality assessment, the fundamental resource control criterion. The first resource control algorithm leveraging this approach is a Measurement Based Admission Control (MBAC) algorithm. Its theoretical underpinning plus performance evaluation are presented in Section 12.5. In Section 12.6 we present the concept of a scheduler that uses measurements or estimations of the current speech quality for making its scheduling decisions. We present a simple *R*-score based scheduler and show that its performance is superior to the channel-oblivious Earliest Deadline First (EDF) scheduler and comparable to the channel-aware MaxSNR scheduler. Finally, we give a short conclusion on this chapter in Section 12.7.

12.2   Quality of Experience versus Quality of Service Assessment

Before entering into technical details, the very first question to answer is how to define QoS for VoIP. While there is a common notion attached to the term *quality*, the actual space for its definition is divided into two parts: subjective and objective. Customers define quality as the overall satisfaction based on subjective assessment in multiple dimensions while engineers tend to express quality in terms of physical and measurable parameters and thus objectively. This is a fundamental discrepancy and frequently leads to notable misunderstandings.

A comprehensive QoS definition and terminology coverage has been given by Gozdecki *et al.* (2003), from which this section is partly borrowed. In brief, and according to

Hardy (2001), a general model divides QoS into three notions: *intrinsic, perceived* and *assessed* QoS. Intrinsic QoS (IQ) is purely technical and evaluates measured and expected characteristics expressed by network parameters such as delay and loss. Perceived QoS (PQ) reflects user satisfaction while using a particular service. It is therefore a subjective measure and the only method to ultimately capture it is to survey human subjects. Assessed QoS (AQ) extends the notion of PQ to secondary aspects such as service price, availability, usability and reliability. Each of this definition can be considered separately but there is a tight interdependence. PQ is a function of IQ and is an element of AQ. Nevertheless, they are commonly considered in isolation and interesting enough, the IETF as well as the ITU-T and ETSI direct their focus on different definitions and neither of them considers AQ to a full extent (Gozdecki *et al.*, 2003).

The IETF lastingly coined the notion of QoS adhering to IQ with the introduction of the IntServ (Braden *et al.*, 1994) and DiffServ (Blake *et al.*, 1998) frameworks. Consequently, a later published IETF QoS definition reads 'A set of service requirements to be met by the network while transporting a flow' (Crawley *et al.*, 1998). In contrast, the ITU-T and ETSI jointly define QoS as 'the collective effect of service performance which determine the degree of satisfaction of a user of the service', expressed the first time in ITU-T (1993) and clearly compliant with PQ. On top of this, the ITU-T recently released the Quality of Experience (QoE) framework ITU-T (2004) in which an explicit distinction has been made between QoS and QoE. In this document, QoS expresses the 'degree of objective service performance' and QoE the 'overall acceptability of an application or service, as perceived subjectively by the end user'. According to these definitions, QoS is equal to IQ and QoE is equal to AQ. Notwithstanding, QoE is prevailingly associated with SQ and as SQ is an element of AQ, we adopt this definition together with QoS for IQ for the remainder of this work.

So far it has been shown that looking at service quality calls for a careful distinction based on its assessment. While QoS evaluation is deemed straightforward and merely a matter of measuring physical parameters, it appears much more complicated (and tedious) for QoE as it involves humans. However, QoE is the ultimate measure and in order to harness it in systems the interdependence between QoS and QoE is to be exploited. This can be achieved by observing that QoE is a function of QoS which can be expressed by mapping physical parameters to user ratings. This quality assessment method based on mapping measured parameters, such as delay or loss, to a QoE scale, such as the Mean Opinion Score (MOS), is called Instrumental Quality Assessment (IQA) (Raake, 2006b).

IQA is central to our work and an elaborate coverage in the context of VoIP will be presented in Section 12.3. For now the reader should note that there is considerable volume of work related to quality assessment and its applications, see Janssen *et al.* (2002), Raake (2006a), Takahashi *et al.* (2006) and Markopoulou *et al.* (2003). In a mobile environment, the relationship of QoS and QoE for a Skype call over a Universal Mobile Telecommunications System (UMTS) Internet access was investigated by Hoßfeld and Binzenhöfer (2008). General investigations on the exponential interdependency between QoS and QoE for different voice codecs can be found in Hoßfeld *et al.* (2008, 2007).

At last IQA paved the way for an emerging area in research and development: resource management based on QoE. Traditionally, resource management was mostly concerned with QoS but frameworks are increasingly considering QoE as the ultimate performance metric. Some examples are given by Sengupta *et al.* (2006) who present parameterization guidelines to improve VoIP quality for WiMAX.

12.3   Methods for Speech Quality Assessment

No matter how a voice service is implemented, either analog or digital, over a circuit-switched or packet network, it essentially means speech transmission and the ultimate service quality depends on how uttered information is being understood by communication participants. Henceforth, the only appropriate quality assessment method for voice services is that of subjective speech quality assessment. Given the adopted terminology it therefore relates to QoE.

12.3.1   Auditory Quality Assessment

As briefly outlined in the previous sections subjective quality assessment involves surveying human subjects. Methods falling into this category are classified as auditory methods and share a basic commonality: in controlled experiments human subjects listen to speech samples subject to varying impairment in space and time and record the perceived quality.

The outcome of an auditory method is highly individual and controlled by a multitude of features anchored in human physiology. Human perception depends on spectral and temporal processing capabilities of the auditory system, on echoic, short-term and long-term auditory memory and speech comprehension, intelligibility and communicability. Finally, it is influenced by the ability to restore missing sounds by way of analyzing context, a feature known as the 'picket fence effect' in analogy to a visual modality: whilst watching a landscape through a picket fence, the fence interrupts the view periodically but the landscape is seen to continue as the human brain is padding missing pieces from memory. Apparently beneficial at first sight, this feature can severely impair perception if padded pieces are selected from alleged context composed of the actual plus extracts of previous context information stored in memory. In an advanced state, subjects even try to anticipate future or missing information and if it does not match the perceived version, they likely classify it falsely as distorted, wrong or even entirely missing.

There is a whole science behind auditory test methods. They are divided in utilitarian and analytic tests. The former aim at directly comparing the quality of different speech communication systems while the latter try to reveal the perceptual features underlying speech quality. Utilitarian methods are subdivided into listening quality, comprehension and listening and talking tests according to different stimuli, context and other features; see Raake (2006b) for a complete treatise. Central to any of these methods is the question on how to scale ratings: absolute, relative, discrete or continuous. The well-known MOS, for example (illustrated in Figure 12.3) is a five-point Absolute Continuous Rating (ACR) scale. Its name is derived from the fact that it expresses the average over a set of individual ratings obtained in an controlled experiment. It is frequently used in methods aiming at capturing time varying quality impairment by recording the instantaneous quality over time, typically with a slider over the ACR scale; cf. Watson and Sasse (1998). As shown later in this chapter, time varying quality impairment plays a central role in speech quality assessment. Nevertheless, it is frequently neglected due to its alleged complexity.

12.3.2   Instrumental Quality Assessment

In many cases it is desirable to evaluate speech quality on demand, in real-time and without human involvement. Accordingly, much effort has been spent in developing alternatives to auditory tests, so-called instrumental methods or IQA. The principle of these methods is to correlate physical and measurable magnitudes with quality as perceived by a human subject. While such unique relationships exist in theory, it has to be noted that, so far, it remains impossible to establish them in practice, even for very simple applications.

Nevertheless, recent advances disclosed methods with considerable precisions and consistency. In particular, signal based methods achieve accurate and reliable results based on the relatively well-understood signal processing performed by human. Examples are Perceptual Evaluation of Speech Quality (PESQ) (ITU-T, 2001) and Perceptual Speech Quality Measure (PSQM) (ITU-T, 1996). Either of these methods compares a clean reference signal with the same signal after been processed by the system under test. In a final step the estimated deviation is mapped to a rating scale, such as the five-point ACR, where this mapping is purely empirical and the result of a large number of auditory tests.

The alternative to signal-based methods are so-called parameter-based methods. These methods require instrumentally measurable magnitudes which are evaluated in a parametric model. The most popular method of this category is the E-Model (ITU-T, 1998), which is the basic component of the presented framework and is presented in detail in Section 12.4. In addition to its distinct internal rationale, it maps the result to a subjective rating based on a purely empirically found function, just like a signal-based method.

12.4 Continuous Speech Quality Assessment for VoIP

As stated in Section 12.1, the objective of this work is to devise the 'identification and implementation of efficient resource control components capable of achieving VoIP QoS support in WiMAX systems'. At this point in time, however, the reader should note the error in this statement. In accordance with the preceding argumentation it must be called 'VoIP QoE support'. The immediate conclusion is that a purpose-built framework requires a component capable of continuously assessing VoIP QoE levels. Furthermore, the overall objective is to assure VoIP QoE levels by means of resource control and thus the second requirement on this framework is the delivery of precise QoE estimates in real-time as a criterion for resource assignment. Obviously this can only be an instrumental method and precludes signal-based methods since there is no space and time for transmitting reference signals over the radio interface.

12.4.1 VoIP Components and their Impact on Speech Quality

Speech is a slowly varying analog signal over time and its frequency components are limited to the lower 4 kHz band. Owing to linguistic structures a speech signal alternates between talk spurts and silence periods. The origin of talk spurts are typically syllables which themselves are phoneme sequences containing one vocalic sound. Average durations of talk spurts range from 300 to 400 ms while silence periods range from 500 to 700 ms. In order to transmit speech over a packet network several pre- and post-processing steps are required. In Figure 12.1, each of these processing steps is represented by a single logical building block and the complete set makes a basic end-to-end VoIP system.

Any of the elements of the VoIP system has an impact on the speech quality perceived by the receiver. First, the analog signal is converted into a digital signal. This step is commonly assumed to be negligible with respect to the overall quality. Additive noise is canceled

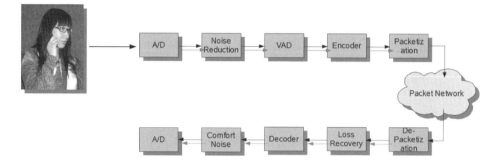

Figure 12.1 Basic Components of a VoIP System.

next before the signal enters the Voice Activity Detector (VAD). The VAD prevents the transmission of silence and thus influences the durations of talk spurts and silence periods. The sojourn times in either state are roughly exponentially distributed with a tendency to longer tails Jiang and Schulzrinne (2000) but the mean is entirely controlled by the VAD's sensitivity. Modern VADs elongate talk spurts by a so-called hangover time in order to prevent speech clipping. Small hangover times result in shorter talk spurts and vice versa. Altogether, VAD explicitly and implicitly impacts on packet loss and its distributions and thus speech quality.

Talk spurts are encoded by one of the very many voice codecs. The simplest and most well-known is Pulse Code Modulation (PCM). It is standardized by the ITU-T where it is named G.711. This encoder produces a 64 kbps digital signal and implies some level of entropy due to its discrete quantization. This is the main reason behind its inherent impact on speech quality.

The digital signal, or bit stream, is then packetized into equal sized packets. For each talk spurt the continuous bit stream therefore results in a periodic sequence of packet emissions where the period is determined by the packet length. Deciding on a proper packet size is crucial with respect to overall efficiency, that is, the tradeoff between transport overhead and the actual payload.

Voice packets are transported over the IP network using the common mechanisms. They might be treated with priority by DiffServ or IntServ implementation in some network access segments but when it comes to the Internet backbone, they most likely share the same fate as any other best-effort traffic, cf. Markopoulou *et al.* (2003). Packets might be delayed, reordered, jostled (jittered) and eventually dropped. Obviously, this part of the VoIP system therefore potentially has the major impact on the overall quality.

Once it arrives at the end system, the speech stream is extracted from the packets. Lost information is identified, for example by means of sequence numbers, and algorithms such as Forward Error Correction (FEC) or Packet Loss Concealment (PLC) recover or mask it to some extent. In a next step the modified bit stream is decoded and depending on the deployed VAD, some comfort noise is added. This is to account for the artificial silence introduced by the VAD: absolute silence is perceived by humans as odd and is likely to be falsely interpreted as the system malfunctioning. In a final step the digital system is re-converted into an analog system and played out by a speaker device.

12.4.2 Continuous Assessment of Time-varying QoE

The *de-facto* parametric IQA method is the E-Model. Its development started with a study conducted by the ETSI which later turned into a standardization activity by the ITU-T (ITU-T, 1998). Its original application domain is network planning and one of the questions we answer with this work is whether it lends itself as a tool for on-demand QoE estimation as input for online resource control.

The E-Model is an IQA method for mouth-to-ear transmission quality assessment based on human perception and is defined as

$$R = R_0 - I_s - I_d - I_e + A. \tag{12.1}$$

In (12.1), R denotes the psychoacoustic quality score defined in [0, 100]. It is an additive, nonlinear quality metric based on a set of impairment factors, namely R_0, I_s, I_d, I_e, and A. It assumes that underlying sources of degradation can be transformed onto particular scales and expressed by an impairment factor. These functional relations are found prevailingly empirical. The classes of degradation are as follows.

- Noise and loudness effects are represented by R_0. These effects originate from basic environmentally inflicted signal-to-noise ratios, such as those induced by reflections and interference in rooms or noise on the line.

- The *simultaneous impairment factor* (I_s) denotes speech signal impairment such as PCM quantizing distortion or VAD hangover times.

- Impairment due to information delay, such as transmission delay or echo, is represented by the *delayed impairment factor* (I_d).

- Degradation due to information loss is expressed by the *equipment impairment factor* (I_e). It covers terminal internal information loss such as low-bit rate coding but also losses caused by lossy transport media such as IP networks.

- The *Advantage Factor* (A) quantifies the user's tolerance with respect to quality degradations if these are perceived as inherent to a feature that otherwise increased system utility or convenience. For instance, cellular-phone subscribers expose higher tolerance to noise than fixed line users as quality degradations are perceived as a natural consequence of ubiquitous telephone access over radio interfaces.

The E-Model's particular appeal lies in its simplicity. To assess the speech quality of a VoIP call, one simply has to measure parameters such as delays and losses, map them onto degradation scales, and sum all factors in order to yield the final score R. Obviously, degradation functionals are therefore central to the E-Model. In Figure 12.2 one such functional is plotted for the equipment impairment factor I_e. Recalling the definition of I_e, it has to be noted that there is a functional for each and any codec as they are differentially sensitive to losses. The depicted functional plots packet loss over quality degradation for the G.711 codec. More of these mappings can be found in Janssen *et al.* (2000), Markopoulou *et al.* (2003) and Raake (2006b).

As mentioned in Section 12.3.2, auditory tests commonly quantify speech quality on an ACR scale. This also applies to the E-Model, whose *R*-score scales on a 100-point ACR.

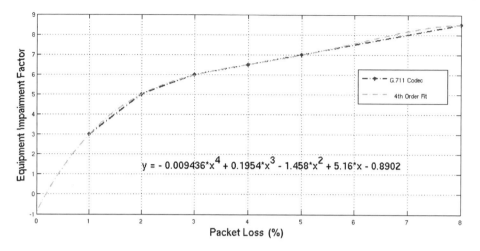

Figure 12.2 Nonlinear relation between the packet loss ratio and the equipment impairment factor (I_e).

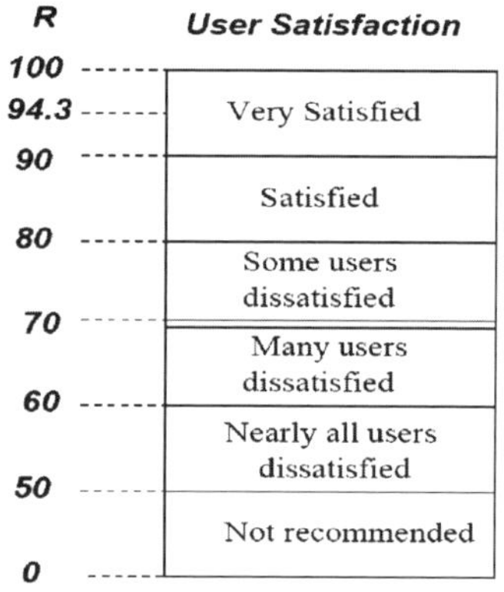

Figure 12.3 Mapping average user satisfaction (MOS) to the R-score.

The globally established ACR scale, however, is the MOS and a translation from the R-score to MOS has been introduced as a result of extensive auditory tests. It can be found in ITU-T (1998) and is depicted in Figure 12.3.

The E-Model was used in various setups in numerous previous works; see, for example, Cole and Rosenbluth (2001), Janssen *et al.* (2002), Markopoulou *et al.* (2003), Meddahi *et al.* (2003), Raake (2006a), Sengupta *et al.* (2006), Takahashi *et al.* (2006). Many of these works discuss weaknesses and propose modifications where the two foremost points are the E-Model's additivity and that packet loss processes in IP networks are far more complex than what is captured by simple loss ratios. How to account for these phenomena is the subject of the following section.

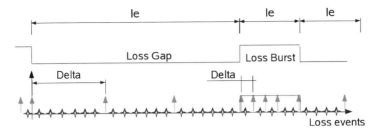

Figure 12.4 A series of consecutive periods of different microscopic loss behaviors, that is packet loss ratio and distribution, together form a macroscopic loss profile. There are two alternating microscopic loss behaviors, loss gaps and loss bursts. Stars represent VoIP packets and arrows indicate packet loss events. The distance between them is called delta. If delta is larger (smaller) than 16, and the model is in gap (burst) state, it remains is this state. Otherwise it changes from burst (gap) to gap (burst) state. In the event of an transition the impairment factor is calculated for the abandoned state. Over time, this leads to a series of I_e values.

12.4.3 Instationary Quality Distortion and Human Perception

The quality of speech is by far most influenced by information loss, that is, I_e (Kostas *et al.*, 1998, Markopoulou *et al.*, 2003, Raake, 2006b). The functional relation expressing this fact is nonlinear and the most relevant part for a G.711 codec is depicted in Figure 12.2. Using a simple fourth-order least-squares fit, the function reads

$$I_e = -0.009436x^4 + 0.1954x^3 - 1.458x^2 + 5.16x - 0.8902. \qquad (12.2)$$

However, measuring packet loss and mapping it to I_e is insufficient. In particular, if speech is delivered over IP networks by means of VoIP, the final quality is prevailingly determined by the packet loss distribution. Intuitively, single packet losses are always preferable over loss bursts. This is exactly the difference between IQ and SQ since by taking averages, as with IQ, such details are inherently ignored. Furthermore, packet loss distributions themselves are frequently instationary over a call's life time and instantaneous as well as ultimate quality assessment by humans exhibits strong correlation with this characteristic Raake (2006b, Chapter 4).

To account for this phenomenon we divide the packet loss process into periods with different loss behaviors, as proposed by Clark (2001) and refined by Markopoulou *et al.* (2003) and Raake (2006a,b). In particular, we adopt the principles proposed by Clark (2001) but modified for our purpose. Essentially, this packet loss driven model defines two alternating states of microscopic loss behavior, loss gap and loss burst state, with respect to the distance of packet loss events, cf. Figure 12.4. According to Clark (2001), the model remains in the (loss) gap state as long as there is a minimum of 16 successfully received packets between two loss events (delta, δ). Otherwise there is a transition from (loss) gap to (loss) burst. The idea behind staying in a gap state under this condition is that modern loss recovery algorithms can handle isolated packet loss relatively well. In the case of a transition to burst state, the model remains in this state until 16 packets are successfully received between the latest and the previous loss event.

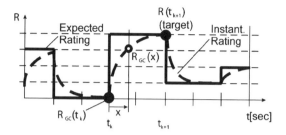

Figure 12.5 The expected rating (solid line) associated with either loss or gap state. The true, delayed perception (dashed line) by humans is indicated as an exponential decay or rise of the R-score with respect to a state transition. (Source: Raake (2006a).)

Upon the detection of any state transition the loss ratio for the previous state is used to calculate the corresponding impairment level, I_e, using the relation depicted in Figure 12.2, resulting in a time series of I_e values with respect to states. However, before these values can be used to compute R, there is another feature, inherent to human perception, which has been integrated into this model, the delayed perception (or acceptance) of quality change.

Naturally, humans tend to perceive a quality change rather continuously and not instantaneously at state transitions. A further distinction has to be made between transitions from good to bad and vice versa. So do humans confirm a change from good to bad much faster than the other way around. Generally, this feature can be modeled by an exponential function, similar to a transistor saturation curve, with specific time constants. It is shown in Figure 12.5 by the dashed lines.

Given $I_{e,g}$ and $I_{e,b}$, the impairment linked to gap or burst, I_1 is the estimated instantaneous impairment level at the change from burst to gap condition and I_2 equals the level at the return from gap to burst. In mathematical terms, I_1 and I_2 can be expressed as

$$I_1 = I_{e,b} - (I_{e,b} - I_2)e^{-b/\tau_1}, \tag{12.3}$$

$$I_2 = I_{e,g} + (I_1 - I_{e,g})e^{-g/\tau_2}. \tag{12.4}$$

Here g and b denote the sojourn time in gap or burst state and τ_1 and τ_2 are the time constants, respectively. Typical values are $\tau_1 = 9$ s and $\tau_2 = 22$ s (Raake, 2006b). A proper combination of (12.3) and (12.4) yields an expression for I_2 independent from I_1:

$$I_2 = I_{e,g}(1 - e^{-g/\tau_2}) + I_{e,b}(1 - e^{-b/\tau_1})e^{-g/\tau_2}. \tag{12.5}$$

Using (12.5) we are now in a position to calculate the average impairment level over a certain time, for example, for the life time of a call. Therefore, we first calculate *average* gap and burst length, $\bar{b}$ and $\bar{g}$, as well as the *average* impairment levels, $\bar{I}_{e,g}$ and $\bar{I}_{e,b}$. Putting these in (12.5) and integrating it over one burst and gap yields the *average* impairment level for a certain loss profile of certain length. It reads

$$\bar{I}_e = \frac{1}{\bar{b} + \bar{g}} \times [\bar{I}_{e,b} \times \bar{b} + \bar{I}_{e,g} \times \bar{g} + \tau_1 \times (\bar{I}_{e,b} - I_2)$$
$$\times (e^{\bar{b}/\tau_1} - 1) - \tau_2 \times (\bar{I}_{e,b} - (\bar{I}_{e,b} - I_2)$$
$$\times e^{-\bar{b}/\tau_1} - \bar{I}_{e,g}) \times (e^{\bar{g}/\tau_2} - 1)]. \tag{12.6}$$

Eventually, by replacing I_e in (12.1) with $\bar{I}_e$ and using proper values for the remaining parameters, one can evaluate the subjective quality for a single call by this parametric IQA method called *integral quality by time averaging* (Raake, 2006b).

12.5 Speech Quality Aware Admission Control for Fixed IEEE 802.16 Wireless MAN

12.5.1 IEEE 802.16d Background and the Deployment Scenario

The IEEE 802.16d standard, officially called 802.16-2004 with reference to its release date, defines an air interface for fixed Broadband Wireless Access (BWA). In doing so it specifies several Physical (PHY) layers and a common Medium Access Control (MAC) layer on top of them. Target deployment is fixed Non-Line-of-Sight (NLOS) within the 2 to 11 GHz frequency band, either in Point-to-Multipoint (PMP) or in mesh mode. In PMP, a central Base Station (BS) controls all traffic interactions between Subscriber Stations (SSs) and itself and all traffic is either sent from a single SS to the BS, called the Uplink (UL), or from the BS to one or many SSs, called the Downlink (DL).

The MAC is connection oriented in order to support QoS, an essential future BWA requirement. There are several types of connections, each unidirectional and between two MAC instances. These connections serve different purposes such as MAC management and signaling with several priorities but also for data transport. Connections are identified by an unique Connection Identifier (CID).

In addition to connections, IEEE 802.16d defines the concept of Service Flows (SFs). A SF itself is defined as unidirectional transport service with predetermined QoS characteristics, that is, QoS parameters. Each SF is mapped to a single connection and has to be served by an UL (or DL) scheduler such that QoS requirements are met. This is being done by a so-called scheduling service which is related to the QoS parameters associated with the respective SF. It should be noted that the standard defines scheduling services but does not define any explicit scheduler for them. It is left to manufactures to select and implement a scheduler which meets the respective requirements. In this respect, IEEE 802.16d is in line with concepts known from DiffServ, which specifies Per Hop Behaviors (PHB); see, for example, Davie *et al.* (2002), but not how to implement them.

One of the envisioned deployments of IEEE 802.16d is to deliver VoIP services in different granularities. As IEEE 802.16 is connection oriented the spectrum ranges from a single VoIP call to VoIP aggregates. We focus on aggregates as single VoIP calls are rather typical for scenarios involving mobile terminals. As this work is in the context of the European research project 'WiMAX Extensions for Isolated Research Data Networks' (WEIRD)[1], the deployment we have in mind is a real deployment scenario defined by WEIRD. In this scenario a remote monitoring station, in the role of a SS, is connected by a BS to a central unit. In reality this is a Forest Fire Monitoring Station (FFMS) somewhere in the mountains connected to a Coordination Center (CC) in a nearby city. In this case VoIP services are used to support the personnel in the FFMS in reporting and for coordination of forest fire prevention activities by the CC. In order to do so, this scenario defines a dedicated, preprovisioned SF with a certain, fixed capacity in either direction. For more details on this

---

[1] See http://www.ist-weird.eu.

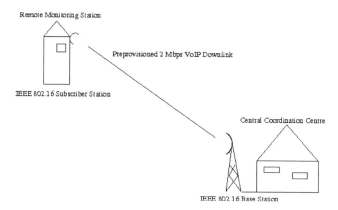

Figure 12.6 The example deployment scenario as defined in the EU IST FP6 IP Project 'WEIRD'. A remote monitoring station (FFMS) is connected with a CC via a preprovisioned IEEE 802.16 SF for VoIP services.

scenario we point the readers to WEIRD Consortium (2007). The scenario is illustrated in Figure 12.6.

12.5.2 The Principle of Admission Control and its Application to VoIP

Admission Control (AC) is the most important mechanism for QoS provisioning on an aggregate level. In other words, if a provider decides to exploit statistical multiplexing gain within a single traffic class, AC regulates the traffic intensity by controlling the number of active flows such that a certain QoS objective is met. In the context of our VoIP scenario, this means an AC function controls the VoIP traffic arriving at the aggregation point, that is, the BS in DL direction or the SS in UL direction and destined to either the FFMS or the CC, such that a certain VoIP quality is assured.

Derived from the definition presented by Bohnert and Monteiro (2007), AC can be generally defined as

$$\chi_k \begin{cases} \geq 0 & \text{admit flow } k, \\ = 0 & \text{reject flow } k, \end{cases} \tag{12.7}$$

where χ_k denotes the admission criterion for the requesting flow k and is defined as

$$\chi_k = \max\{Q(N+1) - Q', 0\}. \tag{12.8}$$

Here we assume that $Q(n)$ expresses the level of QoS for n admitted sources, that is, the traffic aggregate, and is a monotonically decreasing function in n, while Q' is the target QoS. The computation of $Q(N+1)$ reads as

$$Q(N+1) = Q(N) - \Delta_{QoS}^K \tag{12.9}$$

where Δ_{QoS}^K denotes the QoS degradation inflicted on the aggregate by the characteristics of flow k, if the latter would be accepted.

Equipped with the expressions derived in the previous sections we can formulate an admission criterion based on QoE (speech quality). Therefore, we replace $Q(N+1)$ in (12.8) with $R(R_0, I_s, I_d, I_e, A)$ and set $R_0 - I_s = 94$, the default value with respect to inherent features of the G.711 codec. Further, I_d is set to an upper bound determined by the buffer length ω and the link capacity C; see Cox and Perkins (1999) for details. Beyond this bound, packet delay translates into packet loss and is captured by I_e. The respective equation for I_d reads

$$I_d = 4 + 1 * (\omega/C). \tag{12.10}$$

Combining all pieces and further assuming the worst case, that is, we set A to zero, we obtain

$$\bar{R}_T = 94 - 4 + 1 * (\omega/C) - \bar{I}_e(T). \tag{12.11}$$

In this equation the parameter T in $\bar{I}_e(T)$ indicates that the average impairment factor for time-varying speech quality assessment has been calculated over a window of T seconds. This is to account for an inherent feature of Measurement Based Admission Control (MBAC) algorithms, which generally estimate a QoE/QoS criterion over a limited window. Eventually, we can express the admission criterion as follows:

$$\chi_k = \max\{\bar{R}_T - R', 0\}. \tag{12.12}$$

It has to be noted that the criterion in (12.12) slightly differs from that in (12.8) as we put $Q(N)$ ($\bar{R}_T$) in place of $Q(N+1)$. This is due to the difficulty in expressing and quantifying Δ_{QoS}^K without a precise traffic model. As we show, this has little or no impact but we are currently investigating alternatives and their merit.

Furthermore, by using this setup speech quality is assessed on an aggregate level with a method that was originally designed to assess individual call quality. Whether this makes sense is discussed in the following; see Section 12.5.4.2. At least from a model point of view there is little difference in computing $\bar{I}_e$ on an aggregate or call level. What is required in either case are the loss ratio, burst length and gap length. The single difference is the number of packets received (or lost) to trigger state transition, which is 16 for a single call, cf. Section 12.4.3.

In order to translate this trigger threshold to an aggregate level we apply a simple intuitive approach. The AC algorithm knows at any time the number of admitted flows N. By assuming that VoIP traffic can be modeled by a standard exponential on/off model with an average sojourn time in the on (talk) state of 300 ms and a mean off (silence) of 600 ms (Markopoulou *et al.*, 2003), we know that each flow is active (on) for roughly a third of its life time. We further assume that any contribution as well as impact on an individual call scales linearly with the number of calls aggregated. In other words, the contribution of any call as well as the impact on it equals in average that of any other call. On the basis of this assumptions we set the number of packets received (or lost) to trigger state transitions to $16 \times N \times 0.33$.

12.5.3   Experimental Setup and Parameterization

In order to evaluate the concept and performance of the algorithm, we implemented it in the NS-2 framework[2]. The basic scenario has been already described in Section 12.5.1 and

---

[2]See http://www.isi.edu/nsnam/ns.

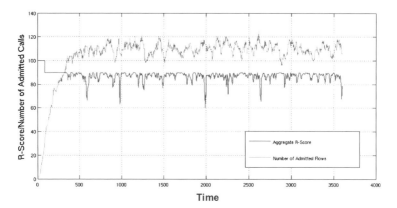

Figure 12.7   The number of admitted flows (upper curve) at the time of an admission request. The aggregate R-score is estimated (lover curve) which serves as admission criterion whenever a new call arrives. For this simulation R' in (12.12) was set to 85 (MOS: satisfied) and this target has been closely achieved over for the time the system remains in a steady state.

complies with an evaluation scenario defined by WEIRD. In this scenario a preprovisioned SF is set for VoIP. By definition this implies a contracted and assured capacity at any time and if there is any channel instability, it is accounted for by proper planning, compensated for by scheduling decisions or any other mechanism. We further assume no packet loss over the air interface by means of retransmissions or appropriate network planning. In such an scenario the UL and DL AC function, placed in the SS or BS are equivalent and allow for a reduction in the simulation setup to a single server queue with fixed capacity.

The preprovisioned link capacity of the respective SF, called the minimum reserved traffic rate in IEEE 802.16 QoS terminology, has been set to 2 Mb/s and the buffer has a length of 30 packets. Call arrivals follow a Poisson process with mean arrival time of 2 s and the holding time is exponentially distributed with mean 210 s.

VoIP traffic was generated by a G.711 coder with voice fames of 20 ms length. The standard exponential on/off model is used to model talk and silent periods where average sojourn time in the on state is 300 ms and mean off time 600 ms (Markopoulou *et al.*, 2003).

Admission control is implicit, cf. Mortier *et al.* (2000), and new calls are detected at the first packet arrival. The algorithm's window length, the past time over which speech quality is assessed, is set to 300 s in order to cover calls of average length. All simulations run for 3600 simulated seconds and the first 500 s are discarded to evaluate the system in a steady state.

12.5.4 Performance Results

12.5.4.1 Admission Control Accuracy

One of fundamental problem of MBAC is precision and only a few algorithms tackle this issue. Hence, we first investigate how closely the algorithm approaches a demanded QoE objective.

Table 12.1 Results of the simulative performance evaluation for different target R-scores, i.e. speech quality levels.

R'	$\bar{R}_{T,\mu}$	R_σ	$R_{\min}$	$R_{\max}$	$t^{\max}_{R<R'}$
80	84.67	4.44	63.36	89.94	31,73
82	85.55	4.16	68.37	89.95	29.66
84	86.53	3.57	66.04	89.98	28.12
86	87.64	2.77	67.05	89.97	37.76
88	88.63	1.71	77.88	89.97	43.22

For the first simulation R' in (12.12) was set to 85 and as shown in Figure 12.7, this target was achieved for most of the time. Skipping transient state the average estimated R-score ($\bar{R}_{T,\mu}$) for the remaining time was 86.82, standard deviation $R_\sigma = 3.89$, $R_{\min} = 59.97$ and $R_{\max} = 98.89$. In addition, we computed the longest continuous period below R', $t^{\max}_{\bar{R}_T<R'}$ and found a value of 39.87 s. We repeated this simulation for different R' in the range [80, 90], which maps on MOS to 'Satisfied'. The results are listed in Table 12.1.

While this results indicate a relative consistent performance, the AC appears a bit too conservative for lower R' values. Perhaps more important in the context of traffic aggregation and statistical QoE is that the average R-score was slightly above R' for all simulations. Among the remaining parameters, $t^{\max}_{R<R'}$ certainly holds the most interesting information. At first sight the maximum duration seems relatively large compared with an average holding time of 210 s. However, the maximum alone does not tell us much and in Figure 12.8 we plot the Cumulative Distribution Function (CDF) of the times $\bar{R}_T$ remained below R', denoted by $t_{\bar{R}_T<R'}$.

This figure further indicates consistency as the curves are very similar. For the whole range of R', the average time $\bar{R}_T$ remains below R' is approximately 10 s and the probability that $t_{\bar{R}_T<R'}$ is larger than 20 s is roughly 0.2. This qualifies the large value for $t^{\max}_{R<R'}$.

12.5.4.2 QoE Performance on the Call Level

Speaking in general terms, what has been achieved by now is an algorithm that can statistically guarantee a predefined application layer metric. However, how meaningful is this metric on the call level? Can we assume that an R-score measured and maintained on aggregate level applies to individual calls too?

In order to find this out we ran the same set of simulations as before, selected randomly 100 consecutively admitted calls and recorded their loss process. We then used the same IQA but with an adjusted state transition trigger to evaluate single call QoE, see Section 12.5.3. The question we tried to answer is how many calls receive the contracted QoE. We therefore assessed the QoE for each call's total life time and Figure 12.9 plots the CDF of these calls R-scores. The figure shows that for each QoE target R' maximally around 5% of calls are rated below $R = 80$, which is the lower threshold for 'satisfied' on the MOS scale, cf. Figure 12.3. Taking the first simulation, depicted in Figure 12.7, as an example this means that approximately 6 calls out of 110 concurrently admitted calls on average would be affected by lower QoE than contracted. Yet some of these calls fall still in the range

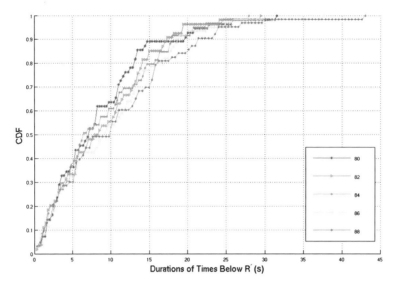

Figure 12.8 CDF of $t_{\bar{R}_T < R'}$. The probability that $t_{\bar{R}_T < R'}$ is larger than $1/10$ of the holding time, in other words that the quality is below the requested one for one-tenth of a calls life time, is approximately 0.2.

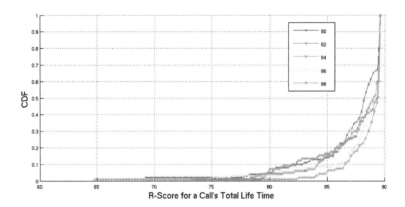

Figure 12.9 CDF of single call quality for a set of randomly recorded calls for all simulations. For each simulation, less than 5% of calls fall below $R = 80$.

$R = [70, 80]$ which maps to MOS 'some users dissatisfied', meaning that some of these may still be rated as 'satisfied'.

Finally, from Figure 12.7 we can draw conclusions with respect to configuration and QoE versus resource utilization tradeoff. If an operator aims at making sure that less than 2% of calls fall below $R = 80$ (MOS: satisfied), it should set $R' = 88$. Obviously, the higher the QoE demands, the lower the network utilization. Hence, an operator has to tradeoff between user satisfaction and resource utilization. It appears to us that configuring $R' = 84$ seems a good tradeoff since only 5% of calls experience QoE below MOS 'satisfied' while roughly half of them are still in the range of MOS 'some users dissatisfied'.

The conclusions of this work are manifold. On top of the list we found that the E-Model lends itself as a metric for QoE control by MBAC. The necessary computations are simple and do not add much burden on equipment. This opens the door to a new domain in VoIP QoS control, namely based on speech quality, the only reliable quality assessment method for VoIP. In support of this statement we found that the algorithm exhibits consistent and accurate behavior for a whole range of configurations. Probably the most intriguing conclusion, and somewhat specific to our setting, is that with a slightly modified measurement procedure we could apply the model on an aggregate level without compromising call level speech quality.

12.6   The Idea of an *R*-score-based Scheduler

In the previous part of the chapter the concept of a call admission control for VoIP services in a fixed WiMAX environment was introduced and its capability to keep the speech quality in terms of the *R*-score above a certain desired threshold was demonstrated by means of simulations. In this section, another novel concept is shortly discussed: to include measurements of the instantaneous speech quality for scheduler decisions in a mobile environment. The idea of this *R*-score-based scheduler is to handle temporary overload situations by maintaining a speech quality as good as possible. In a mobile packet-switched network, call admission control has to take into account both the on/off characteristics of VoIP calls and the temporal variations of the channel quality which again leads to a varying cell capacity. Guaranteeing a high speech quality by providing enough resources to transmit all packets even in temporary overload situations requires a rather conservative call admission control leading to a bad utilization. An aggressive call admission control, however, will achieve a high utilization while having periods of temporary overload where inevitably packets have to be dropped. The *R*-score scheduler will keep track of the instantaneous speech quality per call and consider this value in its decisions for scheduling and dropping packets. In the following, we provide a very first sketch of an *R*-score-based scheduler and evaluate it for the DL scheduling of a single mobile WiMAX cell with a number of VoIP calls large enough to cause a temporary overload situation. The IEEE 802.16 MAC layer is strictly connection oriented and specifies detailed QoS parameters such as maximum latency, minimum reserved traffic rate and maximum sustained traffic rate. Hence, the first goal of the scheduler is to meet these requirements. However, we focus on the situation where the scheduler will not be able to fulfill these requirements for all ongoing VoIP connections, and since all VoIP connection have identical QoS parameters we need another metric for distinguishing the precedence of the packets. The *R*-score measurements will fill this gap and serve as an additional metric.

12.6.1   Scenario

We consider a single cell mobile WiMAX deployment without inter-cell interference. Mobiles are moving around in a 1.5 km square with the BS at the center. Sectorization is not considered. The BS operates in Frequency Division Duplex (FDD) mode with a Partially Used Subchannelization (PUSC) subchannel allocation on a 1.25 MHz band. Assuming a frame length of 5 ms, results in 104 slots subdivided into 4 subchannels with 26 slots each.

Table 12.2 Overview of Modulation and Coding Schemes (MCS).

MCS	Bits/slots	Number of slots/packet	Number of blocks/packet	Required SNR
QPSK 1/2	48	35	6	8.7
QPSK 3/4	72	23	6	10.6
16QAM 1/2	96	18	6	15.8
16QAM 3/4	144	12	6	17.5
64QAM 1/2	144	12	6	21.9
64QAM 2/3	192	9	9	23.9
64QAM 3/4	216	8	8	25.1

Ignoring control traffic and assuming that only one-third of the slots is reserved for VoIP traffic, we end up with 35 slots for VoIP transport per frame. With the G711 speech codec and 20 ms framing, a VoIP packet has a size of 1648 bits including RTP, UDP, IP and MAC header. Depending on the Modulation and Coding Scheme (MCS) a VoIP packet occupies 35 slots with QPSK and 1/2 coding, and 8 slots with 64QAM and 3/4 coding. QPSK with 1/2 coding is the most robust MCS, repetition coding is not considered. The BS receives feedback on the channel quality per mobile in terms of the mean Signal-to-Noise Ratio (SNR) averaged over all subcarriers. In the simulation evaluation, we also use the mean SNR for determining frame errors on an Additive White Gaussian Noise (AWGN) channel. Consequently, in the simulation the BS has perfect knowledge of the channel without feedback delay. The MCS is selected in order to keep the loss probability for an entire VoIP packet below 1%. Since a VoIP packet does not fit into a single coding block, it comprises several coding blocks and, consequently, the product of the frame error rates of the coding blocks has to be below the target packet loss probability of 1%. The rather low packet loss rate of 1% is required since unacknowledged mode is chosen for VoIP traffic transport. Table 12.2 gives an overview of the MCS with required SNR, number of coding blocks and number of slots.

The task of the scheduler is now to decide which packets to transmit in a frame. We do not support fragmentation or packing so a VoIP packet corresponds to a MAC PDU and has to be transported as a whole. Consequently, one frame is just enough to transmit a single QPSK 1/2 VoIP packet. In order to avoid the transmission of too delayed packets we drop packets after a threshold t_{drop}.

12.6.2   The Most Simple *R*-Score Scheduler

The term R-score scheduler relates to any scheduler that uses concurrent measurements of the R-score for making its scheduling decisions. In general, there are two different ways how the BS might obtain R-score measurements: first, the BS can measure the R-score based on its own transmissions. Second, the mobile can measure the R-score based on arriving and missing packets, and signal the R-score to the BS. Both kinds of R-score measurements are based on a series of received and lost packets as described in Section 12.4.2. In fact, the computational complexity of continuously updating the R-score is very low since it is mostly just equal to increasing a counter. Only phase transitions require a more complex operation.

In the following, we estimate the R-score at the BS. We assume that an intact and in order series of VoIP packets arrives at the BS. Then, we assume that a packet that is scheduled for transmission is received independent of its frame error rate. Only those packets are marked as lost that are not scheduled but dropped when the threshold t_{drop} is exceeded. Based on this sequence of transmitted and dropped packets the R-score is evaluated. Please note that this R-score is not equal to the R-score experienced at the MS but more optimistic since erroneous packet transmissions are not included.

The R-score scheduler sorts the packets in an ascending order of the measured R-score. The packets of the mobile with lowest R-score are transmitted first, that is, we could also denote the scheduler as *least R-score first*. We add two more criteria to this order of packets. First, the kth packet of a mobile receives a penalty of $k - 1$. Second, mobiles with a SNR of less than 6 dB are scheduled last since here the packet loss probability exceeds 20% even for the most robust MCS. Consider this example: mobile A has an R-score of 84.2, two packets to transmit, and an SNR of 15 dB. Mobile B has an R-score of 83.6, also two packets to transmit, and an SNR of 25.6 dB. Mobile C has an R-score of 65, one packet to transmit and an SNR of 4 dB. Then, the packets are scheduled in the following order:

Mobile	Packet	R-score	SNR	Metric
B	1	83.6	25.6	(0, 83.6)
A	1	84.2	15	(0, 84.2)
B	2	83.6	25.6	(0, 84.6)
A	2	84.2	15	(0, 85.2)
C	1	65	4.5	(1, 65)

Please note that this scheduler is still a more or less channel oblivious scheduler since the only information it takes into account is that a channel is currently too bad to be used. The packets are reordered after every frame since the R-score or the SNR might have changed, new packets may have arrived or packets that are too old may have been dropped.

12.6.3 Performance Evaluation

In this section we intend to demonstrate the potential of R-score-based schedulers by comparing our *most simple R-score scheduler* with a channel-oblivious indexscheduler!Earliest Deadline First (EDF)Earliest Deadline First (EDF) scheduler that in this scenario degrades to a simple First In First Out (FIFO) scheduler and a channel-aware MaxSNR scheduler that transmits the packets in order of a decreasing SNR. For the MaxSNR scheduler we also use a penalty of $k - 1$ for the kth packet of one mobile. For the EDF scheduler we also schedule packets with an SNR below 6 dB last.

In the following, we compare the three schedulers in a scenario with 25 mobiles and two different multi-path channel profiles: ITU Ped. A (PA) and ITU Ped. B (PB). The other simulation parameters are equal in the two scenarios and listed in the Table 12.3.

For evaluating the quality of a scheduler we observe the distribution of the R-scores experienced at the subscriber stations at the end of a 60 s simulation run. A scheduler works well if no or only few VoIP calls experience bad quality. This means that we want to achieve a homogeneous R-score which is as high as possible. Let us first study the R-score in a scenario where only a single user is present and a VoIP packet is scheduled as soon as it arrives without considering the channel quality. We observe the resulting R-score with 30 independent mean SNR traces for both the ITU Ped. A and the ITU Ped. B channels. Figure 12.10 shows the

Table 12.3 Scenario parameters.

Parameter	Value
Shadow fading	
Standard deviation	$\sigma = 8$ dB
Decorrelation distance	$X_c = 50$ m
Mobile velocity	$v = 3$ km hour^{-1}
Path loss	COST231
$PL(d) = 147.06 + 35.74 * \log_1 0(d[\text{km}])$	
Transmit power	$T_x = 20$ W
Antenna gain	$G = 10$ dB
VoIP call	
On phase	Exp(300 ms)
Off phase	Exp(600 ms)

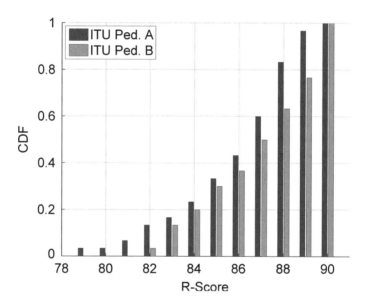

Figure 12.10 Achievable *R*-score without capacity constraints.

CDF of the *R*-scores that are achieved for single users without any capacity constraints. They represent the upper bound of what is achievable by the schedulers. We can see that the PA profile leads to somewhat lower *R*-scores than the PB profile. In particular, with PB we have in 7 of the 30 cases a maximum *R*-score of 90 while we achieve this *R*-score only once with PA. Observing the low *R*-score values, PB has a minimum of 82 whereas PA may lead to values as low as 79. Still, all of these values correspond to an acceptable or good speech quality.

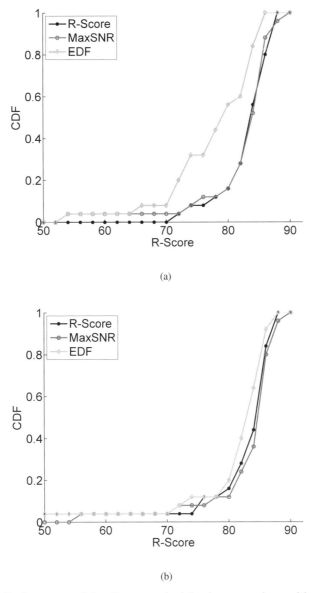

(a)

(b)

Figure 12.11 Performance of the *R*-score scheduler in comparison with channel-oblivious EDF and channel-aware MaxSNR scheduler: (a) ITU Ped. A; (b) ITU Ped. B.

Now, let us compare the impact of the different schedulers when we have a system with light overload, such as with 25 mobiles. Figure 12.11 shows the CDF of the R-score for the three schedulers with Figure 12.11(a) presenting the results for the PA multi-path channel profile and Figure 12.11(b) presenting the results for the PB profile. First, we can observe that in the light overload situation the R-score degrades down to 50 for some of the users while others still experience an excellent speech quality with an R-score close to 90. This holds for both channel profiles. If we observe the performance of the three schedulers in the PB profile, we can see that the three schedulers show marginal differences: only the EDF scheduler performs a little worse than the others. Figure 12.11(a) confirms this observation. Here, the EDF scheduler is clearly worse than the R-score and MaxSNR scheduler. At first glance, we also note that MaxSNR and R-score scheduler show an almost identical performance as in the PB case. If we have a closer look, we detect that for the high R-scores the performance is the same. For the low R-scores, however, the MaxSNR yields one R-score of below 55 while the R-scores obtained by the R-score scheduler are all above 70. This means that at least in this special case the R-score scheduler outperforms the MaxSNR scheduler and is able to avoid the strong degradation of the speech quality of the worst call.

In general, we can state that the 'most simple R-score scheduler' seems to be clearly better than the channel-oblivious EDF or FIFO scheduler. In addition, the performance of the channel-oblivious R-score scheduler is comparable and at least in one case better than the performance of the channel-aware MaxSNR scheduler which is quite remarkable. As a summary from the first experiences with an R-score scheduler, we conclude that the results are quite promising and encourage the further development of QoE aware schedulers not only for VoIP but also for other services such as Video on Demand (VoD), etc. The current simulation results are of course only examples and have to be further confirmed. In particular, the impact of different parameters such as the shadowing model, the velocity, the dropping threshold t_{drop} and many more has to be investigated. Furthermore, the simulation scenario should be more realistic in the sense that, for example, we should not assume a perfect packet arrival process at the BS, perfect knowledge of the channel or the unacknowledged mode. On the other hand, we also feel that there is still a lot of potential for improving the scheduler by combining different metrics for the scheduling decision, in particular current speech quality, current channel quality and urgency of the packet.

12.7   Conclusion

In this chapter we have presented the idea to utilize concurrent measurements of the instantaneous QoE for an improved resource control. We showed two examples of this QoE-based resource control at the scenario of the VoIP service in a WiMAX network. In the first example, a measurement-based admission control scheme for a fixed WiMAX deployment was proposed. The admission decision depends on a certain threshold for the aggregated R-score of ongoing calls. In the second example, a scheduling scheme for the DL of a mobile WiMAX BS is proposed. The scheduler uses the R-score as ordering metric and transmits the packets in a least R-score first fashion.

In both examples the proposed R-score-based resource control schemes show great promise through preliminary results. A more sophisticated evaluation of their performance is required and there is also room for optimization. Furthermore, using the R-score as a QoE metric for speech quality based resource control is only the first and probably most simple example for QoE-based resource control. Of great interest in the future will be QoE-based resource control schemes for different service composites including speech, video or gaming applications.

References

Blake S., Black D., Carlson M., Davies E., Wang Z. and Weiss W. (1998) *An Architecture for Differentiated Services, Request for Comments 2475*, IETF.

Bohnert, T.M. and Monteiro, E. (2007) Multi-class measurement based admission control for a QoS framework with dynamic resource management. *Journal of Network and Systems Management,* **15**(2), 219–240.

Bohnert, T.M., Monteiro, E., Curado, M., Fonte, A., Moltchanov, D., Koucheryavy, Y. and Ries, M. (2007) Internet quality of service: a bigger picture. *Proceedings of the First OpenNet QoS Workshop 'Service Quality and IP Network Business: Filling the Gap'*, Diegem, Belgium.

Braden, R., Clark, D. and Shenker, S. (1994) Integrated services in the Internet architecture: An Overview, *Request for Comments 1633*, IETF.

Clark, A. (2001) Modeling the effects of burst packet loss and recency on subjective voice quality. *Proceedings of the IP Telephony Workshop.*

Cole, R.G. and Rosenbluth, J.H. (2001) Voice over ip performance monitoring. *ACM SIGCOMM Computer Communication Review*, **31**(2), 9–24.

Cox, R. and Perkins, R. (1999) Results of a Subjective Listening Test for G.711 with Frame Erasure Concealment, *Committee contribution T1A1.7/99-016.*

Crawley, E., Nair, R., Rajagopalan, B., and Sandick, H. (1998) A framework for QoS-based routing in the internet, *Request for Comments 2386*, IETF.

Davie, B., Chany, A., Baker, F., Bennet, J.C.R., Benson, K., Bedeuc, J.Y.L., Chui, A., Courtney, W., Davari, S., Firoiu, V., Kalmanek, C., Ramakrishnan, K. and Stiliadis, D. (2002) An Expedited Forwarding PHB (Per-Hop Behavior), *Request for Comments 3246*, IETF.

Eklund, C., Marks, R.B., Ponnuswamy, S., Stanwood, K.L. and van Waes, N.J.M. (2006) Wireless-MAN Inside the IEEE 802.16 Standard for wireless metropolitan networks. *IEEE Standards Wireless Networks Series* (1st edn). IEEE Press, Piscataway, NJ.

Gozdecki, J., Jajszczyk, A. and Stankiewicz, R. (2003) Quality of service terminology in IP networks. *IEEE Communications Magazine* **41**(3), 153–159.

Hardy W.C. (2001) *QoS Measurement and Evaluation of Telecommunications Quality of Service*. John Wiley & Sons Ltd, Chichester.

Hoßfeld, T. and Binzenhöfer, A. (2008) Analysis of skype voIP traffic in UMTS: End-to-end QoS and QoE measurements. *Computer Networks.*

Hoßfeld, T., Hock, D., Tran-Gia, P., Tutschku, K. and Fiedler, M. (2008) Testing the IQX hypothesis for exponential interdependency between QoS and QoE of voice codecs iLBC and g.711. *Proceedings of the 18th ITC Specialist Seminar on Quality of Experience*, Karlskrona, Sweden.

Hoßfeld, T., Tran-Gia, P. and Fiedler, M. (2007) Quantification of quality of experience for edge-based applications. *Proceedings of the 20th International Teletraffic Congress (ITC20)*, Ottawa, Canada.

IEEE (2005) Part 16: Air Interface for Fixed Broadband Wireless Access Systems, Amendment 2, *IEEE Standard for Local and Metropolitan Area Networks*, IEEE.

ITU-T (1993) Terms and Definitions Related to Quality of Service and Network Performance Including Dependability, *ITU-T Rec. E.800*, ITU-T.

ITU-T (1996) *Objective Quality Measurement of Telephone-Band (300-3400 Hz) Speech Codecs.* International Telecommunication Union, Geneva, Switzerland.

ITU-T (1998) The Emodel: a computational model for use in transmission planning, *ITU-T Recommendation G.107,* ITU-T.

ITU-T (2001) *Perceptual Evaluation of Speech Quality (PESQ), an Objective Method for End-to-end Speech Quality Assessment of Narrowband Telephone Networks and Speech Codecs*, International Telecommunication Union, Geneva, Switzerland.

ITU-T (2004) Definition of quality of experience, *ITU-T Rec. D.197*, ITU-T.

Janssen, J., De Vleeschauwer, D. and Petit, G.H. (2000) Delay and distortion bounds for packetized voice calls of traditional pstn quality. *Proceedings of the First IP-Telephony Workshop (IPTel 2000)*, Berlin, pp. 105–110.

Janssen, J., De Vleeschauwer, D. D., Buechli, M. and Petit, G.H. (2002) Assessing voice quality in packet-based telephony. *IEEE Internet Computing*, **6**(3), 48–56.

Jiang, W. and Schulzrinne, H. (2000) Analysis of on–off patterns in VoIP and their effect on voice traffic aggregation. *Proceedings of the Ninth IEEE International Conference on Computer Communication Networks*.

Kostas, T.J., Borella, M.S., Sidhu, I., Schuster. G.M., Brabiec, J. and Mahler, J. (1998) Real-time voice over packet-switched networks. *IEEE Network*, **12**, 1827.

Markopoulou, A.P., Tobagi, F.A. and Karam, M.J. (2003) Assessing the quality of voice communications over internet backbones. *IEEE Transactions on Networking*, **11**(5), 747–760.

Meddahi, A., Afifl, H. and Zeghlache, D. (2003) Packet-E-model: E-model for wireless VoIP quality evaluation. *Proceedings of the 14th IEEE Proceedings on Personal, Indoor and Mobile Radio Communications. (PIMRC 2003)*, vol. 3, pp. 2421–2425.

Mortier, R., Pratt, I., Clark, C. and Crosby, S. (2000) Implicit admission control. *IEEE Journal on Selected Areas in Communications*, **18**(12), 2629–2639.

Raake, A. (2006a) Short- and long-term packet loss behavior: towards speech quality prediction for arbitrary loss distributions. *IEEE Transactions on Audio, Speech, and Language Processing*, **14**(6), 1957–1968.

Raake, A. (2006b) *Speech Quality of VoIP*, John Wiley & Sons Ltd, Chichester.

Sengupta, S., Chatterjee, M., Ganguly, S. and Izmailov, R. (2006) Improving *R*-score of VoIP streams over WiMAX. *Proceedings IEEE International Conference on Communications (ICC '06)*, vol. 2, Istanbul, Turkey, pp. 866–871.

Takahashi. A, Kurashima, A. and Yoshino, H. (2006) Objective assessment methodology for estimating conversational quality of VoIP. *IEEE Transactions on Audio, Speech and Language Processing* **14**(6), 1984–1993.

Teitelbaum, B. and Shalunov, S. (2002) Why premium IP service has not deployed (and probably never will), *Technical Report, Internet2*.

WEIRD Consortium (2007) D.2.2 System Scenarios, Business Models and System Requirements (version 2), *Project Deliverable*, WEIRD Consortium.

Watson, A. and Sasse, M.A. (1998) Measuring perceived quality of speech and video in multimedia conferencing applications. *International Multimedia Conference, Proceedings of the Sixth ACM International Conference on Multimedia*, Bristol, UK. ACM, pp. 55–60.

13

VoIP over WiMAX

Rath Vannithamby and Roshni Srinivasan

13.1 Introduction

Voice over Internet Protocol (VoIP) provides an alternative to the telephone service offered by the traditional Public Switched Telephone Network (PSTN) by using an IP network to carry digitized voice. Packet switched air interfaces that support flat IP architectures have now made it possible to run VoIP applications over wireless technology.

Compression/Decompression (CODEC) techniques for VoIP transform audio signals into digital bit streams. While preserving voice quality, speech samples are further compressed to produce bit streams of 8–12 kbps that are carried over the IP network. The compressed speech sample is then is transmitted using the Real-time Transport Protocol (RTP) over the User Datagram Protocol (UDP) over the Internet Protocol (IP).

VoIP over wireless networks is affected by the choice of CODEC and packet loss, delay and jitter. Fluctuating channel conditions typically cause packet loss and increased latency. In order to keep mouth-to-ear round trip latencies to reasonable levels of 250–300 ms, the delay budget for transmission over the air interface is 50–80 ms. The CODEC, jitter buffer and backbone account for the remaining delay. Channel aware scheduling with Quality of Service (QoS) differentiation, Hybrid Automatic Repeat Request (HARQ) and dynamic link adaptation are used to keep delays within acceptable limits. Jitter buffers are used to compensate for delay jitter experienced by packets due to network congestion, timing drift or route changes.

Third Generation (3G) systems such as 1xEV-DO (Ericson *et al.*, 2006) and High Speed Packet Access (HSPA) (Yavuz *et al.*, 2006) and Fourth Generation (4G) standards such as 802.16e (IEEE, 2006, WiMAX Forum, 2006), Long Term Evolution (LTE) (3GPP, 2007) and Ultra Mobile Broadband (UMB) (3GPP2, 2007) have several features that have been optimized for VoIP support. Fine-grained resource multiplexing to pack many small VoIP

WiMAX Evolution: Emerging Technologies and Applications   Edited by Marcos D. Katz and Frank H.P. Fitzek
© 2009 John Wiley & Sons, Ltd

packets in the available spectrum results in significant overhead to signal resource allocation. This inherent trade-off has led to several innovative techniques to improve VoIP capacity. This chapter provides an overview of VoIP over WiMAX. It includes a detailed description of essential and advanced mobile WiMAX features that support VoIP as well as system simulation results with VoIP capacity based on the mobile WiMAX system profile (WiMAX Forum, 2007).

The chapter is structured as follows. Section 13.2 describes the essential features in the 802.16e standard that support VoIP. Enhanced features including support for persistent scheduling are described in Section 13.3. System simulation results to determine mobile WiMAX VoIP capacity are described in Section 13.4. Finally, Section 13.5 provides a summary and concludes the chapter.

13.2   Features to Support VoIP over WiMAX

802.16e provides a number of features to support VoIP. Prioritization of delay-sensitive VoIP traffic is achieved through the classification of flows into scheduling classes. Voice activity detection and Extended Real-Time Polling Service (ertPS) conserve air link resources during periods of silence. HARQ and channel aware scheduling are used reduce transmission latency over the airlink. Protocol header compression is supported to transport the speech sample efficiently.

13.2.1   Silence Suppression using ertPS

Mobile WiMAX supports QoS requirements for a wide range of data services and applications by mapping those requirements to unidirectional service flows that are carried over Uplink (UL) or Downlink (DL) connections. Table 13.1 describes the five QoS classes, Unsolicited Grant Service (UGS), Real-Time Polling Service (rtPS), ertPS, Non-Real-Time Polling Service (nrtPS) and Best Effort (BE) service, used to provide service differentiation by the Medium Access Control (MAC) scheduler.

In the absence of silence suppression, service requirements for VoIP flows in 802.16e (IEEE, 2006) are ideally served by the UGS, which is designed to support flows that generate fixed size data packets on a periodic basis. The fixed grant size and period are negotiated during the initialization process of the voice session.

Service flows such as VoIP with silence suppression generate larger data packets when a voice flow is active, and smaller packets during periods of silence. rtPS is designed to support real-time service flows that generate variable size data packets on a periodic basis. rtPS requires more request overhead than UGS, but supports variable grant sizes. In conventional rtPS, a bandwidth request header is sent in a unicast request opportunity to allow the Subscriber Station (SS) to specify the size of the desired grant. The desired grant is then allocated in the next UL subframe.

Although the polling mechanism of rtPS facilitates variable sized grants, using rtPS to switch between VoIP packet sizes when the SS switches between the talk and silent states introduces access delay. rtPS also results in MAC overhead during a talk spurt since the size of the VoIP packet is too large to be accommodated in the polling opportunity, which only accommodates a bandwidth request header. The delay between the bandwidth request and subsequent bandwidth allocation with rtPS could violate the stringent delay constraints

Table 13.1 QoS classes supported by mobile WiMAX.

Service	Description	QoS parameters
UGS	Support for real-time service flows that generate fixed-size data packets on a periodic basis, such as VoIP without silence suppression	Maximum sustained rate Maximum latency tolerance Jitter tolerance
rtPS	Support for real-time service flows that generate transport variable size data packets on a periodic basis, such as streaming video or audio	Minimum reserved rate Maximum sustained rate Maximum latency tolerance Traffic priority
ertPS	Extension of rtPS to support traffic flows such as variable rate VoIP with Voice Activity Detection (VAD)	Minimum reserved rate Maximum sustained rate Maximum latency tolerance Jitter tolerance Traffic priority
nrtPS	Support for non-real-time services that require variable size data grants on a regular basis	Minimum reserved rate Maximum sustained rate Traffic priority
BE	Support for best-effort traffic	Maximum sustained rate Traffic priority

of a VoIP flow. rtPS also incurs a significant overhead from frequent unicast polling that is unnecessary during a talk spurt.

The ertPS scheduling algorithm improves upon the rtPS scheduling algorithm by dynamically decreasing the size of the allocation using a grant management subheader or increasing the size of the allocation using a bandwidth request header. The size of the required resource is signaled by the MS by changing the Most Significant Bit (MSB) in the transmitted data. The state transitions for ertPS are shown in Figure 13.1.

13.2.2 HARQ

In addition to link adaptation through channel quality feedback and adaptive modulation and coding, HARQ is enabled in 802.16e using the 'stop and wait' protocol, to provide a fast response to packet errors at the Physical (PHY) layer. Chase combining HARQ is implemented to improve the reliability of a retransmission when a Packet Data Unit (PDU) error is detected. A dedicated Acknowledgment (ACK) channel is also provided in the uplink for HARQ ACK/Negative Acknowledgment (NACK) signaling. UL ACK/NACKs are piggybacked on DL data. A multi-channel HARQ operation with a small number of channels is enabled to improve the efficiency of error recovery with HARQ. Mobile WiMAX also provides signaling to allow asynchronous HARQ operation for robust link adaptation in mobile environments.

The one-way delay budget for VoIP on the DL or the UL is limited between 50 and 80 ms. This includes queuing and retransmission delay. Enabling HARQ retransmissions for error

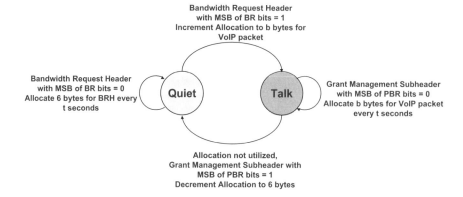

Figure 13.1 Extended rtPS state transitions.

recovery significantly improves the ability of the system to meet the stringent delay budget requirements and outage criteria for VoIP.

13.2.3   Channel Aware Scheduling

Unidirectional connections are established between the BS and the MS to control transmission ordering and scheduling on the mobile WiMAX air interface. Each connection is identified by a unique Connection Identification (CID) number. Every MS, when joining a network, sets up a basic connection, a primary management connection and a secondary management connection. Once all of the management connections are established, transport connections are set up. Traffic allocations on the DL and the UL are connection based, and a particular MS may be associated with more than one connection.

In every sector, the Base Station (BS) dynamically schedules resources in every Orthogonal Frequency Division Multiple Access (OFDMA) frame on the UL and the DL in response to traffic dynamics and time-varying channel conditions. Link adaptation is enabled through channel quality feedback, adaptive modulation and coding and HARQ. Resource allocation on the DL and UL in every OFDMA frame is communicated in Mobile Application Part (MAP) messages at the beginning of each frame. The DL-MAP is a MAC layer message, which is used to allocate radio resources to Mobile Stations (MSs) for DL traffic. Similarly, the UL-MAP is a MAC layer message used to allocate radio resources to the MSs for UL traffic. The BS uses information elements within the DL-MAP and UL-MAP to signal traffic allocations to the MS.

The BS scheduler also supports resource allocation in multiple subchannelization schemes to balance delay and throughput requirements with instantaneous channel conditions. For permutations such as Partially Used Subchannelization (PUSC), where subcarriers in the subchannels are pseudo-randomly distributed across the bandwidth, frequency diversity offers robustness at high mobility. Frequency selective scheduling gains can be exploited with contiguous permutations such as Adaptive Modulation and Coding (AMC) since the subchannels may experience different quality.

13.2.4   Protocol Header Compression

The speech payload from the Adaptive Multi-Rate (AMR) vocoder (3G, 2007) operating at 12.2 kbps is 33 bytes every 20 ms in the active state and 7 bytes every 160 ms in the inactive state. This payload is typically carried over RTP, UDP and IP. Protocol headers associated with RTP, UDP and IP constitute 40 bytes with IPv4 and 60 bytes with IPv6. Excluding the 6 byte MAC header and 2 byte HARQ Cyclic Redundancy Check (CRC), it can be seen that a significant portion of the VoIP packet transmitted over the air interface includes protocol overheads. The fraction of overhead from protocol headers is even greater for VoIP packets carrying speech samples from CODECs operating at lower bit rates (7.95 kbps) such as Enhanced Variable Rate Codec (EVRC) or G.729.

To reduce the protocol header overhead, header compression techniques are typically used for VoIP. With Robust Header Compression (ROHC) (Bormann *et al.*, 2001), the protocol headers are compressed to about 3–4 bytes prior to transmission. Mobile WiMAX enables header compression with support for ROHC. The typical 802.16e VoIP packet size for AMR and G.729 CODECs is shown in Table 13.2.

13.3   Enhanced Features for Improved VoIP Capacity

In this section, we describe the characteristics of VoIP traffic that can be exploited to support efficient scheduling of VoIP packets and the associated control signaling. Enhanced features that provide significant improvement in VoIP capacity are also described in detail.

13.3.1   VoIP Traffic Characteristics

There are several characteristics of VoIP traffic that make VoIP packet scheduling challenging: (a) VoIP packets are small in size; (b) number of VoIP users supported in a given frequency band is large compared with the number of high data rate users that can be supported; (c) the packet inter-arrival time is roughly constant; and (d) speech includes periods of silence for roughly half the time and activity during the rest of the time.

The fact that the VoIP packet size is small makes the ratio of the resources needed for transmitting control information to schedule VoIP to the resources needed for actual VoIP traffic transmission much higher than that observed in data-only systems. Moreover, the high number of VoIP users supported in a given frequency band also adds to the total overhead required to transmit the control information related to the VoIP resource allocation.

In supporting high data rate applications, the focus is on optimizing the throughput, but in supporting VoIP the focus shifts towards delay sensitivity and minimizing the control overhead associated with the VoIP resource allocation.

13.3.2   Dynamic Resource Allocation for VoIP

To support VoIP in an OFDMA system, VoIP packets need to be scheduled on the DL and the UL within a fixed delay bound every time a packet arrives at the BS and at the MS, respectively. The OFDMA resources in frequency and time as well as transmit power and transmission mode need to be specified in each allocation. Furthermore, the MS identification and HARQ transmission related information also need to be specified. All this information

Table 13.2 802.16e VoIP packet size with AMR and G.729 CODECs.

Description	AMR without header compression IPv4/IPv6	AMR with header compression IPv4/IPv6	G.729 without header compression IPv4/IPv6	G.729 with header compression IPv4/IPv6
Voice payload 20 ms (aggregation interval)	7 bytes for inactive 33 bytes for active	7 bytes for inactive 33 bytes for active	0 bytes for inactive 20 bytes for active	0 bytes for inactive 20 bytes for active
Protocol headers (including UDP checksum)	40 bytes/ 60 bytes	3 bytes/ 5 bytes	40 bytes/ 60 bytes	3 bytes/ 5 bytes
RTP	12 bytes		12 bytes	
UDP	8 bytes		8 bytes	
IPv4/IPv6	20 bytes/ 40 bytes		20 bytes/ 40 bytes	
802.16e generic MAC header	6 bytes	6 bytes	6 bytes	6 bytes
802.16e CRC for HARQ	2 bytes	2 bytes	2 bytes	2 bytes
Total VoIP packet size	55 bytes/ 75 bytes for inactive 81 bytes/ 101 bytes for active	18 bytes/ 20 bytes for inactive 44 bytes/ 46 bytes for active	0 bytes/ inactive 68 bytes/ 88 bytes for active	0 bytes/ inactive 31 bytes/ 33 bytes for active

is sent using a robust Modulation and Coding Scheme (MCS), thereby consuming additional resources. In 802.16e, control information associated with resource allocation is signaled through MAP elements. Compressed MAPs can be used with subMAPs to reduce MAP overhead. The compressed MAP header is coded with the most robust MCS and subMAPs can be coded with higher order MCSs. Although compressed MAPs and subMAPs conserve resources compared to conventional MAPs, MAP overhead associated with the larger number of allocations for VoIP can be considerably high.

Dynamic scheduling for every VoIP packet incurs a significant amount of MAP overhead. The motivation for persistent scheduling comes from the fact that the VoIP traffic is periodic and generates constant size packets. As the name suggests, persistent scheduling conserves resources by persistently allocating resources that are required periodically. We discuss two different ways of persistently allocating the resources namely individual persistent

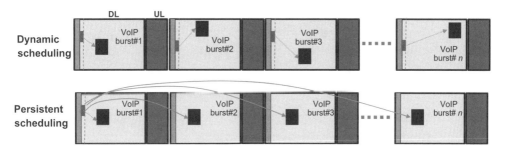

Figure 13.2 MAP assignment with dynamic scheduling and persistent scheduling.

scheduling and group scheduling. Individual persistent scheduling was developed (Bourlas *et al.*, 2008) and incorporated into the IEEE 802.16 Rev2 addendum (IEEE, 2008a). Group-based enhanced persistent scheduling techniques are currently in discussion in the IEEE 802.16 TGm working group for potential inclusion in the IEEE 802.16m standard.

13.3.3 Individual Persistent Scheduling

The basic idea behind individual persistent scheduling is that a user is assigned a set of resources for a period of time and the necessary information for the packet transmission are sent only once at the beginning of the assignment. For the rest of the period of allocation, the MS is assumed to know all of the information for data reception on the DL and data transmission on the UL. Note that the allocation period can be infinite. In other words, persistent scheduling is in effect until updated.

Figure 13.2 compares the operation of dynamic and persistent scheduling operation. In the case of dynamic scheduling, a MAP element is required to specify resource allocation information every time a VoIP packet is scheduled. On the other hand, in the case of persistent scheduling, resource allocation information is sent once in a persistent MAP element and not repeated in the subsequent frames. The additional resource that becomes available due to MAP overhead reduction can be used to increase VoIP capacity.

13.3.3.1 Resource allocation/deallocation for talk spurts/silence periods

As discussed earlier, VoIP users switch between talk spurts and silence. On the average, users in a typical VoIP call will be in either mode for a duration of the order of a second.

Every time a user goes into a talk spurt, resources need to be allocated with all of the information necessary to identify the allocation. The resource is allocated periodically with persistent scheduling as long as the user is in the active state. Similarly, every time the user goes into the silence mode, resources need to be deallocated. Since the frequency of allocation and deallocation of resource for conversational voice (50% voice activity factor) is typically once every 250 WiMAX frames (1.25 s), the overhead associated with a persistently scheduled allocation is small compared with the overhead in dynamic scheduling.

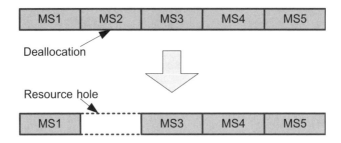

Figure 13.3 Example of a resource hole when MS2 is deallocated.

13.3.3.2 Link Adaptation/MCS Changes

In a mobile environment, the channel conditions are time varying. In order to be spectrally efficient, the MCS used for data transmission and reception needs to be adapted according to channel variations. Adjustment in MCS requires changes in the amount of allocated resources. As a result, every time the MCS needs to be adapted, the BS needs to deallocate or allocate a persistently scheduled resource. Depending on the frequency at which the MCS changes, signaling the changing persistent allocation could result in considerable overhead. Consequently, for fast link adaptation, individual persistent scheduling is not recommended, since the overhead involved in adapting to the channel variations will defeat the purpose of persistent scheduling.

13.3.3.3 HARQ Retransmission

Depending on the operating point chosen by the vendor or the network operator for initial transmissions, HARQ retransmission rates are typically in the range of 10–30%. Allocation of resources for HARQ retransmissions is an important consideration. Although persistent scheduling for HARQ retransmissions may be possible, dynamic allocation of resources is recommended.

13.3.3.4 Resource Holes

One issue with individual persistent scheduling is the associated resource packing inefficiency in the data portion of the frame. As different users transition between active spurts and periods of silence, or MCS changes occur due to link adaptation, resources need to be deallocated and allocated dynamically in addition to scheduling persistent allocations. Every time a resource is deallocated it may or may not be possible to find a user with the same resource request. If there is no match, holes can be created in the data region.

Figure 13.3 shows an example of the creation of resource allocation holes. Hole creation can also defeat the purpose of using persistent scheduling for efficient resource allocation.

13.3.3.5 Implicit and Explicit Allocation

Deallocation and allocation does not always need to be explicit. It is possible to implicitly allocate/deallocate resources to reduce the MAP overhead, for example user N_1 is allocated

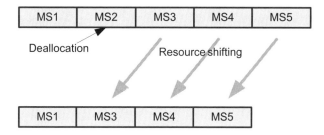

Figure 13.4 Example of resource shifting when MS2 is deallocated.

at location L_1 with MCS_1. When the user is deallocated and another user N_2 with the same MCS need to be allocated, the BS can simply allocate user N_2 at location L_1. User N_1 interprets this allocation to user N_2 as a deallocation of resources, thereby eliminating the need for explicit signaling of the deallocation for user N_1. The broadcast nature of the 802.16e MAP offers this advantage and provides a mechanism to further reduce the overhead.

13.3.3.6 Resource Shifting/Repacking

A mechanism to remove the holes that have been created using individual persistent scheduling is resource shifting or repacking. Basically, the BS broadcasts the size and location of the holes/empty spaces. Using this information, the remaining user allocations can be shifted upwards to account for the holes. Broadcasting information related to the empty space results in some overhead. When the benefit of making the resources available may be higher than the cost of broadcast overhead, it is desirable to perform the shifting operation. It is also possible to tie the shifting operation with each of the deallocation operations, that is, every time there is a deallocation and there is no allocation in its place, all allocations following the deallocated space are automatically shifted up.

Figure 13.4 shows an example of a resource shifting mechanism to pack allocations more efficiently and make more resources available.

13.3.3.7 Reliable MAP Reception

Since persistent scheduling allocates resources not only for the current frame, but also for future frames, the impact of losing MAP information associated with persistent scheduling is much higher than it is with dynamic scheduling. Hence, it is important to make the MAP transmission reliable. A MAP ACK channel can be used to ensure that the MS has received the persistent allocation. This ACK channel is very similar to the HARQ ACK channel. One issue with this approach is that the ACK channels increase overhead in the uplink. In order to reduce the ACK channel overhead in the uplink, shared NACK channels can be used to acknowledge subsequent MAP reception, that is, a MAP ACK channel is used for the initial persistent scheduling assignment, and a shared MAP NACK channel is used for the subsequent assignments. In the case of the shared MAP NACK channel, multiple users use the same NACK channel resource. If the BS receives a MAP NACK signal from one or more users, the incorrect reception of the MAP is detected. The BS can either retransmit the

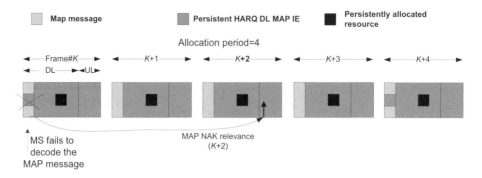

Figure 13.5 Example of MAP error handling operation.

MAP information to all users sharing the NACK channel or intelligently retransmit the MAP information to only those users who signaled the loss of the MAP information.

For the latter case, the BS can monitor the HARQ ACK, etc. for the MSs sharing the MAP NACK channel. Figure 13.5 shows an example of the operation of the error recovery procedure.

13.3.4   Group Scheduling

Group scheduling employs a persistent scheduling allocation intended not just for one user, but for a group of users. Groups can be generated based on the similarities observed in users' channel conditions, the codec used, etc. Once users are assigned to a group, the location of allocations for individual users is not fixed in the OFDMA resource, but the relative position in the group can be fixed if all of the users are active. If the group carries both active and silent users, a bitmap is needed to specify which users are active and which users are inactive. Typically, users with the same MCS or type of codec are grouped together. Knowledge of the allocation size and the bitmap is sufficient to identify the location of the allocation in the frame. This mechanism allows for complete resource packing and is very efficient in resource utilization. Frequent regrouping due to MCS changes and link adaptation, however, could result in significant overhead.

Several efficient grouping mechanisms are currently being discussed in IEEE 802.16m working group and the details are still under development. The performance improvements from these mechanisms on VoIP capacity are promising.

13.4   Simulation Results

Bidirectional VoIP capacity is measured in active users/sector. The VoIP capacity on the DL or the UL is defined to be the maximum number of MSs that can be supported while ensuring that 98% (97%) of packets are successfully received within the delay bound of 50 ms (80 ms) for 98% (95%) of the MSs. In order to determine VoIP capacity, the dynamics of VoIP traffic and the air interface are modeled in system level simulations.

A typical 19 cell network topology with wrap-around is used to model the system. MSs are dropped independently with uniform distribution throughout the system. In addition to accounting for wireless channel variations from path loss and shadow fading in system simulations, mobiles are assigned channel models to simulate fast fading corresponding to a mixed mobility scenario with 60% ITU Pedestrian-B at 3 km hour^{-1}, 30% ITU Vehicular-A at 30 km hour^{-1} and 10% ITU Vehicular A at 120 km hour^{-1}. Fading signal and fading interference are computed from each MS into each sector and from each sector to each mobile for each simulation interval. Channel quality feedback delay and Packet Data Unit errors are modeled and packets are retransmitted as necessary. Asynchronous, nonadaptive, chase combining HARQ is modeled by explicitly rescheduling a packet after a two-frame HARQ feedback delay period.

Each mobile in the system corresponds to an active VoIP session. A series of VoIP packets is generated using a simplified on/off Markov model of the AMR vocoder operating at 12.2 kbps and a voice activity factor of 50% (40%) to model the speech source dynamics. During active speech, the 44 byte VoIP packets are generated every 20 ms, and in periods of inactivity, 18 byte packets are generated every 160 ms. The packet size includes the vocoder payload, compressed RTP/UDP/IP header, 802.16e generic MAC header and CRC.

System parameters are configured according to the WiMAX Forum Wave II System Profile (Release 1.0) (WiMAX Forum, 2007) for PUSC. A TDD test scenario with frequency reuse of 1 corresponding to the baseline configuration and simulation assumptions in Section 2 of the 802.16m evaluation methodology (IEEE, 2008b) is used for performance evaluation. A 2×2 antenna configuration on the downlink and a 1×2 configuration on the uplink were modeled. Adaptive Multiple Input Multiple Output (MIMO) Switching (AMS) between open loop transmit diversity (matrix A) and spatial multiplexing (matrix B) was implemented at the BS. Collaborative Spatial Multiplexing (CSM) was implemented on the uplink. Transmission of MAP signaling was limited to one transmit antenna and two receive antennas.

MAP overhead is dynamically modeled in the simulations using compressed MAPs and three sub-MAPs. Channel aware scheduling and link adaptation are used in the simulations to model the VoIP packet allocations and the corresponding MAP overhead that change from frame to frame. The downlink overhead includes overhead for uplink assignments. In addition, MAP errors were taken into consideration.

A system bandwidth of 10 MHz was assumed for all simulations. The 5 ms Time Division Duplex (TDD) frame was split with 23 Orthogonal Frequency Division Multiplex (OFDM) symbols for the DL, and 24 symbols for the UL when dynamic scheduling was used. In the case of persistent scheduling using the baseline IE defined in IEEE (2008a) and Bourlas *et al.* (2008), 20 symbols were allocated to the DL , while the UL was allocated 27 symbols. Since the system is UL limited, the overhead reduction from persistent scheduling was used to increase the available UL resource and thereby increase overall bidirectional VoIP capacity. Table 13.3 provides a comparison of VoIP capacity with dynamic and persistent scheduling. The system simulation results for two definitions of outage criteria show a reduction in MAP overhead from about 10 symbols to about 6 symbols and a 15% improvement in system capacity. It must be noted that features such as AMC and closed loop MIMO that are currently supported in IEEE (2008a) and expected to provide additional capacity gains have not been simulated.

Table 13.3 Simulation results.

	VoIP capacity with dynamic scheduling (users/sector)	VoIP capacity with persistent scheduling (users/sector)
40% voice activity factor 80ms delay budget	190	220
50% voice activity factor 50ms delay budget	135	155

Results corresponding to two cases are presented in Table 13.3. One set corresponds to a more stringent delay budget of 50 ms, user and system outage criterion of 2% and a voice activity factor of 50%. The other set corresponds to a more relaxed delay requirement of 80 ms, user outage criterion of 3%, system outage criterion of 5% and a voice activity factor of 40%. When the outage criteria are stringent, voice quality can be enhanced at the expense of resource utilization. The results clearly illustrate this trade-off between voice quality and the VoIP capacity of the system.

The persistent scheduling techniques and the associated capacity gains discussed in this chapter are somewhat independent of the delay requirement and the outage criteria for VoIP. Persistent scheduling techniques do not affect the delay sensitive transmission of VoIP packets; persistent scheduling only affects the efficiency with which control information associated with VoIP packet transmission is signaled. As discussed earlier, the control signaling overhead associated with VoIP is very high compared with high throughput data applications. The capacity gain shown in Table 13.3 is a direct result of the control signaling overhead reduction for VoIP through individual persistent scheduling. The performance of group based scheduling techniques is not discussed here since the technology is still under development in the standard.

13.5   Conclusion

Support for VoIP is essential for operators to offer telephony services over wireless networks. This chapter provides an overview of WiMAX features and the functionalities they offer to enable VoIP service. In addition to these features, efficient resource allocation schemes that reduce that high MAP signaling overhead involved in supporting VoIP are described. 802.16e-based systems support a dynamic resource allocation scheme for scheduling delay-sensitive traffic such as VoIP. Simulation results show that persistent scheduling, a mechanism introduced in the 802.16 Rev2 addendum (IEEE, 2008a), reduces MAP overhead by 30–45% and increases the bidirectional WiMAX VoIP capacity by about 15% compared with the dynamic scheduling approach. By optimizing VoIP equipment, the wireless infrastructure and resource allocation, voice capacity can be increased significantly.

Acknowledgements

The authors would like to thank their colleagues, Apostolos Papthanassiou and Shailender Timiri in the Mobile Wireless Group and Shweta Shrivastava in the Corporate Technology Group at Intel Corporation, as well as Mohammad Mamunur Rashid, a student intern from the University of British Colombia, for fruitful discussions in defining and developing simulation models.

References

3G (2007) *3G Technical Specification 26.071, Mandatory speech CODEC speech processing functions; AMR speed CODEC; General description*, July 2007.

3GPP (2007) TS 36.201 V8.1.0, *Technical Specification Group Radio Access Network Evolved Universal Terrestrial Radio Access (E-UTRA); LTE Physical Layer - General Description* (Release 8), November 2007.

3GPP2 (2007) C.S0084 v2.0, *Ultra Mobile Broadband (UMB) Air Interface Specification*, September 2007.

Bormann, C. *et al.* (2001) *Request for Comments 3095, RObust Header Compression (ROHC): Framework and four profiles: RTP, UDP, ESP, and uncompressed*, July 2001.

Bourlas, Y., Etemad, K., Fong, M. Lavi, N., Lim, G., Lu, J. McBeath, S. and Oh, C. (2008), IEEE 802.16maint-08/095r4, Persistent Allocation, *IEEE 802.16 Broadband Wireless Access Working Group*.

Ericson, M., Wanstedt, S. and Ericson, P.J. (2006) Effects of simultaneous circuit and packet switched voice traffic on total capacity. *Proceedings of the 2006 Spring IEEE Vehicle Technology Conference*, May 2006.

IEEE (2006) 802.16e-2005, *IEEE Standard for Local and Metropolitan Area Networks – Part 16: Air Interface for Fixed and Mobile Broadband Wireless Access Systems - Amendment 2: Physical and Medium Access Control Layers for Combined Fixed and Mobile Operation in Licensed Bands and Corrigendum 1*, February 28, 2006.

IEEE (2008a) P802.16Rev2/D4, *Draft Standard for Local and Metropolitan Area Networks – Part 16: Air Interface for Fixed and Mobile Broadband Wireless Access Systems*, May 2008.

IEEE (2008b) 802.16m-08/004r1, *Project IEEE 802.16m Evaluation Methodology*, http://www.ieee802.org/16/tgm/docs/80216m-08_004r2.pdf, May 2008.

WiMAX Forum (2006) *Mobile WiMAX - Part I: A Technical Overview and Performance Evaluation*, White Paper, http://www.wimaxforum.org/news/downloads/Mobile_WiMAX_Part1_Overview_and_Performance.pdf, August 2006.

WiMAX Forum (2007) Mobile System Profile, Release 1.0 Approved Specification (Revision 1.4.0: 2007-05-02), http://www.wimaxforum.org/technology/documents.

Yavuz, M., Diaz, D., Kapoor, R., Grob, M., Black, P., Tokgoz, Y. and Lott, C. (2006) VoIP over cdma2000 1xEV-DO, *IEEE Communications Magazine*, **44**(2), 88–95.

14

WiMAX User Data Load Balancing

Alexander Bachmutsky

14.1 Introduction

Load balancing is an important topic, and it has started to appear frequently on the radar of many WiMAX vendors. The WiMAX Forum does not discuss load balancing functionality formally at any level, and one can barely find that term in the latest specification, but the increasing number of contributions has raised the topic as a reason for the requested change. Some discussions are concentrated on signaling load balancing, others on mobile subscribers balancing between different network elements in the WiMAX network.

There are many aspects for load balancing. One way to achieve it is to perform R3/R4/R6 handovers when appropriate to create more evenly loaded paths. Another way is to terminate the traffic at different points in the network by means of local breakout at different levels. Yet another is to perform network level load balancing. The latter two are described here.

14.2 Local Breakout Use for Load Balancing

Local breakout is another controversial topic that received little attention in the standard. It was in fact raised multiple times on different occasions by different Telecommunication Equipment Manufacturers (TEMs), but was not very well accepted by the operators. The main reason, of course, is a revenue split between visited and home operators in the roaming scenarios. However, that is not what we discuss here, because from a load-balancing point of view the roaming traffic is a relatively small percentage of the total traffic, so that topic can wait until some bigger problems are solved.

WiMAX Evolution: Emerging Technologies and Applications   Edited by Marcos D. Katz and Frank H.P. Fitzek
© 2009 John Wiley & Sons, Ltd

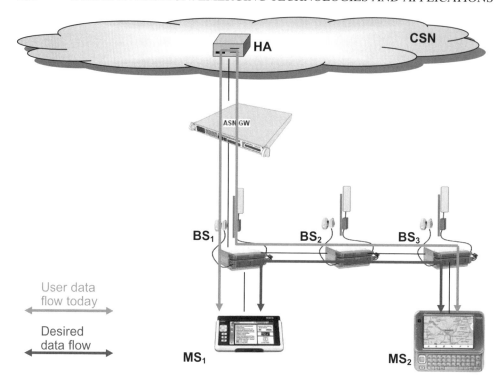

Figure 14.1 Existing and desired data flow for chained BSs.

One of inefficiencies in the current WiMAX architecture is that the entire traffic has to go through a Connectivity Services Network (CSN). Some problems are similar to the roaming case with a revenue split, this time it is needed between Access Service Network (ASN) and CSN operators, and we acknowledge that these problems have to be solved before bringing a comprehensive solution. On the other side, there are many cases when ASNs and CSNs are owned by the same operator, so there are no charging issues, but the inefficiency still persists. Our focus will be on this less-complex scenario.

How many times do we see mobile calls between employees calling each other for a meeting, lunch or any other common activity? Based on our experience, it happens very often driven by the fact that these calls are frequently virtually free of charge driven by either employer unlimited subscriptions or special operator discounts; for example, US AT&T customers can call within the AT&T mobile network without limits and without being charged. Similar deals are often applicable between family members or friends.

14.2.1   Local Breakout at the Base Station Level

Calls are often made between parties located at the same Base Station (BS) or a pair of chained BSs as shown in Figure 14.1.

The entire traffic between such users has to go twice over the link between BS and ASN Gateway (ASN-GW) and twice over the link between ASN-GW and Home Agent (HA) in the CSN. It causes more link load and, of course, much higher latency. It would be much more efficient if such traffic could avoid being sent to the ASN-GW as indicated by the 'Desired data flow'. A few things are required, however, to make this happen.

(a) The BS should be an IP device as opposed to just a L2 network element. Not all vendors have implemented their BSs as IP devices from a user data transport point of view, because with current standard all IP decisions can be made by the ASN-GW.

(b) The BS should be capable of performing IP forwarding tasks.

(c) The BS has to build a forwarding table for all active subscribers on all chained BSs; it can be done through WiMAX control plane messages or standard routing protocols.

(d) The mobile subscriber's IP address might not be unique, because different HAs can have an overlapping IP addressing space; this is why the BS has to know to what HA each side belongs. It is potentially too big a task for the BS to know all HAs and their address spaces; therefore, the recommendation would be to perform a local breakout at the BS only when both sides belong to the same HA. This still means that the BS should know the HA for every mobile subscriber, and the best way is to deliver that information from ASN-GW as a part of mobile subscriber network entry.

(e) The BS might be required to perform counting functionality, and some part of this is already defined today as an optional implementation.

14.2.2 Local Breakout at the ASN-GW Level

When a local breakout at the BS level is not feasible (no IP forwarding information is available because parties are not on the same or chained BSs, parties belong to different HAs, etc.), there can be second level of a local breakout: at the ASN-GW level as shown in Figure 14.2.

As shown in the figure, the ASN-GW can also connect directly to the Internet through the site router bypassing the CSN. In addition to latency and HA load, more optimized desired paths save significant throughput between the site router and CSN.

It is easier to implement ASN-GW-based local breakout because the ASN-GW is defined as an IP device, it has IP forwarding functionality and in most cases it supports routing protocols and virtual routing. All of this creates a perfect infrastructure for the distribution of mobile subscriber IP addresses with their corresponding anchor and/or serving ASN-GW(s).

Depending on the policy and traffic load on different interfaces and network elements, local breakout at the BS and/or ASN-GW level(s) can achieve better traffic load balancing and better Quality of Service (QoS).

14.3 Network-level Load Balancing over Tunneled Interfaces

This chapter concentrates on pure data plane load balancing at the lower L2 and L3 levels, states the current problem and provides some ways to solve it.

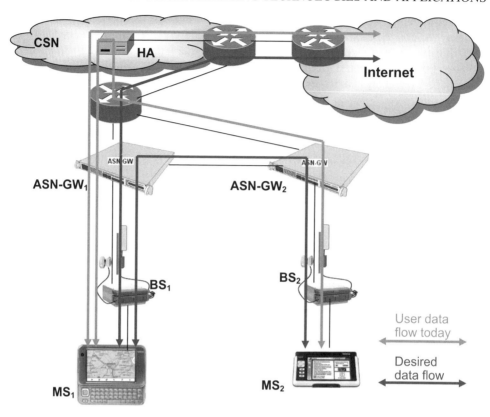

Figure 14.2 Local breakout at the ASN-GW level.

Let us assume that we have data plane traffic to/from mobile subscribers with an aggregated total rate of B_{total} at one of WiMAX reference points R3, R4, R6 or R8. Let us also assume that the traffic has to pass through N physical links with rate B_i ($i = 1, \ldots, N$) each, where $B_i < B_{total}$. A practical example of that would be 5 Gbps traffic sent/received through 10 parallel Gigabit Ethernet links of 1 Gbps each.

The problem is not new and is being solved today using two load-balancing mechanisms: Ethernet Link Aggregation (LAG, sometimes called trunking) standardized in IEEE 802.3ad and Equal Cost Multi-Path (ECMP) with few IETF RFCs covering the topic (for example, RFC 2992 for IP ECMP and RFC 4928 for Label ECMP). The control interface for both implementations is different: Link Aggregation Control Protocol (LACP) for LAG and routing protocols (such as Open Shortest Path First (OSPF)) for ECMP, but the underlying load-balancing mechanism is very similar: extract fields from Layer 2 and/or Layer 3 and/or Layer 4 (UDP/TCP) headers and make link selection based on that information. Both schemes also have the same major assumption: the sender has to preserve the packet order for the same application session with the same priority (sometimes referred to as a 'conversation' or 'flow'). While the definition of application differs depending on the extracted information, it makes one of the simplest modes achieved through a round-robin link selection practically useless. The reason for this is that even in parallel links connected between two physical devices we cannot always ensure that the receiving order of packets will be the same as a sending order, because every link has its own queues and QoS, and every packet has a potentially different size causing different transmit times for all previous messages in corresponding interface queues. Some link types will also cause delayed transmit (such as Ethernet congestion control) or end-to-end retransmits in the case of error detection. This means that we have to send all packets for the same application over the same physical link. The usual implementation is to hash the extracted information and select a link based on the hash value.

One of important properties of load-balancing schemes is in making the treatment of application/flow in the network consistent, because load balancing can be applied multiple times between different network elements while passing messages from source to destination. On the other side, it is frequently impossible to ensure exactly the same configuration in all devices, especially if the packet passes through multiple networks that belong to different transport providers. For example, if one device has a better knowledge of 'real' flows by interpreting some application-level flow identifications, it can apply that knowledge to the local load-balancing scheme for LAG or ECMP, but it will be hard or impossible to enforce the same treatment in already deployed Layer 2 or Layer 3 switches and routers.

14.3.1 Is WiMAX Special for the Case of Traffic Load Balancing?

All of the above is fairly generic and can be applied to most networks. So what is unique in WiMAX networks?

Actually, the problem is not WiMAX-specific. It is the same problem for any tunneled traffic, but WiMAX is one of the first to suffer because it brings multi-gigabit traffic into the network while higher capacity interfaces (10 Gbps and higher) are too expensive or nonexistent.

It is the easiest to explain the problem using the traffic on R3 reference point for a Mobile IPv4-based WiMAX network. In MIPv4 architecture (RFC 3344 - IP Mobility Support for IPv4) all user traffic has to flow through Foreign Agents (FAs) and HAs. All user packets are encapsulated between FAs and HAs into a tunnel (we use IP-in-IP in our example, but Generic Routing Encapsulation (GRE) or any other tunnel will have the same issues). This means that from the point of view of the network devices between FAs and HAs the entire user traffic for all mobile subscribers is a single IP flow between the FA IP address and the HA IP address. There is no differentiation at Layer 4, because the entire traffic has a single IP-in-IP protocol. From the LAG and ECMP points of view there is only a single conversation/flow. The entire traffic can be represented by the diagram in Figure 14.3 (the assumption in this example is the multi-gigabit traffic with Gigabit Ethernet interfaces).

Having a single conversation would mean that switches and routers will not be able to perform load balancing and will use a single link to transfer the data, and a significant portion of that data (everything above 1 Gbps) will be dropped. Of course, this is absolutely unacceptable.

14.3.2 Analysis of Possible Solutions

It is obvious that this problem has to be solved. There are multiple layers of possible solutions, from the physical layer to the application layer; some are WiMAX-independent, others are WiMAX-specific.

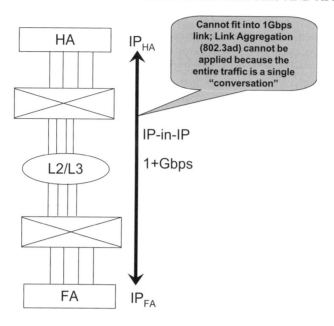

Figure 14.3 Tunnel as a single conversation.

14.3.2.1   Physical Layer: Higher Link Capacity

One of the simplest solutions is to increase link throughput on all reference points. This would mean using, for example, 10 Gbps interfaces. The problem is that these interfaces are still relatively expensive, and such capability would increase the WiMAX deployment costs, exactly the opposite of what WiMAX operators try to achieve to compete with other mobile networks.

Another aspect of this solution is that it cannot solve the problem when traffic is above that throughput. Assuming WiMAX success, user payload on the R3 reference point will quickly surpass 10 Gbps forcing operators and equipment manufacturers to use even higher capacity links: 40, 80 or 100 Gbps, etc. It will be very hard to synchronize the expansion of WiMAX networks and the availability and cost effectiveness of these interfaces.

14.3.2.2   Change IEEE and IETF Standards to Cover Tunneled Traffic

The idea here is to include additional information for load-balancing decisions. The problem is that the scope of that addition depends on the type of tunnel used.

The proposal is to define LAG profiles, similar to what is defined for Robust Header Compression (ROHC). A number of relevant profiles to be defined are as follows.

  (a) Ethernet profile: include Ethernet header information (see Figure 14.4).

  (b) IPv4/v6 profile: include IPv4/v6 header plus optionally Ethernet profile (see Figure 14.5).

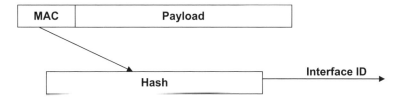

Figure 14.4 Ethernet LAG profile.

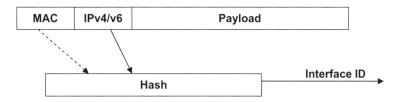

Figure 14.5 IPv4/v6 LAG profile.

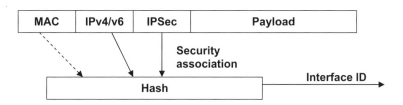

Figure 14.6 IPSec LAG profile.

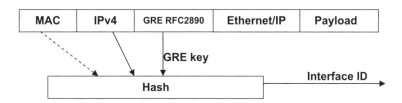

Figure 14.7 GRE RFC2890 LAG profile.

(c) Internet Protocol Security (IPSec) profile: include security association (see Figure 14.6).

(d) GRE RFC2890 profile (with GRE key, used for R4/R6/R8 interfaces): include GRE key (see Figure 14.7).

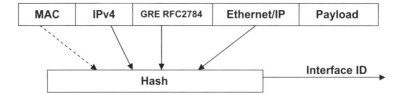

Figure 14.8 GRE RFC2784 LAG profile.

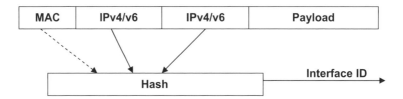

Figure 14.9 IP-in-IP LAG profile.

(e) GRE RFC2784 profile (without GRE key, used optionally for R3 interface): include inner packet headers (Ethernet Layer 2 to cover WiMAX Ethernet CS, inner IP for encapsulated IP data). (See Figure 14.8.)

(f) IP-in-IP profile (used for R3 interface): include inner packet IP header (see Figure 14.9).

Generally speaking there is a need for more profiles (Layer 2 Tunneling Protocol (L2TP), GPRS Tunneling Protocol (GTP), Secure Sockets Layer (SSL), Stream Control Transmission Protocol (SCTP), etc.), but they are not related to WiMAX; therefore, we do not concentrate on those.

Standards update would be the best possible solution in the long term, but it will take time for standardization and even more time for an implementation by major switch and router vendors and corresponding product upgrades in the field.

Meanwhile, we cannot wait for that to happen, marking this solution as impractical for the next few years.

14.3.2.3 Create Multiple Tunnel End-points

This solution is more application-specific. While creating multiple tunnel end-points is a generic concept, the traffic load balancing between these end-points can depend on the application. The following is an example for R3 reference point load balancing in Mobile IP (MIP)-based WiMAX deployment.

Let us enable multiple IP addresses in FAs that force different IP-in-IP tunnels between HAs and FAs (see Figure 14.10); the HAs would think that it is connected to multiple FAs. Every FA–HA tunnel still cannot handle more than the actual capacity of a single physical link; therefore, the proposal is to use dynamic subscriber assignment to these tunnels.

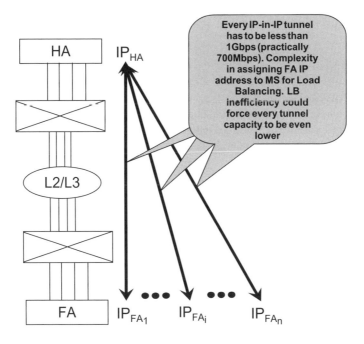

Figure 14.10 MIPv4 traffic load balancing.

When FA sends its advertisement message to the mobile subscriber, it has to select one of the FA IP addresses for the corresponding IP-in-IP tunnel. Such a selection can be based, for example, on a simple or weighted round-robin algorithm; weight can be calculated taking into account the load of the current tunnel and profile of the mobile subscriber (if available at the time the decision is made). Since the traffic pattern might change at any point of time, such an algorithm has to force a lower threshold than the maximum physical link capacity to accommodate traffic bursts. In the case of Gigabit Ethernet links, the recommendation is not to exceed about 700–800 Mbps of traffic per interface.

14.3.2.4 WiMAX-specific Solutions

Creating WiMAX-specific solutions might be not the preferred way to handle the problem, but it can be the best solution in the short term.

One element of the solution could be to eliminate tunnels. That is achieved, for example, in Simple IP deployments that do not use tunnels on R3. When there is no outer tunnel applied, all devices will be able to perform standard LAG or ECMP. Of course, not every operator is happy with such a solution. For example, Sprint is promoting the MIP solution, while Clearwire claims that MIP is not needed at all, and Simple IP is the perfect choice for WiMAX network. The main difference between these operators is that Sprint already has the MIP infrastructure deployed while Clearwire does not and is very reluctant to spend a significant budget on such a deployment. One serious disadvantage of Simple IP is that it requires ASN-GW anchoring after initial network entry for the lifetime of the active session: no CSN anchored mobility. While there are some proposals to introduce the ASN-GW reanchoring into Simple IP, these proposals are similar to MIP and are based on some type of tunnel between ASN-GW and CSN. Of course, any tunneling would bring back our main problem of a link load balancing. On the other side, keeping the initial ASN-GW anchored can be enough for some locations, but creates a serious limitation for others. Take as an example ASN-GWs serving airports, train stations, bus stations, ports, theatres and similar venues where a large number of subscribers would enter the network. In many cases at such locations a number of terminals entering the network will be larger than the number of terminals leaving it or even going idle, causing the ASN-GW to overload at some point of time.

One way to eliminate a tunnel on an R3 interface is to use standard routing protocols: when a mobile subscriber joins the network at a particular ASN-GW, its address is advertised to CSN router(s) and other ASN-GWs; when the mobile subscriber leaves that ASN-GW for any reason, the address is withdrawn. You could consider the scheme as being very similar to MIP registration/deregistration, it is just performed using a routing protocol instead of MIP. There is a concern that the scheme would create a very large number of routes (possibly millions), because there is a need for one route per mobile subscriber. The concern is valid, but on the other side that is exactly what happens in HAs: one entry per mobile subscriber. Routing even has the built-in advantage of automatic summarization (route aggregation) that will reduce the number of routes. Also, mobile subscriber IP addresses can be allocated from the ASN-GW dependent pool, and all terminals still served by the initial ASN-GW are highly aggregated (exact routes for relocated terminals would be placed in the forwarding table ahead of address pool subnets or address ranges). Some routing protocols (such as Border Gateway Protocol (BGP)) can scale even today to millions of routes; some (such as OSPF) are not scalable enough, but their limitations involve calculations of the best route rather than the table sizes, and route calculation in the case of WiMAX should not be very complex, because the WiMAX forwarding infrastructure is a tree without loops. More studies are needed to understand whether routers can really replace HAs.

The load on R4 interfaces can be viewed as less acute, because less traffic is going between ASN-GWs, but we would say that it is very uncomfortable limitation for WiMAX deployment to have at most 1 Gbps (in the case of Gigabit Ethernet links) between every pair of ASN-GWs. This is an unnatural restriction. Also, in the case of Simple IP deployment, we do not have reanchoring, and as a byproduct of that definition the traffic over R4 will be higher than in the MIP case. We can use a similar solution to our proposal for multiple FA addresses: multiple tunnels between each pair of ASN-GWs; with this capability ASN-GWs will assign terminals statically or dynamically to one of these R4 tunnels.

It is hard to avoid mentioning IPv6 as one of the possible methods to eliminate tunnels: if all MSs are assigned routable IPv6 address, and if neither the mobile subscriber nor server uses a flow label, we can use this flow label instead of a GRE key. However, that would only be valid with many 'ifs', and we also have to remember that IPv6 is not recommended for use together with LAG: ECMP has to be used.

Another solution for our original link load problem is a network resource load balancing. For example, a FA can choose the HA with the least loaded R3 interface. The BS can select the ASN-GW taking into account the load on all interfaces for all connected ASN-GWs. In the latter approach we would propose to create in WiMAX what we call network resource management functionality, which is similar to radio resource management but applicable

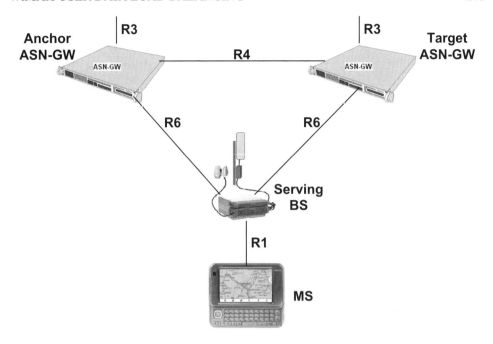

Figure 14.11 Anchor relocation with help from a BS.

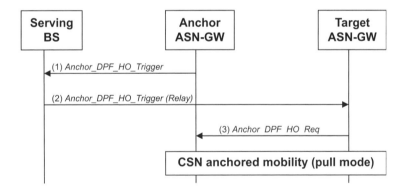

Figure 14.12 Anchor relocation trigger through a BS relay.

between network elements on the WiMAX landline side instead of mobile subscribers and wireless connectivity.

One further way to improve our loads is to move subscribers to less-loaded network elements or connections. This includes moving subscribers from one FA IP address to another within the same ASN-GW or between ASN-GWs. In the first case we would only advertise the new Care-of-Address (CoA) to either MIPv4 client in the mobile subscriber or Proxy Mobile IP (PMIP) client that resides in the ASN-GW without any mobility event. In the

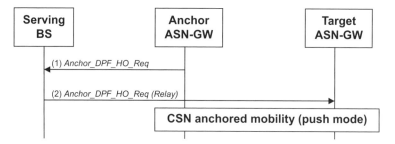

Figure 14.13 Anchor relocation push through a BS relay.

second case the ASN-GW would speed-up changing the mobile subscriber status from active to idle mode (this can be rejected by the terminal if it is in the middle of data transfer), or force the mobile subscriber exit the network, or move the mobile subscriber to another ASN-GW. For the latter action we do not have a generic standard-based mechanism. The only existing method is CSN anchored mobility, but it is applicable today only when anchor and serving ASN-GWs are not co-located. In the general case current anchor ASN-GW does not have any information about potential target ASN-GWs. It is possible to discover potential targets (discussed in Chapter 21), or broadcast a request to all other ASN-GWs (always an undesirable option), or request that from the serving BS. For example, we can reuse the message *Anchor_DPF_HO_Req* and/or *Anchor_DPF_HO_Trigger* and send it over R6; if the receiving serving BS has connectivity to another ASN-GW (the request is rejected otherwise), it will reply with a corresponding target ASN-GW ID or simply redirect the message to that target ASN-GW (similar to the current relay functionality in ASN-GW). In the former case, the anchor ASN-GW would initiate the 'regular' CSN anchored mobility procedure; in the latter case the target ASN-GW would treat it as a start of CSN anchored mobility, and further handling would be similar in both cases. The network diagram looks as shown in Figure 14.11.

The message exchange in pull mode would be as shown in Figure 14.12 and the procedure is similar in push made as shown in Figure 14.13.

Finally, an additional solution can be adding Layer 4 information to our WiMAX tunnels. This proposal takes into account the fact that majority of existing LAG and ECMP implementations in routers/switches support UDP/TCP ports as part of their load-balancing algorithm. We can modify WiMAX standard to add the UDP header between outer and inner IP headers. Instead of IP-in-IP it will be IP-UDP-IP; instead of IP-GRE-IP it will be IP-UDP-GRE-IP or even IP-UDP-IP is no sequence numbers are used and we can use UDP port(s) in the same way as we use the GRE key at present. This solution will, however, increase the packet overhead, and the most sensitive use case is, of course, a Voice over IP (VoIP) session.

14.4 Conclusions

We have touched on only a few aspects of user data load balancing, and great benefits of the described mechanisms should be studied in more detail by the WiMAX community and hopefully quickly adopted for many deployments worldwide making WiMAX a more efficient and more cost-effective solution that will attract mobile subscribers.

15

Enabling Per-flow and System-wide QoS and QoE in Mobile WiMAX

Thomas Casey, Xiongwen Zhao, Nenad Veselinovic, Jari Nurmi and Riku Jäntti

15.1 Introduction

The 802.16 standard offers a possibility for very high utilization of radio resources and good Quality of Service (QoS). QoS has been defined as the ability of the network to provide a service at an assured service level (Kilkki, 1999, Soldani *et al.*, 2006) and has undoubtedly been one of the most attractive features of WiMAX. QoS within one Base Station (BS) in WiMAX is provided by connection-oriented Medium Access Control (MAC) and agile scheduling, which have been popular targets of research. While it is important to guarantee proper QoS, an even more vital issues is how the end-user perceives and experiences the services. This has often been referred to as Quality of Experience (QoE) (Kilkki, 1999, Soldani *et al.*, 2006) and although it is a subjective measure, it will in the end determine how satisfied the user is.

Furthermore the expansion from a fixed WiMAX network to a mobile WiMAX network, with a cellular infrastructure, introduces new challenges to providing QoS as a user is not necessarily static or nomadic, but is likely to be roaming around the network while engaging in a session, thus resulting in Handovers (HOs) between BSs. For example, it is a very annoying experience to be dropped out of an ongoing call due to insufficient resources in a target BS. On the other hand, overlapping cells introduce new possibilities to enhance QoS in the form of load balancing between BSs. Therefore, in a mobile WiMAX access network

WiMAX Evolution: Emerging Technologies and Applications   Edited by Marcos D. Katz and Frank H.P. Fitzek
© 2009 John Wiley & Sons, Ltd

QoS and QoE also become system-level issues, a point of view that has not, so far, received much attention.

In this chapter we discuss different aspects of QoS and QoE on a per-flow level (within one BS) but also on a system-wide level (within a cluster of BSs). The chapter is outlined as follows. In Section 15.2 we conduct an overview of the environment for guaranteeing QoS and QoE both within a single BS and system-wide and introduce the corresponding tools. In Section 15.3 we discuss different aspects of the MAC packet scheduler in charge of per-flow scheduling in the BS and look at it in particular from the QoE point of view. In Section 15.4 we deal with system-level QoS and QoE issues in a mobile WiMAX access network and, finally, in Section 15.5 we draw conclusions.

Nomenclature

δ Hysteresis parameter.

BQ Base quantum size.

$d_{\max}$ Maximum delay allowed.

$F_{\max}$ Maximum radio resource fluctuation.

F_{sys} Average radio resource fluctuation in the system.

G Guard band.

$G_{\mathrm{nrt,ho}}$ Guard band for non-real-time rescue handovers.

$G_{\mathrm{rt,ho}}$ Guard band for real-time rescue handovers.

$G_{\mathrm{rt,new}}$ Guard band for new real-time flows.

$h_{\max}$ Maximum handover rate allowed.

L Average load.

$L_{\max}$ Maximum average load in the system to conduct load balancing.

Q_i Quantum size for the ith service flow.

R Reserved resources.

$r_{\max}$ Maximum packet dropping rate allowed.

T Triggering threshold for load balancing.

T_R Resource reservation-based triggering threshold.

$T_{U,\max}$ Maximum for the resource utilization-based triggering threshold.

$T_{U,\min}$ Minimum for the resource utilization-based triggering threshold.

T_U Resource utilization-based triggering threshold.

U Resource utilization.

U_{new} New resulting resource utilization after a load-balancing handover.

w_i Weight for the ith best-effort service flow

15.2 Overview

How can QoS guarantees given to a user be fulfilled both within a single BS and on system-level in a mobile WiMAX access network? A good starting point is to approach the problem based on the fundamental objective of teletraffic theory which is to determine the relation between the following three components:

1. offered traffic (user needs);

2. system capability and time varying resources;

3. agreed QoS.

The fundamental question is *How can the time varying mobile WiMAX system resources be utilized both on the flow and system level to meet the QoS needs of the offered traffic load?*

WiMAX has been characterized as a system that has the ability to dynamically adjust to the current traffic and radio channel environment and thus utilize radio resources in an efficient manner. We will discuss four main functionalities that enable this: the MAC scheduler and admission control within a single BS and load balancing and HO prioritization on the system level.

The main orchestrator within one BS is the MAC scheduler, whose job it is to take into consideration the incoming traffic buffered in queues and the available time varying radio resources and, based on this, decide the order in which packets are sent to ensure the fulfillment of the QoS guarantees made. An admission control element is used to estimate and predict the long-term average utilization of an incoming flow and the corresponding need for resource reservation and is thus used to protect the ongoing connections and the scheduler from congestion.

Most of the research so far has been concentrated on the per-flow scheduling issues within a single BS, but a lot can also be done on a system level within a cluster of BSs with overlapping coverage. Overloading in a BS can be alleviated with HO-based load balancing which can be thought of as a kind of system-level scheduling of flows. HO drops can be minimized with HO prioritization (that is, prioritizing HO request over new flow requests) which serves as a kind of system-level admission control scheme. Figure 15.1 summarizes both per-flow and system-wide level issues that relate to these functionalities and their environment. In the following sections we briefly go through each point of this triangle both on the level of one BS and a cluster of BSs.

15.2.1 Incoming Traffic

The traffic needs of a single flow can vary a great deal and can be characterized, for example, in terms of delay, jitter, throughput and packet drop and loss rate (ITU-T, 2001). For example, conversational services (e.g. Voice over Internet Protocol (VoIP)) require strict delay and jitter

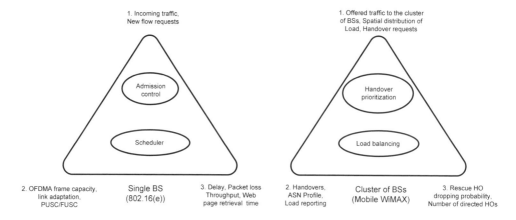

Figure 15.1 The teletraffic triangle model applied to one BS and to the whole system.

requirements but can tolerate some packet loss. Streaming services with play-out buffers, on the other hand, tolerate a little more delay. Elastic Transmission Control Protocol (TCP)-based traffic is then again delay tolerant but requires error-free delivery of data. The holding times of traffic flows, activity periods within the traffic flows (e.g. Web page-retrieval) and their throughput can also vary a great deal.

The resulting traffic mix of individual flows fed to a BS can therefore be very diverse with various needs and thus requires prioritization of traffic. An admission control check based on an estimation of resource consumption has to be made for each new incoming service flow to ensure that there are enough resources for the MAC scheduler to meet the needs of the existing and new traffic flows.

On a system level, unbalanced spatial distribution of users can in a worst case result in one BS serving most of the incoming traffic fed to a cluster of BSs. Also when a user is moving from BS to BS, HO requests are made when entering a new BS. In general, dropping an existing service flow is considered as worse than blocking a new one and thus there is a need to differentiate and prioritize these HO requests over new flow requests.

15.2.2   System and Resources

Mobile WiMAX offers a flexible Orthogonal Frequency Division Multiple Access (OFDMA)-based frame structure and the possibility for simple and efficient frequency reuse planning. The OFDMA frame is divided into a Downlink (DL) and Uplink (UL) subframe which are further divided into slots in the time (OFDMA symbols) and frequency dimensions (subchannels). The capacity of a single slot is controlled by a link adaptation functionality that changes the Modulation and Coding Scheme (MCS) used in the link between a Mobile Station (MS) and a BS according to current channel conditions which in turn depends, for example, on the MSs distance from the BS and the velocity of the MS.

Frequency reuse in mobile WiMAX can be handled with fractional frequency reuse which enables MSs, that are located in the middle of the cell close to the BS, to utilize all of the subchannels (Fully Used subcarrier (FUSC)) in the frequency block. The MSs located in the cell edges can utilize only a part (e.g. one-third) of the subchannel set (Partially Used subcarrier (PUSC)) in order to mitigate interference. Hence, depending on the location of the MS the channel capacity might vary considerably.

Both admission control and the MAC scheduler have a very challenging task in estimating and utilizing these time varying resources in order to fulfill the needs of the incoming traffic flows.

HOs in mobile WiMAX are heavily related to system-wide QoS. It is good to make a clear distinction between two kinds of HO:

1. *directed HOs*, which can be used proactively by an overloaded BS to distribute traffic load to other BSs thus enhancing the possibility to fulfill the QoS guarantees made for the MSs; and

2. *rescue HOs*, that occur as a result of a deteriorating radio signal when a MS drifts away from a Serving BS (SBS) towards a Target BS (TBS).

Hence, in terms of QoS, a HO is on the other hand a tool that enables load balancing and more efficient usage of radio resources, but on the other hand a burden for which the system needs to be prepared (with rescue HO prioritization).

Critical issues for load balancing are, for example, how to discover which terminals are in an overlapping area and are likely to remain there and when to trigger load balancing (i.e. not too early to eliminate unnecessary handovers but no too late to avoid inefficient resource utilization). The challenge with HO prioritization is to reserve an optimal amount of resources for incoming HOs that both eliminates unnecessary HO drops but is not too conservative.

In mobile WiMAX, the Access Service Network (ASN) architecture consists of one or more BSs and ASN Gateways (ASN-GWs) which can be organized according to different ASN profiles. Radio Resource Management (RRM) within an ASN can be completely distributed (i.e. located in the BS) or centralized (divided between the ASN-GW and BSs) but the distributed mode seems to gain more popularity due to easier scalability. The location of RRM affects, for example, the way BSs report their current loading situations and how and who makes the decision to start load balancing.

15.2.3 QoS and QoE

The 802.16e standard defines QoS parameters such as Maximum Sustained Traffic Rate (MSTR), Minimum Reserved Traffic Rate (MRTR), Maximum Latency (ML) and Traffic Priority (TP) that contain basic information on the QoS needs of a particular service flow. Based on these parameters the MAC scheduler provides five different scheduling services: the Unsolicited Grant Service (UGS), mainly targeted at VoIP, that provides a constant bit rate, the Extended Real-Time Polling Service (ertPS) that adds Voice Activity Detection (VAD) support to UGS, the Real-Time Polling Service (rtPS) intended for streaming video and audio with a MRTR and ML guarantee, the Non-Real-Time Polling Service (nrtPS) that provides a MRTR with no delay guarantees targeted for critical TCP-based applications that require a minimum data rate and the Best Effort (BE) scheduling service that gives no guarantees at all of the service level.

There are also other ways of defining QoS. For example the ITU-T E.800 standard (ITU-T, 1994) defines three performance indicators: accessibility (the ability for a user to obtain a service), retainability (the ability to keep the service going) and integrity (level of service delivery without disturbances once the service has been obtained)[1].

One example of accessibility is flow (call) blocking probability which represents the proportion of calls that have to be blocked in order to prevent over congestion. It should be kept under a certain limit (e.g. less than 5% probability that a call is blocked in the network) and can be differentiated in relation to the scheduling service classes and traffic priorities.

QoS parameters can also be defined on the system level. As discussed before, when moving through a mobile WiMAX access network there is the possibility that the next target BSs drops the flow (retainability) or reduces the provided MRTR (integrity) due to insufficient resources. In traditional cellular networks this has been characterized with a QoS parameter called HO dropping probability (e.g. should be less than 2%).

Also, when conducting a HO (be it rescue or directed) the flow might become subject to a 'ping–pong' effect where it is handed over back and forth between two adjacent BSs. Such an effect is especially harmful for delay-sensitive flows such as VoIP (due to HO interruption delays) and should thus be limited.

Fulfilling all traffic QoS parameters will not necessarily guarantee a satisfied user and QoE. For example, requirements and expectations can vary between different WiMAX deployment scenarios and use cases. Thus, true QoE can be provided by gaining an understanding of what are the user needs, what affects a user's perception of services, what is the deployment scenario and by applying this information to the system and individual QoS measures (Soldani *et al.*, 2006). One example of this could be to move from optimizing a scheduler algorithm in relation to single QoS parameters (e.g. delay) to optimizing it in relation to a use case and the corresponding experience (e.g. in relation to a full Web-page retrieval). Another example could be to differentiate HO prioritization in relation to traffic types and their needs: for example, the reduction of MRTR for a Web connection is not such a bad experience compared with a situation where a voice call has to be dropped due to insufficient resources.

15.3   Per-flow-based QoS and QoE

In the previous section, the basic concepts of flow level QoS and QoE were introduced, and the classification of user traffic into five different traffic classes (i.e. UGS, ertPS, rtPS, nrtPS and BE) was discussed. In the MAC layer, a very important issue is how to schedule the MAC protocol data units, and fill the created DL and UL frames. This section discusses how the WiMAX traffic can be scheduled based on priority orders in order to satisfy the QoS requirements. Meanwhile, the QoE performance measure of user satisfaction distribution is studied by optimizing the scheduler parameter(s). The following sections describe and show the considered scheduling algorithm and the scheduler optimization process by using QoE as a performance measure.

---

[1] To avoid confusion, it should be noted that in original ITU recommendations, what is nowadays known as QoS was defined as Grade of Service (GoS) and QoS in turn referred to the current concept of QoE.

15.3.1 MAC scheduler considerations

There have been many proposals on how to implement the key element, the packet scheduler, in IEEE 802.16e-based systems. Kwon *et al.* (2005) considered a utility function based scheduler (with proportional fair scheduler as a special case), which takes into account the application real-time requirements with priority scheduling and instantaneous radio channel conditions by effective resource allocation. The solution was validated in a mixture of VoIP, File Transfer Protocol (FTP) and Hypertext Transfer Protocol (HTTP) traffic. A lower complexity alternative to this proposal has been proposed by Sayenko *et al.* (2006), in the form of Weighted Round Robin (WRR) scheduler. This solution was evaluated in a mixture of VoIP and FTP traffic. A combination of Deficit Round Robin (DRR) for DL scheduling and WRR for the UL scheduling was proposed by Cicconetti *et al.* (2006) and evaluated in a case of pure Web-browsing and a mix of voice, video and Web-browsing traffic. Rath *et al.* (2006) proposed the opportunistic DRR (O-DRR) scheduler for the UL to balance fairness of bandwidth allocation between different users with delay constraints of the application.

In this section, we propose a new priority scheduler shown in Figure 15.2 for the prioritized traffic to meet with the QoS requirements in WiMAX. The scheduler is a combination of a strict priority scheduler that schedules all incoming traffic, a WRR scheduler able to assign resource blocks proportional to weights associated to the service flow queues and the slightly more advanced, yet simple, Deficit WRR (DWRR) scheduler that uses a deficit counter for each flow and is thus able to handle packets of variable size without knowing their mean size. The priority orders of the scheduler are as follows as shown in Figure 15.2.

1. Schedule the MAC management and control messages, Automatic Repeat Request (ARQ)/Hybrid ARQ (H-ARQ) Acknowledgements (ACKs) and Negative ACKs (NACKs) and retransmission.

2. Schedule UGS traffic when available.

3. Schedule ertPS traffic when available.

4. Schedule rtPS traffic for DL and UL by using DWRR and WRR, respectively.

5. Schedule nrtPS traffic for DL and UL by using DWRR and WRR, respectively.

6. Schedule BE using DWRR and WRR when the traffic flows contend for the left bandwidth after scheduling items (1)–(5) traffic in DL and UL, respectively.

As shown in Figure 15.2 the rtPS, nrtPS and BE traffic flows are scheduled by DWRR and WRR in the DL and UL, respectively. When selecting a scheduling algorithm for WiMAX the most important requirements are: (1) QoS guarantees (i.e. throughput and delay) should be met; (2) the algorithm should be simple to be implemented. The WRR and DWRR schedulers meet both of these requirements and are better than simple Round Robin (RR) scheduling that can be unfair if different flows use different packet sizes.

In the proposed algorithm the following input parameters are used to calculate weights for the DWRR scheduler in the DL: the MSTR for every service flow (bytes per second) and the base quantum size (BQ) (slots). The DWRR weight for the ith service flow is calculated as $w_i = \text{MSTR}_i/\text{MSTR}_{\max}$, and the quantum size assigned for the ith service flow is calculated

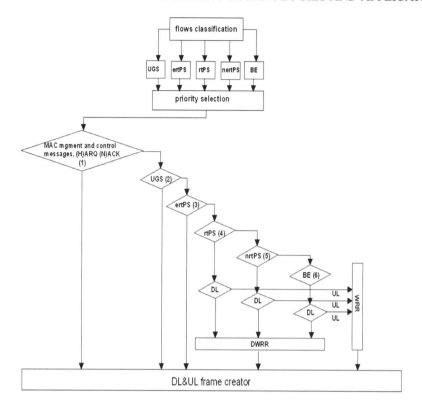

Figure 15.2 Traffic scheduler for WiMAX.

as $Q_i = w_i \times BQ$ (slots). The initial value of the deficit counter is DC_i (slots). The input parameters for the WRR scheduler in the UL are the same as in the DL, and the same formulas can be used to obtain the WRR weight and the scheduled amount of traffic in the UL. The reason that WRR is used in the UL direction is that the packet sizes of the service flows are not known in the MAC scheduler residing in the BS, and thus the deficit counter cannot be updated.

15.3.2   Scheduler Optimization Based on the QoS and QoE Measures

The proposed scheduling algorithm is evaluated by NS2. The simulation parameters are described in Table 15.1. As discussed in the previous section the network can also be optimized in terms of QoE and here we will take the Web-page retrieval time (described in Table 15.2) as a performance measure for the HTTP users. The scheduling algorithm is evaluated by selection of the basic quantum and basic scheduled traffic in the DL and UL, and the corresponding change in user's QoE.

In order to validate that the QoS framework works as required, we compare the following two simulation cases. First, we map all DL traffic (VoIP, FTP and HTTP) on the BE service class. Second, we map VoIP traffic on the ertPS and UGS service classes, and FTP and

Table 15.1 Simulation parameters.

Parameter	Value
Number of BSs	1
Number of MSs	Variable
Bandwidth	10 MHz
Fast Fourier Transform (FFT) size	1024
Frame length	5 ms
DL:UL ratio	2:1
Traffic mix	10% VoIP with VAD
	10% VoIP without VAD
	5% FTP with MSTR 256 kbps
	5% FTP with MSTR 1024 kbps
	35% HTTP with 256 kbps
	35% HTTP with 1024 kbps

Table 15.2 Definition of QoE performance measure.

Metric	Mean Web-page download time (seconds)
Very satisfied	≤ 2
Satisfied	$>2, \leq 4$
Unsatisfied	$>4, \leq 10$
Very unsatisfied	>10

HTTP traffic on the BE class. The results are presented in Figures 15.3 and 15.4 respectively, in terms of VoIP packet delay distribution with the total number of registered users as a parameter.

The QoS performance measures for VoIP are packet delay and jitter. It can be seen that in the first case, the maximum packet delay of VoIP traffic increases with increased number of users. This is what is expected, since VoIP traffic is treated the same as FTP and HTTP within the BE class, despite its real-time behavior. In the second case, the maximum packet delay of VoIP traffic stays under control, that is, under 10 ms, since it has been given adequate priority treatment for UGS and ertPS service classes.

The quantum size is expected to have a major impact on the performance of the scheduler in terms of the BE service class. This effect is illustrated in Figure 15.5, which present user satisfaction distribution versus the quantum size of the DWRR DL scheduler, with different number of users registered to the BS. As expected, if the quantum size is too small, the percentage of unsatisfied HTTP users among those registered with the BS is rather large. This is due to the fact that every user is being allocated only a small portion of the resources at every scheduling round, resulting in large amount of fragmentation overhead, which in turn reduces the overall throughput for BE users. The value of approximately 220 slots in the DL seems to give close to optimal performance regardless of the number of registered users.

It can be expected, however, that the values will change with different DL:UL ratios, and with different proportion of real-time application users in a traffic mix. The corresponding

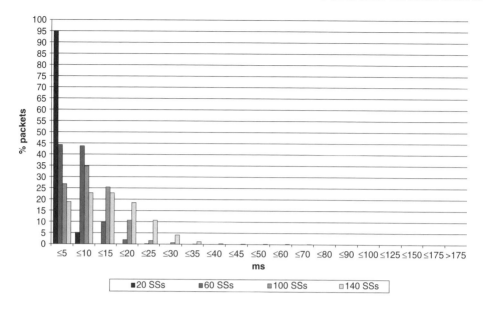

Figure 15.3 Packet delay distribution of VoIP packets when all traffic is mapped onto the BE class (MCS: QPSK1/2).

UL results are available in Veselinovic *et al.* (2008). The simulations also showed that the HTTP user satisfaction is affected not only by the choice of the parameter of the DL but also of the UL scheduler. This is due to the fact that DL TCP throughput produces a fairly proportional amount of TCP ACKs.

The operator may in the end optimize the system by, for example, minimizing the number of unsatisfied users, or maximizing the number of very satisfied users, etc. The optimal parameter (e.g. quantum size, DL:UL ratio, etc.) will then depend on the operator choice.

There remain many ways to further optimize and enhance the scheduler that are worth further study. User grouping and dynamic tuning of the DL:UL ratio and the borders of PUSC and FUSC zones (in the DL subframe) are two examples of this. The IEEE 802.16e standard offers the possibility to reduce control overhead caused by DL-MAP messages by grouping the resource allocation information of users that use the same MCS. In the presented simulations, a fixed DL:UL ratio was used but since the DL and UL ratio of real traffic is expected vary it would be better to dynamically tune the ratio to further improve the QoS and QoE[2]. Furthermore, in order to better allocate the resources in a frame, the borders of PUSC and FUSC zones can be tuned dynamically to increase radio resource utilization. In addition, from a QoE point of view, flow level scheduling principles combined with channel info can provide better performance (e.g. in terms of overall FTP download time) than DRR or proportional fair scheduling (Aalto and Lassila, 2007).

---

[2]This, however, might require a centralized architecture to avoid UL–DL interference.

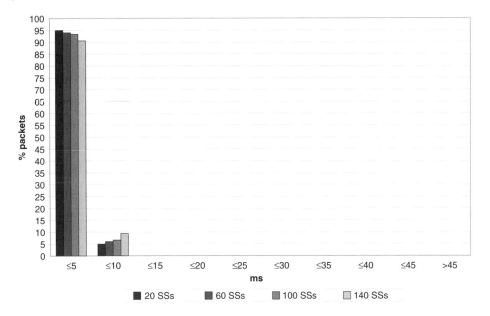

Figure 15.4   Packet delay distribution of VoIP packets when traffic is mapped onto the corresponding priority class (MCS: QPSK1/2).

15.4   System-wide Tools for Enabling QoS and QoE

Both load balancing with directed HOs and the prioritization of rescue HOs form an important part of system-wide QoS. As depicted in Figure 15.6 load balancing can distribute the resource utilization load (U_1, U_2 and U_3) among a group of BSs more evenly and thus free resources for the MAC scheduler to fulfill the guarantees made to the existing service flows and also for the admission control to admit rescue HO requests and new service flows. However, this is not always the case since the amount of resources that can be freed is heavily dependent on the number of terminals residing in the overlapping areas between BSs. Therefore, in some cases load control and prioritization of more important traffic needs to be done and thus HO guard bands (G), can be reserved for the incoming rescue HOs. In the following two sections we go through some key issues for both of these system-wide QoS and QoE enablers.

15.4.1   Load Balancing

The WiMAX Forum network architecture defines a framework for BSs to communicate their loading state to each other and if needed use BS-initiated directed HOs to force MSs residing in overlapping areas to switch their connection from highly loaded ('hot-spot') BSs to lightly loaded BSs. In the following we briefly go through the background behind this framework, present a simple load-balancing algorithm for mobile WiMAX and corresponding initial results and last take a look at some advancements that can be made. Our focus here is on the wireless interface but it should be noted that load balancing is often also needed in the

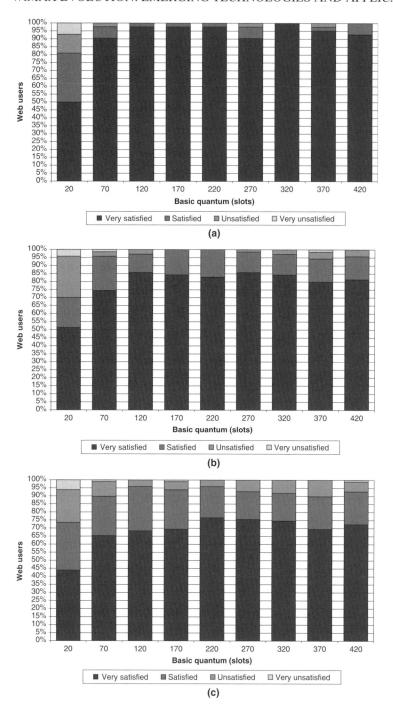

Figure 15.5 QoE of HTTP users versus DL base quantum size of the DWRR scheduler for (a) 60, (b) 100 and (c) 140 registered users.

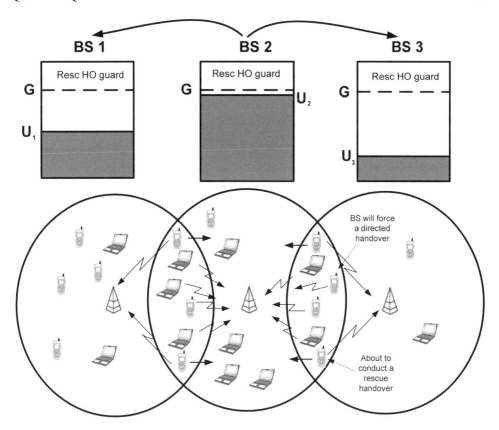

Figure 15.6 Load balancing with directed HOs and rescue HO prioritization.

backhaul part of a WiMAX access network. These challenges are discussed in more detail in Chapter 14.

15.4.1.1 Background

Load-balancing schemes that try to solve the hotspot problem can roughly be divided into *resource allocation* schemes and *load distribution* schemes (Kim *et al.*, 2007). The idea behind balancing the system load with resource allocation is to bring the resources (unoccupied frequencies) to where most of the users are located. Such channel borrowing or Dynamic Channel Allocation (DCA) (Cox and Reudink, 1973), however, needs centralized management and although in principle could also be possible in mobile WiMAX, the actual deployment of such a scheme would be difficult at least in the early stages of mobile WiMAX (especially with a distributed architecture).

With load distribution the goal is to direct the traffic to where the resources are (e.g. with HOs) and it will most likely be the way system wide load balancing will be conducted in mobile WiMAX. So far there have been just a few papers on load balancing with HOs in IEEE 802.16e-based systems. Lee and Han (2007) proposed an inter-cell! HO scheme for load balancing where MSs react to congestion and themselves initiate a HO to a less-congested sector. In this scheme load balancing is not controlled by the network and thus load balancing cannot be done in a controlled manner. Moiseev *et al.* (2006) proposed an advanced, network controlled, HO-based load-balancing algorithm that tries to find the optimal MS–BS association set to both minimize and balance the utilization of common resources in the whole system. In order to be deployed this algorithm would require an extensive amount of signaling and preferably a centralized architecture.

Many mobile WiMAX vendors are likely to implement a more distributed profile (Profile C) of the ASN since it enables better interoperability between BS and central ASN-GW nodes from different vendors and thus results in good scalability. In this case the load-balancing logic will reside in the BS and it will be our working assumption from now on.

One of the load-balancing components in mobile WiMAX is the Spare Capacity Report (SCR) procedure with which a BS can report its loading situation to its peers. In this report spare capacity is described with UL and DL available radio resource indicators that describe the average ratios of nonassigned UL and DL resources to the total usable UL and DL radio resources. Resources are defined by slots which are a good indicator of resource utilization since they describe the used resources not just in terms of traffic throughput, but also in relations to the MCSs used based on the channel conditions experienced by the MSs. Since all measurements are reported in percentages, comparisons can also be made between BSs with different capacities. The SCR can also include the total number of UL and DL slots in the frame so that the BS initiating load balancing has an idea of the actual amount of available resources in the target BS.

Based on the SCR, a threshold can be set that indicates when a BS is in an overloaded state. If it is exceeded a BS can initiate directed HOs for the MSs that reside in the overlapping areas provided that they have been discovered or order MSs most likely residing in the overlapping areas to scan neighboring target BSs. This cell reselection can be conducted before, after or both before and after load-balancing HOs are initiated. The trade-off is that preliminary scanning results in a high number of periodically occurring scans that are especially bad for real-time connections such as VoIP but on the other hand conducting scanning after the load-balancing decision might make load balancing too slow to react to load changes and result in, for example, new flow blocking.

The resulting HOs for the MSs can be conducted in many ways. The IEEE 802.16e specification offers a wide range of options on how to conduct HOs that can be used to meet the requirements of different traffic and mobility characteristics a user might have. For example, a moving VoIP user needs a reliable HO scheme with short interruption whereas a static Web-browsing session a can tolerate several seconds of HO interruption. Based on the needs of the service flow, different preliminary association levels to the TBS can be used to speed up ranging and different types of handover execution mechanisms such as optimized hard handover, Fast Base Station Switching (FBSS) and Macro Diversity Handover (MDHO) can also be used (Dong and Dai, 2007). Reliability, however, comes with the cost of higher HO overhead and general consumption of processing resources.

15.4.1.2   A Load-balancing Algorithm for Mobile WiMAX

To demonstrate some key issues of load balancing in mobile WiMAX we present a simple load-balancing algorithm and show some corresponding initial results. Velayos *et al.* (2004)

proposed a directed HO-based load-balancing scheme for a WLAN AP cluster that serves as a good starting point for an algorithm since it features similar load level reports sent between Access Points (APs) every Load Balancing Cycle (LBC). The reports received from neighbors are used locally in each AP to compute an average load level, L, in the system. The scheme defines three possible loading states: *underloaded*, *balanced* and *overloaded* for each AP. The AP is in an underloaded state if its load U is under the average L and in an overloaded state if its load U exceeds a threshold computed with the equation $T = L + \delta L$. In this equation δ characterizes the size of a hysteresis margin (i.e. how much traffic unbalance will be tolerated in the system) used to avoid the 'ping–pong' effect resulting either from conducting too many directed HOs or from highly varying traffic. Using such a hysteresis offers a simple way to dynamically adjust to the current loading situation in the whole system instead of using a single fixed threshold. Manual configuration of the δ parameter can, however, be challenging. When δ is set large the 'ping–pong' effect is avoided but it can result in call blocking and lower QoS. Then again setting a small hysteresis margin guarantees balance in the system and efficient resource utilization but might cause the 'ping–pong' effect.

A modified version of the scheme tailored to mobile WiMAX is presented in Figure 15.7. In the scheme a SCR will be broadcasted every LBC and its length will also be used as the averaging time. Since spare capacity in the SCR is reported both for DL and UL a worst-case approach will be used for the resource utilization calculations and, hence, the final resource utilization value for BS i is

$$U_i = \max\left(\frac{U_{\text{UL},i}}{U_{\text{UL,tot}}}, \frac{U_{\text{DL},i}}{U_{\text{DL,tot}}} \right) \tag{15.1}$$

where $U_{\text{UL},i}$ and $U_{\text{DL},i}$ are the number of assigned slots and $U_{\text{UL,tot}}$ and $U_{\text{DL,tot}}$ the total number of slots during the averaging interval in the UL and DL[3].

Based on the SCRs the BS will compute L and T and determine its loading state which will in turn define the way incoming requests for new service flows and rescue and directed HOs will be treated.

When a AP goes into the overloaded state it starts conducting directed HOs to the APs that are in the underloaded state[4]. If an AP is in the balanced state, below the threshold T and above the average L, or in the overloaded state it can accept both rescue HO and new service flow request as long the flow is also admitted by admission control[5].

When load balancing is triggered the BS will initiate directed HOs for MSs that reside in overlapping areas or order the MSs to scan neighboring BSs, depending on whether overlap discovery has already been conducted. When resolving the overlapping terminals measurements of the pilot signal strength, round trip delay and channel variation could be used and possibly also mobility prediction techniques could be utilized to narrow down the set of candidate MSs. It might also be beneficial to try to identify and concentrate on nomadic MSs that are likely to reside in the overlapping area throughout their session (i.e. not going to conduct a rescue HO back to the congested BS).

---

[3]Note, that SCR reports UL and DL *spare capacity* to the other BSs.

[4]Bounding values (T_{min}, T_{max} and L_{max}) for the triggering threshold T and for the average resource utilization L can be used to ensure that load balancing is not triggered unnecessarily (e.g. when the system is only lightly loaded or too overloaded) or to make sure that load balancing is initiated before call blocking occurs.

[5]Different from the original scheme, BSs in mobile WiMAX have an admission control function protecting the existing connections and, hence, new flows (and rescue HOs) can be accepted also in the overloaded state.

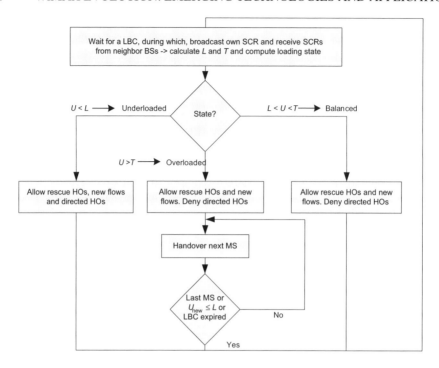

Figure 15.7 Logic for the load-balancing scheme in mobile WiMAX.

After starting the HOs the next MS in the list (or possibly a group of MSs) will be handed over until either there are no more MSs in the overlapping areas that can be handed over[6], the new resulting resource utilization, U_{new}, is equal or below the average L, or the end of the LBC has been reached.

Simulations of the scheme in a simplified system model of an ASN with a cluster of three adjacent BSs indicate a clear need to optimize the size of the margin. In the simulations BS 2 in the middle is overloaded and as a result triggers load balancing to the two neighboring BSs (BS 1 and BS 3) overlapping with BS 2. The corresponding configuration is presented in Table 15.3 (more details of the configuration can be found from Casey (2008) and Casey *et al.* (2008)).

In the simulations all MSs remain static and MSs residing in the overlapping areas far away from the BS will use a more robust MCS. Also a more robust MCS will be used in the UL than in the DL. The simulated traffic is a mixture of VoIP (both with and without silence detection) and TCP-based elastic HTTP and full-buffer FTP traffic served with the BE service. A one-second LBC is used and resource utilization is measured and reported only for the non-BE VoIP service flows meaning that BS initiated load-balancing HOs are not conducted for MSs with BE service flows but they are used as background traffic.

---

[6]In the original scheme only one MS is handed over during a LBC. However, in mobile WiMAX, where MSs can be handed over in a more rapid rate, we can handover as many MSs as necessary as long as some kind of an estimation of the resulting new resource utilization (U_{new}) is made.

Table 15.3 Simulator configuration.

System configuration	
System channel bandwidth	10 MHz
DL/UL subframe ratio	Fixed (2:1)
Traffic profile	
VoIP (UGS) and with VAD (ertPS)	25% and 25%
FTP and HTTP (BE)	25% and 25%
Poisson arrival process	Average 1.2 s
MS distribution	
MSs dropped to the system	400
BS 2 overload	200%

With this configuration and traffic profile the UL subframe is clearly the bottle neck and hence determines the final resource utilization value.

Figure 15.8 presents non-BE and BE resource utilizations for when the load-balancing scheme is not used and where it is used (with a 10% margin). One can observe that the smaller the difference between the resource utilizations of the BSs, the more balanced the system is.

When load balancing was not used the system became unbalanced since directed HOs were not conducted, and as a result admission control had to block 19 new VoIP flows in the congested BS 2. When the load-balancing scheme was used the non-BE VoIP load was distributed to the other BSs and no new calls had to be blocked in BS 2. As can be seen load balancing was triggered three times at about 16, 76 and 95 seconds. In addition, when no load balancing was used, the VoIP traffic in BS 2 was prioritized over the TCP ACKs resulting in a notable decrease in the UL BE throughput which in turn led to a large drop in the BS 2 DL BE throughput (not shown here). However, when load balancing was used UL BE throughput reduced evenly for all BSs, BE connections in BS 2 had enough bandwidth to send the ACKs and as a result the BE DL resource utilization in BS 2 did not drop dramatically.

Figure 15.9 exhibits the trade-off between setting the hysteresis margin too high or too low. As the size of the hysteresis margin increases, the number of blocked calls increases or stays the same and the number of directed HOs conducted decreases or stays the same. In this particular scenario a 0 or 5% hysteresis margin seems to be too small since it results in a HO-based 'ping–pong' effect but at the other end a 30% or 40% hysteresis margin too large since it results in new call blocking and inefficient system wide utilization of resources. With a 10% and 20% hysteresis margin no new call blocking or unnecessary HOs occurred.

From Figure 15.10 we can see in more detail how the scheme behaves. The larger the hysteresis the longer the load-balancing scheme will wait before reacting to the traffic increase and distributing the load so that the system is balanced again. With a 0% hysteresis margin the system is very balanced but this comes with the expense of ping–pong HOs coming back from BS 1 and BS 3 (steep increases in the resource utilization of BS 2).

Both 10% and 20% hysteresis were large enough to avoid the ping–pong effect and small enough to avoid flow blocking. Still it might be better to set the hysteresis to a larger value, for example, 20%, due to the fluctuations caused by a varying channel and MCSs changes not modeled here. In general the problem with the manually set hysteresis margin is that as

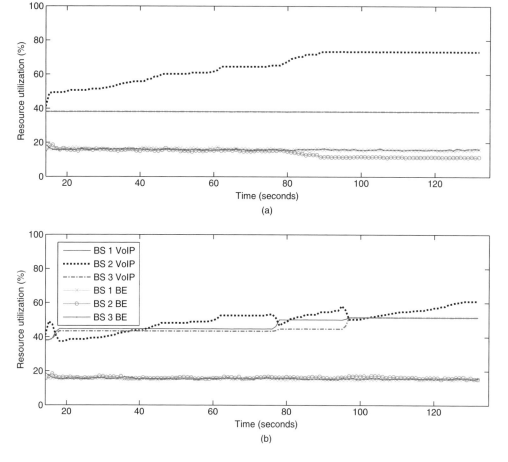

Figure 15.8 UL resource utilizations (a) without and (b) with load balancing.

the average load of the system increases, the triggering threshold can grow too high, meaning that the QoS of the existing connections can degrade and new calls can be blocked before load balancing is initiated. Thus, it is beneficial to set an upper limit (T_{max}) to when load balancing is initiated and with this particular environment it could be, for example, around 74%.

In this rather static environment using an upper limit would be a sufficient solution but as traffic becomes more fluctuating and MSs more mobile, a better way to trigger load balancing would be to tune the hysteresis margin dynamically.

15.4.1.3 Possible Enhancements

Here we briefly discuss enhancements that could be made to the basic algorithm in terms of automatic computation and tuning of the triggering threshold and the use of multiple triggering thresholds to avoid the negative effects of fluctuating traffic.

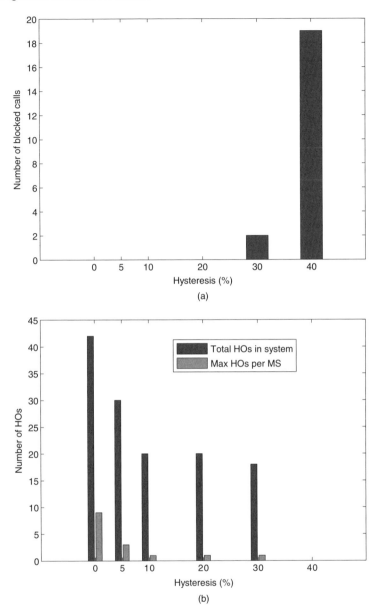

Figure 15.9 Number of blocked calls (a) and number of directed HOs (b) with different hysteresis margins.

Triggering can be dynamically adjusted based on the current traffic characteristics of the system to both deal with premature reaction to variable traffic but also to avoid long delays and packet drops by the BS that occur if the threshold is large and load balancing is triggered too late. The SCR, discussed earlier, can contain a radio resource fluctuation field

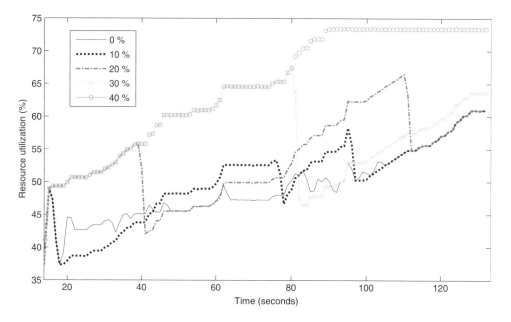

Figure 15.10 Resource utilization of BS 2 with different hysteresis margins.

that describes the degree of fluctuation in channel data traffic throughputs for the reporting BS and can thus be used as a basis when setting the triggering threshold.

The fluctuation value ranges from a minimum 0 that could correspond to a traffic mix of UGS-based VoIP connections with steady channel conditions to a maximum 255 that, in turn, could correspond to a traffic mix of highly varying mobile traffic sources with rapidly changing channel conditions.

As a basis to automatically compute the triggering threshold two boundary values $T_{U,\min}$ and $T_{U,\max}$ could be set. The lower boundary value $T_{U,\min}$ includes a minimum hysteresis margin required to avoid the ping–pong effect resulting from one BS initiating and another accepting too many load-balancing HOs. We call this the *HO-based ping–pong effect* and call the ping–pong effect caused by general resource utilization fluctuation the *fluctuation based ping–pong effect*[7].

Here $T_{U,\min}$ can be set in relation to the average system load L and average system radio resource fluctuation F_{sys}, and will increase as F_{sys} increases; F_{sys} can be calculated based on the values received from the SCR of other BSs thus describing the overall fluctuating nature of the incoming traffic.

The upper bound reference value $T_{U,\max}$ is based on the reliability and performance of the scheduler and denotes the maximum value for the triggering threshold after which the service of the existing connections starts to degrade. So an estimation of the new resource utilization threshold can be computed every LBC as a function of the above-mentioned variables: $T_U = f(T_{U,\min}, T_{U,\max}, F_{\text{sys}})$. One example of a simple way to compute the threshold would be

---

[7]Similar ping–pong effects can also be seen in a signal-based HO decision.

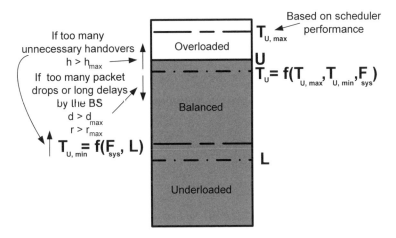

Figure 15.11 Automatic triggering threshold tuning.

with the following equation

$$T_U = T_{U,\min} + (T_{U,\max} - T_{U,\min})\frac{F_{\text{sys}}}{F_{\max}}, \qquad (15.2)$$

where $F_{\max}$ is the maximum fluctuation value 255. As can be seen, as the system fluctuation F_{sys} increases the size of the hysteresis margin increases so that the system does not react prematurely to the varying traffic[8]. Both the lower boundary value $T_{U,\min}$ and resulting threshold T_U can be reactively tuned in relations to a maximum allowed HO rate per MS ($h_{\max}$) as depicted in Figure 15.11 (i.e. tuned up to avoid ping–pong HOs). The resulting threshold can also be tuned in relations to maximum values for the number of dropped packets ($r_{\max}$) and overlong delays ($d_{\max}$) (i.e. tuned down to trigger load balancing earlier).

The increase of fluctuation in resource utilization can also be relieved by increasing the averaging interval used to measure the resource utilization. However, this has to be done with care as it might make the system too slow to react to varying traffic. Yet another way of dealing with fluctuation is to set a hysteresis margin also in time. Soldani *et al.* (2006) described a method where load balancing is triggered only after resource utilization has been over the triggering threshold for a certain period of time.

An even more enhanced way to deal with the fluctuating environment is to use multiple thresholds to differentiate load balancing triggering in relations to the traffic types served. Although traditionally, with rather static traffic conditions, the higher priority connections have been handed over first to the less-congested cell, in a fluctuating environment, such as a mobile WiMAX access network, it might actually be beneficial to handover the delay sensitive connections last. In this way the delay sensitive connections (e.g. UGS-based VoIP) avoid unnecessary HOs and the delay tolerant connections (e.g. nrtPS-based FTP), not so sensitive to HO interruption delays, have a chance to react to the load increase and get higher bandwidth from a less-congested BSs. Traffic prioritization within the classes could still be

---

[8]Note that this is just one example of a way to dynamically compute the threshold and a more elaborate formula could be the target of future research.

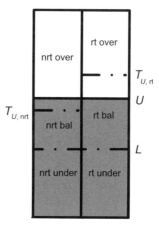

Figure 15.12 Multiple threshold triggering.

used so that for example a higher traffic priority nrtPS FTP connection would be handed over before a lower traffic priority nrtPS FTP connection, so that it would have access to more bandwidth.

Figure 15.12 presents the basic idea of the scheme with two general traffic classes real-time (rt) and non-real-time (nrt). To make the rt connections most robust against traffic fluctuation we will set the load-balancing triggering threshold $T_{U,\text{rt}}$ for the rt-class to a higher value than $T_{U,\text{nrt}}$ for the nrt-class. This can either be done with manual configuration of fixed thresholds, with fixed hysteresis values in relations to average resource utilization or with a more advanced algorithm that computes and automatically tunes the thresholds in relations to monitored differentiated parameters in a similar manner to that we described earlier[9]. If time hysteresis's are used they can also be set to a longer value for the delay sensitive classes to further protect them from premature reaction.

In the example in Figure 15.12, load balancing would be initiated for the nrt class and directed HOs would be conducted to the TBSs in the nrt underloaded state. If the load increase would be only temporary the delay and HO sensitive rt connections would be spared from an unnecessary HO. Furthermore if after a period of time, one of the TBSs load would temporarily increase, the nrt connections would be handed over back to the original cell. This 'visit' would be beneficial to the nrt connections because they had access to a larger amount of bandwidth than what they would have had in the original BS[10].

One advantage of using the multiple threshold approach is that since the UGS and ertPS connections usually reserve and use less bandwidth than rtPS and nrtPS connections, handing over rtPS or nrtPS connections releases more resources in the congested BS. Furthermore the UGS and ertPS based flows require only a certain guaranteed rate and do not benefit from the extra bandwidth available in a less congested BS as much as rtPS and nrtPS connections do. In addition, with the multiple threshold approach ready lists of MSs in overlapping areas

---

[9]In mobile WiMAX we could assume the maximum allowed handover rates $h_{\text{nrtPS}} > h_{\text{rtPS}} > h_{\text{ertPS}} = h_{\text{UGS}}$.

[10]It should be noted that when TCP-based delay tolerant connections are handed over many times the WiMAX system has to conduct proper MAC context transfer or buffering of packets during the HO so that no packets are lost and TCP congestion avoidance is not triggered.

could be kept only of the delay tolerant MSs not sensitive to scanning and thus scanning for the delay sensitive connections could be minimized.

15.4.2   HO Prioritization

Since the proactive use of load balancing with directed HOs will only improve the possibility to fulfill QoS guarantees, load control functions are also needed. Load control can be conducted for example by blocking new service flow requests, by lowering the QoS of existing flows (and restoring it later) or if possible by conducting vertical HOs to other systems[11]. As a last resort ongoing flows can be dropped but should be done in a controlled and prioritized manner.

The focus in this section will be on how HO prioritization over new flow requests could be conducted in mobile WiMAX, how traffic differentiation in HO prioritization could be done to enable better QoE in the mobile WiMAX access network and finally how load balancing could complement HO prioritization and how the two could be combined.

15.4.2.1   Background

HO prioritization is usually conducted by reserving some of the resources (i.e. a guard band) for incoming rescue handovers and can be roughly classified to two categories: *fixed guard band schemes* and *dynamic guard band schemes*. In fixed guard band schemes the guard band is fixed and defined manually in network planning. With dynamic guard channel schemes the idea is to tune the guard channel dynamically based on, for example, the number of ongoing calls in neighboring cells, the estimation of the channel holding times and the number of HOs to and from the BS. The biggest advantage that the dynamic schemes introduce is more efficient resource utilization without compromising the QoS requirements. Complexity that results from required information exchange between BSs and logic are on the other hand a disadvantage.

Roughly speaking it seems that there are two approaches to dynamic HO resource reservation. The threshold can be adjusted based on information recorded only locally or based on both local information and information exchanged between adjacent BSs. The information exchange-based schemes can further be divided into those that just try to estimate the resources needed in each BS for incoming HOs and those that enable explicit HO resource reservation per-connection for the entire route that each MS will traverse (the latter can become quite expensive unless the mobility patterns of the MSs are known).

In terms of the existing WiMAX Forum network architecture both approaches, local and information exchange-based, are in principle possible but the local approach seems more feasible due to its simplicity and distributed nature. The WiMAX Forum network architecture does not currently provide a framework for such information exchange between BSs which means that new messages would have to be introduced to enable information exchange-based dynamic reservation which would in turn result in lower scalability. Therefore, due to the distributed, but also flexible nature of mobile WiMAX, local dynamic guard band adaptation

---

[11] An approach called Directed Retry (DR) (introduced by Eklundh (1986)) could in principle also be used by the BS to direct blocked flows to another BS and even direct the users that are not in an overlapping area to the nearest BS with most free capacity (this kind of Network Directed Roaming, proposed by Balachandran *et al.* (2002), however, requires cross-layer design).

for example, based on the arriving rescue HO rate and the corresponding traffic characteristics seems natural.

One thing that should be noted is that, in order to make rescue HO prioritization work while conducting load-balancing triggered directed HOs, it would be necessary to distinguish between the two HO types while conducting preparatory HO signaling between a serving BS and a target BS. Rescue HOs should be prioritized over load-balancing HOs (as is done in the load-balancing scheme discussed earlier) and it could be accomplished by using the extra bits in the HO type field of the HO_req message defined in the current WiMAX Forum access network specification (WiMAX Forum, 2008).

15.4.2.2   HO and Traffic Prioritization in Mobile WiMAX

As mobile WiMAX carries many types of traffic it makes sense to further differentiate prioritization of traffic by setting multiple guard bands to protect, for example, new service flows with higher priority over lower priority service flows (e.g. new UGS VoIP service flow over the new nrtPS FTP service flow) and also incoming HOs of a higher priority QoS class over HOs of a lower priority QoS class (e.g. UGS VoIP service flow HO over a nrtPS FTP service flow HO).

There have been few proposals for such schemes in the context of mobile WiMAX. Tsang *et al.* (2007) proposed a dynamic call admission control scheme which featured a quadra-threshold bandwidth reservation algorithm that reserves guard bands for each of the WiMAX service classes and a common guard band for all incoming rescue HOs. Ge and Kuo (2007) introduced an advanced dynamic bandwidth quasi-reservation scheme that reserves bandwidth for both HO real-time traffic and potential new real-time traffic based on probabilistic estimations of HO and new flow arrivals. The scheme also enables non-real-time service flows to temporarily occupy the quasi-reserved bandwidth to mitigate the inefficiency of full-bandwidth reservation. Wang *et al.* (2007) proposed an admission control policy for non-preprovisioned (MS initiated) service flows that uses a multiple guard channel scheme to give higher priority to HO connections and new real-time connections of such as UGS, rtPS and ertPS. It also introduced a proportional bandwidth-borrowing scheme that enables incoming higher priority flows to borrow some bandwidth from existing connections of lower priority service classes (i.e. from existing rtPS/ertPS and nrtPS flows) in case congestion occurs.

The advantage of using this kind of differentiation is that the higher priority traffic can experience even better QoS. For example, it is far more irritating to be dropped out of a conversation than to wait a little longer for your FTP download and therefore it is good to have the possibility to ensure that HOs for non-real-time connections are dropped[12] before dropping HOs for real-time connections.

To illustrate this functionality Figure 15.13 presents a simple multiple guard band concept based on two basic traffic classes (non-real-time and real-time) presented by Chen *et al.* (2005) where three dynamic guard bands are used to maintaining the relative priorities of different types of traffic.

---

[12]Their QoS can also be lowered or in case of BE service the flow can even be queued for the whole cell visit.

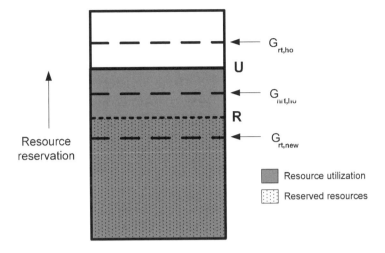

Figure 15.13 Multiple-threshold bandwidth reservation (Chen *et al.*, 2005).

As resource reservation increases[13], the resources reserved after the guard band for new real-time connections has been passed, can be used by new rt connections and by nrt and rt handovers. In the same way the resources reserved after the guard band for non-real-time HOs can be only used by nrt and rt handovers. Finally, the resources reserved after the guard band for real-time HOs can only be used by rt real-time HOs. All new nrt connections will be blocked after the new real-time connection guard band has been passed which will happen in the example in Figure 15.13.

This scheme works locally both by estimating initial values for the thresholds based on instantaneous mobility and traffic load situation and by further adapting the thresholds according to instantaneous QoS measures such as dropped handovers and blocked new calls and thus fits in well with the WiMAX Forum network architecture.

It should be noted that the case presented in Figure 15.13 is only one example and that an operator can configure the thresholds as it sees best, that is, it can also prioritize new calls of a higher class over HO requests of a lower class (as is defined, for example, by Wang *et al.* (2007)). This is especially important with emergency services that should have access to the network at all times, but also for example in the case where higher class users are paying a lot more for their services[14]. In order to reduce complexity that results from using many thresholds, an operator can use a single threshold for a combination of service classes.

---

[13]It should be noted that in guard band schemes, comparisons are made in terms of the *reserved resources, R, of the BS, not the used resources, U*, as is done with load balancing. Reserved resources correspond to slots reserved (in the long term) in order to fulfill at least the MRTR guarantees made to the served MSs in the BS whereas resource utilization corresponds to instantaneous slot utilization that also contains excess traffic (i.e. everything assigned by the MAC scheduler after MRTR until MSTR) and can therefore be higher than resource reservation (as shown in the example in Figure 15.13) especially if many of the bursty traffic flows are active. Vice versa, if many of the service flows are not active resource reservation can be higher.

[14]Furthermore, a guard band can also be reserved for the BE services flows, which have no MRTR, in order to provide them access to at least some bandwidth.

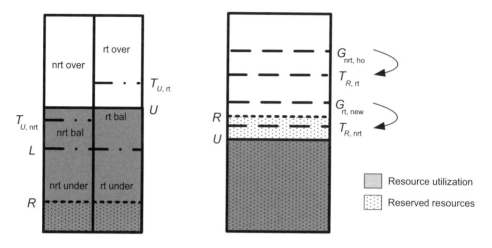

Figure 15.14 Resource utilization- and reservation-based multiple threshold triggering.

15.4.2.3   Combined Scheme with Load Balancing

In the load-balancing schemes discussed earlier load balancing is triggered in relations to high resource utilization. It is, however, possible, especially if guard bands are used to prioritize incoming traffic and if many of the admitted service flows are not very active, that load balancing will not be triggered (i.e. resource utilization based triggering threshold is not exceeded) before a reserved guard band is exceeded and hence admission control can unnecessarily start to block flows. Therefore, it would be beneficial to be able to trigger load balancing also in relation to the reserved guard bands.

In order to trigger load balancing before a guard band is reached, but not too early to avoid unnecessary HOs, a *resource reservation based triggering threshold* can be computed based on, for example, the average arrival rate of new slot reservations and the corresponding average holding times of slots and also in relations to the rate at which the load-balancing scheme can release slots. After computing such a reservation-based triggering threshold, it can be further reactively tuned in relations to experienced call blocking and HO rate[15]. As in resource utilization-based triggering these parameters could be differentiated correspondingly.

Putting all of the pieces together, an example of the final framework with the rt and nrt classes is presented in Figure 15.14. It features differentiated thresholds $T_{U,\text{nrt}}$ and $T_{U,\text{rt}}$ for resource utilization-based load balancing, differentiated guard bands $G_{\text{nrt,ho}}$ and $G_{\text{rt,new}}$ and the corresponding resource reservation-based load-balancing triggering thresholds $T_{R,\text{nrt}}$ and $T_{R,\text{rt}}$.

Although the examples for both of the schemes have been presented with the nrt and rt example classes, the schemes can be extended to all of the scheduling services that mobile WiMAX supports by adding more thresholds or used as such for aggregated groups of scheduling services[16] to simplify the scheme and limit the number of thresholds. All in

---

[15] Also possibly in relation to bandwidth borrowing.

[16] For example, UGS, ertPS and rtPS use real-time and nrtPS and BE use non-real-time.

all this kind of a framework has the ability to react to the loading situation on the level[17] that is at the time most critical and can thus maximize the efficient use of resources and minimize unnecessary packet drops and flow blocking. At the same time it is able to avoid handover drops in a differentiated way resulting in a mobile WiMAX access network that has an improved capability to meet the needs of users and provide better QoE on a system level.

15.5   Conclusions

In this chapter we have discussed different ways of enabling QoS and QoE in a mobile WiMAX system. The issue has been discussed both on the view point of per-flow scheduling within a single BS and on the system-level within a cluster of BSs.

The BS MAC scheduler was discussed and evaluation cases were presented where the MAC scheduler was optimized in relation to the service experienced in a Web-page retrieval procedure. The system-level QoS tools load balancing and HO prioritization, were also covered and new enhancements were proposed to enable differentiation of QoS and better user experiences.

As discussed in some of the other chapters of the book, requirements and expectations can vary between different WiMAX scenarios and use cases. Thus, when designing and deploying a WiMAX access network, the important question to ask, is not necessarily how to ensure that the requirement of a certain QoS parameter is fulfilled but rather, what are the needs of the users. Then, once there is a good understanding on what kind of an experience a user of a service (e.g. UGS or BE) would typically like to have in the particular WiMAX deployment, the parameters of the access network can be optimized and differentiated in relations to these needs to enable better QoE for the end user.

Acknowledgements

The authors of Chapter 15 are grateful to Mr Jose Pradas, Mr Keijo Hyttinen and Dr Jaakko Talvitie for their contributions to this work.

References

Aalto, S. and Lassila, P.E. (2007) Impact of size-based scheduling on flow level performance in wireless downlink data channels. *Proceedings of the 20th International Teletraffic Congress*, pp. 1096–1107.

Balachandran, A., Bahl, P. and Voelker, G.M. (2002) Hot-Spot congestion relief in public-area wireless networks. *Proceedings of the 4th IEEE Workshop on Mobile Computing Systems and Applications*, pp. 70–80.

Casey, T. (2008) *Base Station Controlled Load Balancing with Handovers in Mobile WiMAX*. Master's thesis Department of Communications and Networking, Helsinki University of Technology.

Casey, T., Veselinovic, N. and Jäntti, R. (2008) Base station controlled load balancing with handovers in mobile WiMAX. *Proceedings of the 19th IEEE International Symposium on Personal, Indoor and Mobile Radio Communications*.

---

[17] Packet-level resource utilization versus flow level resource reservation.

Chen, X., Li, B. and Fang, Y. (2005) A dynamic multiple-threshold bandwidth reservation (DMTBR) scheme for QoS provisioning in multimedia wireless networks. *IEEE Transactions on Wireless Communications* **4**(2), 583–592.

Cicconetti, C., Lenzini, L., Mingozzi, E. and Eklund, C. (2006) Quality of service support in IEEE 802.16 networks. *IEEE Network* **20**(2), 50–55.

Cox, D. and Reudink, D. (1973) Increasing channel occupancy in large-scale mobile radio systems: Dynamic channel REassignment. *IEEE Transactions on Vehicular Technology* **22**(4), 218–222.

Dong, G. and Dai, J. (2007) An improved handover algorithm for scheduling services in IEEE 802.16e. *Proceedings of the IEEE Mobile WiMAX Symposium, 2007*, pp. 38–42.

Eklundh, B. (1986) Channel utilization and blocking probability in a cellular mobile telephone system with directed retry. *IEEE Transactions on Communications* **34**(4), 329–337.

Ge, Y. and Kuo, G.S. (2007) Dynamic bandwidth quasi-reservation scheme for real-time services in IEEE 802.16e networks. *Proceedings of the IEEE Wireless Communications and Networking Conference, 2007*, pp. 1700–1705.

ITU-T (1994) *ITU-T Recommendation E.600 - Terms and Definitions Related to QoS and Network Performance Including Dependability*.

ITU-T (2001) *End-user multimedia QoS categories - Recommendation G.1010*.

Kilkki, K. (1999) *Differentiated Services for the Internet*. Macmillan Technical Publishing, USA.

Kim, D., Sawhney, M. and Yoon, H. (2007) An effective traffic management scheme using adaptive handover time in next-generation cellular networks. *International Journal of Network Management* **17**, 139–154.

Kwon, T., Lee, H., Choi, S. *et al.* (2005) Design and implementation of a simulator based on a cross-layer protocol between MAC and PHY layers in a WiBro compatible IEEE 802.16e OFDMA system. *IEEE Communications Magazine* **43**(12), 136–146.

Lee, S.H. and Han, Y. (2007) A novel inter-FA handover scheme for load balancing in IEEE 802.16e system. *Proceedings of the IEEE 65th Vehicular Technology Conference*, pp. 763–767.

Moiseev, S., Filin, S., Kondakov, M. *et al.* (2006) Load-balancing QoS-guaranteed handover in the IEEE 802.16e OFDMA network. *Proceedings of the Global Telecommunications Conference, 2006*, pp. 1–5.

Rath, H.K., Bhorkar, A. and Sharma, V. (2006) An opportunistic DRR (O-DRR) uplink scheduling scheme for IEEE 802.16-based broadband wireless networks. *Proceedings of the IETE International Conference on Next Generation Networks (ICNGN)*.

Sayenko, A., Alanen, O., Karhula, J. and Hämäläinen, T. (2006) Ensuring the QoS requirements in 802.16 scheduling. *Proceedings of the 9th ACM International Symposium on Modeling, Analysis and Simulation of Wireless and Mobile Systems*, pp. 108–117.

Soldani, D., Cuny, R. and Li, M. (2006) *QoS and QoE Management in UMTS Cellular Networks*. John Wiley & Sons Ltd, Chichester, England.

Tsang, K.F., Lee, L.T., Tung, H.Y. *et al.* (2007) Admission control scheme for mobile WiMAX networks. *Proceedings of the IEEE International Symposium on Consumer Electronics, 2007*, pp. 1–5.

Velayos, H., Aleo, V. and Karlsson, G. (2004) Load balancing in overlapping wireless LAN cells. *Proceedings of the IEEE International Conference on Communications, 2004*, Vol. 7, pp. 3833–3836.

Veselinovic, N. *et al.* (2008) QoS of experience in 802.16e. *Proceedings of the 11th International Symposium on Wireless Personal Multimedia Communications*.

Wang, L., Liu, F., Ji, Y. and Ruangchaijatupon, N. (2007) Admission control for non-preprovisioned service flow in wireless metropolitan area networks. *Proceedings of the 4th European Conference on Universal Multiservice Networks*, pp. 243–249.

WiMAX Forum (2008) *WiMAX Forum Network Architecture (Stage 3: Detailed Protocols and Procedures - Release 1, Version 1.2)*.

Part VI

WiMAX Evolution and Future Developments

16

MIMO Technologies for WiMAX Systems: Present and Future

Chan-Byoung Chae, Kaibin Huang and Takao Inoue

16.1 Introduction

Multiple input multiple output (MIMO) technology is a key breakthrough in wireless communication. By using multiple antennas, MIMO technology multiplies throughput without requiring additional frequency bandwidth, enhances link reliability through spatial diversity and enlarges the coverage area by increasing the transmission range. These features have motivated extensive research on developing MIMO theory and techniques in the last decade. One of the main goals of the IEEE802.16 standard (WiMAX)[1] is to deliver last-mile wireless broadband access as an alternative to cable and Digital Subscriber Line (DSL). Achieving this goal critically hinges on the application of MIMO for enhancing throughput and link reliability as expected in the last-mile of wireless access, for applications such as high-quality video streaming. In this chapter we summarize single-user MIMO techniques that have been adopted in IEEE 802.16e, and multiuser MIMO techniques that are being considered for IEEE802.16m (WiMAX evolution).

MIMO techniques can be grouped into two categories: single-user and multiuser MIMO. As illustrated in Figure 16.1(a), a single-user MIMO system is a point-to-point link between a Base Station (BS) and one scheduled Mobile Station (MS), where multiple antennas are employed at both ends. For single-user MIMO systems, MIMO techniques help to increase user throughput by either supporting multiple data streams or improving link reliability. In Section 16.2, we discuss various single-user MIMO techniques that have been included in the IEEE802.16e standard.

Multiuser MIMO techniques represent the latest developments in MIMO theory. As shown in Figure 16.1(b), multiuser MIMO enables a BS to transmit multiple data streams to multiple

[1]WiMAX, the World Interoperability for Microwave Access, is based on the IEEE802.16 standard. Hereafter, we use the IEEE802.16 for consistency.

WiMAX Evolution: Emerging Technologies and Applications   Edited by Marcos D. Katz and Frank H.P. Fitzek
© 2009 John Wiley & Sons, Ltd

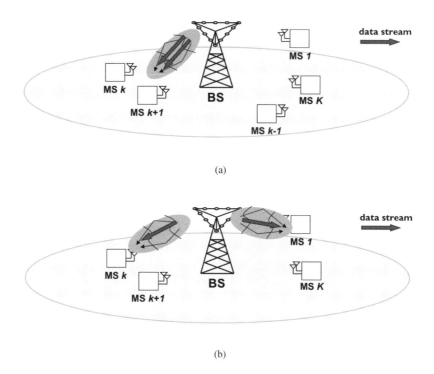

(a)

(b)

Figure 16.1 (a) Single-user MIMO transmission. The BS selects only one user. (b) Multiuser MIMO transmission. The BS selects multiple users and shares the radio resource.

MSs in the same time and frequency slot. The broadcast capability of multiuser MIMO relies on the efficient use of degrees of freedom in the virtual space created by multiple antennas. Multiuser MIMO techniques are usually integrated with scheduling algorithms to maximize downlink throughput. In Section 16.3, we discuss a nonlinear multiuser technique called vector perturbation, followed by linear multiuser techniques in Section 16.4.

Throughout this chapter, we use the following notation. Upper case and lower case boldface are used to denote matrices $\mathbf{A}$ and vectors $\mathbf{a}$, respectively. If $\mathbf{A}$ denotes a complex matrix, and $\mathbf{A}^{\mathrm{T}}$, $\mathbf{A}^*$, $\mathbf{A}^{-1}$, and $\mathbf{A}^\dagger$ denote the transpose, conjugate transpose, inverse and pseudo-inverse of $\mathbf{A}$, respectively; $[\mathbf{A}]_u$ denotes the uth column of matrix $\mathbf{A}$; $\|\mathbf{A}\|_F$ denotes the Frobenius norm of matrix $\mathbf{A}$ and adj($\mathbf{A}$) denotes the adjoint matrix of $\mathbf{A}$; $\mathbb{E}$ denotes expectation.

16.2   IEEE802.16e: Single-user MIMO Technologies

IEEE802.16 aims to support the development of broadband wireless metropolitan area networks (WMANs). The standard IEEE 802.16e was finalized in September 2005 (IEEE, 2006) and now the IEEE 802.16m (IEEE 802.16e evolution) is in progress (IEEE, 2007).

The main standardization goal of the IEEE802.16e was to provide mobility functions for seamless communications but several MIMO solutions were also discussed at the standards based on the IEEE802.16d (fixed WiMAX) (IEEE, 2006).

In the IEEE802.16e standard meetings, various MIMO options were discussed for inclusion in the standard. In this section, we describe a representative subset of MIMO options adopted in the IEEE802.16e. There are several Mobile Application Parts (MAPs) for control message but we consider the extended normal MAP adopted in March 2005. The extended normal MAP supports open-loop and closed-loop MIMO techniques. For the MIMO options, up to four transmit antennas are used at the BS and up to two transmit antennas at the MS. In this chapter, we focus on the downlink MIMO techniques as it will provide a rich overview of advanced MIMO techniques.

16.2.1 Open-loop Solutions

There are two main categories of open-loop single-user MIMO transmission technologies. *Spatial multiplexing* techniques transmit parallel data streams to increase data rate, by exploiting the spatial dimension of wireless propagation channels. *Diversity* techniques are considered to improve the link reliability. There are fundamental tradeoffs between diversity and multiplexing Zheng and Tse (2003), and practical algorithms for switching between these techniques have been proposed by Chae *et al.* (2004), Heath and Paulraj (2005). In practice, the channel knowledge can be fed back to the transmitter (for *closed-loop* techniques) to improve the performance thanks to precoding, at the cost of higher complexity.

A total of seven open-loop techniques were adopted through technical meetings. For two transmit antenna systems, the classic Alamouti Space–Time Block Code (STBC) was included for rate one (Alamouti, 1998). To achieve a rate of two and a diversity order of two, constellation rotation-based STBC was adopted, but this requires more complex nonlinear decoders such as a sphere decoder or Maximum Likelihood Decoder (MLD) at the MS.

In IEEE802.16d, there were no MIMO techniques for a three transmit antenna configuration. For IEEE802.16e, two techniques called Coordinate Interleaved STBC (CI-STBC) were adopted in July 2004 for rate one and two. To provide a better understanding, we briefly introduce the solutions. Note that CI-STBC can also be used for four transmit antenna systems but simple antenna circulations are used in the IEEE802.16e to provide backward compatibility with the IEEE802.16d system.

Let the complex transmit symbols be x_1, x_2, x_3, x_4, which take values from a square Quadrature Amplitude Modulation (QAM) constellation. Let $s_i = x_i e^{j\theta}$ for $i = 1, 2, 3,$ $\ldots, 8$, where $\theta = \tan^{-1} \frac{1}{3}$ and we obtain

$$\tilde{s}_1 = s_{1I} + j s_{3Q}, \quad \tilde{s}_2 = s_{2I} + j s_{5Q},$$
$$\tilde{s}_3 = s_{3I} + j s_{1Q}, \quad \tilde{s}_4 = s_{4I} + j s_{2Q}, \quad (16.1)$$

where the subscripts I and Q stand for in-phase and quadrature-phase, respectively. The BS maps the new symbols $\tilde{s}_i$ over two time symbols and two subcarriers as follows:

$$\mathbf{A} = \begin{bmatrix} \tilde{s}_1 & -\tilde{s}_2^* & 0 & 0 \\ \tilde{s}_2 & \tilde{s}_1^* & \tilde{s}_3 & -\tilde{s}_4^* \\ 0 & 0 & \tilde{s}_4 & \tilde{s}_3^* \end{bmatrix}. \quad (16.2)$$

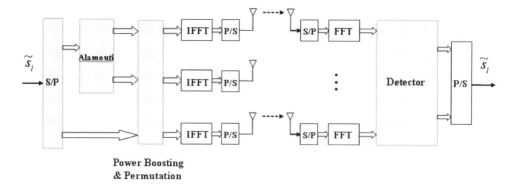

Figure 16.2 Transmitter and receiver structures for Matrix **A** and Matrix **B** for three transmit antenna systems.

Thus, the transmit symbols $\tilde{s}_1$ and $\tilde{s}_2$ are transmitted through subcarrier f_1 and transmit antennas 1 and 2 over time 1 and 2, while $\tilde{s}_3$ and $\tilde{s}_4$ are transmitted through subcarrier f_2 and transmit antennas 2 and 3 over time 1 and 2.

For rate two transmission, the BS uses the following mapping matrix,

$$\mathbf{B} = \begin{bmatrix} \sqrt{\tfrac{3}{4}} & 0 & 0 \\ 0 & \sqrt{\tfrac{3}{4}} & 0 \\ 0 & 0 & \sqrt{\tfrac{3}{2}} \end{bmatrix} \begin{bmatrix} \tilde{s}_1 & -\tilde{s}_2^* & \tilde{s}_5 & -\tilde{s}_6^* \\ \tilde{s}_2 & \tilde{s}_1^* & \tilde{s}_6 & \tilde{s}_5^* \\ \tilde{s}_7 & -\tilde{s}_8^* & \tilde{s}_3 & -\tilde{s}_4^* \end{bmatrix}, \tag{16.3}$$

where the definition for the remaining variables are given by

$$\tilde{s}_5 = s_{5I} + js_{7Q}, \quad \tilde{s}_6 = s_{6I} + js_{8Q},$$
$$\tilde{s}_7 = s_{7I} + js_{5Q}, \quad \tilde{s}_8 = s_{8I} + js_{6Q}. \tag{16.4}$$

Since each symbol s_i experiences more independent channels (than a simple Alamouti-based approach), better diversity gains are attained. The mapping matrices **A** and **B** are permuted based on the mapping subcarrier index to avoid a power imbalance. Figure 16.2 illustrates the transmitter and receiver structures for three transmit antenna systems.

For rate three transmission, simple spatial multiplexing is used as follows:

$$\mathbf{C} = \begin{bmatrix} x_1 \\ x_2 \\ x_3 \end{bmatrix}. \tag{16.5}$$

Note that original complex transmit symbols x_i are used for the rate three transmission while the coordinated interleaved symbols $\tilde{s}_i$ are used for the rate one and two transmissions.

For four transmit antenna systems, simple concatenated Alamouti codes were adopted for rate one and two cases even though CI-STBC can also be used because of backward compatibility. The mapping matrices $\mathbf{A}$, $\mathbf{B}$ and $\mathbf{C}$ are given by

$$\mathbf{A} = \begin{bmatrix} x_1 & -x_2^* & 0 & 0 \\ x_2 & x_1^* & 0 & 0 \\ 0 & 0 & x_3 & -x_4^* \\ 0 & 0 & x_4 & x_3^* \end{bmatrix}, \tag{16.6}$$

$$\mathbf{B} = \begin{bmatrix} x_1 & -x_2^* & x_5 & -s_7^* \\ x_2 & x_1^* & x_6 & -x_8^* \\ x_3 & -x_4^* & x_7 & -x_5^* \\ x_4 & x_3^* & x_8 & x_6^* \end{bmatrix}, \tag{16.7}$$

$$\mathbf{C} = \begin{bmatrix} x_1 \\ x_2 \\ x_3 \\ x_4 \end{bmatrix}. \tag{16.8}$$

It is notable that full diversity cannot be achieved through the matrices $\mathbf{A}$ and $\mathbf{B}$, so circulated versions of the matrices are used to achieve better diversity gain (see IEEE (2006)).

16.2.2 Closed-loop Solutions

Aiming to improve open-loop MIMO systems, several closed-loop solutions with a good performance-complexity trade-off were proposed Chae *et al.* (2008c), Gore and Paulraj (2002), Heath *et al.* (2001), Shim *et al.* (2005). Most closed-loop MIMO techniques exhibit good link performance in low-mobility environments, but link performance deteriorates considerably in rapidly changing channels.

In the IEEE802.16e, there are two categories of closed-loop solutions: (i) antenna grouping/selection; (ii) beamforming/precoding. Exploiting only a few feedback bits was shown to significantly enhance the link quality. As an example, for a four transmit antenna case, antenna grouping techniques to minimize Frobenius norm of each grouping pair were proposed for rate one (Chae *et al.*, 2008c) and as well as to minimize the received Mean Square Error (MSE) for rate two (Shim *et al.*, 2005), based on the mapping matrices $\mathbf{A}$ and $\mathbf{B}$. Diversity gain can be further enhanced through antenna selection methods but their performance deteriorates in high mobility environments. In contrast, antenna grouping shows good performance with relatively low complexity in a wide range of mobility environments.

When more than two bits are available for the feedback channel, Grassmannian beam-forming and precoding can be used (Love and Heath, 2005, Love *et al.*, 2003). The MSs calculate their MSE of the received signal or capacity to choose the codeword that the BS will use in data transmission and feed back the codeword index to the BS. Instead of storing all the codewords, efficient codewords construction methods were included in IEEE802.16e.

We summarize all MIMO solutions adopted in the IEEE802.16e in Table 16.1.

16.2.3 Limitations

From the point of view of information theory, supporting multiple users in the spatial domain is necessary to enhance the cell throughput (Gesbert *et al.*, 2007). As can be seen from Table 16.1, however, no multiuser MIMO solution exists in the IEEE802.16e standard. This

Table 16.1 MIMO solutions in the IEEE802.16e. OL: open loop; CL: closed loop; FDFR: full diversity full rate; AC: antenna circulation; AG: antenna grouping; AS: antenna selection; BF: beamforming; CR: constellation rotation.

#BS antennas	2 Tx	3 Tx	4 Tx	Comments
OL Div.	Alamouti	FDFR STC	AC	Spatial rate $(N_s) = 1$
Hybrid	CR STC	FDFR STC	AC	$N_s = 2$
SM	STC $w/R = 2$	STC $w/R = 3$	STC $w/R = 4$	$N_s = N_t$
CL Ant. grouping	AS for $R = 1$	AG for $R = 1, 2$	AG for $R = 1, 2$	Grouping/selection
Ant. selection		AG for $R = 1, 2$	AG for $R = 1, 2, 3$	index feedback
Beamformimg	Grassmanian BF	Grass. precoding	Grass. precoding	Codebook based
precoding	for $R = 1$	for $R = 1, 2$	for $R = 1, 2, 3$	BF/precoding

is mainly due to the feedback and control channel overheads. In the next section, we explain several possible techniques that can improve the system performance.

16.3   IEEE802.16m: Evolution Towards Multiuser MIMO Technologies – Part I. Nonlinear Processing

The MIMO broadcast channel achieves high capacity on the downlink by coordinating the transmissions to multiple users simultaneously (Caire and Shitz, 2003, Vishwanath *et al.*, 2003, Viswanath and Tse, 2003, Weingarten *et al.*, 2006, Yu and Cioffi, 2004). The optimal transmit strategy according to information theory is Dirty Paper Coding (DPC), which achieves the capacity region (Weingarten *et al.*, 2006). Since DPC does not directly lead to a realizable transmission strategy (Caire and Shitz, 2003) there has been substantial interest in developing practical transmission strategies. In this section and 16.4, we summarize several nonlinear and linear multiuser MIMO techniques that approach the performance of DPC.

One *nonlinear* technique for multiuser MIMO is *vector perturbation* (Hochwald *et al.*, 2005) where the transmit symbol is intentionally modified in a channel dependent way so as to minimize the transmit energy (but improve the received Signal-to-Noise Ratio (SNR)). The benefits of vector perturbation such as diversity gain and asymptotic sum-rate achieving performance are now known (Stojnic *et al.*, 2006, Taherzadeh *et al.*, 2005, Windpassinger *et al.*, 2004b). There are still, however, many difficulties to be overcome for the practical use of a vector perturbation technique. In January 2007, a contribution on vector perturbation (Vetter *et al.*, 2008) was proposed by the IEEE 802.16m Task Group indicating some interest in using vector perturbation. In this section, we give an introductory presentation on the current state of vector perturbation techniques.

16.3.1   System Model

This section introduces the basic building blocks of the vector perturbation multiuser MIMO system and the typical underlying assumptions that are often made in the literature. The purpose of this section is to understand the key building blocks, namely, (1) channel

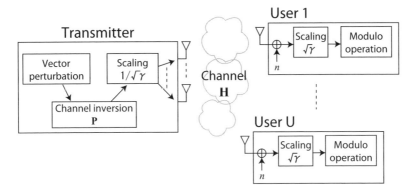

Figure 16.3 Block diagram of vector perturbation system.

inversion, (2) vector perturbation, and (3) modulo operation, where (1) and (2) are at the transmitter side and (3) is at the receiver side.

Let us consider the downlink communication of a multiuser MIMO system. We note that our discussion is not strictly confined to downlink scenarios. It is a matter of convenience to consider downlink communications because we often find cases where the BS is equipped with multiple antennas servicing multiple MSs with a limited number of antennas due to device size, etc.

Suppose that the BS has N_t transmit antennas and each $U \leq N_t$ noncooperating MS has single receive antenna as shown in Figure 16.3. The noncooperating MS means that each MS does not know the propagation channel, *Channel State Information (CSI)*, between the BS and all the other MSs. Instead, we assume that the BS has the CSI for all of the MSs, either through the CSI feedback mechanism or by channel estimation in the Time Division Duplex (TDD) reciprocal channel. The BS intends to transmit U complex baseband symbols for each user. We denote the collective complex transmit symbols as a vector, $\mathbf{s} = [s_1 \cdots s_U]^\mathrm{T}$. Each s_1 through s_U are intended for users 1 through U. The equivalent baseband input–output relationship, assuming perfect synchronization, sampling and linear memoryless channel, is

$$\mathbf{y} = \mathbf{Hx} + \mathbf{n}. \tag{16.9}$$

The output $\mathbf{y}$ is a $U \times 1$ vector of collected output symbol at each of the MSs. This is purely for notational convenience and it should not be confused with the single-user MIMO system where $\mathbf{y}$ is the received signal vector for one user. The $U \times N_t$ composite channel matrix is $\mathbf{H}$ where each row of $\mathbf{H}$ represents the vector MISO channel between the BS and each MS. The transmit signal $\mathbf{x}$ is a precoded symbol with unit transmit energy constraint, $\|\mathbf{x}\|^2 = 1$, obtained from $\mathbf{s}$, which we shall elaborate on below. Finally, the additive noise term $\mathbf{n}$ represents various system noises (e.g. thermal noise, circuit noise, etc.) introduced into the signal chain. The entries of $\mathbf{n}$ are assumed to be complex Gaussian independent and identically distributed (i.i.d.) with variance N_0 according to $\mathcal{CN}(0, N_0)$.

Given the general system model (16.9), the most intuitive approach to decouple the available channel to all of the serviced users is the *channel inversion* technique. The channel inversion technique was proposed and analyzed in detail by Peel *et al.* (2005). The basic principle of channel inversion, as the name implies, is to precode the complex data vector **s** by the pseudo-inverse of the channel matrix, $\mathbf{P} = \mathbf{H}^*(\mathbf{H}\mathbf{H}^*)^{-1}$. We use pseudo-inverse to cover the possibility that U may be less than N_t. The transmit symbol becomes

$$\mathbf{x} = \frac{\mathbf{Ps}}{\sqrt{\gamma}}, \tag{16.10}$$

where $\gamma = \|\mathbf{Ps}\|^2$ is the normalization factor to maintain unit transmit energy. The received signal for the uth MS can be written as

$$y_u = \frac{1}{\sqrt{\gamma}}s_u + n_u. \tag{16.11}$$

Therefore, precoding the data by the pseudo-inverse of the channel results in a parallel Gaussian channel without inter-user interference (uth user's received signal does not depend on any other user's signal). The received SNR for the uth user thus becomes

$$\text{SNR}_u = \frac{\|s_u\|^2}{\gamma N_0}. \tag{16.12}$$

The unfortunate implication of channel inversion is that the received SNR is scaled by a channel dependent factor γ. To see this, consider the eigenvalue decomposition of the channel matrix product, $\mathbf{H}\mathbf{H}^* = \mathbf{V}\Lambda\mathbf{V}^*$, where $\mathbf{V}$ is the matrix of eigenvectors and Λ is a diagonal matrix with ordered eigenvalues, λ_u for $k = 1, \ldots, U$, along the diagonal. Using the eigenvalues and eigenvectors, we may rewrite the scaling factor γ as

$$\gamma = \sum_{k=1}^{U} \frac{1}{\lambda_u} |\langle \mathbf{f}_u, \mathbf{s} \rangle|^2, \tag{16.13}$$

where $\langle \cdot, \cdot \rangle$ denotes the inner product of vectors and $\mathbf{f}_u$ is the uth eigenvector of $\mathbf{V}$. When the channel is ill-conditioned, the smallest eigenvalue may become very small leading to a very large γ (Edelman, 1989). Thus, conditioned on the channel, some users may see a very poor signal degrading the sum rate performance. The next component, vector perturbation, helps to reduce the effect of the channel and improve the sum rate performance of the system.

16.3.2   Vector Perturbation

Vector perturbation (Hochwald *et al.*, 2005) was proposed to overcome the aforementioned effects of γ in channel inversion. The basic idea is to perturb the data symbols individually by a scaled integer such that the scaling factor γ is minimized. By minimizing γ, the received SNR is maximized, thus resulting in better Bit Error Rate (BER) performance. But how exactly do we perturb the data?

The perturbations are performed for real and imaginary components of the complex data symbol **s**. For this reason, we now modify our notation to consider the real and imaginary components separately. Using the notation from Windpassinger *et al.* (2004a), we define an equivalent real-valued system model. Let $\Re$ and $\Im$ denote the real and imaginary parts of a complex variable, respectively. Then, (16.9) can be rewritten as

$$\begin{bmatrix} \Re\mathbf{y} \\ \Im\mathbf{y} \end{bmatrix} = \begin{bmatrix} \Re\mathbf{H} & -\Im\mathbf{H} \\ \Im\mathbf{H} & \Re\mathbf{H} \end{bmatrix} \begin{bmatrix} \Re\mathbf{x} \\ \Im\mathbf{x} \end{bmatrix} + \begin{bmatrix} \Re\mathbf{n} \\ \Im\mathbf{n} \end{bmatrix}. \tag{16.14}$$

We use the subscript $(\cdot)_r$ to denote the real vectors and real matrices obtained this way. The system model in (16.9) becomes a $2U$-dimensional real-valued expression $\mathbf{y}_r = \mathbf{H}_r \mathbf{x}_r + \mathbf{n}_r$. The received signal for user u will be denoted by $\mathbf{y}_{r,u}$ where uth and $(U+u)$th rows are taken from $\mathbf{y}_r$.

In a vector perturbation system, the transmit symbols are perturbed by a scaled integer vector as $\mathbf{x} = (\mathbf{P}_r/\sqrt{\gamma})(\mathbf{s}_r + \tau \ell_r)$ where $\gamma = \|\mathbf{P}_r(\mathbf{s}_r + \tau \ell_r)\|^2$ is now the normalization factor to maintain the unit transmit energy, τ is some fixed scalar and ℓ_r is a $2U$-dimensional integer perturbation vector. The scalar τ is chosen large enough so that the receiver may apply the modulo function to recover the transmit symbol (to be elaborated on in the next section). Hochwald *et al.* (2005) suggested that the scalar parameter τ be chosen according to $\tau = 2(|c|_{\max} + \Delta/2)$, where $|c|_{\max}$ is the absolute value of the constellation symbol(s) with largest magnitude and Δ is the distance between the constellation points. The perturbation vector ℓ_r is chosen to minimize γ as

$$\ell_r = \arg \min_{\ell_r' \in \tau \mathbb{Z}^{2U}} \|\mathbf{P}_r(\mathbf{s}_r + \tau \ell_r')\|^2. \tag{16.15}$$

This form of optimization problem is known as a $2U$-dimensional integer lattice least squares problem over the scaled integer lattice, $\tau \mathbb{Z}^{2U}$. This optimization problem is known to have exponential complexity using the sphere encoder (Hassibi and Vikalo, 2005, Windpassinger *et al.*, 2004a). Therefore, the search for the optimal perturbation is in general quite involved. It is also worthwhile to note that the perturbation must be found for every transmit vector $\mathbf{s}_r$, so the search must take place at the symbol rate. If the channel matrix does not change, for say over a frame period, perturbation corresponding to each $\mathbf{s}_r$ can be stored so that when the same $\mathbf{s}_r$ appears, perturbation is found through a lookup table.

There are two main approaches in the literature for finding the optimal and sub-optimal perturbation vector in (16.15). The well-known sphere decoding algorithm can be used to find the exact solution (Boccardi and Caire, 2006), or approximately using the Lenstra–Lenstra–Lovász (LLL) lattice reduction algorithm (Lenstra *et al.*, 1982, Windpassinger *et al.*, 2004a). Other approaches which have appeared in literatures but are unfortunately beyond the scope of this chapter, are the Minimum Mean Squared Error (MMSE)-based vector perturbation (Kim *et al.*, 2006) and integer relations based lattice reduction using Brun's algorithm (Seethaler and Matz, 2006) (the Very-Large-Scale Integration (VLSI) implementation of this algorithm appears in Burg *et al.* (2007)).

Windpassinger *et al.* (2004a) proposed an approximation technique using lattice reduction and Babai approximation. It has been proven that lattice reduction using the polynomial time LLL algorithm and Babai approximation achieves maximum diversity (Taherzadeh *et al.*, 2005). For the lattice reduction preprocessing, the columns of the precoding matrix $\mathbf{P}_r$ are viewed as a basis of a $2U$-dimensional lattice. The goal is to transform the lattice basis $\mathbf{P}_r$ into a more orthogonal basis $\tilde{\mathbf{P}}_r$ for the same lattice. The LLL algorithm (Lenstra *et al.*, 1982) is used on $\mathbf{P}_r$ to obtain $\tilde{\mathbf{P}}_r = \mathbf{P}_r \mathbf{B}$ where $\tilde{\mathbf{P}}_r$ is the lattice reduced basis with approximately orthogonal columns and $\mathbf{B}_r$ is a $2U \times 2U$ unimodular matrix, that is, it has integer entries with $\det(\mathbf{B}_r) = \pm 1$. Using the LLL algorithm, the matrix $\mathbf{P}_r$ is transformed into a 'nearly orthogonal' matrix $\tilde{\mathbf{P}}_r$ such that a simple rounding operation can be performed to find the approximate perturbation vector ℓ_r by

$$\tilde{\ell}_r = -\mathbf{B}^{-1} Q_{\tau \mathbb{Z}^{2U}} \{\mathbf{B} \mathbf{s}_r\}, \tag{16.16}$$

where we have used $Q_{\tau\mathbb{Z}^{2U}}\{\cdot\}$ to denote the component-wise rounding of a $2U$-dimensional vector to a scaled lattice $\tau\mathbb{Z}^{2U}$.

16.3.2.1   Modulo Operation at the Receiver

In contrast to relatively complex transmit preprocessing, one of the benefits of vector perturbation is its simplicity at the receiver. Assuming that the scaling factor γ and τ are known at the receiver, the effect of perturbation is effectively removed by the *modulo operator* at the receiver. The received signal for user u with vector perturbation is

$$\mathbf{y}_{r,u} = (1/\sqrt{\gamma})(\mathbf{s}_{r,u} + \tau\ell_{r,u}) + \mathbf{n}_{r,u}. \tag{16.17}$$

A modulo operator is defined as $f_\tau(\mathbf{y}_{r,u}) = \mathbf{y}_{r,u} - \lfloor(\mathbf{y}_{r,u} + \tau/2)/\tau\rfloor\tau$, where the function $\lfloor\cdot\rfloor$ returns the largest integer less than or equal to the argument. The modulo function removes the effect of perturbation and the detector observes the receive symbols $\hat{\mathbf{s}}_u = f_\tau(\sqrt{\gamma}\mathbf{y}_{r,u})$ for demodulation.

16.3.3   Performance of a Vector Perturbation System

A comparison of BER performance of vector perturbation with exhaustive search over integer lattice $\{\pm2, \pm1, 0\}$, lattice reduction-aided vector perturbation and channel inversion is shown in Figure 16.4. For this simulation, $N_t = U = 4$, and 4-QAM constellation were used for all of the transmissions. The channel $\mathbf{H}$ was assumed to have unit variance complex Gaussian i.i.d. entries according to $\mathcal{CN}(0, 1)$. Perfect CSI was assumed at the transmitter. As discussed above, channel inversion results in diversity order of one because the channel is decoupled into a parallel independent Gaussian channel. The search-based vector perturbation in contrast results in full diversity performance, however, at the cost of performing extensive search for each transmit symbol. The lattice reduction-aided vector perturbation results in a slight loss over the exhaustive search-based vector perturbation due to approximations made in lattice reduction. The lattice reduction technique results in more efficient perturbation computation compared with exhaustive search thus providing a balanced performance and complexity for practical applications.

16.4   IEEE802.16m: Evolution Towards Multiuser MIMO Technologies – Part II. Linear Processing

To further reduce complexity, there has been considerable interest in linear multiuser MIMO techniques that avoid the need for nonlinear DPC-like processing (Chae *et al.*, 2008a,b, Choi and Murch, 2003, Farhang-Boroujeny *et al.*, 2003, Joung and Lee, 2007, Mazzarese *et al.*, 2007, Pan *et al.*, 2004, Peel *et al.*, 2005, Sharif and Hassibi, 2005, Spencer *et al.*, 2004, Wong, 2006). For linear multiuser MIMO, each data symbol of a scheduled user is transmitted over multiple antennas at the BS after multiplication with a set of complex coefficients. The vector grouping these coefficients is usually normalized and called a transmit *beamforming vector*. There exist three popular methods for designing multiuser beamforming vectors, namely *zero-forcing beamforming*, *orthogonal beamforming* and *coordinated beamforming*, which are discussed in the following sections. To simplify our discussion, we consider a downlink

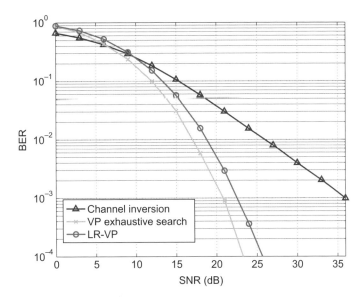

Figure 16.4 BER performance of the vector perturbation system.

where a BS employs an antenna array and MSs use single antenna for Zero-forcing Beamforming (ZFBF) and Orthogonal Beamforming (OGBF). We later consider more than one receive antenna for coordinated beamforming.

16.4.1   Linear Multiuser MIMO: Perfect Channel State Information

16.4.1.1   Zero-forcing Beamforming (ZFBF)

The downlink channel for each user can be represented by a vector of complex channel coefficients, referred to as a *channel vector*. ZFBF constrains the beamforming vector of a scheduled users to be orthogonal to those of all other scheduled users in the vector space. Provided that perfect multiuser CSI is available at the BS, ZFBF ensures no interference between scheduled users. Let N_t denote the number of antennas at the BS and U represents the number of scheduled users served by the BS. Moreover, the $N_t \times 1$ vector channel for the uth scheduled user and the corresponding transmit BF vector are represented by $\mathbf{h}_u$ and $\mathbf{f}_u$, respectively. Then ZFBF applies the constraint $\mathbf{f}_n \perp \mathbf{h}_m$ for all $m \neq n$. We can group the channel vectors of scheduled users as a $N_t \times U$ matrix $\mathbf{H} = [\mathbf{h}_1, \mathbf{h}_2, \ldots, \mathbf{h}_U]$. The rank of $\mathbf{H}$ gives the number of *spatial degrees of freedom* in the multi-antenna downlink channel. This value gives the maximum number of multiuser MIMO users. This fact can be verified for ZFBF by considering the beamforming constraint and using linear algebra. If the users are located sufficiently far away, scheduled users' channel vectors are usually independent. In this case, the number of spatial degrees of freedom is equal to N_t. In other words, in theory, linear multiuser MIMO can potentially support as many simultaneous users as the number of antennas at the BS.

For downlink using linear multiuser MIMO, a data symbol[2] received at the uth MS is given as

$$y_u = \frac{P}{U}\mathbf{f}_u^*\mathbf{h}_u x_u + \sum_{\substack{m=1 \\ m \neq u}}^{U} \frac{P}{U}\mathbf{f}_m^*\mathbf{h}_u x_m + z_u, \tag{16.18}$$

where z_u represents a sample of the additive Gaussian white noise process. Under the ZFBF constraint with perfect transmit CSI, the second term on the right-hand-size of (16.18) is equal to zero, indicating no multiuser interference. Hence, the downlink throughput can be written as

$$R_z = \sum_{u=1}^{U} \mathbb{E}\left[\log_2\left(1 + \frac{P}{U\sigma^2}|\mathbf{h}_u^*\mathbf{f}_u|^2\right)\right], \tag{16.19}$$

where P is the total transmission power and σ^2 represents the variance of z_u. Usually, ZFBF is designed to achieve high throughput by maximizing the number of scheduled users, namely $U = N_t$. In contrast, Time Division Multiple Access (TDMA) schedules a single user and applies beamforming for maximizing this user's link reliability. Using MIMO terminologies, multiuser MIMO and TDMA achieve *multiplexing gain* and *diversity gain*, respectively. This difference between multiuser MIMO using ZFBF and TDMA using single-user beamforming is illustrated by the following example.

Example 16.1 In this example, the downlink channel coefficients $\mathbf{h}_1$, $\mathbf{h}_2$, ..., $\mathbf{h}_{N_t}$ are i.i.d. $\mathcal{CN}(0, 1)$. For TDMA, one user is scheduled in each slot and the single data stream is transmitted using Maximum Ratio Transmission (MRT), mentioned earlier. For high SNRs and using Lapidoth and Moser (2003, Equation 209), the TDMA ergodic throughput is obtained as

$$R(\text{TDMA}) \approx \mathbb{E}\left[\log_2\left(P\sum_{u=1}^{N_t}|h_u|^2\right)\right], \quad \text{SNR} \gg 1, \tag{16.20}$$

$$= \log_2 P - \zeta + \sum_{u=1}^{N_t-1}\frac{1}{u}, \tag{16.21}$$

where ζ denotes Euler's constant ($\zeta = 0.577\ldots$). For linear multiuser MIMO using ZFBF and having N_t scheduled users, note that the term $|\mathbf{h}_u^*\mathbf{f}_u|$ in (16.19) is $\mathcal{CN}(0, 1)$ since $\mathbf{f}_u$ is independent of $\mathbf{h}_u$ under the ZFBF constraint. Thus, from (16.19) and Lapidoth and Moser (2003, Equation 209), the multiuser MIMO throughput is

$$R(\text{MU-MIMO}) \approx \mathbb{E}\left[N_t \log_2\left(\frac{P}{N_t}|h_1|^2\right)\right], \quad \text{SNR} \gg 1, \tag{16.22}$$

$$= N_t \log_2 P - \log N_t - \zeta. \tag{16.23}$$

For high SNRs, $R(\text{TDMA})$ in (16.21) and $R(\text{MU-MIMO})$ in (16.23) can be further approximated as $\log_2 P$ and $N_t \log_2 P$, respectively. Thus, the throughput of linear multiuser MIMO is about N_t times higher than that of TDMA for high SNRs.

---

[2]The time index of a data symbol is omitted to simplify notation.

It is well known that the zero-forcing constraint potentially causes small receive SNRs for downlink data streams if the angles between channel vectors of scheduled users are small. Furthermore, the inaccuracy of CSI at the BS called *transmit CSI* leads to residual multiuser interference even if ZFBF is applied. Both problems can be alleviated by exploiting *multiuser diversity* as elaborated on in Section 16.4.3.

16.4.1.2 Orthogonal Beamforming (OGBF)

In the presence of many users, it is possible and desirable to schedule multiuser MIMO users whose channel vectors are close to orthogonal. This leads to small interference and thus increases throughput. Given small interference, the main function of transmit beamforming is to maximize receive SNRs through MRT. Specifically, an optimal beamforming vector is closely aligned with the channel vector of the associated users in vector space. As a result, the optimal beamforming vectors of scheduled users are also close to orthogonal. For downlink multiuser MIMO with a large number of users, an arbitrary set of OGBF vectors can achieve the same throughput scaling law as DPC. This motivates OGBF, where the transmit beamforming vectors for scheduled multiuser MIMO users are constrained as $\mathbf{f}_1 \perp \mathbf{f}_2 \perp \cdots \perp \mathbf{f}_U$, based on the notation introduced in the preceding section. If $U = N_t$, $\{\mathbf{f}_1, \mathbf{f}_2, \ldots, \mathbf{f}_U\}$ forms a basis of the vector space with N_t complex dimensions, where N denotes the codebook size.

The expression for linear multiuser MIMO using OGBF is derived as follows by using simple geometry. To illustrate, consider multiuser MIMO downlink with $N_t = N = 3$ and real channel coefficients. Figure 16.5 illustrates OGBF vectors for scheduled multiuser MIMO users $\mathbf{f}_1, \mathbf{f}_2, \mathbf{f}_3$ and the channel vector of the 1st user $\mathbf{h}_1$. From Figure 16.5 and (16.18), the signal-to-interference-and-noise ratio (SINR) of the 1st user is

$$
\begin{aligned}
\text{SINR}_1 &= \frac{(P/N_t)|\mathbf{f}_1^*\mathbf{h}_1|^2}{\sigma^2 + (P/N_t)|\mathbf{f}_2^*\mathbf{h}_1|^2 + (P/N_t)|\mathbf{f}_3^*\mathbf{h}_1|^2} \\
&= \frac{(P/N_t)\|\mathbf{h}_1\|^2 \cos^2 \theta_1}{\sigma^2 + (P/N_t)\|\mathbf{h}_1\|^2 \sin^2 \theta_1(\cos^2 \theta_2 + \cos^2 \theta_3)}.
\end{aligned}
\tag{16.24}
$$

As observed from Figure 16.5, $\theta_2 + \theta_3 = 90°$ and hence $(\cos^2 \theta_2 + \cos^2 \theta_3) = 1$. Thus, it follows from (16.24) that

$$
\text{SINR}_1 = \frac{(P/N_t)\|\mathbf{h}_1\|^2 \cos^2 \theta_1}{\sigma^2 + (P/N_t)\|\mathbf{h}_1\|^2 \sin^2 \theta_1}.
\tag{16.25}
$$

By generalizing the above example, the ergodic throughput for multiuser MIMO using OGBF is obtained as

$$
R_o = \sum_{u=1}^{N_t} \mathbb{E}\left[\log_1\left(1 + \frac{(P/N_t)\|\mathbf{h}_u\|^2 \cos \theta_u}{\sigma^2 + (P/N_t)\|\mathbf{h}_u\|^2 \sin \theta_u}\right)\right],
\tag{16.26}
$$

where $\theta_u = \angle(\mathbf{f}_u, \mathbf{h}_u)$.

We can observe from (16.26) that the angles $\{\theta_n\}$ should be minimized to increase throughput. In other words, OGBF vectors must be geometrically aligned with the channel vectors of corresponding scheduled users. Nevertheless, it is difficult to find a set of OGBF

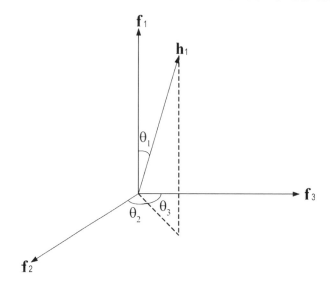

Figure 16.5   OGBF vectors and the channel vector of the 1st user in $\mathbb{R}^3$ for downlink multiuser MIMO.

vectors that aligned with channel vectors so that throughput is maximized. We can find an approximation of the optimal set of OGBF vectors by performing an exhaustive search over a sequence of randomly generated orthonormal bases, denoted as $\{\mathcal{V}^{(m)}\}_{m=1}^{M}$ with $\mathcal{V}^{(n)} = \{\mathbf{f}_1^{(m)}, \mathbf{f}_2^{(m)}, \ldots, \mathbf{f}_{N_t}^{(m)}\}$. Specifically, the throughput achieved by the exhaustive search is

$$R_o^\star = \max_{1 \leq m \leq M} \sum_{u=1}^{N_t} \mathbb{E}\left[\log_1\left(1 + \frac{(P/N_t)|\mathbf{h}_u^*\mathbf{f}_u^{(m)}|^2}{\sigma^2 + (P/N_t)\sum_{\substack{n=1 \\ n \neq u}}^{N_t}|\mathbf{h}_u^*\mathbf{f}_n^{(m)}|^2}\right)\right] \tag{16.27}$$

and the corresponding transmit beamforming vectors belong to the orthonormal set $\mathcal{V}^{(m^\star)}$ where $m^\star$ is the maximizing index in (16.27). Although inefficient, the above approach for finding OGBF vectors is simple. Moreover, higher computational complexity is more affordable at the BS where the power consumption requirement is less stringent than the MS.

16.4.1.3   Coordinated Beamforming

ZFBF and OGBF described above only work for one receive antenna per user. Throughput performance, however, can be further improved using jointly optimized transmitter and receiver beamforming vectors when each MS has more than one receive antenna (Chae *et al.*, 2006a,b, Farhang-Boroujeny *et al.*, 2003, Pan *et al.*, 2004, Schubert and Boche, 2004). These approaches perform close to the sum capacity but require an iterative algorithm to compute the transmit and receive beamformers. In this section, we summarize a noniterative linear multiuser MIMO technique called *coordinated beamforming* (Chae *et al.*, 2008a,b, Mazzarese *et al.*, 2007).

   Consider a multiuser MIMO system with two antennas at the transmitter and N_r receive antennas for each of U users. Here we specifically focus on the case where the transmitter is equipped with two transmit antennas[3]. In this case, the channel between the transmitter and the uth user is represented by an $N_r \times N_t$ matrix. Let $\mathbf{x}_u$ denote the transmit symbol of the uth user, and $\mathbf{n}_u$ be the additive white Gaussian noise vector observed at the receiver. Let $\mathbf{f}_u$ denote the unit-norm transmit beamforming vector and $\mathbf{w}_u$ the unit-norm receive combining vector for the uth user. The signal at the uth user after receiver combining is

$$y_u = \mathbf{w}_u^* \mathbf{H}_u \mathbf{f}_u x_u + \mathbf{w}_u^* \mathbf{H}_u \sum_{l=1, l \neq u}^{U} \mathbf{f}_l x_l + \mathbf{w}_u^* \mathbf{n}_u. \tag{16.28}$$

The subspace decomposition algorithm presented in Farhang-Boroujeny *et al.* (2003) iteratively leads to a choice of the transmit and receive beamformers that achieve the property

$$\mathbf{w}_u^* \mathbf{H}_u \mathbf{f}_l \begin{cases} = 0, & u \neq l, \\ > 0, & u = l. \end{cases} \tag{16.29}$$

Therefore, there is no inter-user interference by applying this algorithm in an ideal channel. It was noted in Farhang-Boroujeny *et al.* (2003) that although the iterative algorithm seemed to converge in most cases, it could not be guaranteed. Note that Maximal Ratio Combining (MRC) is assumed at the MS, given by $\mathbf{w}_u = \mathbf{H}_u \mathbf{f}_u$. This is a reasonable design since it achieves the sum rate very close to capacity under the zero interference constraint.

Thus, the effective channel of user u, which includes the effect of the receiver matched filter, is $\mathbf{f}_u^* \mathbf{R}_u$ where $\mathbf{R}_u = \mathbf{H}_u^* \mathbf{H}_u$. The U transmit beamformers are initialized to some random vectors $\mathbf{f}_{u,1}$, $u = 1, \ldots, U$. Then the following two operations are repeated with increasing i (iteration index) until a stopping criterion is met

$$\tilde{\mathbf{H}}_i = [(\mathbf{f}_{1,i}^* \mathbf{R}_1)^{\mathrm{T}} \cdots (\mathbf{f}_{U,i}^* \mathbf{R}_U)^{\mathrm{T}}]^{\mathrm{T}}, \tag{16.30}$$

$$\mathbf{F}_{i+1} = \tilde{\mathbf{H}}_i^{-1}, \tag{16.31}$$

where $\mathbf{F}_{i+1} = [\mathbf{f}_{1,i+1} \cdots \mathbf{f}_{U,i+1}]$, and $\mathbf{f}_{u,i+1}$ is the transmit beamforming column–vector for the uth user at the $(i + 1)$th iteration, without normalization. A stopping criterion based on convergence can be used, but a fixed number of iterations, for example 20 with four transmit antennas, is sufficient in most cases to obtain a stable solution.

In Chae *et al.* (2008a), the authors derived a closed-form expression for the transmit beamforming vectors. As shown in Chae *et al.* (2008a), we can exactly solve

$$\begin{bmatrix} \mathbf{f}_1 & \mathbf{f}_2 \end{bmatrix} = [[(\mathbf{f}_1^* \mathbf{R}_1)^{\mathrm{T}} (\mathbf{f}_2^* \mathbf{R}_2)^{\mathrm{T}}]^{\mathrm{T}}]^{-1}. \tag{16.32}$$

The cases where the convergence fails occur with probability zero, and are thus not considered in practice. The closed-form expression allows us to solve the problem of slow iteration cases, and offers a computationally attractive solution.

As shown in Chae *et al.* (2008a) that the algorithm will converge to the same solution if we define

$$\mathbf{R}_u = \frac{\mathbf{H}_u^* \mathbf{H}_u}{\|\mathbf{H}_u\|_F^2} = \begin{pmatrix} R_{u,11} & R_{u,12} \\ R_{u,21} & R_{u,22} \end{pmatrix}, \tag{16.33}$$

---

[3]For more than two antenna cases, see Chae *et al.* (2008a).

where $\|\mathbf{H}_u\|_F^2$ is the squared Frobenius norm of the 2×2 complex matrix $\mathbf{H}_u$. Let us define the following notation

$$\mathbf{A}_1 = \begin{pmatrix} R_{1,22} & -R_{1,12} \\ -R_{1,12}^* & R_{1,11} \end{pmatrix}, \tag{16.34}$$

$$\mathbf{A}_2 = \begin{pmatrix} R_{2,11} & R_{2,12} \\ R_{2,12}^* & R_{2,22} \end{pmatrix}. \tag{16.35}$$

With the matrices $\mathbf{A}_1$ and $\mathbf{A}_2$, we define $\mathbf{G}$, a, b, $\mathbf{t}_1$ and $\mathbf{t}_2$ as

$$\mathbf{G} = \mathbf{A}_1 \mathbf{A}_2 = \begin{pmatrix} g_{11} & g_{12} \\ g_{21} & g_{22} \end{pmatrix}, \tag{16.36}$$

$$a = (g_{11} - g_{22})/2, \tag{16.37}$$

$$b = \sqrt{(a^2 + g_{21}g_{12})}, \tag{16.38}$$

$$\mathbf{t}_1 = \begin{pmatrix} a - b \\ g_{21} \end{pmatrix} \quad \text{and} \quad \mathbf{t}_2 = \begin{pmatrix} a + b \\ g_{21} \end{pmatrix}. \tag{16.39}$$

Finally, the transmit beamformers for user 1 and 2 are given by

$$\mathbf{f}_1 = \frac{\mathbf{t}_1}{\sqrt{(a-b)^2 + g_{21}^2}}, \tag{16.40}$$

$$\mathbf{f}_2 = \frac{\mathbf{t}_2}{\sqrt{(a+b)^2 + g_{21}^2}}. \tag{16.41}$$

The derivation of the closed-form expression in the two transmit antenna case, explained by Chae *et al.* (2008a), gives insights as to why the algorithm does not converge, or converges in some cases very slowly.

16.4.2   Linear Multiuser MIMO: Limited Feedback

Linear multiuser MIMO requires transmit CSI for adapting beamforming to time-varying channels. For most systems including the IEEE802.16, transmit CSI is acquired through feedback by the receiver that performs channel estimation. In the IEEE802.16 systems, CSI feedback relies on finite-rate feedback channels. Thus, efficient CSI quantization is required, which is related to an active research field known as *limited feedback*, focusing on the design of efficient CSI feedback techniques (Love and Heath Jr., 2006, Love *et al.*, 2003).

The most common approach for limited feedback relies on codebook-based quantization for reducing CSI feedback into a small number of bits for each feedback instant. Designing codebooks is a popular theme in the area of limited feedback. The most important finding related to the codebook design is its equivalence with the classic geometry problem of packing on the Grassmannian or the Stiefel manifold (Love *et al.*, 2003, Mukkavilli *et al.*, 2003). Beamforming and precoding codebooks designed based on such a relationship have been demonstrated to achieve low-rate CSI feedback with near optimal performance. Let $\mathcal{C}$ denote the codebook of N $N_t \times 1$ unitary vectors that is designed based on the criterion of packing on the Grassmannian manifold. Then at the uth MS, the CSI is quantized as

$$\hat{\mathbf{h}}_u = \max_{\mathbf{f} \in \mathcal{C}} |\mathbf{f}^* \mathbf{h}_u|^2. \tag{16.42}$$

The codebook index of $\hat{\mathbf{h}}_u$ is sent back from the uth MS to the BS, where the CSI feedback requires $\log_2 N$ bits. Based on limited feedback, the BS obtains quantized multiuser CSI and uses it for computing transmit beamforming vectors.

Limited feedback results in CSI inaccuracy in transmit CSI. Consequently, if all MSs have single antennas, MSs can no longer be decoupled even if zero-forcing transmit beamforming is applied. Specifically, limited feedback results in residual interference for linear multiuser MIMO. From (16.18), the ergodic throughput for linear multiuser MIMO is given as

$$\tilde{R} = N_t \mathbb{E}\left[\log_2\left(1 + \frac{(P/N_t)|\mathbf{h}_u^* \mathbf{f}_u|^2}{\sigma^2 + (P/N_t) \sum_{\substack{n=1 \\ n \neq u}}^{N_t} |\mathbf{h}_u^* \mathbf{f}_n|^2} \right) \right]. \tag{16.43}$$

The above expression holds for ZFBF, OGBF and coordinated beamforming. Nevertheless, they differ in both the CSI quantizer codebook design and the computation of beamforming vectors, resulting in different throughput. For ZFBF, the single-user Grassmannian codebooks (see, e.g., Love *et al.* (2003), Mukkavilli *et al.* (2003)) are suitable for CSI quantization in (16.42). For OGBF, the quantizer codebook $\mathcal{C}$ in (16.42) must consist of multiple sets of orthonormal vectors. This facilitates OGBF where a particular orthonormal vector set is selected as beamforming vectors and the associated users are scheduled. For ZFBF, transmit beamforming vectors are computed from quantized feedback CSI such that the beamforming vector $\mathbf{f}_n \perp \tilde{\mathbf{h}}_m$ for all $m \neq n$. This computation involves inversion of the matrix $[\tilde{\mathbf{h}}_1, \tilde{\mathbf{h}}_2, \ldots, \tilde{\mathbf{h}}_{N_t}]$ and normalization of the columns of the resultant matrix.

It has been found that limited feedback places an upper-bound on downlink throughput, which is independent of transmission power (Ding *et al.*, 2007, Jindal, 2006). A relevant observation is that constraining the throughput loss due to limited feedback requires more CSI feedback for larger transmission power. These two observations are quantified in the following example for ZFBF and the rich scattering environment. Similar results in this example can be also obtained for OGBF.

Example 16.2 We consider downlink where N_t users are scheduled for multiuser MIMO using ZFBF. The downlink channel is narrow-band and channel coefficients are i.i.d. $\mathcal{CN}(0, 1)$. Each MS sends back CSI using limited feedback. The codebook $\mathcal{C}$ used for quantizing CSI as in (16.42) consists of randomly generated unitary vectors which are i.i.d. and isotropic. Given these assumptions, the throughput in (16.19) is upper-bounded as (Jindal, 2006)

$$R_z \leq N_t \left[1 + \frac{B + \log_2 e}{N_t - 1} + \log_2(N_t - 2) + \log_2 e \right], \tag{16.44}$$

where $Q = \log_2 M$ is the number of feedback bits per user. The upper-bound in (16.44) is independent of the transmission power P and is a function of Q. This implies that the residual interference caused by CSI inaccuracy prevents throughput from continuously growing with increasing transmission.

The throughput loss due to limited feedback can be measured by the throughput difference between the cases of perfect CSI and limited feedback, which is define as $\Delta R_z = R_z - \tilde{R}_z$

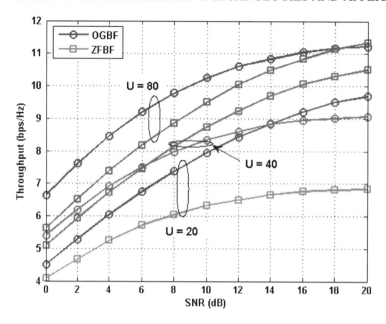

Figure 16.6 Throughput comparison between OGBF and ZFBF for an increasing SNR; the codebook size $N = 64$ and the number of transmit antennas $N_t = 4$.

with R_z and $\tilde{R}_z$ given in (16.19) and (16.43), respectively. Given above assumptions, ΔR_z can be upper-bounded as (Jindal, 2006)

$$\Delta R \leq N_t \log_2(1 + P2^{-B/(N_t-1)}).$$ (16.45)

As suggested by (16.45), scaling up Q with P at the rate of $Q = (N_t - 1) \log_2 P$ is sufficient to contain the throughput loss. In other words, higher throughput places a more stringent requirement on the accuracy of feedback CSI.

Figure 16.6 compares the throughput of OGBF and ZFBF for an increasing SNR. The number of transmit antennas is $N_t = 4$ and the codebook size is $N = 64$. As observed from Figure 16.6, for the number of users $U = 20$, OGBF achieves lower throughput than ZFBF over the range of SNR under consideration ($0 \leq$ SNR ≤ 20 dB). Nevertheless, for larger numbers of users ($U = 40$ or 80), OGBF outperforms ZFBF for a subset of the SNRs. Specifically, the throughput versus SNR curves for OGBF and ZFBF crosses at SNR $= 7$ dB for $U = 40$ and at SNR $= 18$ dB for $U = 80$. The above results suggest that in the practical range of SNR, OGBF is preferred to ZFBF only if the user pool is sufficiently large.

For the case where the MS has more than one receive antenna, coordinated beamforming can be used. In this case, the feedback strategy is a little bit different. Since the transmitter needs the matched channel matrices $\{\mathbf{R}_u\}_{u=1}^{U}$, and not just its subspace information as with ZFBF and OGBF, direct quantization exploiting the symmetry of $\mathbf{R}_u$. One example is to quantize $\mathbf{R}_u$ as follows:

$$R_{u,11} = \alpha_u \quad \text{and} \quad R_{u,22} = 1 - \alpha_u, \tag{16.46}$$

$$R_{u,21} = \beta_u e^{j\varphi_u} \quad \text{and} \quad R_{u,12} = R^*_{u,21}, \tag{16.47}$$

where $0 \leq \alpha_u, \beta_u \leq 1$ and $0 \leq \varphi_u < 2\pi$. By choosing α_u, β_u and φ_u uniformly distributed in these intervals, the users can directly quantize the channel using a finite number of bits. We choose the following quantization method. Parameters α_u and β_u are quantized using Q_1 bits, while φ_u is quantized using Q_2 bits:

$$\alpha_u, \beta_u \in \left[\frac{1}{2^{Q_1+1}}, \frac{1}{2^{Q_1+1}} + \frac{1}{2^{Q_1}}, \ldots, 1 - \frac{1}{2^{Q_1+1}} \right],$$

$$\varphi_u \in \left[-\pi, -\pi + \frac{2\pi}{2^{Q_2}}, \ldots, \pi - \frac{2\pi}{2^{Q_2}} \right].$$

Quantization loss can be further reduced using non-uniform quantization (see Chae *et al.* (2008a) for more information).

Figure 16.7 illustrates the sum rate of closed-form coordinated beamforming with limited feedback, the sum capacity and the largest single-user closed-loop MIMO capacity with perfect channel state information, when the number of users in the cell is the same as the number of transmit antennas. In this case, no scheduling algorithm is needed. As can be seen from Figure 16.7, the throughput of coordinated beamforming is quite close to the sum capacity even with limited feedback $Q_1 = 6$ bits. It also has about 3.3 bps/Hz gap gain against the TDMA approach.

16.4.3 Linear Multiuser MIMO: Multiuser Diversity

As we saw in Example 16.2, the CSI inaccuracy inherent in limited feedback causes residual interference between multiuser MIMO users even if ZFBF is applied (Jindal, 2008, Yoo *et al.*, 2007). Multiuser interference becomes a bottleneck for increasing multiuser MIMO throughput at high SNRs or for a relatively larger number of multiuser MIMO users. This places a more stringent requirement on CSI accuracy for multiuser MIMO than that for point-to-point MIMO. To acquire highly-accurate CSI without incurring excessive overhead, limited feedback can be integrated with *multiuser diversity*.

Multiuser diversity refers to the degrees of freedom due to independent fading in different users' channels. This concept was introduced by Knopp and Humblet (1995) for single-antenna downlink, where scheduling the user with the largest channel gain is shown to maximize capacity. The downlink throughput gain contributed by multiuser diversity is called *multiuser diversity gain*. For MU-MIMO, this gain is achieved by scheduling users with not only large channel gains but also small mutual interference. Thereby the stringent requirement on CSI accuracy is relaxed and as a result CSI feedback overhead decreases. In the following section, scheduling algorithms that exploit multiuser diversity are discussed separately for ZFBF and OGBF.

16.4.3.1 ZFBF with Limited Feedback

A greedy search scheduling algorithm has been designed by Yoo *et al.* (2007) for the zero-forcing multiuser MIMO discussed in the preceding section, which provides a suboptimal

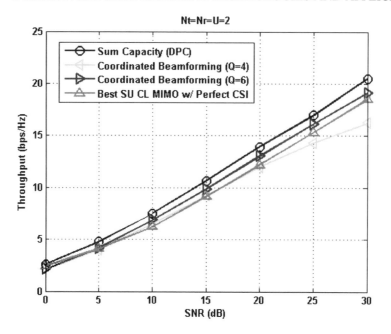

Figure 16.7 Sum rates versus SNR for (2, 2, 2) scenario with two transmit antennas at the BS, two receive antennas at the MS and two users in a cell.

approach for increasing the downlink throughput. Note that finding an optimal set of multiuser MIMO users involves an exhaustive search, and is hence impractical unless the number of users is very small. The greedy-search algorithm uses both the channel shape, defined based on feedback CSI from the uth user as $\hat{\mathbf{s}}_u = \hat{\mathbf{h}}_u / \|\hat{\mathbf{h}}_u\|$. Furthermore, this algorithm assumed perfect feedback of the SINR estimated at each user using that of OGBF in (16.27)

$$\mathrm{SINR}_u \approx \frac{(P/N_t)\|\mathbf{h}_u\|^2 \cos\theta_u}{\sigma^2 + (P/N_t)\|\mathbf{h}_u\|^2 \sin\theta_u}. \tag{16.48}$$

The steps for this algorithm are summarized as follows.

1. Select the user with the largest SINR and record down the user index as u_1. Initiate the index set of scheduled users as $\mathcal{A} = \{u_1\}$.

2. Find a subset of unscheduled users whose quantized channel shapes form sufficiently large angles with those of scheduled users. The above subset is obtained as

$$\mathcal{I} = \{1 \le u \le U : u \notin \mathcal{A}; \ |\hat{\mathbf{s}}_u^* \hat{\mathbf{s}}_k|^2 \le \eta \text{ for all } k \in \mathcal{A}\} \tag{16.49}$$

where $0 < \eta < 1$ is a predesigned threshold.

3. If $\mathcal{I} \ne \emptyset$ and $|\mathcal{A}| < N_t$, schedule the user in the set $\mathcal{I}$ with the largest SINR. The index sets $\mathcal{A}$ and $\mathcal{I}$ are augmented as $\mathcal{A} = \mathcal{A} \cup \{\arg\max_{u \in \mathcal{I}} \mathrm{SINR}_u\}$ and $\mathcal{I} = \mathcal{I} - \{\arg\max_{u \in \mathcal{I}} \mathrm{SINR}_u\}$.

4. If $\mathcal{I} \neq \emptyset$ and $|\mathcal{A}| < N_t$, go to step 2. Otherwise, schedule the users in $\mathcal{A}$ for downlink multiuser MIMO and compute their beamforming vectors using the zero-forcing method.

Despite being fixed by Yoo *et al.* (2007), the threshold η in step 2 should be optimized for different numbers of users for the following reasons. A small value of η can result in too few multiuser MIMO users and hence a loss in multiplexing gain. However, a large value of η can lead to poor orthogonality between the channels of scheduled users and cause strong interference.

16.4.3.2 OGBF with Limited Feedback

The following suboptimal scheduling algorithm has been proposed by Huang *et al.* (2009) for linear multiuser MIMO using limited feedback and orthogonal beamforming. A similar design has also been developed in the industry and is called *Per-User Unitary and Rate Control* (PU2RC) (Samsung Electronics, 2006). First, each member of the codebook $\mathcal{C}$, which is a potential beamforming vector, is assigned a user with the maximum SINR. Consider an arbitrary vector, for instance $\mathbf{f}_n^{(m)}$, which is the nth member of the mth orthonormal subset $\mathcal{V}^{(m)}$ of the codebook $\mathcal{C}$. This vector can be the quantized channel shapes of multiple users, whose indices are grouped in a set defined as $\mathcal{I}_n^{(m)} = \{1 \leq u \leq U : \hat{\mathbf{s}}_u = \mathbf{f}_n^{(m)}\}$, where $\hat{\mathbf{s}}_u$ is the uth user's quantized channel shape given in (16.42). From (16.42), $\mathcal{I}_n^{(m)}$ can be equivalently defined as

$$\mathcal{I}_n^{(m)} = \{1 \leq u \leq U \mid d(\mathbf{s}_u, \mathbf{f}_n^{(m)}) < d(\mathbf{s}_u, \mathbf{f}) \text{ for all } \mathbf{f} \in \mathcal{C} \text{ and } \mathbf{f} \neq \mathbf{f}_n^{(m)}\}. \qquad (16.50)$$

Among the users in $\mathcal{I}_n^{(m)}$, $\mathbf{f}_n^{(m)}$ is associated with that providing the maximum SINR, which is feasible since the SNRs are known to the BS through feedback. The index $(i_n^{(m)})$ and SINR $(\xi_n^{(m)})$ of this user associated with $\mathbf{f}_n^{(m)}$ can be written as

$$i_n^{(m)} = \arg \max_{u \in \mathcal{I}_n^{(m)}} \text{SINR}_u \quad \text{and} \quad \xi_n^{(m)} = \max_{u \in \mathcal{I}_n^{(m)}} \text{SINR}_u, \qquad (16.51)$$

where the index set $\mathcal{I}_n^{(m)}$ and the function SINR_u are expressed respectively in (16.50) and (16.48). In the event that $\mathcal{I}_n^{(m)} = \emptyset$, the vector $\mathbf{f}_n^{(m)}$ is associated with no user and the maximum SINR $\xi_n^{(m)}$ in (16.51) is set to zero. Second, the orthonormal subset of the codebook that maximizes throughput is chosen, whose index is $m^\star = \arg \max_{1 \leq m \leq M} \sum_{n=1}^{N_t} \log(1 + \xi_n^{(m)})$. Thereby, the users associated with this chosen subset, specified by the indices $\{i_n^{(m^\star)} \mid 1 \leq n \leq N_t\}$, are scheduled for simultaneous transmission using beamforming vectors from the $(m^\star)$th orthonormal subset.

The above scheduling algorithm does not guarantee that the number of scheduled users is equal to N_t, the spatial degrees of freedom. For a small user pool, the number of scheduled users is smaller than N_t. This is desirable because finding N_t simultaneous users with close-to-orthogonal channels in a small user pool is unlikely. In this case, having fewer scheduled users than N_t reduces interference and leads to higher throughput. As the total number of users increases, the number of scheduled users converges to N_t.

Based on the above scheduling algorithm, the ergodic throughput for multiuser MIMO using OGBF is given as

$$R = \mathbb{E}\left[\max_{1 \leq m \leq M} \sum_{n=1}^{N_t} \log\left(1 + \max_{u \in \mathcal{I}_n^{(m)}} \frac{(P/N_t)\|\mathbf{h}_n\|^2 \cos \theta_n}{\sigma^2 + (P/N_t)\|\mathbf{h}_n\|^2 \sin \theta_n} \right) \right].$$

(16.52)

16.4.3.3 Multiuser Diversity Gain

Using scheduling algorithms that exploit multiuser diversity leads to performance gain, known as *multiuser diversity gain*. For the design criterion of maximizing throughput, this gain is reflected in the growth of throughput with the number of users. For linear multiuser MIMO, it is difficult to characterize multiuser diversity gain as simple functions of the number of users and other system parameters such as the number of feedback bits per user. The common approach for quantifying multiuser diversity gain is to consider an asymptotically large number of users and derive the asymptotic throughput scaling laws. Such analysis usually relies on simplified channel models and uses mathematical tools including extreme value theory (Sharif and Hassibi, 2005, Yoo *et al.*, 2007) or the uniform convergence in the weak law of large number (Huang *et al.*, 2009). We illustrate the throughput scaling laws for linear multiuser MIMO in the following example.

Example 16.3 We consider the same system and channel models as in Example 16.2. Furthermore, multiuser MIMO users are scheduled from a total of U available users. The operational SNR range can be partitioned into three regimes: *low SNRs* where interference is negligible with respect to noise, *normal SNRs* where noise and interference have comparable power, and *high SNRs* where noise is negligible with respect to interference. The following scaling laws for different SNR regimes have been shown to hold for ZFBF (Yoo *et al.*, 2007) and OGBF (Huang *et al.*, 2009, Sharif and Hassibi, 2005):

$$\lim_{U \to \infty} \frac{\tilde{R}}{N_t \log \log U} = 1 \quad \text{low and normal SNR regimes,}$$

(16.53)

$$\lim_{U \to \infty} \frac{\tilde{R}}{(N_t/(N_t - 1)) \log U} = 1 \quad \text{high SNR regime.}$$

The identical scaling laws indicate that ZFBF and OGBF lead to identical asymptotic performance. As observed from the above scaling laws, R scales with U much faster (logarithmically) in the high SNR regime than in other regimes.

Simulation results are obtained based on the same system model as in Example 16.3. In Figure 16.8, the throughput of multiuser MIMO using OGBF is compared with that using ZFBF for an increasing number of users. The number of the transmit antenna is $N_t = 4$ and the SNR is 5 dB. Moreover, the codebook sizes $N = \{4, 8, 16, 32\}$ for channel shape quantization are considered. As in Yoo *et al.* (2007), the threshold 0.25 is applied in the greedy-search scheduling for ZFBF. Figure 16.8(a) and (b) show respectively the small ($1 \leq U \leq 35$) and the large ($1 \leq U \leq 200$) user ranges. As observed from Figure 16.8(a), for a given codebook size (either $N = 16$ or 64), OGBF achieves higher throughput than ZFBF for a relatively large number of users but the reverse holds for a smaller user pool. Specifically, in Figure 16.8(a), the throughput curves for OGBF and ZFBF cross at $U = 19$

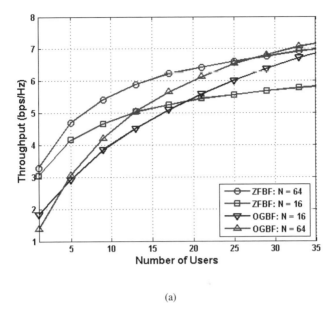

(a)

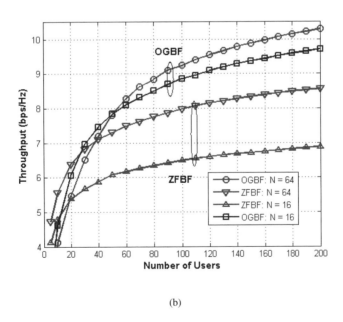

(b)

Figure 16.8 Throughput comparison between OGBF and ZFBF for an increasing number of users U, SNR $= 5$ dB, and the number of transmit antennas $N_t = 4$: (a) small numbers of users; (b) large numbers of users.

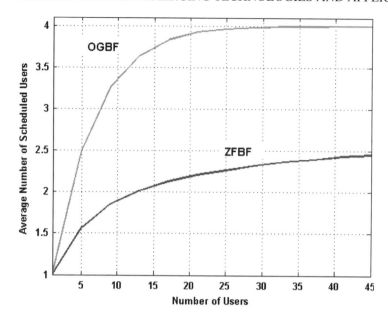

Figure 16.9 The average numbers of scheduled users for OGBF and ZFBF for SNR = 5 dB, and the number of transmit antennas $N_t = 4$.

for $N = 16$ and at $U = 27$ for $N = 64$. For a sufficiently large number of users, OGBF always outperforms ZFBF in terms of throughput as shown in Figure 16.8(b). Furthermore, compared with ZFBF, OGBF is found to be more robust against CSI quantization errors. For example, as observed from Figure 16.8(b), for $U = 100$, the throughput loss for OGBF due to the decrease of the codebook size from $N = 64$ to 16 is 0.3 bps/Hz but that for ZFBF is 1.5 bps/Hz. The above observations are explained below. In summary, these observations suggest that OGBF is preferable to ZFBF for a large user pool but not for a small one.

To explain the observations from Figure 16.8, the average numbers of scheduled users for OGBF and ZFBF are compared in Figure 16.9 for an increasing number of users. It can be observed from Figure 16.9 that OGBF tends to schedule more users than ZFBF. First, for a small number of users, interference between scheduled users cannot be effectively suppressed by scheduling, and hence more simultaneous users result in smaller throughput. This explains the observation from Figure 16.8(a) that OGBF achieves lower throughput than ZFBF due to more scheduled users. Second, for a large user pool, the channel vectors of scheduled users are close-to-orthogonal and interference is negligible. Therefore, a larger number of scheduled users leads to higher throughput. For this reason, OGBF outperforms ZFBF for a large number of users as observed from Figure 16.8(b). Last, with respect to ZFBF, the better robustness of OGBF against CSI quantization errors is mainly due to the joint beamforming and scheduling. Note that beamforming and scheduling for ZFBF are performed separately (Yoo *et al.*, 2007).

16.5 Conclusion

In this chapter, we introduced single-user MIMO techniques that have been included in IEEE 802.16e and multiuser MIMO techniques being considered for IEEE 802.16m. For single-user MIMO, we discussed open-loop techniques including classic, concatenated and constellation rotation-based STBCs, and closed-loop techniques including antenna grouping, antenna selection and limited feedback beamforming and precoding. Multiuser MIMO solutions were not adopted in IEEE 802.16e due to potentially overwhelming feedback and control overhead. Nevertheless, with increasing demands for high throughput and the recent progresses in multiuser limited feedback, multiuser MIMO techniques will certainly become a key physical-layer component for IEEE 802.16m. In the second part of this chapter, we discussed a nonlinear multiuser MIMO technique, namely vector perturbation and lattice reduction-aided precoding, and linear techniques including ZFBF, OGBF and coordinated beamforming. Furthermore, we presented multiuser limited feedback for reducing channel feedback overhead and multiuser diversity algorithms for enhancing the throughput of multiuser MIMO.

In this chapter, we did not consider another popular multiuser MIMO solution called block diagonalization, which supports multiple data streams for each user given multiple antennas available at MSs. In our discussion, we also omitted practical factors such as CSI feedback delay and channel estimation errors. In summary, with multiuser MIMO techniques finding their way into the IEEE802.16m, there is no doubt that MIMO will continue to play a key role in supporting high-rate access in the next-generation of WiMAX systems.

References

Alamouti, S.M. (1998) A simple transmit diversity technique for wireless communications. *IEEE Journal of Selected Areas in Communications*, **16**, 1451–1458.

Boccardi, F. and Caire, G. (2006) The p-sphere encoder: peak-power reduction by lattice precoding for the MIMO Gaussian broadcast channel. *IEEE Transactions on Communications*, **54**(11), 2085–2091.

Burg, A., Seethaler, D. and Matz, G. (2007) VLSI implementation of a lattice-reduction algorithm for multi-antenna broadcast precoding. *Proceedings of the IEEE International Symposium on Circuits and Systems (ISCAS)*, pp. 673–676.

Caire, G. and Shitz, S.S. (2003) On the achievable throughput of a multi-antenna Gaussian broadcast channel. *IEEE Transactions on Information Theory*, **43**, 1691–1706.

Chae, C.B., Heath, Jr., R.W. and Mazzarese, D. (2006a) Achievable sum rate bounds of zero-forcing based linear multi-user MIMO systems. *Proceedings of the Allerton Conference on Communications Control and Computers* pp. 1134–1140. Available at http://users.ece.utexas.edu/~rheath/papers/2006/allerton/.

Chae, C.B., Katz, M., Suh, C.H. and Jeong, H.S. (2004) Adaptive spatial modulation for MIMO-OFDM. *Proceedings of the IEEE Wireless Communications and Networks Conference*, **1**, 87–92.

Chae, C.B., Mazzarese, D. and Heath, Jr., R.W. (2006b) Coordinated beamforming for multiuser MIMO systems with limited feedforward. *Proceedings of the Asilomar Conference on Signals, Systems and Computers*, pp. 1511–1515.

Chae, C.B., Mazzarese, D., Jindal, N. and Heath, Jr., R.W. (2008a) Coordinated beamforming with limited feedback in the MIMO broadcast channel. *IEEE Journal of Selected Areas in Communications*, **26**, 1505–1515.

Chae, C.B., Mazzarese, D., Jindal, N. and Heath, Jr., R.W. (2008b) A low complexity linear multiuser MIMO beamforming system with limited feedback. *Proceedings of the Conference on Information Science and Systems,* pp. 418–422.

Chae, C.B., Shim, S. and Heath, Jr., R.W. (2008c) Space–time block codes with limited feedback using antenna grouping. *IEICE Transactions on Communications,* **E91-B**, 3387–3390.

Choi, L. and Murch, R.D. (2003) A transmit preprocessing technique for multiuser MIMO systems using a decomposition approach. *IEEE Transactions Wireless Communications,* **2**, 773–786.

Ding, P., Love, D.J. and Zoltowski, M.D. (2007) Multiple antenna broadcast channels with shape feedback and limited feedback. *IEEE Transactions on Signal Processing,* **55**, 3417–3428.

Edelman, A. 1989 Eigenvalues and Condition Numbers of Random Matrices. *PhD thesis,* Massachusetts Institute of Technology.

Farhang-Boroujeny, B., Spencer, Q. and Swindlehurst, A.L. (2003) Layering techniques for space–time communications in multi-user networks. *Proceedings of the IEEE Vehicle Technology Conference,* **2**, 1339–1342.

Gesbert, D., Kountouris, M., Heath, Jr., R.W., Chae, C.B. and Salzer, T. (2007) Shifting the MIMO paradigm: From single user to multiuser communications. *IEEE Signal Processing Magazine,* **24**, 36–46.

Gore, D.A. and Paulraj, A. (2002) MIMO antenna subset selection with space–time coding. *IEEE Transactions on Signal Processing,* **50**, 2580–2588.

Hassibi, B. and Vikalo, H. (2005) On the sphere-decoding algorithm I. expected complexity. *IEEE Transactions on Signal Processing,* **53**(8), 2806–2818.

Heath, Jr., R.W. and Paulraj, A. (2005) Switching between spatial multiplexing and transmit diversity based on constellation distance. *IEEE Transactions on Communications,* **53**(6), 962–968.

Heath, Jr., R.W., Sandhu, S. and Paulraj, A. (2001) Antenna selection for spatial multiplexing systems with linear receivers. *IEEE Communication Letters,* **5**, 142–144.

Hochwald, B., Peel, C. and Swindlehurst, A. (2005) A vector-perturbation technique for near-capacity multiantenna multiuser communication-part II: perturbation. *IEEE Transactions on Communications,* **53**(3), 537–544.

Huang, K., Andrews, J.G. and Heath, Jr., R.W. (2009) Performance of orthogonal beamforming for SDMA with limited feedback. *IEEE Transactions on Vehicle Technologies.*

IEEE (2006) IEEE P802.16e-2006 draft standards for local and metropolitan area networks part 16: Air interface for fixed broadcast wireless access systems. *IEEE Standard 802.16e.*

IEEE (2007) IEEE P802.16m-2007 draft standards for local and metropolitan area networks part 16: Air interface for fixed broadcast wireless access systems. *IEEE Standard 802.16m.*

Jindal, N. (2006) MIMO broadcast channels with finite-rate feedback. *IEEE Transactions on Information Theory,* **52**(11), 5045–60.

Jindal, N. (2008) Antenna combining for the MIMO downlink channel. *IEEE Transactions on Wireless Communications,* **7**, 3834–3844.

Joung, J. and Lee, Y.H. (2007) Regularized channel diagonalization for multiuser MIMO downlink using a modified MMSE criterion. *IEEE Transactions on Signal Processing,* **55**, 1573–1579.

Kim, E.Y., Kim, E.Y. and Chun, J. (2006) Optimum vector perturbation minimizing total MSE in multiuser MIMO downlink. *Proceedings of the IEEE International Conference on Communications* (ed. Chun, J.), vol. 9, pp. 4242–4247.

Knopp, R. and Humblet, P. (1995) Information capacity and power control in single-cell multiuser communications. *Proceedings of the IEEE International Conference on Communications,* **1**, pp. 331–335.

Lapidoth, A. and Moser, S.M. (2003) Capacity bounds via duality with applications to multiple-antenna systems on flat-fading channels. *IEEE Transactions on Information Theory,* **49**(10), 2426–67.

Lenstra, A.K., Lenstra, H.W. and Lováz, L. (1982) Factoring polynomials with rational coefficients. *Mathematische Annalen*, **261**(4), 515–534.

Love, D.J. and Heath, Jr., R.W. (2005) Limited feedback unitary precoding for spatial multiplexing. *IEEE Transactions on Information Theory*, **51**, 2967–2976.

Love, D.J. and Heath Jr., R.W. (2006) Feedback techniques for MIMO channels. *MIMO Antenna Technology for Wireless Communications*. CRC Press Boca Raton, FL.

Love, D.J., Heath, Jr., R.W. and Strohmer, T. (2003) Grassmannian beamforming for multiple-input multiple-output wireless systems. *IEEE Transactions on Information Theory*, **49**, 2735–2747.

Mazzarese, D., Chae, C.B. and Heath, Jr., R.W. (2007) Jointly optimized multiuser beamforming for the MIMO broadcast channel with limited feedback. *Proceedings of the IEEE International Symposium on Pers., Indoor and Mobile Radio Comm.*, pp. 1–5.

Mukkavilli, K.K., Sabharwal, A., Erkip, E. and Aazhang, B. (2003) On beamforming with finite rate feedback in multiple antenna systems. *IEEE Transactions on Information Theory*, **49**(10), 2562–79.

Pan, Z., Wong, K.K. and Ng, T.S. (2004) Generalized multiuser orthogonal space-division multiplexing. *IEEE Transactions on Wireless Communications*, **3**, 1969–1973.

Peel, C.B., Hochwald, B.M. and Swindlehurst, A.L. (2005) A vector-perturbation technique for near capacity multiantenna multiuser communication – part I: channel inversion and regularization. *IEEE Transactions on Communications*, **53**, 195–202.

Samsung Electronics (2006) Downlink MIMO for EUTRA *3GPP TSG RAN WG1 # 44/R1-060335*.

Schubert, M. and Boche, H. (2004) Solution of the multiuser downlink beamforming problem with individual SINR constraints. *IEEE Transactions on Vehicle Technologies*, **53**, 18–28.

Seethaler, D. and Matz, G. (2006) Efficient vector perturbation in multi-antenna multi-user systems based on approximate integer relations. *Proceedings of the European Signal Processing Conference (EUSIPCO)*.

Sharif, M. and Hassibi, B. (2005) On the capacity of MIMO broadcast channel with partial side information. *IEEE Transactions on Information Theory*, **51**, 506–522.

Shim, S., Kim, K. and Lee, C. (2005) An efficient antenna shuffling scheme for a DSTTD system. *IEEE Communications Letters*, **9**, 124–126.

Spencer, Q., Swindlehurst, A.L. and Haardt, M. (2004) Zero-forcing methods for downlink spatial multiplexing in multiuser MIMO channels. *IEEE Transactions on Signal Processing*, **52**, 462–471.

Stojnic, M., Stojnic, M., Vikalo, H. and Hassibi, B. (2006) Asymptotic analysis of the gaussian broadcast channel with perturbation preprocessing. *Proceedings of the IEEE International Conference on Acoustics, Speech and Signal Processing* (ed. Vikalo, H.), vol. 4, pp. IV–IV.

Taherzadeh M, Mobasher A and Khandani A (2005) LLL lattice-basis reduction achieves the maximum diversity in MIMO systems. *Proceedings of the IEEE International Symposium on Information Theory*, pp. 1300–1304.

Vetter, H., Sun, Y., Ponnampalam, V., Aoki, T., Kikuchi, S. and Mori, H. (2008) Proposal on multi-user mimo precoding considerations of IEEE802.16m. *IEEE 802.16m Standards Contribution – IEEE C802.16m-08/058r1*.

Vishwanath, S., Jindal, N. and Goldsmith, A. (2003) Duality, achievable rates, and sum capacity of Gaussian MIMO broadcast channels. *IEEE Transactions on Information Theory*, **49**, 2658–2668.

Viswanath, P. and Tse, D. (2003) Sum capacity of the vector Gaussian broadcast channel and uplink-downlink duality. *IEEE Transactions on Information Theory*, **49**, 1912–1921.

Weingarten, H., Steinberg, Y. and Shitz, S.S. (2006) The capacity region of the Gaussian multiple-input multiple-output broadcast channel. *IEEE Transactions on Information Theory*, **52**, 3936–3964.

Windpassinger, C., Fischer, R. and Huber, J. (2004a) Lattice-reduction-aided broadcast precoding. *IEEE Transactions on Communications*, **52**(12), 2057–2060.

Windpassinger, C., Fischer, R., Vencel, T. and Huber, J. (2004b) Precoding in multiantenna and multiuser communications. *IEEE Transactions on Wireless Communications*, **3**(4), 1305–1316.

Wong, K.K. (2006) Maximizing the sum-rate and minimizing the sum-power of a broadcast 2-user 2-input multiple-output antenna system using a generalized zeroforcing approach. *IEEE Transactions on Wireless Communications*, **5**, 3406–3412.

Yoo, T., Jindal, N. and Goldsmith, A. (2007) Multi-antenna broadcast channels with limited feedback and user selection. *IEEE Journal on Selected Areas in Communications*, **25**(7), 1478–91.

Yu, W. and Cioffi, J. (2004) Sum capacity of Gaussian vector broadcast channels. *IEEE Transactions on Information Theory*, **50**, 1875–1892.

Zheng, L. and Tse, D.N.C. (2003) Diversity and multiplexing: a fundamental tradeoff in multiple-antenna channels. *IEEE Transactions on Information Theory*, **49**(5), 1073–1096.

17

Hybrid Strategies for Link Adaptation Exploiting Several Degrees of Freedom in WiMAX Systems

Suvra Sekhar Das, Muhammad Imadur Rahman and Yuanye Wang

17.1 Introduction

The wireless channel condition is constantly changing, both in time and frequency. This is especially evident in cases when there is relative motion between transmitter and receiver, and when multipaths exist between them. The signal can experience high fluctuation in very short intervals of time and frequency. This makes it difficult to achieve high data rates while maintaining a target Bit Error Rate (BER)/Frame Error Rate (FER) constraint. Thus, the variation in channel gain, which is termed as channel dynamics, has to be well exploited to maximize the data rate. One of the techniques to achieve this is Link Adaptation (LA), where system parameters are adapted according to the channel condition. This allows us to take advantage when channel conditions are favorable while optimizing bit and power resources during poorer conditions.

LA is a technique used to introduce a real-time balancing in the link budget in order to increase the spectral efficiency of a system over fading channels (Chow *et al.*, 1995, Chung and Goldsmith, 2001, Goldsmith, 2005, Hayes, 1968). Parameters such as transmitted power level, modulation scheme, coding rate or any combination of these can be adapted according

WiMAX Evolution: Emerging Technologies and Applications   Edited by Marcos D. Katz and Frank H.P. Fitzek
© 2009 John Wiley & Sons, Ltd

to the channel conditions. Channel conditions are estimated at the receiver and the Channel State Information (CSI) is sent to the transmitter to adapt the transmission accordingly.

Nearly all communication systems require some target BER not to be exceeded. It is reasonable to assume an average BER or an instantaneous BER (or FER) as a constraint. One assumption that we make is that we have knowledge of the instantaneous CSI for each subcarrier. With this knowledge, transmission parameters can be changed at the transmitter.

When no adaptation is done, the performance is highly dependent on the channel condition. In cases when the channel gain is poor, the system presents low values of Signal-to-Noise Ratio (SNR) and results in poor BER performance. On the other hand, when channel gain is very high, this high gain is not efficiently exploited hence the throughput is not optimized. In order to improve the system performance, adaptation of transmitting bits per symbol and/or power is required. With LA, a low modulation level or high transmit power will be used when the channel is in deep fade and vice versa, so that the received SNR is just enough to provide the required BER target. Some reference points of SNR are required to 'switch' from one modulation level to another. Theoretically these reference points correspond to the SNR-BER curve under Additive White Gaussian Noise (AWGN) channel conditions.

Defining the thresholds for switching between LA parameters is an interesting area of research. One possible criterion for switching points is described by Siebert and Stauffer (2003). A different approach can be found in the work of Song *et al.* (2002), where the adaptation process is performed considering the subcarrier fading statistics rather than instantaneous values. The adaptation can be performed individually for each subcarrier as in Tase *et al.* (2005) and Zhen *et al.* (2002), or a certain number of subcarriers can be grouped together to apply the same modulation technique, as described by Hwang *et al.* (2005). An advantage of performing adaptation by using blocks is a reduction in the amount of information sent during the feedback to the transmitter, as well as a reduced calculation complexity. The effects of imperfect CSI in multi-carrier systems with adaptive modulation have been studied by Ahn and Sasase (2002), Leke and Cioffi (1998), Souryal and Pickholtz (2001) and Ye *et al.* (2002).

17.2   Link Adaptation Preliminaries

In this section we describe a basic scenario of how the LA is done in an Orthogonal Frequency Division Multiplying (OFDM) system. A block diagram of an adaptive modulation over one single subcarrier (or subchannel) is shown in Figure 17.1. However, the concept shown in Figure 17.1 can be extended to multi-carrier scenarios by including the channels for the N subcarriers (or N_{sch} subchannels).

We assume that the entire frequency band is divided into N flat fading subcarriers based on the criteria that the symbol duration is shorter than the coherence time of the channel and the bandwidth of each subcarrier is narrower than the coherence bandwidth of the system.

We denote h_k as the channel response for the kth subcarrier. To simplify the study at the beginning, it is assumed that an error-free and instantaneous estimation/feedback of the channel gain is done at the receiver so that the transmitter knows the channel gain perfectly. However, later on we study the effects of delay and noisy estimations of the channel on the system.

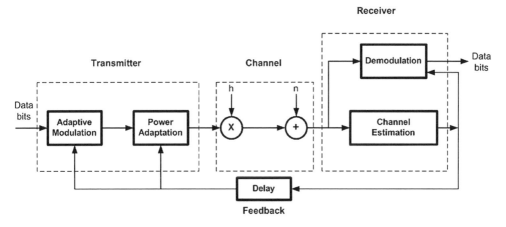

Figure 17.1 Block diagram of a system using LA, that is, adaptive modulation and power adaptation.

AWGN $n(t)$ is assumed with σ_n^2 as its variance. At first, if we consider a constant average power, the average transmitted power will be denoted as $\bar{S}$ and γ_k the corresponding SNR at the kth subcarrier, the relation between these two parameters can be described as (Chung and Goldsmith, 2000)

$$\gamma_k = \bar{\gamma}|h_k|^2 = \frac{\bar{S}|h^k|^2}{\sigma_n^2} \tag{17.1}$$

where $\bar{\gamma}$ is the average SNR.

When power is not considered to be constant we use the notation S_k for the power level at the kth subcarrier, the instantaneous SNR then becomes (Chung and Goldsmith, 2000)

$$\gamma_k = \frac{S_k|h_k|^2}{\sigma_n^2}. \tag{17.2}$$

The process of LA implies the adaptation of the modulation and transmitted power according to the channel conditions. Channel conditions are evaluated for each subcarrier based on Equation (17.2).

17.2.1 Trade-offs and Optimization Target

Owing to the requirements of high data rate for multimedia applications in current systems, one of the main concerns is the optimization of the average spectral efficiency. Depending on the channel conditions, the power required to transmit data with a fixed rate will change with the time. In addition to rate and power, it is necessary to consider that the system must meet some error rate requirement, thus these three parameters are the key to LA.

Ideally, an optimal S_k needs to be found for all subcarriers (or subchannels), for $k \in [1, \ldots, N]$, so that

$$\eta_t \approx \sum_k \mathfrak{R}_k \leq \log_2\left(1 + \frac{S_k}{\sigma_n^2}|h_k|^2\right). \tag{17.3}$$

We omit the time notation from the right-hand side of Equation (17.3) for clarity, without losing any generality. To solve the power problem in this equation, $2^N - 1$ combinations can be found from all S_k, for all $k \in [1, \ldots, N]$, on which the combination that provides highest rate need to be selected. This is a complex optimization problem, which can be solved using the well-known *waterfilling* algorithm as described below (Paulraj *et al.*, 2003). Assuming a Gaussian input signal and perfect CSI feedback with no delay, the optimal power allocation to maximize the capacity (or supported rate) under total power constraint $\bar{S}$ can be solved as

$$S_k = \left(\mu - \frac{\sigma_n^2}{|h_k|^2} \right)^+ ; \quad k = 1, \ldots, N, \tag{17.4}$$

where μ is chosen to satisfy $\sum_{k=1}^{N} S_k = N\bar{S}$, and where

$$(x)^+ = \begin{cases} x, & \text{if } x \geq 0, \\ 0, & \text{if } x < 0. \end{cases} \tag{17.5}$$

To understand the capacity regions, it is easier to study the optimization problem across two subcarriers (or subchannels). This problem can be explained as follows:

$$\Re_1 \leq \log_2 \left(1 + \frac{S_1}{\sigma_n^2} |h_1|^2 \right), \tag{17.6}$$

$$\Re_2 \leq \log_2 \left(1 + \frac{\bar{S} - S_1}{\sigma_n^2} |h_2|^2 \right), \tag{17.7}$$

$$\Re_1 + \Re_2 \leq \log_2 \left[\left(1 + \frac{S_1}{\sigma_n^2} |h_1|^2 \right) \left(1 + \frac{\bar{S} - S_1}{\sigma_n^2} |h_2|^2 \right) \right]. \tag{17.8}$$

Equation (17.8) can be solved to find a closed-form solution for S_1, which will give the optimum rate. The rate region of two subcarriers will be a two-dimensional plot, while capacity regions of more than two subcarriers will be polyhedral. This means that, solving the optimum amount of power for the kth subcarrier when N subcarriers are present in the system, can be very complex, and virtually impossible to solve with any reasonable amount of complexity. Thus, suboptimal algorithms need to be found, which are discussed in the next section.

Once the proper power level and bit allocation is done, then the average spectral efficiency, η can be written in bits per second per Hertz (Chung and Goldsmith, 2000),

$$\eta = \frac{1}{T} \sum_{t=1}^{T} \eta_t = \sum_{t} \frac{1}{BT_s} \sum_{k=1}^{N} \text{ß}_{k,t}, \tag{17.9}$$

where $\text{ß}_{k,t}$ is the number of bits transmitted by the kth subcarrier at the tth OFDM symbol. Here B and T_s are the system bandwidth and OFDM symbol duration, respectively. The power constraint is given by (Chung and Goldsmith, 2000)

$$\frac{1}{N} \sum_{k=1}^{N} S_k \leq \bar{S}. \tag{17.10}$$

17.3   Link Adaptation Algorithms

Based on the requirements mentioned in Section 17.2, we move on to find a suitable LA algorithm for our analysis and studies. After an extensive literature survey, we chose two algorithms for further review.

Simple Rate Adaptation (SRA). This algorithm is described by Toyserkani *et al.* (2004). This algorithm is based on the criteria of maximizing the throughput while keeping the total transmission power constant. It does not necessarily assign the best modulation scheme based on the Carrier-to-Interference Ratio (C/I). The bit loading is performed by comparing the actual channel gain for each subcarrier with pre-defined thresholds. This algorithm is very simple but its throughput performance is not optimized.

Adaptive Power Distribution (APD). APD is an adaptive power distribution algorithm, the objective of which is to improve the spectral efficiency in OFDM systems (Lei *et al.*, 2004). It starts with 0 bits and 0 power for each subcarrier. Then uses an iterative adaptive power distribution that tries to achieve a high throughput while satisfying the transmit power threshold and maintaining a target BER. In each iteration, this algorithm allocates one subcarrier with the 'best' possible bit loading and power level. Iterations are continued until the maximum power threshold is reached or all subcarriers are assigned the highest possible bit level. The price for the high performance achieved is high calculation complexity which increases with the number of subcarriers and maximum power threshold.

To find a balance between complexity and spectral efficiency, a new algorithm is proposed. It is a combination of the two algorithms discussed above: the iterative approach is still taken, but unlike the APD algorithm, instead of starting from 0 power and 0 bits, it starts with equal power for all subcarriers. Then by comparing received SNR with the SNR-lookup table, loaded bits for each subcarrier can be found, and power required for each subcarrier is recalculated. Then an approach similar to the APD algorithm is taken, to find the best bit and power distribution. We call such an algorithms as Simple Adaptive Modulation and Power Adaptation Algorithm (SAMPDA).

Different symbols that are used for describing SAMPDA are as follows:

- P_T, transmit power threshold;

- P_L, loaded power;

- N, number of subcarriers;

- F, the highest modulation level;

- $\psi_{\mathrm{mod}} = [0, 1, \ldots, F]$, usable modulation set;

- $P_{1 \times N}$, vector of power for each subcarrier;

- $k_{1 \times N}$, vector of loaded bits for each subcarrier;

- $M_{1 \times N}$, vector of modulation scheme for each subcarrier;

- k, sequence number of the subcarrier;

- hk, channel frequency response at the kth subcarrier;

- $\Delta P/\Delta k_{1\times N}$, incremental power per incremental bit;

- γ_k, SNR in each subcarrier;

- σ_n^2, noise power in each subcarrier;

- SNR^f, required SNR to maintain the target BER for the fth modulation level;

- NaN, not a number

- k^*, sequence number of the best subcarrier which has the minimum value of $\Delta P/\Delta k$.

17.3.1 SAMPDA Algorithm

SAMPDA works as follows.

- **Step 1: Initialization**

$$P_k = \frac{P_T}{N}$$

- **Step 2: Initial modulation scheme and power calculation**

$$\gamma_k = \frac{P_k hk^2}{\sigma_n^2}$$

$$M_k = f_k \quad \text{where SNR}^{f_k} \leq \gamma_k < \text{SNR}^{f_k+1}$$

$$P_k = \frac{\text{SNR}^{M_k}\sigma_n^2}{hk^2}$$

$$P_L = \sum_{k=1}^{N} P_k$$

$$\frac{\Delta P}{\Delta k}_k = \frac{(\text{SNR}^{M_k+1} - \text{SNR}^{M_k})\sigma_n^2}{2hk^2} \quad \text{if } M_k \neq F$$

$$\frac{\Delta P}{\Delta k}_k = \text{NaN} \quad \text{if } M_k \text{ equals } F$$

- **Step 3: Check the Termination Condition**

If $P_L = P_T$ or $\min(M) = F$, go to Step 6, otherwise continue.

- **Step 4: Iteration starts**

Find the best subcarrier:

$$k^* = \text{argmin}_k \frac{\Delta P}{\Delta k}$$

Recalculate power and modulation scheme for the kth subcarrier:

$$M_{k^*} = M_{k^*} + 1$$

$$P_{k^*} = \frac{\text{SNR}^{M_{k^*}}\sigma_n^2}{h_{k^*}^2}$$

- **Step 5: Check whether the Distributed Power Overflows**

If $\sum_{k=1}^{N} P_k \geq P_T$, exclude the impossible modulations:

$$\frac{\Delta P}{\Delta k}\Big|_{k^*} = \text{NaN}$$

$$M_{k^*} = M_{k^*} - 1$$

$$P_{k^*} = \frac{\text{SNR}^{M_{k^*}} \sigma_n^2}{h_{k^*}^2}$$

go to Step 3.

Otherwise update the parameters:

$$P_L = \sum_{k=1}^{N} P_k$$

$$\frac{\Delta P}{\Delta k}\Big|_{k^*} = \frac{(\text{SNR}^{M_{k^*}+1} - \text{SNR}^{M_{k^*}})\sigma_n^2}{2h_{k^*}^2} \quad \text{if } M_{k^*} \neq F$$

$$\frac{\Delta P}{\Delta k}\Big|_{k^*} = \text{NaN} \quad \text{if } M_{k^*} \text{ equals } F$$

go to Step 3.

- **Step 6: End**

Calculate bits loaded for each subcarrier:

$$\text{B}_k = \log 2(2^{2 \times M_k}) = 2 \times M_k$$

and stop.

After these six steps, the bit rates and power for each subcarrier are stored in the two N length vectors $\text{B}_{1 \times N}$ and $P_{1 \times N}$, which will be used for the transmission. A flow diagram is shown in Figure 17.2 to help understand this LA process.

17.4 Link Adaptation Scenario

In this section, we present the different system related issues that are considered in our investigation throughout the chapter.

17.4.1 Link Adaptation Process

We assume either Time Division Duplex (TDD) or Frequency Division Duplex (FDD) for this study. In case of TDD, the Mobile Station (MS) receives the pilot signals at the Downlink (DL) transmission slot, and then it measures the received SNR based on the used receiver technique. The received SNR is mapped to certain Channel Quality Information (CQI), which is transported back to the Base Station (BS) at the Uplink (UL) time slot. For FDD case, the CQI is reported via the UL frequency.

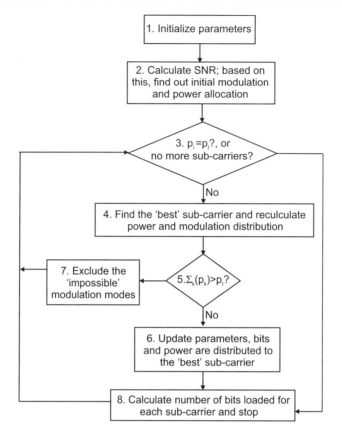

Figure 17.2 Flow diagram of the proposed algorithm.

At the beginning of each adaptation window, the resulting SNR, γ_k, for all subchannels is measured. This SNR is mapped back to the BS as CQI, which is used to decide on the power allocation level, modulation bits and coding rate to be used for that subchannel by using different LA algorithms. Naturally it is understood that for long adaptation windows, the CQI will be somewhat invalid at the end of the adaptation window.

Figure 17.3 describes the LA model. An example of FDD is shown in this figure. We assume that there are six OFDM symbols in one frame (or block). If the processing time (i.e. time required by the MS to calculate the subchannel gains and decide on the LA levels) is negligible compared with the OFDM symbol duration, then there is a delay of four OFDM symbols in the actual estimation of the CQI and its implementation in the LA system. For TDD-based systems, the UL and DL are separated in time rather than frequency. So, the resultant delay in the LA process will depend on the ratio of slots assigned to DL and UL.

17.4.2   System Parameters

We choose following system parameters for our studies, which are based on WiMAX standard.

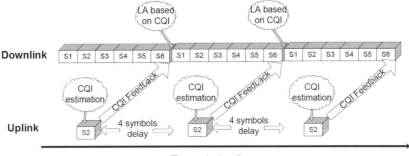

Figure 17.3 LA model; an example of the FDD scenario is shown.

- Convolutional coding with rate 1/3, 1/2 and 2/3 is used, together with 4QAM, 16QAM and 64QAM as the switchable modulation schemes.

- The maximum channel RMS delay spread that is considered in this work is 2 μs. This corresponds to a high-frequency selective channel. We have also used other lower RMS delay spread values, such as 0.5 and 1 μs. The channel power delay profile is taken from Erceg *et al.* (1999).

- Subchannelization is used according to the WiMAX standard. A subchannel consists of 8, 16, 32 or 512 subcarriers. We denote this as *subN* in the subsequent discussions.

- A frame is defined as one subchannel across six consecutive OFDM symbols, thus a frame duration is around 0.6 ms. For example, one frame has 96 Quadrature Amplitude Modulation (QAM) symbols when the subchannel size is 16.

- The LA adaptation window is taken to be 1, 2, 4, 10 or 20 frames.

- The maximum Doppler spread is taken as 50 and 250 Hz for the two cases of our investigations. The corresponding user speeds are 15.43 and 77.14 km h^{-1}, respectively. The corresponding 50% coherence times are 8.5 and 1.7 ms, respectively (Rappaport, 1996).

17.4.3   Frame Structure

The frame structure fundamental to LA being considered in this work is given in Figure 17.4. The minimum unit over which Forward Error Correction (FEC) and interleaving is applied is called a frame (or block). It is a set of consecutive subcarriers which span a successive sequence of OFDM symbols over a period of 0.6 ms. The subchannel size is defined by the number of consecutive subcarriers that make the frame. The frame size can be made to vary in the frequency domain by changing the subchannel size and in the time domain by changing the number of symbols inside one frame. The modulation and coding level is adapted once every adaptation interval, which can be a multiple of the block duration.

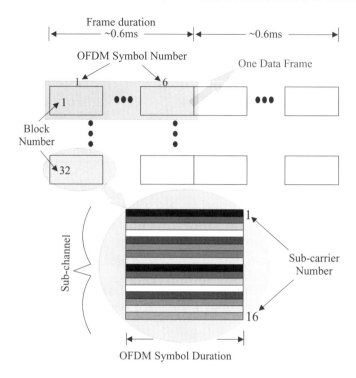

Figure 17.4 LA frame structure; an example of 6 OFDM symbols per frame (or block) and 16 subcarriers per subchannel.

17.5 Role of Power Adaptation in Collaboration with Bit Adaptation

In a system where modulation, channel coding and transmit power across OFDM subchannels are adapted in real time, the adaptation algorithms are usually quite complex. One such algorithm was proposed by Das *et al.* (2007). In this case, the algorithm has to take care of the fact that the total power is constant and a threshold BER, or FER, needs to be met. There are other suboptimal algorithms where different combinations of bit and power adaptation can be used. It is much easier to assign equal power across all OFDM subchannels and then determine the allowable modulation and coding level for any particular subchannel: such a method is called Adaptive Modulation and Coding (AMC). In AMC, the complexity of the LA procedure is reduced compared with AMC with dynamic power allocation. Usually, AMC and power allocation are not suitable to be carried out at the same rate in any wireless system. Several authors previously suggested that, when a point-to-point wireless link is considered, then AMC with dynamic power allocation does not provide any significant improvement in throughput in comparison with AMC-only systems (Chung and Goldsmith, 2001, Czylwik, 1996, Figueiredo *et al.*, 2006, Hunziker and Dahlhaus, 2003). Some of these articles suggest that it is not recommended to perform AMC in collaboration with adaptive power distribution due to excessive system complexity. Rhee and Cioffi (2000) suggested the same for multi-user OFDM systems.

In this section, we study bit and power allocation strategies for multi-antenna assisted OFDM systems. In contrast to the suggestions in the above-mentioned articles, we have found that in some scenarios and in some system conditions, some form of power adaptation along with bit allocation across OFDM subchannels are required together for efficient system performance. For broadband OFDM systems, the channel variations are quite high inside the OFDM system bandwidth, thus dynamic power distribution is advantageous together with AMC. This is in-line with the conclusions made by Bohge *et al.* (2005). Using different LA mechanisms, we have found that, when we cannot find the exact SNR thresholds due to different reasons, such as reduced LA rate, CSI error, feedback delay etc., it is better to fix the transmit power across all subchannels to guarantee the target FER. Then the power allocation will actually act as a safety margin for the impairment to a certain degree. Otherwise, we can use adaptive power distribution to save power, which can be used for other purposes, or we can increase the modulation level to increase the system throughput. These benefits are even more visible when multi-antenna schemes are used in the system.

Our investigation is mainly concentrated on different bit and power allocation rates in OFDM systems. In Section 17.5.1 we study the impact of both adaptations at the same rate and in Section 17.5.2 we investigate the adaptations at different rates.

17.5.1 AMC and Power Adaptation at the Same Rate

In this section, we study bit and power adaptation at the same rate, so the adaptation window for bit and power allocation is always same. In this way, the channel conditions and user mobility have the same impact on both of the allocations.

As we have seen in Section 17.4, when the adaptation window is only one frame long, then the channel conditions are quite static for the Doppler conditions that we are considering here. Thus, we can say that the feedback CQI is almost correct according to the channel conditions when the LA decisions are implemented. Our previous analysis points out that the resultant FER is well below the FER threshold when instantaneous AMC is performed. Thus, we clearly see that some power is wasted when only AMC is performed. When we also adapt the power after the AMC decisions are made, we find that the FER threshold is still maintained. Now, we need to look at the corresponding spectral efficiency curves. Intuitively, when some power is removed after power adaptation, the spectral efficiency should be lower. However, we can observe that the spectral efficiency is almost equal with and without power adaptation. Note that, in this case, the power adaptation essentially means that we are saving some power, as described in Figure 17.5(a) and discussed later.

For the outdated feedback case, we should find new AMC thresholds compared with the original AMC thresholds used in previous simulations. These new threshold margins can be found via Monte Carlo simulations. Again, referring to previous analysis, we can see that the FER threshold is not met when AMC is followed by power adaptation. This is the case when the adaptation window is equal to four frames. In the same scenario, the fixed power based mechanism works well. As we have a slower rate of adaptation, we practically need a new threshold (which is obviously higher than the previous threshold) for the slower adaptation rate. In this case, the extra power available after bit and coding rate allocation compensates for the required additional power when the adaptation is slowed down. The spectral efficiency

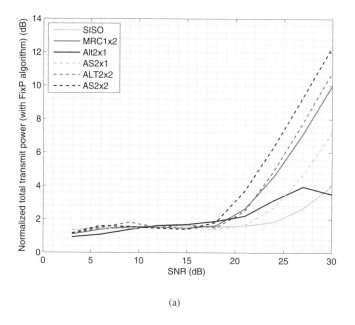

(a)

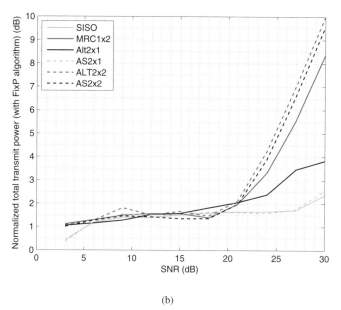

(b)

Figure 17.5   Power savings of different LA rates with and without power adaptation. Ensemble average of the ratio of total transmit power required between AMC, AdaptP and AMC, FixP is plotted against the average pre-SNRs. (a) Instantaneous adaptation (LA per frame). (b) Slower adaptation (LA per four frames).

is almost the same when the power is adapted and the power is fixed for all of the schemes, but the FER threshold is not met when power adaptation follows the AMC procedure. Also, a loss in spectral efficiency is experienced when a wider adaptation window is used.

Figure 17.5(a) and (b) summarize the resulting power saving when AMC followed by a power adjustment is performed, in comparison with the AMC-only scheme. We can see that a significant amount of power can be saved for instantaneous adaptation. When the system SNR is roughly more than 20 dB, the power savings to meet the FER requirement are more than 2 dB. When the average SNR increases, the power saving increases greatly. This result demonstrates that power allocation also has a role to play in WiMAX-like broadband OFDM systems, in contrast to the original claim that power allocation does not bring much benefit when AMC is already performed (Chung and Goldsmith, 2000, 2001). As expected, the power savings becomes irrelevant when the adaptation window is longer. In this case, the power saving is performed, but the threshold FER is not met. Thus, we see a throughput degradation.

The transmit power saved can be used for many other purposes in the system. For example, more bits can be transmitted in a particular subchannel. Thus, a window of opportunity arises when we perform the power adaptation in collaboration with AMC. In a multi-user scenario, the power saved can be allocated to a user with a weaker SNR and the spectral efficiency of the system can be improved in that way.

17.5.1.1   Impact of Different Power Adaptation Algorithms

Figures 17.6, 17.7(a) and 17.7(b) give us the FER, resultant spectral efficiency and power savings respectively, when we use two different power adaptation mechanisms. In these results, APMC refers to our optimal SAMPDA algorithm as developed by Das *et al.* (2007), and AMC with subsequent power adaptation refers to power adaptation applied after AMC decision is made (labeled as 'AMC,AdaptP' in the figures). Under a low-SNR regime, APMC will utilize more power than AMC with power allocation. APMC algorithms (e.g. SAMPDA) do not assign any bit to any weaker subchannel (i.e. withdraws power from the weaker subchannels), rather it concentrates the available power across subchannels where some bits can be transmitted. For the case of AMC with power adaptation, not many bits can be transmitted at low SNR, so throughput is very low (as seen in Figure 17.7(a)), although some power saving is seen in Figure 17.7(b). In the same figures, we can see that APMC is able to transmit significantly higher throughput even at very low average SNR, because APMC allocates most of the available power to some subchannels where some bits can be transmitted. So, we can conclude that APMC utilizes the total available power efficiently. This throughput optimization in APMC is obtained at the price of added complexity in the LA algorithm. Under a high SNR regime, both of these algorithms can save power, but APMC will utilize the available power more efficiently than AMC with adapted power to achieve a better throughput performance.

In these results, we have only shown the FER, throughput and power saving performances for fast (or instantaneous) LA rate, because both AMC with adapted power and APMC can maintain the FER target at fast LA rates, while at a low LA rate, an additional margin is needed.

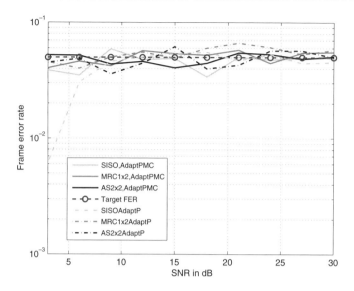

Figure 17.6 FER results of instantaneous LA with different power adaptation (LA perframe).

17.5.2   AMC and Power Adaptation at Different Rates

Conventional wideband systems, such as the Global System for Mobile Communication (GSM) and Wideband Code Division Multiple Access (WCDMA), use a mixture of AMC and Power Control (PC), which can broadly be classified as Adaptive Power Fixed Rate (APFR) methods, where one strives to adapt the power to maintain the required throughput over the duration of communication. On the other hand, systems such as High Speed Downlink for Packet Access (HSDPA), use another class of methods, namely Fixed Power Adaptive Rate (FPAR) methods, where the modulation and coding rate adaptation enabled by the momentary channel quality using fixed power (Larsson, 2007). In our previous discussions, we have studied different combinations of APFR and FPAR-type LA methods. For OFDM-like multi-carrier systems, APFR is not a good solution, because using APFR means we will not be able to exploit the available frequency diversity of the channel. OFDM enables us to exploit the channel gains up to a subcarrier level, and this degree of freedom should be exploited. Thus, a mixture of AMC and PC will always be a preferable solution.

Until now in our discussions, we have used the WiMAX system parameters for our analysis and evaluations for different LA issues. From this point onwards until the end of this chapter, we use another set of parameters taken from the Universal Mobile Telecommunications Systems–Long Term Evolution (UMTS-LTE) standard. The parameters are listed in Table 17.1.

In Section 17.5.1, we studied the role of simultaneous power and bit adaptations (i.e. AMC and PC at the same rate). We have concluded that simultaneous bit and power allocations are needed in some scenarios to ensure power savings at the expense of insignificant throughput reduction. In Section 17.6.3, the results show that, even at lower Doppler frequency values, the system throughput is severely degraded when the LA window is very large. In both of these sections, it is explained that instantaneous (or near instantaneous) AMC and PC is the

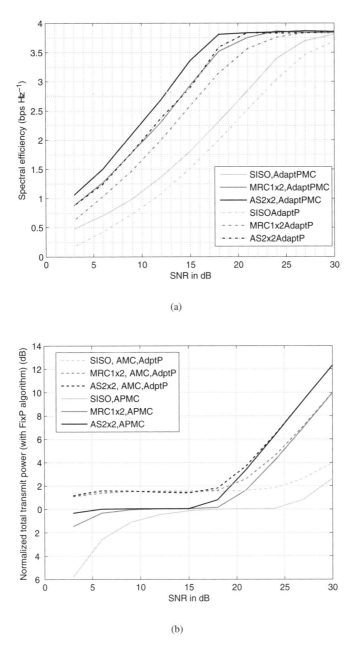

(a)

(b)

Figure 17.7 (a) Spectral efficiency and (b) power saving of instantaneous LA with different power adaptation mechanisms (LA per Frame).

Table 17.1 Parameters from UMTS-LTE Standard.

Carrier frequency	2 GHz
Bandwidth	5 MHz
FFT size	512
Subcarrier spacing	15 kHz
Useful part of OFDM symbol	66.67 μs
Frame duration	0.5 ms
Adaptation window size	[1, 2, 4, 10, 20] frames
Adaptation window durations	[0.5, 1.0, 2.0, 5.0, 10.0] ms
Maximum transmit power	38 dBm

best solution in terms of spectral efficiency. In reality, this requires frequent feedback from the transmitter to receiver; this can be of large amount of traffic compared with the available spectrum and link capacity. This can be very difficult to achieve in FDD systems. In TDD systems, the required CQI can be obtained without using any feedback channel, but it still remains an issue that the transceivers need to perform complex calculations on every frame to find out the possible modulation, coding and power level. This may require high processing power and also large processing times. For some traffic conditions, such as delay-intolerant real-time traffic, this can even be an inefficient scenario. One possible solution can be to use a fast PC while using slow rate control (i.e. AMC). For AMC, the required amount of feedback is much higher compared with PC, because AMC needs multi-level feedback for selected modulation and coding rates, while PC requires only a single bit (power up or power down) of feedback. Thus, In this section, the performance for reduced AMC rates but with fast PC rates is investigated.

Figure 17.8 shows the concept of reduced LA with fast PC rates. The idea is to use a fast PC to compensate for the SNR mismatching loss due to the time domain channel variation caused by slow AMC. The steps are shown in Figure 17.9, and explained below.

1. Assume a slow AMC rate of every T_{amc}(ms), and a fast PC rate of every T_{pc}(ms), $T_{amc}/T_{pc} = K$ and K is an integer. It is understood that k is always greater than one.

2. Within one single data block, the power, modulation and coding rate for each subchannel is found using above-mentioned LA algorithms. These levels are kept constant during the whole LA window.

3. In the beginning of the kth PC window ($1 < k \leq K$), the MS will compare the instantaneous channel gain with the previous PC window. If the difference between the two channel gains is within a certain limit L_{unchg}, no change in bit, power or coding level is performed.

4. The BS collects all of the requirements for each block. In the first step, the BS decreases the power level to those blocks which require a reduction in power. However, for the blocks that require an increase in the transmission power due to worsening channel conditions, the BS needs to consider the total power constraints by taking into account the saved power after bringing down those mentioned before. Assume that the granularity for bringing down the power is always G_{down} (which is typically 20% in

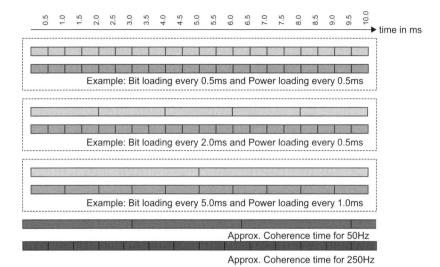

Figure 17.8 Examples of different power and bit allocation rates. For clarity, the coherence time related to 50 and 250 Hz is also shown.

our investigations) and the maximum granularity for increasing the power is G_{up}. The increase in power, G_{up}, can be decided dynamically. It can start from an increase of 20%, and then increase step-by-step until the requirement of the subchannel is met. Thus, the actual increasing in power level, $G_{up,actual}$, can be calculated as

$$G_{up,actual} = \epsilon G_{pu} \quad \text{where } \epsilon = \min\left\{1, \frac{P_{av} + G_{down} \sum_{n=1}^{N_{down}} P_{k-1,n}}{G_{up} \sum_{n=1}^{N_{up}} P_{k-1,n}}\right\}, \quad (17.11)$$

where N_{down} is the number of blocks which need less power and N_{up} is the number of blocks which need more power; $P_{k-1,n}$ is the power assigned for the nth block during the $(k-1)$th PC window; P_{av} is available power before redistributing the power and is calculated using

$$P_{av} = P_T - \sum_{n=1}^{N} P_{k-1,n}; \quad (17.12)$$

P_T is the total power constraint which indicates the up-limit for the total transmission power.

5. Step 3 is then repeated for each PC window within the same AMC window. Steps 2 and 3 are repeated for each AMC window during the whole transmission time.

In this work, we take that $G_{up} = G_{down} = 0.2$. Figure 17.10(a) and (b) show the achievable spectral efficiency for different combinations of AMC and PC rates, for low- and high-diversity conditions, respectively. This analysis is valid when 16 consecutive subcarriers are placed together in one subchannel (i.e. $subN = 16$). For a better understanding of the ratio of time between the adaptation windows and channel characteristics, we can check Figure 17.8. For low-diversity conditions, we can see the following.

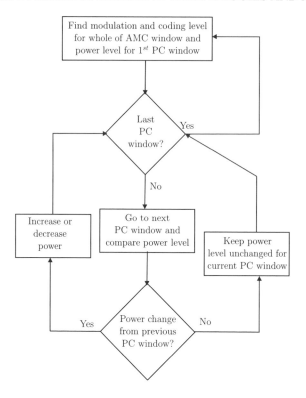

Figure 17.9 Flowchart of link adaptation mechanism when faster power control and slower AMC is used.

1. When the AMC rate is less or equal to 1 ms, there is not much difference in performance compared with cases with faster power control.

2. When AMC is performed every 2 ms, a faster PC rate of every 0.5 ms can improve the performance compared with AMC at 2 ms and PC at 1 ms by ~0.2 dB. It can also be noted that very fast power control (i.e. PC at 0.5 ms) reduces the need for faster AMC, which can be seen from the cases of AMC at 2 ms and PC at 0.5 ms, AMC at 1 ms and PC at 1.0 ms and AMC at 1 ms and PC at 0.5 ms. These three cases are very similar in performance, thus the lower AMC rate can be chosen.

3. When AMC is performed every 5 ms, both PC rates of every 0.5 ms and 1 ms give the same performance. For 50 Hz of Doppler spread, the coherence time is 8.5 ms. So inside the AMC window of 5 ms, we will see some channel variations, thus, PC faster than 1 ms is not beneficial. What is noticeable is that, when PC and AMC are both done at 5 ms, the spectral efficiency performance is degraded largely compared with AMC at 5.0 ms/PC at 1 ms. Once again, this proves that fast PC is striving to mitigate the SNR threshold imbalances caused by slower AMC rate.

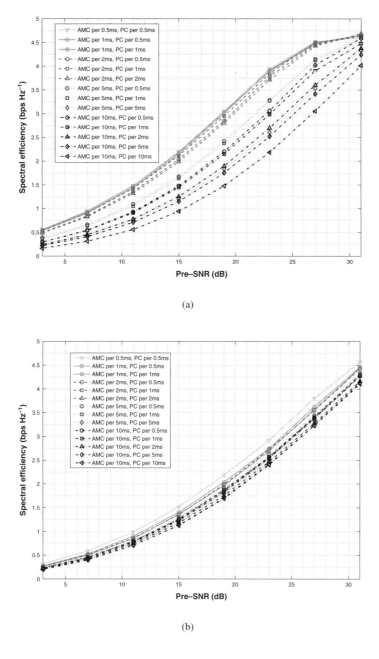

(a)

(b)

Figure 17.10 Spectral efficiency for different AMC and PC Rates. PC rate is always higher than the AMC rate. (a) Doppler 50 Hz, RMS delay spread 0.5 μs. (b) Doppler 250 Hz, RMS delay spread 2.0 μs.

4. Similar trends are seen for the case of AMC at 10 ms. The cases of AMC at 10 ms/PC at 0.5 ms and AMC at 10 ms/PC at 1.0 ms show that the resultant spectral efficiency is improved by almost 50% compared with the spectral efficiency of AMC at 10 ms/PC at 10 ms case, as seen in Figure 17.10.

In essence, decreasing the AMC rate has a severe impact on the system performance, as also seen earlier. In a real scenario, a reduction in the AMC rate may sometimes be required. In those cases, faster PC rates need to be implemented for acceptable level of system performance. For the particular system and user parameters as shown in Figure 17.10(a), PC rate of 1 ms provides good performance for several AMC rate combinations.

Figure 17.10(b) shows the performance when $f_d = 250$ Hz. As can be seen from this figure, with high Doppler frequency, PC gives very limited benefit as compared with low Doppler cases. This result is in accordance with Figure 17.13(b).

To conclude, a reduced AMC rate can be used to reduce the system complexity. At high Doppler frequency, this reduced AMC rate will not affect performance much, while at low Doppler frequency, a large degradation can be observed for low AMC rate. By using PC at a fast rate, the combined performance can be improved at reasonably lower complexity than a fast AMC-only rate.

17.5.3   Overhead Analysis

In our work related to LA, we have mainly considering the gain in throughput obtained by bit and power adaptation, while this gain is obtained at a cost of increased overhead. For a proper system design, the exact amount of overhead needs to be measured, and proper feedback mechanisms also need to be found to minimize the feedback overhead while still maintaining the system performance at a satisfactory level. In this section, we perform an initial evaluation of the required feedback when different rates of AMC and PC are used.

Considering the discussions related to different rates of AMC and PC rates, we need to have a look at the required overhead bits for transporting the channel information between the transmitter and the receiver.

The following parameters are used for overhead estimates:

- N, total number of available subcarriers;

- $subN$, number of subcarriers in one subchannel;

- T_{amc}, AMC interval, that is, $1/T_{amc}$ gives us the AMC rate;

- T_{pc}, power adaptation interval, that is, $1/T_{pc}$ gives us the AMC rate;

- $K = T_{amc}/T_{pc}$, an integer number representing AMC over PC duration;

- PC_{ix}, the index of power adaptation times within one AMC window.

We can divide the overhead into two parts, DL and UL. The total overhead can be written as

$$overhead_{total} = overhead_{ul} + overhead_{dl}. \tag{17.13}$$

17.5.3.1 Overhead in the UL

In the UL, we need to transmit two types of overhead information.

1. We need to transport the CQI from the MS to inform the BS about the possible modulation and coding level that can be supported for each subchannel. Thus, 5 bits (i.e. 2^5 levels of information) for one subchannel will be fed back to the BS for the received SNR level.

2. For faster PC, we need to transport the $PC_{ix} = [2, \ldots, K]$. In this case, only 1 bit is needed per subchannel per T_{pc} to indicate whether the power should go up or down. Note that we are considering the situation when $T_{amc} > T_{pc}$.

So, we can express the overhead in the UL as follows:

$$overhead_{ul} = \left[5 + \left(\frac{T_{amc}}{T_{pc}} - 1 \right) \right] \frac{N}{subN} \frac{1}{T_{amc}}$$
$$= \frac{(T_{amc}/T_{pc} + 4)N}{subN\, T_{amc}}$$
$$= \frac{(K + 4)N}{subN\, T_{amc}}. \tag{17.14}$$

17.5.3.2 Overheads in Downlink

In the DL, the BS needs to inform the MS about the modulation and coding level for each subchannel. If we have 16 possible rate levels, then 4 bits for each subchannel will be transmitted from the BS via the control channels. This can be expressed as

$$overhead_{dl} = 4 \frac{N}{subN} \frac{1}{T_{amc}} = \frac{4N}{subN\, T_{amc}}. \tag{17.15}$$

17.5.3.3 Overhead Calculation Corresponding to Our System Parameters

Using (17.14) and (17.15), we can write that the total system overhead expression as

$$overhead_{total} = \frac{(K + 8)N}{subN\, T_{amc}}. \tag{17.16}$$

Table 17.2 presents the amount of overhead signaling required for different combinations of AMC rate, PC rate and different subchannel size. As can be seen in the table, the highest amount of overhead is required in the case of AMC at 1.0 ms/PC at 1.0 ms, while we have seen earlier that the achievable spectral efficiency is also very high at this adaptation rates. Depending on different channel conditions and different system complexity requirements, we can choose to increase both the AMC and PC rates. We can see that when AMC window is extended from 1.0 to 10.0 ms, we have approximately 80% less overhead requirement for all subchannel sizes.

Table 17.2 Overhead (Mbps) for adapt power LA.

	1.0 ms	2.0 ms		5.0 ms		10.0 ms
T_{amc}						
T_{pc}	1.0 ms	0.5 ms	1.0 ms	0.5 ms	1.0 ms	1.0 ms
subN = 8	0.57	0.384	0.32	0.2302	0.1664	0.1152
subN = 32	0.144	0.096	0.08	0.0575	0.0416	0.0288
subN = 128	0.036	0.024	0.02	0.0144	0.0104	0.0072
subN = 512	0.009	0.006	0.005	0.0036	0.0026	0.0018

17.6   Link Adaptation Considering Several System Issues

As we have described in previous sections, LA schemes adapt transmission parameters according to the channel conditions so that the maximum bit rate is achieved whilst keeping the error rate below the target. Once the values of the adapted parameters are selected, they are kept constant over a region where the channel is relatively flat. OFDM with its fine granularity of the minimum allocation unit as a subcarrier (or subchannel), which experiences flat fading, provides the inherent support needed to exploit the advantages of LA techniques (Das *et al.*, 2007, Toyserkani *et al.*, 2004) in multiple dimensions. The degrees of freedom that can be exploited by LA techniques increase when applied in OFDM systems and this leads to an increase in complexity of the system with the benefit of improved spectral efficiency. LA involves adaptation of the modulation level (M), the FEC rate (C) and the power level (P) at the transmitter as per the channel state information fed back from the receiver. When applied in the OFDM framework, LA additionally includes selection of adaptation interval for M and C, adaptation interval for P, subchannel size (a set of consecutive subcarriers that span a few successive OFDM symbols), and choice of bit and power loading algorithms. Other than the fast fading of the channel gains, the dynamic variation of the channel parameters such as the RMS delay spread, Doppler frequency spread and average SNR conditions, heavily influence the values to be selected for the LA parameters.

   The LA scheme, maximizing the throughput while maintaining a target error rate, is expected to become highly complex when it tries to optimally adapt so many parameters which depend on another large set of varying channel and system conditions. Therefore, hybrid strategies, for example that limit to some of the degrees of freedom by slowly varying some parameters, while using fast adaptation for the others are investigated in this section. The objective is to analyze the tradeoff between spectral efficiency loss and complexity and overhead reduction that can be obtained by the hybrid strategies. When FEC with interleaving is considered in LA, as in this case, the interplay of subchannel size, RMS delay spread, Doppler spread and adaptation window size becomes especially important. This is because, on the one hand (frequency–time) diversity gain is brought by FEC while on the other hand LA with adaptive modulation and power loading exploits diversity in a different way. Therefore, it is very interesting to study the synergy of these techniques, which is also one of the objectives of this section.

   In this section, strategies for simplified LA, which reduce processing complexity but not compromising on throughput significantly, are presented. Results obtained in this perspective are also very important for multi-user resource allocation, since the resource unit to be allocated to one user must be such that there is maximum benefit in terms of overall throughput considering the signaling overhead. This work serves as a first step to such systems.

17.6.1 Subchannelization

In this section the influence of RMS delay spread and Doppler velocity on different subchannel sizes (8, 16, 32 and 64 subcarriers in one subchannel) is investigated. Figure 17.11 shows the spectral efficiency in terms of bits per second per Hertz for a LA system when AMC is done every 2ms while keeping P fixed (i.e. the total available power is divided equally across all subchannels for all frames). It can be seen from the figures that when RMS delay spread is small and user velocity is low, that is, the coherence bandwidth and coherence time are large, the subchannel size of 8 subcarriers has the highest throughput. At this subchannel size, the channel is very flat inside the whole of the subchannel, that is, LA gains will dominate over the frequency–time interleaving gains[1].

Interestingly, at high velocity and high RMS delay spread, that is, small coherence bandwidth and small coherence time, the subchannel size of 8 subcarriers has very similar performance to that of subchannel with 64 subcarriers. This is because of the fact that the channel is already time and frequency selective inside the subchannel space at low T_c and low B_c, even when only 8 subcarriers are included inside one subchannel. In this case, the interleaving gain is much higher compared with the LA gain, thus the spectral efficiency is not improved for different subchannel sizes.

Therefore, the LA performance is optimal in terms of throughput when the subchannel size is small for low-diversity conditions. It can be concluded that for very high velocity and high RMS delay spread conditions, it is better to use a large subchannel size since it will lower the required overhead without reducing the achievable spectral efficiency, while under contrasting conditions of velocity and RMS delay spread, it is suggested to use a small subchannel size. It must be noted that subchannel size selection can be a statistical adaptation in combination with instantaneous adaption of modulation and coding rate.

17.6.2 Fixed Coding Rate

In most of the references cited in this work, it is found that adaptive bit loading is considered without any constraint on user devices. The best bit loading may bring out a situation where more than one coding rate is allocated to one user. Although this might lead to a spectrally efficient system, this might not be feasible since it will put a heavy signal processing burden on the user equipment. Using multiple coding rates simultaneously in a dynamically fashion means that the user equipment needs multiple FEC and decoders, which would increase the complexity prohibitively. Therefore, using only a single FEC coder (i.e only one FEC rate) for one user is highly desirable. In a practical system, the FEC coding rate adaptation window can also be made long enough to find a compromise between performance and signal processing requirements.

---

[1]Time–frequency interleaving inside one subchannel provides time–frequency diversity in any coded OFDM systems, provided that the channel gains vary significantly inside the subchannel duration and subchannel bandwidth.

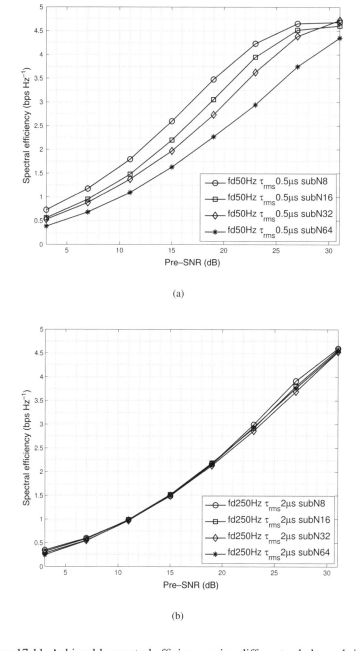

(a)

(b)

Figure 17.11 Achievable spectral efficiency using different subchannel sizes.

Table 17.3 Average SNR thresholds (dB) for a switching coding rate for different RMS delay spread and Doppler conditions.

τ_{rms}	f_d	subN = 8			subN = 16			subN = 32			subN = 64		
FEC		$\frac{1}{3}$	$\frac{1}{2}$	$\frac{2}{3}$	$\frac{1}{3}$	$\frac{1}{2}$	$\frac{2}{3}$	$\frac{1}{3}$	$\frac{1}{2}$	$\frac{2}{3}$	$\frac{1}{3}$	$\frac{1}{2}$	$\frac{2}{3}$
0.5	50	NA	NA	–	NA	NA	–	–	16	24	–	18	27
	150	NA	–	19	–	13	19	–	18	25	–	19	27
	250	NA	–	18	–	16	23	–	17	26	–	20	28
1	50	–	11	16	–	15	24	–	18	27	–	19	28
	150	NA	–	19	–	13	19	–	19	27	–	20	28
	250	NA	–	21	–	16	24	–	17	27	–	20	29
2	50	–	14	23	–	17	26	–	20	27	–	20	28
	150	–	16	24	–	19	26	–	18	27	–	20	28
	250	–	17	24	–	19	27	–	19	27	–	20	27

In this section, we would like to show the impact of channel coding on the LA scenario. Thus we study the following.

1. All three dimensions, that is, the bit, power and code rate are adapted dynamically on a frame-by-frame basis.

2. The total transmit power is equally distributed across all subchannels and coding rate is kept fixed while only modulation is varied on a frame-by-frame basis, that is, fixed P and C and varying M.

Figures 17.12(a) and (b) show the throughput comparison for different channel conditions when different FEC rates are used in the system. In the figures, 'APMC' means adaptive power allocation, modulation and coding simultaneously, using our proposed SAMPDA algorithm. 'APM' is used to denote adaptive bit and power allocation using the same algorithm, while the FEC rate is fixed. Figure 17.12(a) gives us the spectral efficiency results when a low-diversity channel (large B_c and T_c) is used, while Figure 17.12(b) presents the results when high diversity channel is used. In this case, 50 and 250 Hz correspond to 15.43 and 77.14 km h^{-1} respectively, similarly for coherence times of 8.5 and 1.7 ms, respectively.

As seen from the previous section, for the low-diversity case as in Figure 17.12(a), the interleaving gain is nonexistent, thus only coding gain is available. APMC always provides the optimal gain in all SNR ranges. Closer inspection reveals that rate-$\frac{2}{3}$ FEC approaches the performance of APMC when the average SNR is higher than 12 dB. For lower SNR, rate-$\frac{1}{2}$ FEC is very close to the optimal results. This is because, in low-diversity channels, coding gain plays the important role. For SNR even lower than that, rate-$\frac{1}{3}$ performs a little worse in spectral efficiency compared with rate-$\frac{1}{2}$. For high-diversity channels, as seen in Figure 17.12(b), we can now see more gains due to channel coding and interleaving for different ranges of SNR. Up to 18 dB of average SNR, rate-$\frac{1}{3}$ can be used, while rate-$\frac{1}{2}$ can be used for SNR between 18 and 25 dB. Beyond this, rate-$\frac{2}{3}$ can be used.

Figure 17.12(a) and (b) provide us with the results when the subchannel size is 16. As discussed in the previous section, we understand that different subchannel sizes will also

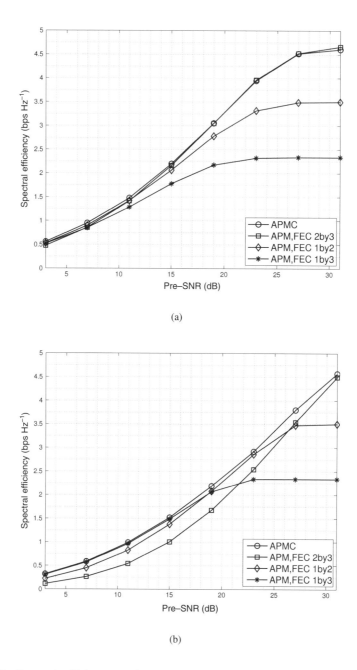

(a)

(b)

Figure 17.12 Spectral efficiency performance comparison for fixed coding with adaptive modulation vs adaptive modulation and coding. (a) Doppler 50 Hz, RMS delay spread 0.5 μs. (b) Doppler 250 Hz, RMS delay spread 2.0 μs.

Table 17.4 SNR lookup table for SISO with different AMC rates.

Coding rate	Modulation								
	4-QAM			16-QAM			64-QAM		
	$\frac{1}{3}$	$\frac{1}{2}$	$\frac{2}{3}$	$\frac{1}{3}$	$\frac{1}{2}$	$\frac{2}{3}$	$\frac{1}{3}$	$\frac{1}{2}$	$\frac{2}{3}$
LA per 1	5dB	8dB	11dB	NaN	14dB	17dB	NaN	21dB	23dB
LA per 2	5dB	9dB	11dB	NaN	15dB	17dB	NaN	21dB	24dB
LA per 4	6dB	10dB	12dB	NaN	16dB	19dB	NaN	22dB	25dB
LA per 10	10dB	13dB	16dB	NaN	19dB	23dB	NaN	25dB	28dB
LA per 20	12dB	16dB	18dB	NaN	22dB	25dB	NaN	28dB	31dB

have an influence on thresholds for choosing different FEC rates. Table 17.3 summarizes the switching thresholds for all of the subchannel sizes, different Doppler frequencies and delay spread values. The dash '–' indicates that the coding rate is the default coding rate to start with, while the SNR values indicate the starting average SNR from where the particular coding rate can be used, and 'NA' indicates that the corresponding coding rate will not be used. In those code rates, the target FER is not achieved even at very high SNR, thus we exclude those code rates.

It can be seen that the performance of different fixed coding rates at different ranges of SNR is not very far from the optimal APMC scheme. If average SNR is used as the threshold for switching between one coding and another, then one user can choose a coding rate based on the average channel SNR information and then use adaptive modulation.

17.6.3 AMC Rate

In Section 17.6.1, we have studied the impact of different size of adaptation space by varying the subchannel size, that is, the adaptation window was varied in the frequency domain. In this section, we investigate the impact of different AMC rates, that is, we study the varying adaptation in the time domain when the subchannel size is fixed. The adaptation window can be varied in time by including a varying number of OFDM symbols inside one LA space. Here, the power is kept constant, while modulation and coding rates are adapted.

Figure 17.13(b) shows the impact of changing the adaptation rate for high-diversity channels. The corresponding SNR look up table is shown in Table 17.4. It can be observed that when the subchannel size is small the adaptation rate has a big impact, but when the subchannel size is very large, then the adaptation rate has hardly any influence. It can also be seen that for low and moderate average SNR levels under these kinds of channel conditions (RMS delay spread of 2 μs at 77.14 km h^{-1}) a large subchannel size can be good enough; in this way, the overhead for signaling the LA modulation and coding level can be minimized. It may also be concluded that with these channel conditions low rate LA with a large subchannel size may be selected; thus, the adaptation window in time and frequency can be quite large. This conclusion is in-line with our understanding from Figure 17.11.

Figure 17.13(a) shows a similar performance comparison but for a low-diversity situation, that is, RMS delay spread of 0.5 μs and 15.43 km h^{-1} velocity. It can be seen that there is a large impact of the decreased adaptation rate as in the earlier case for a subchannel size

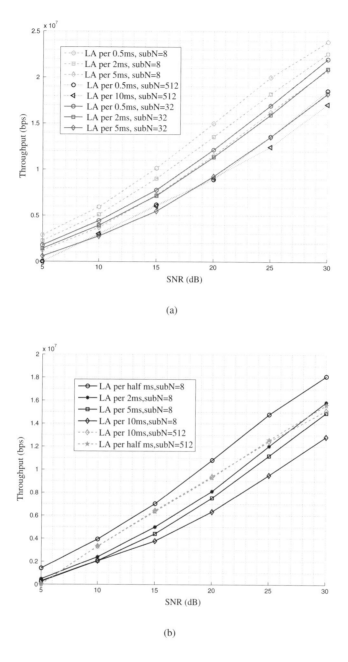

Figure 17.13 Throughput comparison for different bit adaptation rates. (a) Doppler 50 Hz, RMS delay spread 0.5 μs. (b) Doppler 250 Hz, RMS delay spread 2.0 μs.

of 8 subcarriers and also to some extent for a subchannel size of 32. When the subchannel size is made large there is little impact on the adaptation time interval, that is, short-term adaptation in the time domain is not necessary when the adaptation window is large in the frequency domain even under low mobility conditions. Compared with the previous case there is a difference in performance between large and small subchannel sizes.

17.7 Summary

In this chapter, several aspects of link adaptation in OFDM systems have been presented. We now present a short summary of the studies performed in this chapter.

17.7.1 Guidelines for Hybrid Link Adaptation

It is seen that there are a large number of options to maximize the throughput in link-adapted OFDM systems. If all of these options are intended to be optimized in the same rate, then the system becomes very complex since it involves optimization of several parameters. From the wireless channel point of view, Doppler conditions, RMS delay spread, average SNR range, etc., are the statistical measures and the channel gain values for each subchannel in real time are the instantaneous information that provides a number of opportunities for improving the system performance.

Hybrid LA strategies do not exploit all degrees of freedom simultaneously at the same rate. As mentioned, several parameters can be adapted, namely, modulation level (M), the FEC rate (C), power level (P), adaptation interval for M and C, adaptation interval for P, subchannel size and choice of bit and power loading algorithms. Different combinations of slow and fast adaptation can be made between these parameters so that only few parameters are adapted instantaneously using immediate channel gains, while others are adapted statistically, that is, using some average information such as RMS delay spread, Doppler conditions, average SNR, etc.

We can summarize our findings as shown in Figure 17.14. From the results presented it can be said that bit and power loading algorithms can be selected based on the average SNR conditions. For optimum throughput performance, APMC-type algorithm is the best, while it also requires the highest complexity. Thus, suboptimal methods can be taken as shown in this chapter. In general, AMC with adapted power should be used for low-diversity channel (i.e. low RMS delay spread and low Doppler values), while AMC with fixed power algorithms can be used under contrasting conditions. The coding rate can also made quasi-static which adapts based on the average SNR criteria, which includes shadowing loss of the channel. The selection of a subchannel, which has a large influence on the overhead, is dependent on channel time and frequency correlation factors. This suggests that a significant gain can be obtained in case of low Doppler by using a small subchannel size, where as for high velocities, increases the subchannel size does not reduce throughput but decreases the overhead, which can be a great advantage. It must be remembered that each case is dependent on the coherence bandwidth, coherence time of the channel and the subchannel size.

It is seen that using hybrid strategies, that is, using a combination of slow and fast adaptation of different parameters, can simplify the LA process while there is little impact on the spectral efficiency performance. This understanding will give significant information on resource allocation strategies where users may have different channel conditions and then

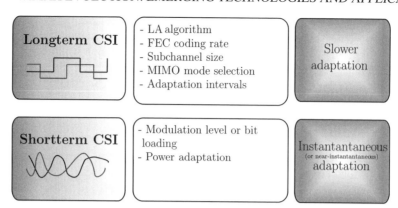

Figure 17.14 Summary of the hybrid LA approach.

selecting different LA parameters for different users will be very important. Results, such as those presented in this chapter, will play a key role in identifying such cases. These results are also expected to trigger investigations into resource allocation algorithms for future broadband wireless systems, such as UMTS-LTE and IMT-Advanced.

17.7.2   Conclusion from Bit and Power Allocation Analysis

We have studied several combinations of bit and power allocation rates, so that the throughput can be optimized without increasing the system complexity. It is popular belief that simultaneous bit and power allocation at the same rate is not very useful in terms of system throughput. In contrast to this popular belief, we have found that in some scenarios and under some system conditions, some kind of power adaptation along with bit allocations across OFDM subchannels are required together for efficient exploitation of the wireless channel. In the case of multi-antenna schemes, the benefits of simultaneous bit and power allocation are even greater compared with bit allocation only schemes. We have tested different LA algorithms in different multi-antenna systems. We have found that, if we cannot find the exact SNR thresholds due to different reasons, such as a reduced LA rate, CSI error, feedback delay, etc., it is better to fix the transmit power across all subchannels to guarantee the target FER. Otherwise, we can use adaptive power distribution to save power, which can be used for other purposes, or to increase the throughput of the system by transmitting a higher number of bits.

Owing to system complexity requirements, if we need to reduce the adaptation rates, then modulation and coding can be adapted at a slower rate if fast power control is applied in the time domain. This provides a satisfactory system performance compared with the required system complexity. We make the following conclusions.

1. If the LA window is quite narrow (i.e. short in the time domain), then we practically waste the transmit power if we fix the amount of power for each subchannel. If we can adjust the power, once the FER threshold is met based on our AMC threshold table, then there will be power saving without losing much of the spectral efficiency.

2. For power-nonadapted cases, if we adapt both FEC and modulation, then by introducing more switchable rates, the difference in wastage power for the cases with and without power adaptation will be smaller, thus less power will be wasted. This can be done by introducing more channel-coding rates. However, this will also bring other system-level complexities.

3 If the LA window is quite wide (i.e. long in the time domain), then faster PC rates can be used to compensate for the mismatch between actual SNR and modulation level assigned to any subchannel.

17.7.3 Future Work

In the future, the impact of different subchannel sizes in collaboration with different bit and power allocation rates can be studied. More importantly, studying these issues in a multi-user scenario will more clearly demonstrate the benefits of the power savings.

References

Ahn, C.J. and Sasase, I. (2002) The effects of modulation combination, target BER, Doppler frequency, and adaptation interval on the performance of adaptive OFDM in broadband mobile channel. *IEEE Transactions on Consumer Electronics*, **48**(1), 167–174.

Bohge, M., Gross, J. and Wolisz, A. (2005) The potential of dynamic power and subcarrier assignments in multi-user OFDM-FDMA cells. *Proceedings of the IEEE GlobeComm'05*, St. Louis, MO, pp. 2932–2936.

Chow, P.S. Cioffi, J. and Bingham, J.A.C. (1995) A practical discrete multitone transceiver loading algorithm for data transmission over spectrally shaped channel. *IEEE Transactions on Communication*, **43**(234), 773–775.

Chung, S.T. and Goldsmith, A.J. (2000) Adaptive multicarrier modulation for wireless systems. *Proceedings of the 34th Asilomar Conference on Signals, Systems and Computers*, Vol. 2, pp. 1603–1607.

Chung, S.T. and Goldsmith, A.J. (2001) Degrees of freedom in adaptive modulation: a unified view. *IEEE Transactions on Communications*, **49**(9), 1561–1571.

Czylwik, A. (1996) Adaptive OFDM for wideband radio channels. *Proceedings of the Global Telecommunications Conference*, Vol. 2, pp. 713–718.

Das, S.S. *et al.* (2007) Influence of PAPR on link adaptation algorithms in OFDM systems. *Proceedings of the Semiannual Vehicular Technology Conference*, Dublin, Ireland.

Erceg, V. *et al.* (1999) A model for the multipath delay profile of fixed wireless channels. *IEEE Journal of Selected Areas in Communications*, **17**(3), 399–410.

Figueiredo, D.V.P., de Carvalho, E., Deneire, L. and Prasad, R. (2006) Impact of feedback delay on rate adaptation for multiple antenna systems. *Proceedings of the IEEE PIMRC'06*, Helsinki, Finland.

Goldsmith, A. (2005) *Wireless Communications*. Cambridge University Press.

Hayes, J. (1968) Adaptive Feedback Communications. *IEEE Transactions on Communications*, **16**(1), 29–34.

Hunziker, T. and Dahlhaus, D. (2003) Optimal power adaptation for OFDM systems with ideal bit-interleaving and hard-decision decoding. *Proceedings of the IEEE ICC'03*, Vol. 5, pp. 3392–3397.

Hwang, Y.-T., Tsai, C.-Y. and Lin, C.-C. (2005) Block wise adaptive modulation for OFDM WLAN systems. *Proceedings of the IEEE International Symposium on Circuits and Systems*, Vol. 6, 6098–6101.

Larsson, P. (2007) Joint power and rate control for delay tolerant traffic in a wireless system. *Proceedings of the IEEE Semiannual Vehicular Technology Conference*, Dublin, Ireland.

Lei, M., Zhang, P., Harada, H. and Wakana, H. (2004) An adaptive power distribution algorithm for improving spectral efficiency in OFDM. *IEEE Transactions on Broadcasting*, **50**(3), 347–351.

Leke, A. and Cioffi, J.M. (1998) Multicarrier systems with imperfect channel knowledge. *Proceedings of the Ninth IEEE International Symposium on Personal, Indoor and Mobile Radio Communications*, Vol. 2, pp. 549–553.

Paulraj, A.J., Nabar, R. and Gore, D. (2003) *Introduction to Space–Time Wireless Communications*, 1st edn. Cambridge University Press, Cambridge.

Rappaport, T.S. (1996) *Wireless Communications Principles and Practice*. Prentice Hall, Englewood Cliffs, NJ.

Rhee, W. and Cioffi, J.M. (2000) Increase in capacity of multiuser OFDM system using dynamic subchannel allocation. *Proceedings of the IEEE Semiannual Vehicular Technology Conference*, Tokyo, Japan, pp. 1085–1089.

Siebert, M. and Stauffer, O. (2003) Enhanced link adaptation performance applying adaptive subcarrier modulation in OFDM systems. *Proceedings of the 57th IEEE Semiannual Vehicular Technology Conference*, Vol. 2, pp. 920–924.

Song, Z., Zhang, K. and Guan, Y.L. (2002) Statistical adaptive modulation for QAM-OFDM systems. *Proceedings of the IEEE Semiannual Vehicular Technology Conference*, Vol. 1, pp. 706–710.

Souryal, M.R. and Pickholtz, R. (2001) Adaptive modulation with imperfect channel information in OFDM. *Proceedings of the IEEE International Conference on Communications*, Vol. 6, pp. 1861–1865.

Tase, H., Ono S. and Hinamoto, T. (2005) Bit-rate maximization for multiuser OFDM systems. *Transactions of the Institute of Electronics Information and Communication Engineering A*, **J88-A**(3), 364–372.

Toyserkani, A.T., Naik, S., Ayan, J. *et al.* (2004) Subcarrier based adaptive modulation in HIPERLAN/2 system. *Proceedings of IEEE ICC'04*, Vol. 6, pp. 3460–3464.

Ye, S., Blum, R. and Cimini, L.J. (2002) Adaptive modulation for variable-rate OFDM systems with imperfect channel information. *Proceedings of the 55th IEEE Semiannual Vehicular Technology Conference*, Vol. 2, pp. 767–771.

Zhen, L. *et al.* (2002) Link adaptation of wideband OFDM systems in multi-path fading channel. *Proceedings of IEEE CCECE*, Vol. 3, pp. 1295–1299.

18

Applying WiMAX in New Scenarios: Limitations of the Physical Layer and Possible Solutions

Ilkka Harjula, Paola Cardamone, Matti Weissenfelt, Mika Lasanen, Sandrine Boumard, Aaron Byman and Marcos D. Katz

18.1 WiMAX in New Scenarios

While the WiMAX systems have been designed for Local Area Network (LANs) and Metropolitan Area Networks (MANs), the system parameters and algorithms have been designed to meet the requirements set by the radio channel in dense urban, suburban and rural areas. This is natural considering the fact that the majority of the potential customers reside in urban, densely populated areas. However, as the specification of the IEEE 802.16e-2005 standard, on which the WiMAX systems are based, is highly flexible, it is natural to raise a question of extending the usage environment from the urban areas to less-populated areas outside the urban environment.

As the WiMAX specification makes it possible to use high data rates for relatively long link distances, the number of applications that could be used in such systems outside the dense urban areas would be vast. This topic has been discussed for example in Information Society Technologies (IST) 6th Framework Programme (FP) project WiMAX Extension to

WiMAX Evolution: Emerging Technologies and Applications Edited by Marcos D. Katz and Frank H.P. Fitzek
© 2009 John Wiley & Sons, Ltd

Isolated Research Data Networks (WEIRD) (WEIRD, 2008), where an extension of WiMAX to isolated research networks was considered. In WEIRD, possible applications such as environmental monitoring, fire prevention and telemedicine were identified, and a network architecture supporting these applications was proposed. In WEIRD, the physical (PHY) and medium access control (MAC) layers of WiMAX were also studied from the viewpoint of extending the use of the standardized systems to uninhabited areas, namely mountainous areas covering application scenarios such as volcano monitoring in Iceland and Italy and fire prevention in Portugal.

The operational environment of the mobile WiMAX systems considered in this section consist of environmental monitoring devices transmitting over fixed WiMAX stations placed in the valleys and slopes of the mountains, as well as communications and data-mining applications used by the scientist visiting the environmental monitoring sites either by foot or relatively slowly moving vehicles. Therefore, we consider here slow mobility and a situation where the Line-of-Sight (LOS) connection between the Mobile Station (MS) and Base Station (BS) might or might not exist. The link distances might be relatively high, up to several kilometers, and the terrain is relatively free from buildings and similar obstacles, but instead characterized by the presence of the mountain slopes and valleys.

One of the major challenges of modeling a communication system in a given environment is choosing a channel model that accurately represents the prevailing transmission conditions affecting the performance of the system. In the next section, several publicly available channel models are reviewed in order to select the correct channel model for the novel WiMAX scenarios. These channel models include the European Cooperation in the Field of Scientific and Technical Research (COST) modes (Correia, 2001, Molisch and Hofstetter, 2006), Third Generation Partnership Project (3GPP) model (3GPP, 2003) and the Wireless World Initiative New Radio (WINNER) Phase I and II channel models (WINNER, 2008). As can be seen from that section, none of the channel models was designed to specifically model mountainous environments. Therefore, the WINNER Phase I channel model developed was used as a starting point, and analytical tools were used for deriving an extension to the existing Matlab channel model in section 18.3 to model the transmission conditions in the mountainous environment (Cardamone *et al.*, 2008).

The IEEE 802.16-2005 standard offers a variety of tools for tailoring the transmission for different operational environments including several levels of modulation, variable Cyclic Prefix (CP) length, variety of error-correction coding rates and types, and different Multiple Input Multiple Output (MIMO) technologies including Space–Time Coding (STC), spatial multiplexing, and beamforming just to mention a few. In the mountainous environment, the dominant features of the channel model are long channel delay spread and highly directive angle-of-arrival of the received signal. Therefore, scarce signal bandwidth must be sacrificed for extensive CP, and multi-antenna techniques are recommended to deal with the spatial properties of the channel. As a result, two beamforming techniques, namely pre- and post-FFT receiver beamforming, are proposed as solutions for the mountainous environments in Sections 18.4.1 and 18.4.2 and their performance is studied with Matlab simulations in Section 18.4.3 (see Harjula *et al.* (2008)).

The mountainous environment may also be challenging for synchronization between the BS and MS. We study the performance of Downlink (DL) timing synchronization in such a scenario in Section 18.5. We use a practical hardware (HW) system in the study and demonstrate various approaches that may be used for synchronization purposes.

18.2 Channel Model for Mountainous Environments

As the radio channel places fundamental limits on the current wireless communications systems, the modeling of the radio channel has been under active study for a relatively long period of time. Starting from Maxwell's well-known equations for the electromagnetic wave propagation from 1861 (Maxwell, 1861) and pioneering work by Bello (1963, 1969) and Kailath (1963), the work has been continuing actively to recent days. While the fundamentals of the channel modeling have been derived for the Single Input Single Output (SISO) channel, the modeling of the MIMO channel has been gaining increased attention in the recent years.

In the following sections, we briefly present some recent standardized or publicly available radio channel models for outdoor environments. The study has been limited to MIMO channel models, and the focus has been on searching for a suitable channel model for modeling the radio environment for mountainous environment. The channel models presented in this section include COST 259 and COST 273 channel modes, 3GPP and 3GPP2 Spatial Channel Models (SCMs), Stanford University Interim (SUI) models, models defined for the IEEE 802.16a standard and the channel model developed in the WINNER Phase I and II projects.

18.2.1 COST 259/273

According to COST (2008), COST is the longest-running instrument supporting co-operation among scientists and researchers across Europe. In the telecommunications field, COST is most probably best known for the standardized channel models, referred to as COST 259 and COST 273 (Correia, 2001, Molisch and Hofstetter, 2006). Sometimes the 3GPP SCM and the WINNER Phase I channel model are considered as the subsets of the very general COST 259.

18.2.2 3GPP/3GPP2 Statistical Channel Model

The 3GPP/3GPP2 SCM (3GPP, 2003) was developed for evaluating different MIMO concepts in outdoor environments at a frequency of 2 GHz with a system bandwidth of 5 MHz. This model is divided into two parts: link-level evaluation for calibration purposes and system-level evaluation for simulations. The link-level calibration model can be implemented either as a physical model or as an analytical model. The first is a nongeometrical stochastic physical model. It describes the wideband characteristics of the channel as a tapped delay line, and all of the paths are assumed to be independent. Each resolvable path is characterized by its own spatial channel parameters, such as power azimuth spectrum which is Laplacian or uniform, angular spread and mean direction, at both the BS and the MS. The parameters are fixed, thus the model shows stationary channel conditions. The speed and the direction of the MS characterizes implicitly the Doppler spectrum. The physical model can be transformed into an equivalent analytical model using the antenna configurations defined in the model itself.

18.2.3   SUI Models and IEEE 802.16a Channel Models

The IEEE 802.16a channel model is based on the SUI channel model (Erceg *et al.*, 2001). Therefore, they share mostly the same features, and they are therefore discussed together in this section. However, there are some differences between the channel models, and these are also pointed out in this discussion.

The common features regard the cell radius of less than 10 km, the system bandwidth from 2 to 20 MHz, the BS height varying from 15 to 40 m, and the use of a fixed user end antenna. The channel models can be used in the 1 to 4 GHz range. The concept of MIMO and directional characteristics of the transmission are not considered in the standard, but subsequent studies have taken these elements into account (COST, 2008).

The IEEE 802.16a channel model is based on a modified version of the SUI channel model, valid for both omni-directional and directional antennas. The second aspect causes an increase in the Ricean factor, but a decrease in the delay spread. The model also includes different path loss models, the K-factor model and introduces an antenna gain reduction factor which considers the reduction of the antenna gain due to a scattering effect.

18.2.4   WINNER Phase I and II Channel Models

The channel model developed in the WINNER Phase I (WINNER, 2008) project in phase 1 is focusing on the beyond 3G radio system using a frequency bands up to 100 MHz for the frequency band between 2 and 6 GHz. The channel model is related to both the COST 259 model and the 3GPP SCM model. The WINNER Phase I channel model is based on seven different propagation scenarios for indoor and outdoor environments. As a 3GPP SCM model, it is divided into link-level simulation and system-level simulation models. The generic channel model is a geometric-based stochastic channel model and most of its features are based on the principle of the 3GPP SCM model. Actually, these aspects are valid for six scenarios; the other scenario is not considered in the generic model since it is a stationary wireless feeder scenario, where transmitted and receiver ends are fixed. It is modeled separately as Clustered Delay Line (CDL) model. Various measurement campaigns provide the parameterization of the scenarios for both LOS and Non-Line-of-Sight (NLOS) conditions.

From the WINNER Phase I channel model, the scenario D1 defined for the flat rural area seems very interesting from the point of view of this work. The possibility of adjusting the center frequency to 3.5 GHz which is the center frequency for mobile WiMAX used in the testbeds in the WEIRD project is a very appealing property. The Matlab code is readily available at the WINNER Web site (WINNER, 2008).

At the time of writing the WINNER Phase II channel models had also became public (WINNER, 2008). According to their Web page, the WINNER Phase II channel model includes the model parametrization based on measurements, 17 model scenarios (including indoor), stochastic models for system-level simulations, fixed CDL models for calibration simulations, network layout visualization tool autocorrelation of large-scale channel parameters and support for arbitrary 3D antenna geometry and field patterns. However, as the work in this section is based on the WINNER Phase I channel model, the reader is advised to look for more information on the WINNER Phase II Web site (WINNER, 2008).

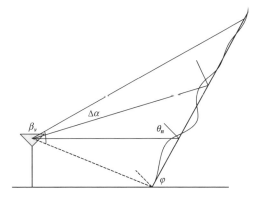

Figure 18.1 Mountain slope geometry.

18.3   Mountainous Scenario and Channel Modeling

18.3.1   Analytical Modeling of the Channel in the Presence of Mountains

An analytical way of modeling the channel in the presence of a mountain slope was presented in Cardamone *et al.* (2008). In mountainous terrain, high peaks of the mountain may be visible to both the transmitter and the receiver and act as large reflectors. The resulting multipath signal components may contain a significant amount of received signal power, and the signal might be significantly delayed relative to the direct signal causing a spread over a wide range of delays. Several studies have adopted a model using the bistatic radar equation and the concept of normalized scattering cross section of the mountain slope to predict the path losses and relative delays of the indirect paths from transmitter via mountain to receiver (Driessen, 2000, Thomson and Carvalho, 1978), and in Cardamone *et al.* (2008) this basic model has been used as a starting point for implementing a channel model for the mountainous environment. For calculating the reflection parameters, the well-known Lambertian scattering model (Lambert, 1760) was used. The general geometry used in the analysis for the Lambertian model in Cardamone *et al.* (2008) can be seen in Figure 18.1.

18.3.2   Extension of the WINNER Phase I Channel Model for the Mountainous Scenario

Generally speaking, the environments that have been modeled in the WINNER I channel model are typically found in urban areas of European and North-American countries, and it is not intended to cover all possible environments and conditions such as mountainous or hilly rural environments. However, by adjusting the initialization parameters the desired kind of channel coefficients matrix can be created. Aiming at being as realistic as possible, we have decided to use an actual environment to model our channels. Our environment considers Mount Vesuvius, a volcano near Naples, in Italy. The details of this environment can be found in the WEIRD project deliverable (Dinis *et al.*, 2006). This mountainous scenario can generally be represented as a rural terrain with a great mountain, acting as a reflector in the background. Since the reflector is obviously fixed, the starting point of the model is deterministic, that is, the powers and the delays are not considered to be extracted from a Probability Distribution Function (PDF), but calculated in a deterministic manner.

One of the partners of the WEIRD project is the Osservatorio Vesuviano, and this institute gave us the maps with the locations of their sensors which are used to monitor the activity of the volcano and for the purposes of fire, eruption and earthquake prevention. By using these maps, we have chosen one fixed sensor as our BS and one of the mobile sensors as the MS. We also had the geographical coordinates of each sensor and of Mount Vesuvius. Thus, we have used this information to compute the distances between the BS and MS and between them and the mountain.

The results of the channel modeling in Matlab can be seen in Figure 18.2, which shows the delay spread of the paths due to the mountainous environment. In this Matlab simulation with WINNER Phase I channel model enhanced with the analytical mountain model, a NLOS path between source and receiver has been assumed. As we can see, the mountain acts as a large reflector resulting in long-delayed echoes which are exponentially distributed. In order to relate the timescale to the WiMAX system, all of the possible WiMAX cyclic prefix sizes are remarked: it simple to see that only a CP of $1/4$ ($22.85\ \mu s$) can be used without affecting the performance of the IEEE 802.16e-2005-based system. On the other hand, using the longest size for the CP means that a great fraction of the power is assigned to the transmission of the CP itself and cannot be used for the data-carrying symbols. In addition, the simulation results with the WINNER Phase I D1 scenario are presented in this Figure 18.2, and the division of power is assumed to be equal between the D1 and the reflections from the mountain.

18.4   Beamforming Algorithms and Simulation

MIMO channels arise in many different scenarios in communications, such as modeling frequency selective channel using transmit and receive filter banks, treating a bundle of twisted pairs in digital subscriber lines as a whole, or when using multiple antennas at both sides of the wireless link (Palomar *et al.*, 2003). This work focuses on MIMO channels arising from the use of multiple antennas at both sides of the wireless link since they offer a significant capacity increase over SISO channels in certain channel conditions (Foschini and Gans, 1998, Telatar, 1999). Good references for the MIMO techniques in general include Giannakis *et al.* (2008) and Goldsmith (2005).

In recent years, the trend in the MIMO studies has been more towards the multiuser MIMO techniques. These systems consist of several BS and MS, each equipped with several antennas and operating in the same frequency bands while sharing the same spatial resources. A good overview of these techniques is given in Gesbert *et al.* (2007), and the same topic has been approached from a viewpoint of spatial diversity in Diggavi *et al.* (2004). However, in this section the focus is in a cell with a single user or several users sharing the spectrum by using Orthogonal Frequency Division Multiple Access (OFDMA), and sharing the properties of the mountainous environment.

The IEEE 802.16-2004 standard (IEEE, 2004) with the IEEE802.16-2005 standard amendment (IEEE, 2005) offers a variety of tools for MIMO processing including STC, spatial multiplexing and beamforming. The WirelessMAN-OFDMA PHY layer definition

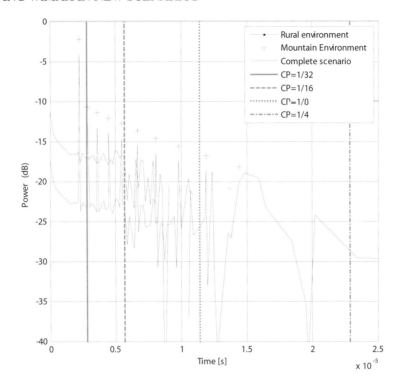

Figure 18.2 Delay spread of the mountainous and WINNER Phase I D1 scenarios.

in IEEE (2005) is the most sophisticated from that point of view while the WirelessMAN-OFDM PHY layer specification includes a smaller set of algorithms for MIMO processing. The WirelessMAN-SCa PHY and WirelessMAN-SC PHY specifications are the least suitable for the multi-antenna transmission.

The support for the MIMO techniques is provided in the IEEE802.16-2005 standard by dividing the transmission frame into several parts, referred to as zones. The first zone is used for single antenna transmission, while the latter zones can be used for Space–Time Coded, spatially multiplexed or beamformed signaling. The latter zones can be also used for some other transmission schemes, such as optional Partially Used Subcarriers (PUSCs) and Fully Used Subcarriers (FUSCs), and Adaptive Modulation and Coding (AMC), some of them including a possibility for spatial processing.

18.4.1   Pre-FFT Receive EVD Beamforming

Pre-FFT receiver beamforming for Orthogonal Frequency Division Multiplexing (OFDM) systems has been introduced in several papers (Budsabathon *et al.*, 2004, Kim *et al.*, Lei *et al.*, 2004a,b). Usually, the criterion for selecting the weights for the beamformer is max-SNR, which can be achieved by computing the dominant eigenvector of the spatial cross-correlation matrix **R** of the received signal vector in time domain, before Fast Fourier Transform (FFT), and using the dominant eigenvector as the weight vector w.

More practically this can be achieved by calculating the inverse of the sample matrix gathered from the different receive antennas (Lei *et al.*, 2004a, Matsuoka and Shoki, 20003).

This technique is applicable for both the OFDM and OFDMA systems presented in the IEEE 802.16-2005 standard. It should be noted anyway that since this technique is capable of maximizing only the total received SNR, it might not be optimal for the Uplink (UL) reception in OFDMA systems, where the subcarriers within an OFDM symbol might originate from several different users and therefore are affected with different channels and thus arrive from different directions. Optimizing the pre-FFT beamformer for OFDMA UL reception might be an interesting task for the future.

18.4.2   Post-FFT Receive EVD Beamforming

Instead of processing the received signal in time domain, the beamforming can be executed in the frequency domain (Alam *et al.*, 2004, Matsuoka and Shoki, 20003). This post-FFT beamformer performs better than the pre-FFT beamformer simply because of the fact that it is capable of maximizing the SNR separately for each of the received subcarriers. The method for calculating the weights is similar as in pre-FFT receiver beamforming, but in this case the matrix $\mathbf{R}$ is describing the spatial cross-correlation of the frequency domain subcarrier instead of the whole received time domain signal. As in pre-FFT beamformer, the dominant eigenvector of $\mathbf{R}$ is used as the frequency domain weight vector w for each subcarrier.

As well as pre-FFT beamforming, the post-FFT beamforming is directly applicable to the OFDM and OFDMA systems presented in the IEEE 802.16-2005 standard. However, because of the capability of processing the data subcarrier-by-subcarrier basis, it is more suitable for multi-user OFDMA system since it can steer the beam towards all of the desired users at the same time instant. The drawback of this method, compared with the pre-FFT method is the increased required computational complexity which will be discussed in more detail later on in this work.

18.4.3   Simulation Results

The simulation results presented in this section were generated using Matlab. The channel model used in the simulations was either WINNER Phase I D1 channel model, or WINNER Phase I D1 channel model extended with the novel channel model presented earlier in this chapter. In the simulations a scaled-down version of the WiMAX uplink transmission parameters were used. The carrier frequency was at 3.5 GHz, the inverse fast Fourier transform (IFFT) size was set to 512, and no error-correction coding was used. The sampling rate used in these simulations was one. Cyclic prefix sizes of 64 and 128 were used, corresponding to 1/8 and 1/4 of the IFFT length. Perfect synchronization was assumed as well as perfect channel estimation, except for the results presented in Figures 18.6 and 18.7, where the effect of imperfect channel estimation was studied. The channel parameters were set to relatively slow mobility, MS speed of 10 km h^{-1} was used, and the channel was kept constant during the transmission of a single OFDM symbol.

The pre- and post-FFT Eigenvalue Decomposition (EVD) beamforming was studied in a channel consisting of the novel channel model and WINNER Phase I D1 scenario (Harjula *et al.*, 2008). The Bit Error Rate (BER) results are presented in Figure 18.3 as a function of E_b/N_0. It can be seen from the figures, that the pre-FFT beamforming improves the system

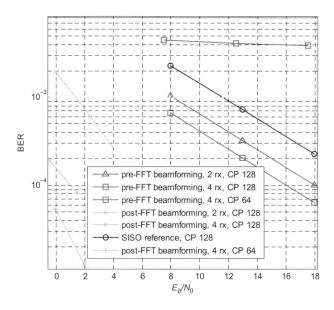

Figure 18.3 E_b/N_0 versus BER for pre- and post-FFT EVD beamforming in a novel mountain model and WINNER I D1 NLOS fading channel.

performance also under these more demanding channel conditions. It can be also seen that pre-FFT technique is not capable of compensating for the too short CP size. The post-FFT results demonstrate superior performance over the pre-FFT technique, and it can be seen that it is capable of providing a much better performance than the pre-FFT beamformer also in the case of CP shorter than the channel impulse response.

The computational complexity of the post-FFT beamforming is high compared with the computational complexity of the pre-FFT beamforming. Most of the complexity is caused by the fact that the EVD of the channel matrix has to be calculated separately for every subcarrier, while in the pre-FFT beamforming only one EVD calculation is required. In order to reduce the computational complexity of the post-FFT beamforming, the clustering of the beamforming weights was studied. In this method, the subcarriers are divided into clusters of arbitrary size, and the beamforming weight is calculated only once for each cluster. Since the frequency response of the channel is correlated between the antennas, the clustering will cause some performance degradation but not destroy the system performance completely.

The calculation of the weight vector for each cluster was done in a simple manner. The subcarrier in the middle of the cluster was selected, and the weight vector was calculated for this subcarrier. This weight vector was then used for all of the subcarriers in the cluster. No interpolation was used to compensate for the difference between the calculated weight vector and the actual weight vector that should be used in the subcarriers close to the edge of the cluster. This simple technique is by no means optimal in any sense, but is somewhat revealing of the lower bound of the performance for this technique.

Simulation results for the clustered post-FFT beamforming are presented in Figure 18.4 and Figure 18.5 for four and two receiver antennas, respectively. The reference scenario is

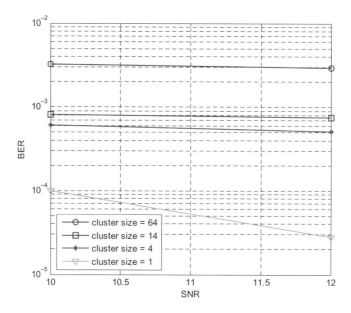

Figure 18.4   SNR versus BER for post-FFT EVD beamforming with clustering and 4 rx antennas in a WINNER I D1 NLOS fading channel.

presented with the line with triangles and it corresponds to the case where the weight vector is calculated separately for each subcarrier. Simulation results for cluster sizes 4, 14 and 64 are presented in the other lines. A cluster size of 4 corresponds to the size of UL tile defined in IEEE (2004), and a cluster size of 14 corresponds to the size of DL cluster in IEEE (2004). A cluster size of 64 is studied to see the performance with a relatively large cluster size. It can be seen from the figures that the post-FFT beamforming is highly sensitive to the clustering of the subcarriers, and the clustering seems to cause an error floor to different BER levels depending on the cluster size. This might be caused by the high-frequency sensitivity of the channel model used in the simulations.

The sensitivity of the pre- and post-FFT beamforming algorithms to the channel estimation errors was also studied. The channel estimation errors were modeled as Gaussian Independent and Identically Distributed (IID) noise, which is the worst-case estimation error assumption for the linear Minimum Mean Squared Error (MMSE)-type channel estimator. A detailed explanation and justification for this kind of approach can be found, for example, in Tölli (2008).

The simulation results for pre- and post-FFT beamforming with imperfect channel estimates are presented in Figure 18.6 and Figure 18.7, respectively. The results are present as BER versus SNR graphs for two receiver antennas, and the different curves indicate the different amount of estimation error as a function of Channel-to-Noise ratio (CNR). In this case the noise refers to the noise used to model the channel estimation error, not the noise used in the AWGN channel. The results indicate that the post-FFT beamformer is far more sensitive to the estimation errors than the pre-FFT beamformer. While the CNR value of 5 dB is enough providing the receiver with pre-FFT beamformer a nondegraded performance, the

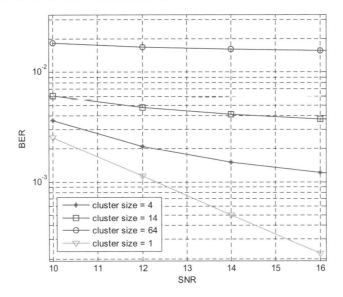

Figure 18.5 SNR versus BER for post-FFT EVD beamforming with clustering and 2 rx antennas in a WINNER I D1 NLOS fading channel.

same amount of error degrades the performance of the receiver with post-FFT beamforming more than 2.5 dB at the BER level of 10^{-3}.

18.5   A Timing Synchronization Study in a Mountain Environment

In Section 18.4, ideal synchronization was assumed in a mountain environment. Here we are interested in timing synchronization performance in the same environment. We concentrate on DL synchronization that enables a MS to set and maintain its time-base to receive the BS's transmission. In addition, we assume that OFDMA system parameters such as the frame period, FFT size and cyclic prefix period are known. In practice, these parameters are identified from the DL transmission. For UL communications, further ranging procedures based on MS transmissions are supported by the IEEE 802.16e-2005 standard to adjust UL transmission parameters. We use a field-Programmable Gate Array (FPGA) hardware environment and channel emulator in the study.

The IEEE 802.16e-2005 preamble starts each frame transmitted by a WiMAX BS. We emphasize the repetitive structure of the preamble consisting of near-identical Synchronization Patterns (SPs) in Figure 18.8. In addition, CP is a copy of the last G samples of the OFDM symbol. A conventional synchronization approach employs repetition in the preamble by correlating received samples that have a distance of L samples. Location of the maximum correlation gives ideal timing for OFDM symbols. A block diagram for this approach is shown in Figure 18.9. In addition to a correlation **C** calculated in the upper signal path, an energy signal **E** is needed to detect whether a signal is present or not. The approach is

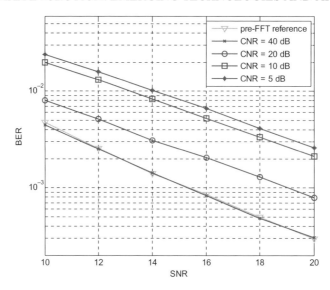

Figure 18.6 SNR versus BER for pre-FFT EVD beamforming with channel estimation errors in a WINNER I D1 NLOS fading channel.

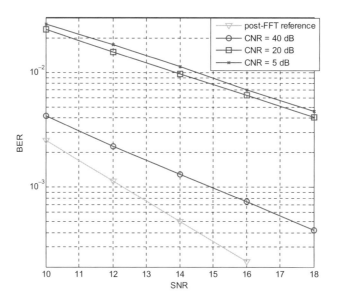

Figure 18.7 SNR versus BER for post-FFT EVD beamforming with channel estimation errors in a WINNER I D1 NLOS fading channel.

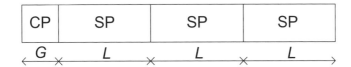

Figure 18.8 Repetition preamble OFDM symbol.

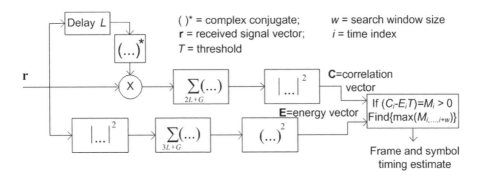

Figure 18.9 Timing synchronization block diagram.

adopted from Schmidl and Cox (1997). Timing decisions are made from a timing metric **M** defined in Figure 18.9. In addition to **C** and **E**, the timing metric uses a threshold T during whole synchronization process as proposed in Lasanen *et al.* (2002). In calculating **M** in Figure 18.9, we do not consider normalizations of **C** and **E** due to different summation periods because this can be taken into account in T in practice.

Timing synchronization has two phases in our model. First we use moving average of individual one-shot timing estimates to define an accurate timing location. For the next one-shot estimate, we wait a little less than a frame period from the current timing point before re-starting procedures shown in Figure 18.9. In this way, we try to minimize the risk of false alarms.

After having the moving average of one-shot timing estimates converged, we go to a tracking state and start updating the moving average in a slightly different way. In the tracking state, we allow the next one-shot timing estimate to be used to update the current moving average if it is inside, for instance, $\pm G/4$ sample windows from the current moving average. In this way, we cancel the effect of very weak estimates that are typical for low SNRs and, hence, deep channel fades. Furthermore, we can use various options in setting up the actual timing synchronization location. We can emphasize the last individual timing estimate and moving average results differently in this process. These options are beyond the scope of the study reported here.

For measurements, we use following frame structure and parameters. Some of these may differ somewhat from those defined in the IEEE802.16e standard. The preamble depicted in Figure 18.8 is the first OFDM symbol of the frame. Then, we have two OFDM symbols to represent frame configuration information. Both symbols have 48 active subcarriers. These are followed by eight data-carrying OFDM symbols that are filled with data and pilots.

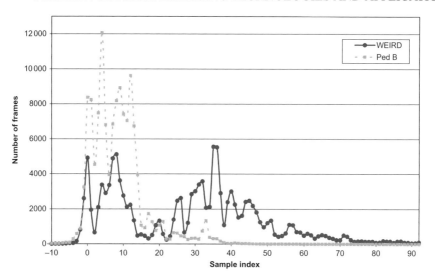

Figure 18.10   Distributions of one-shot estimates for the frame start for WEIRD and pedestrian B channel models with SNR = 10 dB.

The remaining UL period is silent. The frame period is about 2.5 ms. The bandwidth of the OFDM signal is 7 MHz. With an 8 MHz sampling rate, 1024-point (I)FFT is used. The CP period corresponds to 1/4 of FFT size. Therefore, the OFDM symbol period is 160 μs in total with 32 μs CP. We use the largest CP period due to the long delay spread of the channel.

The preamble has been at a level about 4 dB higher than the data OFDM symbols. In measurements, we have used SNRs of about 0 and 10 dB for the preamble when measured from 7 MHz signal bandwidth. In the implementation, we have a 16 MHz sampling rate, that is, oversampling by two is used. Synchronization is performed at this rate meaning that summation periods related to FFT size and cyclic prefix are doubled when compared with the case in Figure 18.9. The number of transmitted frames is 128 000. Mobility speed is about 5 km h^{-1} and a carrier frequency of 2.2 GHz is assumed. We have used threshold of 0.3 corresponding to definitions in Lasanen *et al.* (2002) and Schmidl and Cox (1997) and a search window of 1536. We show histogram-like distributions for the estimated frame start positions in Figures 18.10–18.12. Between measurements cases, we do not have exact calibration for an ideal synchronization point that may also be quite hard to define for a multipath channel. In results, very small negative sample indices or large positive sample indices may indicate that the frame start estimate is somewhat early or late, respectively. The results have been collected with a logic analyzer from the HW.

In Figure 18.10, we show distributions of estimated one-shot frame start positions for both the WEIRD and the ITU pedestrian B channel models with a SNR of 10 dB. The sample area shown corresponds to 20% of the cyclic prefix period. These results suggest that one-shot synchronization would give a high enough accuracy so that both timing inaccuracy and intersymbol interference from the used multipath channel models can be compensated for with a long cyclic prefix. This is true especially with the pedestrian B channel model. On the other hand, some estimates can have values outside the window used in the figure. In addition,

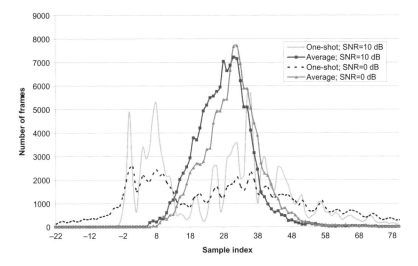

Figure 18.11 Distributions of one-shot and averaged estimates for frame start with SNRs of 0 and 10 dB.

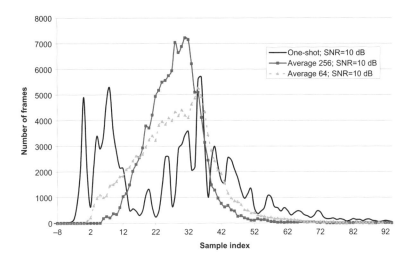

Figure 18.12 Distributions of one-shot and averaged estimates for averaging periods of 64 and 256 frames with SNR = 10 dB.

false alarms or frame misses may take place depending on the threshold value. A natural reason for having a large area for estimation results with the WEIRD channel model is in the long delay spread of the channel.

Next we demonstrate how the deviation of individual timing estimates can be made smaller for timing synchronization. In Figure 18.11, we present frame start estimates that have been averaged over 256 successive frames. We use SNRs of 0 and 10 dB. If the threshold has not been exceeded during the time period of $\pm G/4$ from the current synchronization point, the average has not been updated. For this case, the previous average could be used in timing synchronization. The results show that averaging reduces the variation of estimates considerably. Interestingly, we obtain quite a similar curve with both SNRs via averaging. This may be explained by the limited time period over which the new estimate is accepted for the average. This hypothesis is supported by the fact that the number of frames not exceeding a threshold within the window are about 6500 (5%) and 470 (0.4%) when comparing SNRs of 0 and 10 dB. Furthermore, one-shot results for the lower SNR are dominated less by two peaks that have distances of tens samples. This may also result in a denser average distribution for the lower SNR.

Finally, we briefly study the effect of different averaging periods for the estimates in Figure 18.12. In practice, a short averaging period can be desirable to be able to follow for example, delay changes due to movements of the MS. As expected, averaging over 64 frames gives a much more concentrated plot than one-shot results. The gain for using 256 frames instead of 64 appears more modest in this case.

18.6   Analysis and Conclusions

We have presented the extension of the WiMAX PHY to a mountainous environment. An analytical model for a radio channel for a link operating in the proximity of a mountain slope has been derived, and the channel model was also tested with simulation tools. The simulations demonstrated that the presence of the mountain increases the length of the channel delay spread and also affects the angular distribution of the received signal by concentrating more power in the direction of the mountain slope. These effects might cause performance degradation in WiMAX systems in a form of extended CP length or, in the worst case, even total loss of the connection if the peaks in the channel impulse response exceed the CP length destroying the orthogonality of the subcarriers.

As one possible solution to the problem caused by the presence of the mountain slopes, two MIMO algorithms, namely pre- and post-FFT EVD beamforming, were selected for closer study. The simulation results indicate that both of the techniques were effective for increasing the received signal quality and thereby improving the system performance. On the other hand, only the more complicated and sensitive post-FFT algorithm was capable of dealing with the channel impulse response that was exceeding the CP length.

We studied also DL timing synchronization in the WEIRD channel model. A practical FPGA implementation was used in measurements. We showed that one-shot frame start estimates have a large deviation that is mostly manageable with the long cyclic prefix. Furthermore, we have demonstrated that the deviation can be diminished considerably via averaging and leaving out some weak estimates. These methods may also be used to maintain synchronization during deep channel fades and avoid false alarms in synchronization. Further work could consider including multiple receiver antenna support for synchronization as illustrated, for example, in Czylwik (1999), Schellmann *et al.* (2005) and study these with the beamforming methods discussed above.

The results presented in this chapter indicate that extending the WiMAX use scenarios to more remote locations might place limitations and challenges on the design of the WiMAX

PHY layer algorithms. The use of MIMO techniques, especially beamforming, might offer a solution for providing the end user with reliable operation even in a challenging radio environment.

References

3GPP (2003) 3rd Generation Partnership Project; technical specification group radio access network; spatial channel model for MIMO simulations (release 6), *3GPP Technical Report 25.996, V6.1.0.*

Alam, F., Cheung, B.L.P., Mostafa, R., Newhall, W.G. and Reed, J.H. (2004) Sub-band beamforming for OFDM system in practical channel condition. *Proceedings of the IEEE Semiannual Vehicular Technology Conference*, Vol. 1, pp. 235–239.

Bello, P.A. (1963) Characterization of randomly timevariant linear channels. *IEEE Transactions on Communication Systems*, **11**, 360–393.

Bello, P.A. (1969) Measurement of random time-variant linear channels. *IEEE Transactions on Information Theory*, **15**, 469–475.

Budsabathon, M., Hara, Y. and Hara, S. (2004) Optimum beamforming for pre-FFT OFDM adaptive antenna array. *IEEE Transactions on Vehicular Technology*, **53**, 945–955.

Cardamone, P., Harjula, I., Katz, M. and Mucchi, L. (2008) Proposal of a New WiMAX channel model for mountainous areas. *Proceedings of the IEEE Semiannual Vehicular Technology Conference*.

Correia, L.M. (2001) *Wireless Flexible Personalized Communications – COST259: European Co-operation in Mobile Radio Research*. John Wiley & Sons Ltd, Chichester.

COST 273 Web page, http://www.ftw.at/cost273 (accessed 2008).

Czylwik, A. (1999) Synchronization for systems with antenna diversity. *Proceedings of the IEEE Semiannual Vehicular Technology Conference*, Vol. 2.

Diggavi, S. N., Al-Dahir, N., Stamoulis, A. and Calderbank, A.R. (2004) Great expectations: the value of spatial diversity in wireless networks. *Proceedings of the IEEE*, **92**(2), 219–270.

Dinis, M., Neves, P., Angori, E. *et al.* (2006) Deliverable D2.1; System Scenarios, Business Models and System Requirements. *IST-034622-IP WEIRD deliverable D2.1.*

Driessen, P.F. (2000) Prediction of multipath delay profiles in mountainous terrain. *IEEE Journal on Selected Areas in Communications*, **8**(3), 336–346.

Erceg, V., Hari, K.V.S., Smith, M.S., Baum, D.S. *et al.* (2001) Channel models for fixed wireless applications. Contribution IEEE 802.16.3c-01/29r1, February.

Foschini, G. and Gans, M. (1998) On limits of wireless communications in a fading environment when using multiple antennas. *Wireless Personal Communications*, **6**, 311–335.

Gesbert, D., Kountouris, M., Heath, Jr., R.W. Chae, C.-B. and Sälzer, T. (2007) Shifting the MIMO paradigm. *IEEE Signal Processing Magazine*, September 2007, pp. 36–46.

Giannakis, G.B., Liu, Z., Ma, X. and Zhou, S. (2008) *Space–Time Coding for Broadband Wireless Communications*. John Wiley & Sons Ltd, Chichester.

Goldsmith, A. (2005) *Wireless Communications*. Cambridge University Press, Cambridge, pp. 644.

Harjula, I., Cardamone, P., Katz, M. and Albiero, F. (2008) MIMO Processing for WiMAX in Challenging Radio Environment. *Proceedings of the ICT Mobile Summit 2008*.

IEEE (2004) IEEE standard for local and metropolitan area networks part 16: air interface for fixed broadband wireless access systems, *IEEE Standard 802.16-2004.*

IEEE (2005) IEEE standard for local and metropolitan area networks part 16: air interface for fixed broadband wireless access systems. Amendment 2: physical and medium access control layers for combined fixed and mobile operation in licensed bands. *IEEE Std 802.16e™-2005 and IEEE Std 802.16™-2004/Cor1-2005 (Amendment and Corrigendum to IEEE Std 802.16-2004).*

Kailath, T. (1963) Time-variant communication channels. *IEEE Transactions on Information Theory*, **9**, 233–237.

Kim, C.K., Lee, K. and Cho, Y.S. (2000) Adaptive beamforming algorithm for OFDM systems with antenna arrays. *IEEE Transactions on Consumer Electronics*, **46**(4), 1052–1058.

Lambert, J.H. (1760) *Photometria sive de mensure de gratibus luminis, colorum umbrae*. Eberhard Klett.

Lasanen, M., Rautio, J. and Nissilä, M. (2002) Timing Synchronization of the WIND-FLEX OFDM Prototype. *Proceedings of the IST Mobile Wireless Telecommunications Summit*.

Lei, M., Zhang, P., Harada, H. and Wakana, H. (2004a) Adaptive beamforming based on frequency-to-time pilot transform for OFDM. *Proceedings of the IEEE Semiannual Vehicular Technology Conference*, Vol. 1, pp. 285–289.

Lei, M., Zhang, P., Harada, H. and Wakana, H. (2004b) A combinational scheme of pre-FFT adaptive beamforming and frequency-domain adaptive loading for OFDM. *Proceedings of the IEEE Semiannual Vehicular Technology Conference*, Vol. 1, pp. 290–294.

Matsuoka, H. and Shoki, H. (2003) Comparison post-FFT and pre-FFT processing adaptive arrays for OFDM systems in the presence of co-channel interference. *Proceedings of the PIMRC2003*, Vol. 2, pp. 1603–1607.

Maxwell, J.C. (1861) On physical lines of force. *The Philosophical Magazine*, **XXI** (as quoted in Simpson T.K. (1997) *Maxwell on the Electromagnetic Field*. Rutgers University Press, 1997, pp. 146).

Molisch, A.F. and Hofstetter, H. (2006) The COST 273 channel model. *Mobile Broadband Multimedia Networks*, ed. Correia, L. Academic Press.

Palomar, D.P., Cioffi, J.M. and Lagunas, M.A. (2003) Joint transmit-receive beamforming design for multicarrier MIMO channels: a unified framework for convex optimization. *IEEE Transactions on Signal Processing*, **51**(2), 2381–2401.

Schellmann, M., Jungnickel, V. and Helmolt, C. (2005) On the value of spatial diversity for the synchronisation in MIMO-OFDM systems. *Proceedings of the IEEE PIMRC 2005*, Vol. 1.

Schmidl, T.M. and Cox, D.C. (1997) Robust frequency and timing synchronization for OFDM. *IEEE Transactions on Communications*, **45**(12).

Telatar, I.E. (1999) Capacity of multi-antenna Gaussian channels. *European Transactions on Telecommunications (ETT)*, **10**, 585–595.

Thomson, W. E. and Carvalho, P.F. (1978) VHF and UHF links using mountains as reflectors. *IEEE Transactions on Communications*, **26**(3), 391–400.

Tölli, A. (2008) Resource management in cooperative MIMO-OFDM cellular systems. *PhD dissertation*, University of Oulu, Acta Universitatis Ouluensis, Oulu, pp. 196.

WEIRD project web page, http://www.ist-weird.eu/ (accessed 2008).

WINNER project Web page, http://www.ist-winner.org/ (accessed 2008).

19

Application of Radio-over-Fiber in WiMAX: Results and Prospects

Juan Luis Corral, Roberto Llorente, Valentín Polo,
Borja Vidal, Javier Martí, Jonás Porcar, David Zorrilla and
Antonio José Ramírez

19.1 Introduction

19.1.1 Radio-over-Fiber systems

The increasing demand for broadband communication systems to deliver multimedia content to a growing number of users that must be able to access ICT content in a seamless and ubiquitous manner has pushed the development of several standards, both wireline and wireless, as discussed by Stuckmann and Zimmermann (2007). Wireless access has experienced a high degree of development due to their intrinsic benefits, such as mobility, fast deployment and quick revenue. The proliferation of wireless technologies and the spectrum scarcity has pushed the need to develop efficient means to transport and distribute the wireless signals remotely, avoiding the use of costly equipment to implement the wireless backhaul network in the millimeter-wave band. This technology is called microwave photonics, as in Seeds and Williams (2006), or Radio-over-Fiber (RoF) in a more system-oriented denomination. RoF technology offers a cost-effective scalable and transparent solution to implement transport architectures to distribute wireless signals in both indoor and outdoor environments.

RoF systems comprise broadband optical sources either based on direct or external modulation, a suitable transmission media such as Multi-Mode Fiber (MMF), Singlemode Fiber (SMF) or Plastic Optical Fiber (POF), and equally broadband photodetectors or photoreceivers (Dagli, 1999).

WiMAX Evolution: Emerging Technologies and Applications   Edited by Marcos D. Katz and Frank H.P. Fitzek
© 2009 John Wiley & Sons, Ltd

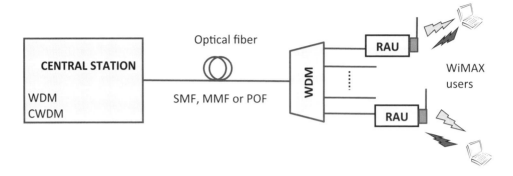

Figure 19.1 Simplified schematic of a RoF system for a WiMAX distribution.

Many techniques have been the subject of research in RoF technology during the last 20 years, including optical signal processing (photonic analogue-to-digital converters, photonic-microwave filters, arbitrary waveform generation), antenna array beamforming, millimeter-wave and terahertz generation systems, or photonic up- and down-converting links for applications such as broadband wireless access networks, electronic warfare and RADAR processing, imaging and spectroscopy or radio-astronomy, as described by Capmany and Novak (2007). In particular, the use of optical fiber links to distribute telecommunication standards is the more successful application of RoF technology. This application is known as Hybrid Fiber Radio (HFR), as in Jager and Stohr (2001). The HFR concept is similar to the distribution of CATV signals over a Hybrid Fiber Coaxial (HFC) network, in which a combination of digital and analog channels is distributed from a central location to many users distributed geographically as described by Darcie and Bodeep (1990) and Wilson *et al.* (1995). HFC implies that last mile connectivity is provided through coaxial cable, whilst in HFR the last mile connection is a wireless link. This is not a minor difference, as the wireless environment is much more hostile than cable which imposes restrictive RoF link performance requirements in terms of linearity, noise and power handling capabilities. These parameters must be engineered to guarantee a Spurious Free Dynamic Range (SFDR) for the whole link high enough to cope with geographical dispersion of users and complex modulation formats used by current wireless standards, such as Orthogonal Frequency Division Multiplexing (OFDM).

A simplified schematic of a HFR system is shown in Figure 19.1. RoF technology allows the required RF signal processing functions to be centralized in a single shared location, the Central Station (CS), and then optical fiber is employed to distribute the RF signals to a set of Remote Access Units (RAUs). This allows important cost savings as the RAUs can be simplified to perform just opto-electronic conversion, filtering and amplification functions. It is possible to use Wavelength Division Multiplexing (WDM) techniques in order to increase capacity and to implement advanced network features such as dynamic allocation of resources. This centralized and simplified RAU scheme allows lower cost system operation and maintenance, which are reflected in major system Operational Expendatures (OPEX) savings, especially in broadband wireless communication systems where a high density of RAUs is necessary.

In both the CS and the RAU, both electro-optical (E/O) and opto-electronic (O/E) conversion of WiMAX signals to/from electrical and optical domains are carried out, mainly using Intensity Modulation Direct Detection (IMDD) techniques. E/O conversion is achieved employing either directly modulated lasers or external modulators and O/E conversion employing photodetectors or photoreceivers, as described by Seeds and Williams (2006). There are several possibilities to transport the RF signal from the CS to the RAU, as discussed by Jager and Stohr (2001). In particular, when the signal is transported directly at the frequency of operation many benefits regarding cost and complexity and also upgradeability arise, as there is no need for complex RF signal processing at the RAU, such as up/down conversion or base-band mux/demux, always required when alternative techniques such as digital-over-fiber or intermediate-frequency-over-fiber are used.

19.1.2 Analog Transmission on Fiber State-of-the-Art

RoF systems have been demonstrated (Hirata *et al.*, 2003) at frequencies up to 120 GHz, being commercially available for inter-satellite and RADAR antenna remoting for frequencies up to 40 GHz and above. The market-driving application of RoF technology has been the transmission of wireless standards over optical fiber links in centralized architectures. The broad bandwidth of the optical fiber and the available devices facilitate independent standards and multiservice operation for existing cellular systems, such as GSM as in Ogawa *et al.* (1992), UMTS as in Persson *et al.* (2006), wireless LAN (WiFi 802.11 a/b/g/n) as in Chia *et al.* (2003), Niiho *et al.* (2004) and Nkansah *et al.* (2006), WiMAX as in Pfrommer *et al.* (2006), and Ultra Wide-Band (UWB), as reported in Llorente *et al.* (2008).

The devices required in RoF technology operate at frequencies used for major wireless standards (GSM, WiFi 802.11 a/b/g, UMTS) and also WiMAX up to 5–6 GHz. Directly modulated semiconductor lasers are preferred as optical transmitter due to lower cost (Qian *et al.*, 2005). At higher frequencies, the required performances can only be satisfied by externally modulated transmitters. For indoor applications where picocell configurations are envisaged, advanced multi-function devices such as waveguide electro-absorption modulators (Wake *et al.*, 1997), or polarization independent asymmetric Fabry–Perot modulators, as described by Liu *et al.* (2000, 2003) can be used as both detector and modulator.

Devices with bandwidth handling capabilities in excess of these required by near-term WiMAX deployments exist commercially. In particular Distributed Feedback (DFB) lasers offer the required bandwidth and performances, but normally at a high cost. Recently, important research efforts have been devoted to the development of low-cost/high-performance transmitters, for instance uncooled lasers (Ingham et al., 2003, Hartman *et al.*, 2003) or Vertical-Cavity Surface Emitting Lasers (VCSELs), (Chia *et al.*, 2003, Persson *et al.*, 2006). Probably, the most restrictive requirement for wireless services provision over RoF systems is the SFDR. Nowadays SFDRs in excess of 100 dB Hz$^{-2/3}$ have been demonstrated experimentally, providing enough dynamic range to be employed in the targeted applications, as discussed by Seeds and Williams (2006).

19.1.3 Market Overview and Technology Forecast

The growth in high-bandwidth radio services such as WiMAX, has led to a renewed focus on the definition of the optimum network infrastructure to transmit signals between central offices and remote antenna units. To provide some indicators regarding WiMAX only, Sprint/Nextel are deploying an 802.16e compliant mobile WiMAX network which will reach 100 million American users by the end of 2008. The estimated revenue for the worldwide WiMAX equipment market will grow to more than \$3.3 billion, and connections will reach \$48 million by 2010, as forecasted by Daily Wireless (2008).

Mobile operators are investigating several radio access interfaces such as HSDPA, LTE or WiMAX, not only from the technology point of view but also taking into account CAPEX and OPEX minimization. It is not clear at this moment which technology will become dominant in the near future, as they are in a different stage of development, so coexistence and compatibility are key requirements in RoF technology. Taking into account this heterogeneous environment, huge efforts have been made to develop converged infrastructures able to cope with increasing bandwidth demand, including dynamic adaptation to changing traffic conditions, offering multi-standard transmission capabilities and providing integration with both the fiber-based transport network and the wireless backhaul.

The required integration of the transport network with the wireless access network is a straightforward application of RoF technologies and systems and is of particular importance in highly sectorized indoor environments. This market is growing very fast as several market analysts have recently predicted. ABI Research, for instance, has estimated the CAGR for this market to be 20% with today's market of \$3.8 billion growing to \$15 billion by 2013. ABI states that distributed-antenna indoor systems will make a significant impact in larger buildings because they offer multi-service broadband capabilities.

The use of RoF technologies is expected to become important in indoor deployments due to the flexibility, multi-standard capabilities, transparency and unlimited bandwidth offered by fiber-based solutions. In the case of WiMAX, recent field trial demonstrations have shown the potential of this technology to increase the coverage of WiMAX deployments using RoF technology. For example, the development of RoF systems adapted to both fixed and mobile WiMAX has been started in the frame of WEIRD project (IST, 2008). The transmission of mobile WiMAX requires a careful engineering of the time division duplexing hardware and different transmission aspects in order to support advanced features such as Multiple Input Multiple Output (MIMO) and beamforming, as described by Harjula *et al.* (2008).

19.2   Optical Transmission of WiMAX Signals

19.2.1   Optical Link Key Elements

The basic elements of any RoF link, as shown in Figure 19.2, are a device capable of up-converting the radio signal to optical frequencies (i.e. E/O conversion), the optical fiber as transmission medium towards the remote facility, and a device performing the recovery of the radio signal from the optical carrier (i.e. O/E conversion).

E/O conversion can be carried out using mainly two approaches: direct modulation of a laser and the use of a CW-laser with an external modulator. External modulation is usually performed using Mach–Zehnder modulators (Blumenthal, 1962) or electro-absorption modulators (Dagli, 1999). It benefits from broad bandwidth, low drive voltage and good linearity. However, owing to the cost of external modulators, direct modulation is preferred for low-cost RoF applications.

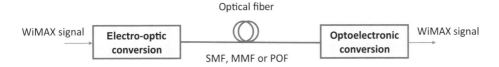

Figure 19.2 Key elements of a RoF link.

In direct modulation, the modulation signal directly changes the intensity of the laser output, implementing a compact and low-cost solution to E/O conversion. The main laser parameters to be considered for direct modulation are modulation bandwidth, optical wavelength and laser efficiency. Since the first proposals of direct modulation (Ikegami and Suematsu, 1967), considerable work has been devoted to increasing the direct modulation bandwidth. Recently, semiconductor lasers at 1550 nm showing modulation bandwidth greater than 30 GHz have been demonstrated by Matsui *et al.* (1997). The most employed laser in direct modulation is the semiconductor diode laser in its various configurations: Fabry–Perot, DFB and VCSEL technologies (Capmany and Novak, 2007).

Fabry–Perot diode lasers are composed of a *p–n* diode junction and an optical waveguide with partially reflecting mirrors at either end (forming a Fabry–Perot cavity, hence the reason for its name), as described by Cox (1997). Fabry–Perot lasers show an optical spectrum with several wavelengths (modes) that can limit the RF bandwidth when combined with fiber dispersion. In addition, it makes the joint operation of several optical sources in the same optical fiber difficult. Therefore, the use of this laser type is limited to short fiber lengths and/or low RF frequencies.

A different laser configuration is the DFB diode laser as described by Agrawal (1997). Its structure is similar to the Fabry–Perot laser but with enhanced wavelength selectivity of the laser cavity provided by an internal filter (a grating written on the active medium which leads to a periodic variation of the mode index). DFB lasers emit a single longitudinal mode. Experimental prototypes have shown direct modulation bandwidths at 1550 nm up to 37 GHz (Bach *et al.*, 2003). Commercial DFB devices can be found with operational bandwiths close to 10 GHz.

Recently, great attention has been focused on a different laser configuration suitable for direct modulation: VCSELs, as described by Persson *et al.* (2006) and Chia *et al.* (2003). These lasers operate in a single longitudinal mode thanks to the use of a very small cavity length (around 1 μm) and show efficient coupling with optical fiber as well as low cost. These are very interesting features for the implementation of low-cost RoF links. VCSELs emit light in a direction normal to the active-layer plane like LEDs. VCSEL technology is not as mature as, for example, DFB and usually present lower output power and bandwidth. In addition, they are mainly available at 850 and 1310 nm although devices at 1550 nm (the usual wavelength for telecommunication applications) are beginning to become commercially available with enhanced direct-modulation bandwidth. There are singlemode and multimode devices commercially available, with the multimode type usually being less expensive.

Commercial VCSELs exhibit bandwidths up to 3 GHz around 1310 and 1550 nm and up to 7 GHz at 850 nm wavelength. In general, semiconductor lasers show wavelength drifts with temperature (about 0.08 nm/°C for a DFB). In order to avoid this undesirable effect, lasers can be cooled to stabilize their operating temperature. This results in a higher cost of the optical source. In the case of RoF applications, low-cost uncooled lasers can be employed if the link is designed to tolerate the wavelength drift, as reported in Ingham et al. (2003) and Hartman *et al.* (2003).

The second key element of a RoF link is the optical fiber. This transmission medium presents many advantages as mentioned in the introduction section which makes fiber a good medium for the transmission of microwave signals in general and WiMAX in particular. Different kinds of fibers have been developed since the 1970s (Agrawal, 1997, Gambling, 2000). Among the most suitable for RoF links are standard SMF, MMF and POF. The main difference is the fiber core radius (typically around 9 μm in SMF and 62.5 μm in MMF) which results in the propagation of a single or multiple modes through the fiber. POF is made of plastic instead of glass and shows a larger core radius than MMF (the insertion loss and dispersion is higher but the cost is lower than glass fibers).

SMF is the fiber of choice in long haul links of optical communication core networks. SMF exhibits huge capacity, with capacity $\times$ distance products up to 41 petabit s^{-1} km^{-1} have been demonstrated (Charlet *et al.*, 2008) using a single SMF.

On the other side, MMF is intended for short links inside buildings, airports, shopping malls and corporate office premises. Owing to its low cost and ease of handling, MMF is already installed in many large buildings with typical lengths of pre-installed MMF up to 300 m. The capacity of MMF is limited by the dispersion between its multiple modes making the bandwidth dependent not only on the fiber length but also on the laser type and launch condition. MMF shows data rates up to 10 Gbps at hundreds of meters.

Finally, a device to perform the O/E conversion and to recover the WiMAX signal from the optical carrier is needed. This is the task of the photodetector (PD) (Agrawal, 1997). The main requirements for PDs in RoF applications are high efficiency in the O/E conversion, large bandwidth and the capability of handling high optical power. High-speed PDs in the 1310–1550 nm bands have been reported with 3-dB bandwidths up to 300 GHz, as reported by Ito *et al.* (2000). Commercial devices up to 100 GHz are available. There is also interest in integrating RF amplifiers with the PD to increase the output power level, as described by Umbach *et al.* (1996). In that case it is called a photoreceiver.

19.2.2   Transmission Performance

A typical RoF link for antenna remoting is shown in Figure 19.3. In the downlink path, the WiMAX signal directly modulates a laser diode. The optical modulated signal travels through an optical fiber span with a length equal to the distance between the Central Station (CS) and the RAU. In the uplink path, the received WiMAX signal from the antenna directly modulates a laser diode and the modulated optical signal is transmitted to the CS through the same optical fiber span.

According to the scheme shown in Figure 19.3, both optical links (downlink and uplink) present a similar configuration. For the downlink and assuming impedance matching is provided at both laser input and photodiode output ($Z_0 = 50\ \Omega$), the detected signal power, $P_{\text{rec\_DL}}$, at the output of the photodiode in the RAU is given by (19.1) (Agrawal, 1997):

$$P_{\text{rec\_DL}}(\text{dBm}) = P_{\text{trx\_DL}}(\text{dBm}) - 2 \cdot L_{\text{opt}}(\text{dB}) + 20 \cdot \log_{10}(\eta \cdot \Re), \qquad (19.1)$$

where $P_{\text{trx\_DL}}$ is the electrical power at the laser input at the CS, L_{opt} are the optical losses from the laser output to the photodiode input, η (W A^{-1}) is the laser efficiency and

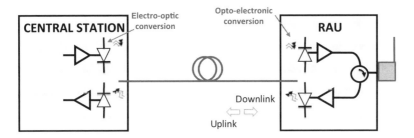

Figure 19.3 RoF link: uplink and downlink operation.

$\Re$ (A/W) is the photodiode responsivity. In a typical scenario ($P_{trx\_DL} = 5$ dBm, $L_{opt} = 8$ dB, $\eta = 0.4$ W A^{-1} and $\Re = 0.7$ A W^{-1}), the detected electrical power is $P_{trx\_DL} = -22$ dBm. In an actual implementation, the maximum detected power at the photodiode output will be limited by the nonlinear performance of the laser diode when modulated by the WiMAX signal.

Considering the main noise contributions, Relative Intensity Noise (RIN) from laser source, shot noise from the photodiode and electrical thermal noise at the photodiode output, the total noise power at the photodiode output is given by

$$N_{rec\_DL} = N_{RIN} + N_{shot} + N_{thermal}$$

$$= 10^{RIN/10}(\eta \overline{P_{opt}})^2 \Delta f R_L + 2q(\eta \overline{P_{opt}} + i_{dark})\Delta f R_L + 4kT\Delta f, \qquad (19.2)$$

where RIN (dB/Hz) is the relative intensity noise of the laser source, $\overline{P_{opt}}$ (W) is the mean optical power impinging the photodiode, i_{dark} (A) is the dark current of the photodiode, Δf (Hz) is the electrical signal bandwidth at the photodiode output, R_L (Ω) is the load resistance at the photodiode output, T (K) is the actual temperature, q (C) is the electron charge (1.6×10^{-19} C) and k is Boltzmann's constant (1.38×10^{-23} J K^{-1}). It is important to point out that the mean optical power at the photodiode input depends basically on the laser bias point as the optical losses of the optical fiber will just modify this value by a couple of decibels for typical scenarios depending on the total fiber length.

In the downlink, the influence of the electrical thermal noise at the laser diode input can be considered negligible if compared with the other noise contributions owing to the relatively high level of the available electrical signal power driving the laser source.

Figure 19.4 shows the respective values of each noise contribution as a function of the mean optical power at the photodiode input. From the results shown it can be stated that the main noise contribution is the thermal noise for low optical powers at the photodiode input and the RIN for high optical powers owing to its dependence on the square of the incident optical power. This behavior is standard for any optical link although the specific values for each noise contribution will depend on the actual parameters of the E/O components in use.

Taking into account that the signal power level at the photodiode output is also proportional to the square of the optical power at the photodiode input, the SNR at the photodiode output will increase with the optical power at the photodiode input reaching its maximum value when the RIN contribution is dominant. At this point the SNR at the photodiode output will remain constant.

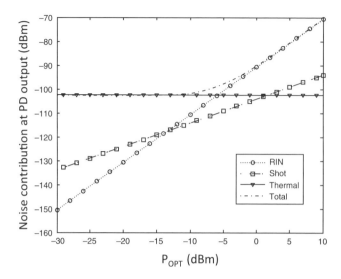

Figure 19.4 Detected power at the photodiode output in Figure 19.3 for three different noise contributions (RIN, shot and thermal) and total noise power as a function of mean optical power at the photodiode input ($T = 290$ K, $R_L = 50$, $\eta = 0.7$ A W^{-1}, $i_{dark} = 40$ nA, RIN $= -140$ dB Hz^{-1}, $\Delta f = 3.5$ MHz).

If a preamplifier is placed at the output of the photodiode the signal (Equation (19.1)) and noise (Equation (19.2)) levels would increase by the gain value of this amplifier. In addition, the thermal noise contribution in Equation (19.2) would further increase its value by the noise figure of the amplifier.

For the uplink, the optical link configuration is similar to the downlink case. The main difference between both cases is the signal and noise electrical levels at the laser diode input. The laser diode at the uplink is driven by the WiMAX signal coming from the antenna with a signal level lower and a noise level higher than the corresponding levels at the downlink case. The signal level at the laser input could be increased by an electrical amplifier stage but the electrical SNR at this point will be reduced. The signal and thermal noise levels at the photodiode output owing to the signal and thermal noise terms at the laser diode input are shown in Equations (19.3) and (19.4). The noise contributions previously considered for the downlink given in Equation (19.3), shown in Figure 19.4, should be added to the additional noise term in Equation (19.4); however, in a typical scenario the two main contributions will be the RIN and the thermal noise term coming from the antenna with relative importance mainly depending on the optical power impinging the photodiode:

$$P_{rec\_UL}(dBm) = P_{trx\_UL}(dBm) - 2 \cdot L_{opt}(dB) + 20 \cdot \log_{10}(\eta \cdot \Re), \qquad (19.3)$$

$$N_{rec\_thermal\_UL}(dBm) = N_{thermal\_UL\_input}(dBm) - 2 \cdot L_{opt}(dB) + 20 \cdot \log_{10}(\eta \cdot \Re). \quad (19.4)$$

Let us now evaluate the maximum reach transmitting WiMAX RoF at the maximum bit rate defined in the IEEE 802.16d standard, that is, 20 mbps. The performance of a WiMAX-on-fiber point-to-point link is analyzed for two electrical/optical conversion

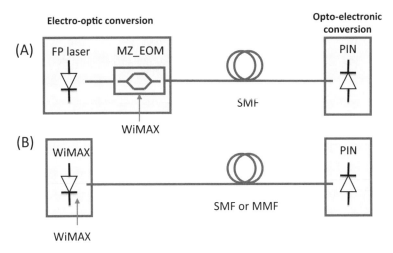

Figure 19.5 WiMAX RoF optical link analysis configuration: (A) Mach–Zehnder external modulation; (B) VCSEL direct modulation.

Table 19.1 WiMAX radio-over-fiber analysis parameters.

VCSEL parameters	Value	PIN parameters	Value	MZ-EOM parameters	Value
Core radius	$2\ \mu$m	Responsivity	0.7 A W^{-1}	E/O BW	20 GHz
Active region thickness	$0.3\ \mu$m	Thermal noise	10 pA Hz^{-1}	V_{π_DC}	5 V
Confinement factor	0.03			V_{π_RF}	5.5 V
				Insertion loss	6 dB
				Thermal frequency shift	-10^{-12} df dJ^{-1}
				Extinction ratio	35 dB

scenarios (external Mach–Zehnder or direct VCSEL modulation) and the two optical media (SMF or MMF). The WiMAX signal is centered at 3.5 GHz and comprises 256 carriers 64-QAM modulated, at 15.625 KHz carrier spacing, with an overall bandwidth of 3.5 MHz. The optical link configuration is shown in Figure 19.5. The MZ-EOM employed is a chirp-free lithium-niobate x-cut modulator. The modulation index has been optimized in every configuration analyzed to maximize WiMAX reach. The analysis is performed by employing a split-step Fourier tool (VPI Photonics, 2006). Table 19.1 summarizes the device parameters employed in the analysis.

The performance of WiMAX transmission is given by the Error Vector Magnitude (EVM). The EVM threshold for WiMAX is 3.1% when 64-QAM modulation per carrier is employed, as described in IEEE (2006). Figure 19.6 shows the EVM obtained when transmitting the WiMAX signal on SMF/MMF optical media for the direct modulation and external modulation cases.

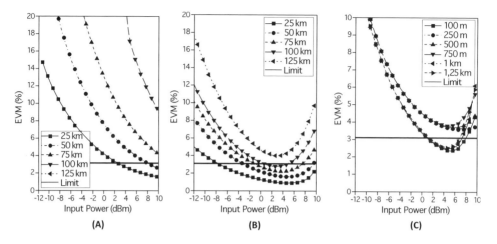

Figure 19.6 WiMAX-on-fiber performance for (A) external modulation on SMF, (B) direct modulation on SSMF and (C) direct modulation on MMF. Horizontal lines indicate the WiMAX EVM threshold.

The results shown in Figure 19.6 indicate that external modulation on SMF gives a maximum reach of 75 km, with 0 dBm optical power launched in the fiber. If MMF is employed as optical transmission media, the maximum reach is 1.25 km also for 0 dBm optical power.

19.3   WiMAX-on-Fiber Applications

19.3.1   Target Applications

WiMAX-on-fiber technology has two main applications today: first, serving the interconnection of WiMAX access points or base stations via a backhaul of point-to-point fiber links and, second, increasing the user coverage deploying a Distributed-Antenna System (DAS) in a given area. These two applications are depicted in Figure 19.7 and described briefly now.

19.3.1.1   WiMAX base-station backhaul

WiMAX access techniques include indoor and outdoor pico- and micro-cells to provide high user capacity density supporting bandwidth-intensive services. WiMAX base stations with integrated backhaul, that is, WiMAX radio itself is employed as a backhaul, can be deployed forming self-connected clusters. This technology is called in-band backhaul. Nevertheless, when the number of users increases, it is necessary to reduce the WiMAX cell size and the spectral reuse in order to deliver an adequate bandwidth per subscriber. In this situation, out of band backhaul is required. Fiber is the best backhaul option in densely populated areas, such as urban areas, when the maximum return on investment from WiMAX deployment can be obtained.

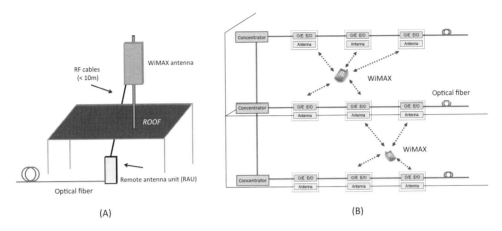

Figure 19.7 WiMAX-on-fiber applications: (A) base station backhaul; and (B) distributed antenna.

19.3.1.2 WiMAX distributed antenna systems

In a DAS several point-source antennas are distributed along a fiber span to provide continuous coverage in heavy-usage areas, such as tunnels, airports, buildings and shopping areas. The antennas transmit the same frequency at different locations to provide a relatively continuous signal-level coverage employing reasonable power levels, thus keeping that signal low to avoid interference. Fiber-based DASs are preferable over coaxial cables in typical applications such as covering a subterranean environment, owing to its lower transmission losses. Fiber-based systems also exhibit greater flexibility. Moving the antenna location in a cable-based DAS means that the whole design must be changed because of noise contributions, impedance mismatching and power levels/amplifier placement restrictions. Available commercial systems, however, are typically limited to frequency ranges between 800 and 2500 MHz. Demonstrations DASs include their deployment to provide uniform wireless coverage in important sportive events such as the 2000 Olympic Games (Rivas and Lopesr, 1998), and the 2006 World Cup, described by Casini and Faccin (2003).

19.3.2 Transmission Impairments

The two applications previously described require the transmission of WiMAX wireless over optical fiber, so suffer from optical analog transmission impairments, which can be summarized as follows.

Group Velocity Dispersion (GVD) originates in the refractive index dependence of the fiber with the wavelength. The different spectral components of a given optical signal propagate at different velocity, broadening the transmitted signal and potentially causing Inter-Symbol Interference (ISI) in digital transmission and severe frequency-selective attenuation in RoF transmissions, as reported by Schmuck (1995). SMF exhibits a GVD value of around 16 ps nm^{-1} km^{-1} at a 1550 nm wavelength. GVD can be compensated employing so-called Dispersion Compensation Fibers (DCFs), which exhibits negative GVD values. Combining transmission through SMF and DCF, the overall effect is mitigated. This approach requires expensive fibers, and it must be implemented at fiber link deployment. Another approach shifts the GVD compensation problem from the optical to the electrical domain by means of electronic compensators/equalizers, as described in Jansen *et al.* (2007). This approach is of special interest in WiMAX-on-fiber applications when OFDMA modulation is employed.

Polarization-Mode Dispersion (PMD) is a complicated process that cannot be solved by link design. PMD becomes important in high bit rate digital transmissions. PMD originates in the fiber core birefringence and is influenced by mechanical (stress, vibration) or environmental (temperature) factors. Birefringence is a characteristic of silica optical fiber originating in the manufacturing process. State-of-the-art fibers are manufactured following enhanced processes achieving a minimal asymmetry in the core. PMD levels are typically lower that 0.1 ps km$^{-1/2}$. PMD compensation in the electrical domain has recently been reported as suitable for OFDMA-based WiMAX (Djordjevic, 2007).

Modal Dispersion is a distortion mechanism present in MMFs. The optical signal transmitted is distorted because of the different propagation velocities of the electromagnetic modes present in the optical media. This effect severely limits the available bandwidth of the MMF. For example, step-index fibers (50μm core) exhibit a bandwidth of 20 MHz · km. The bandwidth of typical off-the-shelf graded-index MMFs (50μm) is close to 1 GHz · km. Fortunately, electronic compensation of modal dispersion has been demonstrated for OFDM communications, being applicable for OFDMA-based WiMAX, as in Lowery and Armstrong (2005).

Another important technology issue in WiMAX-over-fiber is the coexistence problem between different wireless standards when transmitted through fiber. This is an important issue, as fiber and wireless access will coexist in the last mile. The coexistence of WiMAX over wireless signals, for example UMTS, when transmitted through the fiber usually does not pose a problem, as shown by Alemany *et al.* (2008). The different licensed or unlicensed wireless services are allocated at different transmission bands and a careful design of the optical transmission system can minimize interference.

19.3.3   Field Trials

In the framework of the WEIRD project (IST, 2008), two demonstrations of RoF systems for WiMAX signals distribution were carried out. The prototypes employed in these field trials were provided by the company DAS Photonics, member of the WEIRD consortium. In both cases the selected transmission technique was direct modulation of semiconductor lasers.

In a first field trial, a RoF system tailored to a fixed WiMAX base station was tested. The developed prototypes were designed to provide a clean full-duplex transmission path for the WiMAX signal in the 3.5 GHz band by Frequency Division Duplexing (FDD). At both edges of the RoF system a frequency duplexer separates and combines the uplink and the downlink signals. Figure 19.8(A) depicts the testbed architecture.

In Figure 19.8(A) the WiMAX base station (BS) is linked through its RF interface to a Central Station (CS). Two optical fiber coils of 1.1 km connect the CS to the RAU, which was

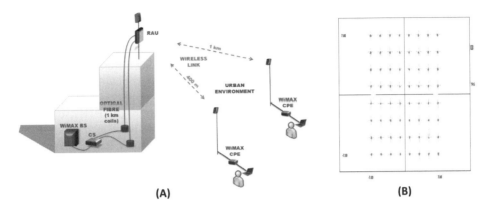

Figure 19.8 (A) WiMAX FDD RoF trial architecture; (B) 64-QAM constellation received after 400 m transmission.

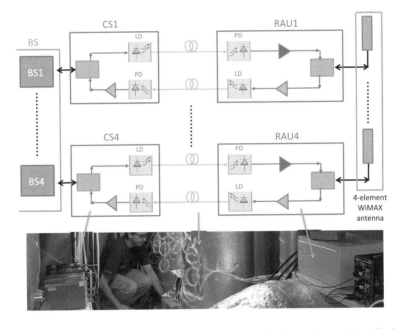

Figure 19.9 Four-channel WiMAX TDD RoF trial schematic and installation.

located on the roof and, in turn, connected through a RF cable to the WiMAX BS antenna. In the surroundings, two CPEs were installed at 1 km and 400 m away from the WiMAX CS and they were linked through a wireless link at 3.5 GHz. The RAU prototype was designed to raise the downlink signal power up to +13 dBm ensuring an EVM better than −31 dBm (maximum EVM allowed for 64-QAM OFDM signals in IEEE 802.16-2004). Figure 19.8(B) shows the demodulated constellation at the receiver in the downlink RAU at 400 m radio distance.

Two applications were tested by the WEIRD consortium over this WiMAX RoF network: remote volcano monitoring using seismic sensors and Voice-over-IP communication. The results of the trial showed that the implemented RoF system provided good quality of service to the WiMAX CPEs in terms of signal strength and SNR, allowing the use of high spectral-efficient modulations combined with high channel coding rates in both uplink and downlink paths.

A second field trial performed within the WEIRD project aimed at the demonstration of a RoF system adapted to mobile WiMAX features. The system was developed in accordance with the mobile WiMAX base station C-WBS (Compact WiMAX Base Station) from Alcatel-Lucent. The C-WBS had four RF (3.4–3.6 GHz) ports that are connected to a four-element panel WiMAX antenna, which allows beamforming, if desired. In this case Time-Division Duplexing (TDD) was employed as a multiple-access technique.

The RoF system implemented in this case consisted of optical and electronic hardware allowing the optical analog transmission of the WiMAX signal between the base station and its antenna. In the case of the C-WBS, four RF signals had to be transmitted. In order to transmit four WiMAX signals in parallel, the selected approach was to produce four independent RoF links as shown in the Figure 19.9. This figure also shows a picture of the RoF system deployed and its integration with the WiMAX C-WBS.

These two field trials validated the RoF WiMAX approach in both FDD and TDD WiMAX operation in a real-world scenario, employing fully commercial equipment.

19.4   Conclusions

In this chapter, RoF systems as an efficient technology to support WiMAX services deployment in base station backhaul and distributed antenna scenarios have been described. Simulation work and field trials that were published recently validate this approach. RoF technology can play a significant role in the widespread adoption of WiMAX services in high user density environments, such as office buildings, government premises and also as backhaul technology for access networks. The key factor for the success of this technology is the optimization of high-performance/low-cost devices to reduce deployment expenses. Recent technological advances in this direction, as discussed in Section 19.2.1, foresee a successful market introduction of WiMAX over RoF systems.

References

Agrawal, G.P. (ed.) (1997) *Fibre-Optic Communication Systems*. John Wiley & Sons Ltd, Chichester.

Alemany, R., Pérez, J., Llorente, R., Polo, V. and Martí, J. (2008) Coexistence of WiMAX 802.16d and MB-OFDM UWB in Radio Over Multi-mode Fiber Indoor Systems. *Proceedings of MWP 2008*.

Bach, L., Kaiser, W., Reithmaier, J.P., *et al.* (2003) Enhanced direct-modulated bandwidth of 37 GHz by a multi-section laser with a coupled-cavity-injection-grating design. *IEE Electronics Letters*, **39**(22), 1592–1593.

Blumenthal, R.H. (1962) Design of a microwave frequency light modulator. *Proceedings of the IRE*, **50**(4), 452–456.

Capmany, J. and Novak, D. (2007) Microwave photonics combines two worlds. *Nature Photonics*, **1**(6), 319–330.

Casini, A. and Faccin, P. (2003) Wavelength Division. Multiplication Technologies for UMTS Radio. Coverage extension by using the radio over fibre technique. *Proceedings of the IEEE International Topical Meeting on Microwave Photonics*, Budapest, pp. 123–128.

Charlet, G., Renaudier, J., Mardoyan, H., *et al.* (2008) Transmission of 16.4 Tbit/s capacity over 2550 km using PDM QPSK modulation format and coherent receiver. *Proceedings of the Optical Fiber Conference 2008*, San Diego, CA, Post Deadline Paper PDP3.

Chia, M., Luo, B., Yee, M.L. and Hao, E.J.Z. (2003) Radio over multimode fibre transmission for wireless LAN using VCSELs. *IEE Electronics Letters*, **39**(15), 1143–1144.

Cox, C.H. (ed.) (1997) *Analog Optical Links. Theory and Practice*. Cambridge University Press, Cambridge.

Dagli, N. (1999) Wide-bandwidth lasers and modulators for RF photonics. *IEEE Transactions on Microwave Theory and Techniques*, **47**(7), 1151–1171.

Daily Wireless (2008) LTE vs Wimax, http://www.dailywireless.org/ (accessed 13 February 2008)

Darcie, T.E. and Bodeep, G.E. (1990) Lightwave subcarrier CATV transmission systems. *IEEE Transactions on Microwave Theory and Techniques*, **38**(5), 524–533.

Djordjevic, B. (2007) PMD compensation in fibre-optic communication systems with direct detection using LDPC-coded. OFDM. *Optics Express*, **15**(7), 3692–3701.

Gambling, W.A. (2000) The rise and rise of optical fibers. *IEEE Journal on Selected Topics in Quantum Electronics*, **6**(6), 1084–1093.

IEEE (2006) WiMAX 802.16 Standard, http://standards.ieee.org/getieee802/802.16.html (accessed 2 October 2008).

Ihgham, J., Webster, M., Wanfar, A., *et al.* (2003) Wide-frequency-range operation of a high linearity uncooled DFB laser for next-generation radio-over-fibre. *Proceedings of the Optical Fiber Conference 2003*, Atlanta, GA, pp. 754–756.

IST (2008) WiMAX extension to isolated research data networks, http://www.ist-weird.eu/ (accessed 2008)

Harjula, I., Ramirez, A., Martinez, F., *et al.* (2008) Practical issues in the combining of mimo techniques and RoF in OFDM/A systems. *Proceedings of The 7th WSEAS International Conference on Electronics, Hardware, Wireless and Optical Conference*, Cambridge, pp. 244–248.

Hartman, P., Webster, M., Wonfor, A., *et al.* (2003) Low-cost multimode fibre-based wireless LAN distribution system using uncooled directly modulated DFB laser diodes. *Proceedings of the Optical Fiber Conference 2008*, Rimini, pp. 804–805.

Hirata, A., Harada, M. and Nagatsuma, T. (2003) 120-GHz wireless link using photonic techniques for generation, modulation, and emission of millimeter-wave signals. *Journal of Lightwave Technology*, **21**(10), 2145–2153.

Ikegami, T. and Suematsu, Y. (1967) Resonance-like characteristics of the direct modulation of a junction laser. *Proceedings of the IEEE*, **55**, 122–123.

Ito, H., Furata, T., Kodama, S. and Ishibashi, T. (2000) INP/INGAAS uni-traveling-carrier photodiode with 310 GHz bandwidth. *IEE Electronics Letters*, **36**, 1809–1819.

Jager, D. and Stohr, A. (2001) Microwave photonics. *Proceedings of the European Microwave Conference*, London, pp. 1–4.

Jansen, S.L., Morita, I., Takeda, N. and Tanaka, H. (2007) 20-GB/s ofdm transmission over 4,160-km SSMF enabled by RF-pilot tone phase noise compensation. *Proceedings of the Optical Fiber Conference*, Anaheim, CA, Post-deadline paper PDP15.

Liu, C., Polo, V., Dijk, F.V., Pfrommer, H., *et al.* (2000) Full-duplex docsis/wirelessdocsis fibre-radio network employing packaged AFPMs as optical/electrical transducers. *Journal of Lightwave Technology*, **25**(3), 673–684.

Liu, C., Seeds, A., Chadha, J., *et al.* (2003) Bi-directional transmission of broadband 5.2 GHz wireless signals over fibre using a multiple quantum-well asymmetric Fabry–Perot modulator/photodetector. *Proceedings of the Optical Fiber Conference 2003*, Atlanta, GA, pp. 738–740.

Llorente, R., Alves, T., Morant, M., *et al.* (2008) Ultra-wideband radio signals distribution in FTTH networks. *Journal of Lightwave Technology*, **20**(11), 945–947.

Lowery, A.J. and Armstrong, J. (2005) 10 Gbit/s multimode fibre link using power efficient orthogonal frequency division multiplexing. *Optics Express*, **13**, 10003–10009.

Matsui, Y., Murai, H., Arahira, S., *et al.* (1997) 30-Ghz bandwidth 1.55 μm strained-compensated INGAALAS-INGAASP MQW laser. *IEEE Photonics Technology Letters*, **9**, 25–27.

Niiho, T., Nakaso, M., Masuda, K., *et al.* (2004) Multi-channel wireless LAN distributed antenna system based on radio-over-fibre techniques. *Proceedings of the IEEE LEOS Annual Meeting.*, Rio Grande, Puerto Rico, pp. 57–58.

Nkansah, A., Das, A., Lethien, C., *et al.* (2006) Simultaneous dual band transmission over multimode fibre-fed indoor wireless network. *IEEE Microwave and Wireless Components Letters*, **16**(11), 627–629.

Ogawa, H., Polifko, D. and Banba, S. (1992) Millimetre-wave fibre optic systems for personal radio communication. *IEEE Transactions on Microwave Theory Technology*, **40**(12), 2285–2292.

Persson, K.A., Carlsson, C., Alping, A., *et al.* (2006) WCDMA radio-over-fibre transmission experiment using singlemode VCSEL and multimode fibre. *IEE Electronics Letters*, **42**(12), 372–374.

Pfrommer, H., Piqueras, M.A., Polo, V., *et al.* (2006) Radio-over-fibre architecture for simultaneous feeding of 5.5 and 41 Ghz WIFI or WiMAX access networks. *Proceedings of the Microwave Symposium Digest, IEEE MTT-S International Volume*, pp. 301–303.

Qian, X., Hartmann, P., Ingham, J.D., *et al.* (2005) Directly-modulated photonic devices for microwave applications. *Proceedings of the IEEE MTT-S Intl Microwave Symp.*, Long Beach (CA) USA.

Rivas, I. and Lopesr, L. (1998) A microcellular DCA scheme using variable channel exclusion zones. *Proceedings of the IEEE Semiannual Vehicular Technology Conference* pp. 1395–1399, Ottawa, Canada.

Schmuck, H. (1995) Comparison of optically millimeter-wave system concepts with regard to chromatic dispersion. *IEE Electronics Letters*, **31**(21), 1848–1849.

Seeds, A.J. and Williams, K.J. (2006) Microwave photonics. *Journal of Lightwave Technology*, **24**(12), 4628–4641.

Stuckmann, P. and Zimmermann, R. (2007) Towards ubiquitous and unlimited capacity communication networks: European research in framework programme 7. *Journal of Lightwave Technology*, **45**(5), 148–147.

Umbach, A., Waasen, S., Auer, U., *et al.* (1996) Monolithic pin-hemt 1.55 μm photoreceiver on INP with 27 Ghz bandwidth. *Electronics Letters*, **32**(23), 2142–2143.

VPI Photonics (2006) VPItransmissionMaker, Version 7.1.

Wake, D., Hansson, D. and Moodie, D.G. (1997) Passive picocell: A new concept in wireless network infrastructure. *Electronics Letters*, **33**(5), 404–406.

Wilson, B., Ghassemlooy, Z. and Darwazeh, I. (eds.) (1995) *Coexistence Study Between UWB and WiMAX at 3.5 GHz Band*. Institute of Electronic Engineers, London.

20

Network Planning and its Part in Future WiMAX Systems

Avraham Freedman and Moshe Levin

20.1   Introduction

Wireless network planning is the art of 'integrating' a given wireless system into an specific geographic and market environment. The task facing the network planners is complex. They have to offer a deployment plan that would answer the operator requirements and business plan, making the best use of the wireless system properties and features in the most cost-effective way. This plan has to be robust enough to accommodate the inherent uncertainty in the information available in the environment, scalable and flexible enough to enable network growth and demand changes. The role of the network planner does not end with providing the best deployment plan. It goes further during the lifetime of a network, and includes network optimization, trouble shooting, scaling and upgrading. Mishra (2007) provides a description of the network planning process for various cellular systems, while Laiho *et al.* (2007) give a detailed analysis of Universal Mobile Telecommunications System (UMTS) planning.

WiMAX (Andrews *et al.*, 2007, Eklund *et al.*, 2006) offers robustness and flexibility in a level exceeding any other system. The IEEE 802.16 standard (IEEE (2004) amended in IEEE (2005)) was developed to provide a variety of options to the manufacturer and the operator. The WiMAX Forum had narrowed down the range of possibilities by providing interoperability profiles that enable economy of scale (WiMAX Forum, 2006). Still, a set of options and features provide the system with enhanced capacity, robust operation and interference mitigation within a variety of equipment cost ranges.

New technologies may offer capabilities that may render some of the traditional planning tasks obsolete. For example, frequency planning is not as fundamental a task in Orthogonal Frequency Division Multiple Access (OFDMA) and Code Division Multiple Access

WiMAX Evolution: Emerging Technologies and Applications   Edited by Marcos D. Katz and Frank H.P. Fitzek
© 2009 John Wiley & Sons, Ltd

(CDMA) system as it has been for Time Division Multiple Access (TDMA) systems. Nevertheless, WiMAX represents a careful step-back from the Direct Sequence CDMA (DS-CDMA) concept of 'no planning and optimization': it provides several mechanisms that allow the network to control the RF channel by generating different levels of isolation among customers. Consequently, deployment planning and optimization are important tasks, and probably will always remain so. Similarly, system configuration in the general sense, including frequency planning, fractional frequency planning, power planning, frame planning, etc., will be the bread and butter of planners of current WiMAX systems. Whereas system designers seek new ways and new means to improve system performance in a given environment, it is the planner's task to make use of the additional capabilities and plan the most cost-effective deployment for the actual environment at hand.

The systems described in this book provide much more. Base station coordination, self organizing networks, cognitive radios and other techniques put intelligence at the base station level. A central optimization entity will still be needed, to plan, for example, the base stations' locations and to serve as a hub to gather and distribute information across the network, however a lot of the system monitoring and optimization activities will take place in a distributed manner across the network, thus allowing for fast real-time response and improved performance.

The wireless network operator seeks better control of the network, higher network quality, better use of resources, lower maintenance costs, faster responses to problems and adaptability to changing reality: the actual service demand, traffic load and propagation environment.

The main theme in our vision of the future wireless planning and optimization is the new concept of a network model. Current wireless networks are installed and managed using reports generated by the network planning department: the *Plan*. The main limitation of the *Plan* is that it is not dynamically updated; therefore the decisions taken, which are based on it, are not optimal.

The *Model* is a constantly updated version of the *Plan*, escorting the network throughout its life cycle. Appropriate processes and interfaces should be provided to facilitate simple and accurate dynamic integration of network information: configuration and traffic demand patterns. Such a model will allow the user to perform planning and optimization on an accurate database. A key success factor for efficient modeling is accurate estimation of the traffic demand: geographical distribution and resources used. One solution to that might be radio location techniques, applied to information collected from the network.

One might be tempted to claim that planning is not important. The technology is 'smart' enough to autonomously optimize itself and find real-time solutions for real-time problems. Our answer is that wise planning will generate better infrastructure over which smart optimization will be able to operate. On top of that, planning as a methodological process has two distinctive properties that cannot be met by any online optimization mechanism:

(i) planning for a future scenario, e.g. what-if and sensitivity analysis;

(ii) optimization that is influenced by arbitrary external requirements such as preferring to give a certain area of the network greater preference than its relative traffic impact.

In this chapter we describe the process of planning a wireless network to cover large metropolitan or rural areas. We define the various stages of the process, and its role in

WiMAX deployment. The impact of WiMAX system features on network planning is described and, in particular, we discuss the role network planning is going to take in future WiMAX systems deployments. Finally the concept of the *Model* is described as a key to the integration of planning information into a system.

20.2   The Network Planning Process

Figure 20.1 is a schematic network planning process, with the purpose of planning a new wireless system in a given area: a green-field deployment planning. A planning process can be described as comprising three main phases. The first is the information-gathering stage, the second is the planning itself and the third is the planning testing, verification and optimization stage, which, as a matter of fact, lasts throughout the whole lifetime of a system. The process is by no means linear. Feedback from the third stage can, and always does, affect the gathered information and may change the planning. Other feedback loops are inherent in the process, mainly due to the limitation of gathering full and reliable information on the deployment environment.

20.2.1   Data Collection

The information the planner has to use is of three main types.

Geographical information includes all of the geographical data of the area to be covered, terrain ground heights, land use maps and maps of the location and shapes of buildings and their heights. The latter type of information is very important when planning wireless systems in cities and urban areas, although it is not always easy to come by. One should also note that the availability and accuracy of geographical information varies greatly from place to place and project to project. As part of the information the planner needs for an area, a list of potential locations for base stations should also be included or created by the planner.

Customer information includes the market data, namely the expected number of users, a map of expected user density across the area, expected traffic profile of the users, etc. While traditional cellular networks assumed a single type of user with a single type of service, and third-generation systems introduced several types of service, modern systems enable us to define different types of users, each with their own usage profile, service-level agreements, etc.

Technology information refers the parameters of the wireless system to be deployed. Those parameters include base station and terminals transmission powers and power control capabilities, receiver sensitivity under various operation conditions, rejection of interference capability (adjacent channel attenuation, net filter discrimination against external interference, etc.), total capacity, adaptive modulation and coding support, antenna system description, which includes the radiation pattern of the antenna, total antenna gain and cross polarization. Modern cellular systems also include antenna arrays with a variety of spatial techniques, such as diversity, beamforming, interference cancellation and spatial multiplexing. Thus, the parameters of the array and the applicable spatial technique to the base stations and terminals should be determined.

An important part of the technology information is a specification of the spectrum to be used, which includes the number of the frequency channels (at the system bandwidth) available for deployment. Information about external systems operation, regulatory limitations

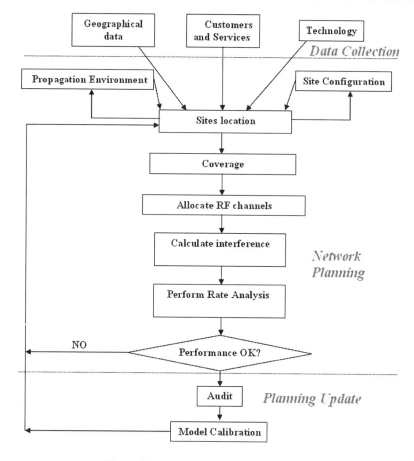

Figure 20.1 Network Planning Process.

on the usage of spectrum and power in different parts of the planned area should also be included.

20.2.2   Network Planning

20.2.2.1   Basic Terms and Concepts

Before actual planning starts, some basic terms and concepts should be mentioned.

Coverage Database   What would be the coverage area to plan for? In the case of a fixed wireless access system (a common and viable design for WiMAX) the planning should concentrate on the actual location of the Subscriber Station (SS) antennas, which may very well be the rooftops. Each such point serves as a reference point for which the planner calculates the Received Signal Strength (RSS), estimates the interference level and derives the Signal-to-Interference-plus-Noise Ratio (SINR) and the radio performance. In addition,

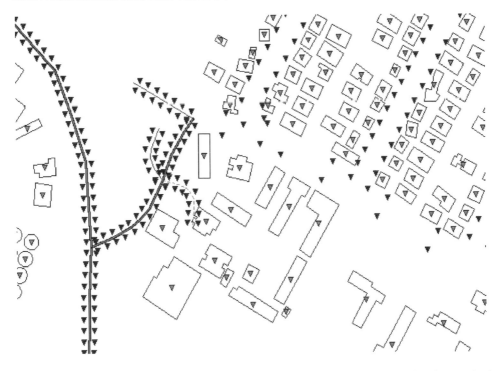

Figure 20.2 Indoor (grey triangles) and outdoor (black triangles) reference points in a typical urban area.

each such point serves as a traffic node, which generates or sinks traffic. For a mobile system, the coverage is typically calculated for an equally spaced grid of points, each representing an area called a bin. Advanced planning tools support a three-dimensional coverage database, which also defines bins over building floors at different heights. Each bin center is a reference point, as described above. Traffic can also be assigned to it, but it is not a real traffic source, but rather a virtual source, as users will not always be present within each bin. Traffic assigned to each bin is taken in proportion to the total area to which it belongs. Another approach would be to define the set of reference points where possible subscribers can be found according to the geographical features, such as along streets and within buildings. This approach has an inherent advantage as it is more focused around the subscribers' real locations, reducing the computation load for the planning tool on the one hand, and providing more accurate statistics on the other. An example of this approach is given in Figure 20.2, where such points were distributed uniformly along streets and on every floor within each building in a given area.

Propagation and Channel Models   Prediction of the signal strength is a basic procedure for estimation the reference signal, as well as the interference. As exact solution of the Maxwell equation is impossible, a large variety of models are available (see Parsons (1992), Blaunstein (1999), COST (1999), and ITU (2007a,b)) and references therein), ranging from empirical models, based on a set of experiments and tests made in various conditions, and physical models, which perform the prediction based on a simplified model of the propagation environment. The more elaborate models, such as ray tracing, take into account reflection, diffraction and scattering, however these models usually require a very long computation time for complete region coverage. Model selection depends on the nature of the area, but also on available information and available computation time. Different models should be applied to different points in the coverage area according to the land use and the location of the users (indoor, outdoor, rooftop level, street level, etc.)

System performance is not determined only by the RSS or SINR at the receiver. It also depends on the channel and the receiver used, and it is random in nature, as the channel fades in an unpredictable manner. A statistical model should be applied in order to predict the actual performance in a required availability level. Sophisticated tools use different models for different reference points to provide a better picture of the environment.

Range-limited or Capacity-limited Deployment   The number of base stations required to cover a given area is determined either by the size of the area to be covered, relative to the area that can be covered by a single base station, or by the total traffic that needs to be carried divided by the capacity a single base station can carry. In the first case, the deployment is said to be range limited, while in the second case is a capacity limited deployment. Planning is different in the two cases. In the first case the focus of the planner is to increase, as much as possible, the range each base station can reach. In the second case, a base station will not cover to its full range, which may lead to interference between base stations. The focus of the planner is to mitigate this interference.

Supply-driven or Demand-driven Deployment   These are two different approaches to the planning process. In the first approach planning will be made for the projected system size in the future. Actual system deployment will be derived out of this master plan, and can be scaled up until it reaches the final system size. This approach ensures optimized long-term operation, and provides a path for system growth. On the other hand, if the predictions are not accurate, planning and deployment may have to be redone. The demand driven approach starts with the minimum system configuration needed to provide for the existing demand. For a fixed wireless access system this would mean deploying only the base stations in limited regions where demand exists. For a mobile system this would mean deployment only a range limited system at the first stage and later addition of cells to provide capacity where needed. Cell splitting is a good example of this approach. Demand driven deployment is more flexible, however, it results in an unoptimized system, as each growth has to be made based on the constraints imposed by existing deployment and an overall network optimization is never performed at any stage.

Base Station Configuration   A base station's range and capacity depend on its configuration, namely the height of the antenna above ground and the antenna directivity, which implies the number of sectors deployed. Tall antenna masts provide wider coverage and longer range and thus are widely used for range limited deployments, while installation below rooftops limits the base station range. For capacity limited deployments, this is a preferred option as the interference between cells is reduced. Multi-sector base stations (four, six and even more) can be found in fixed wireless access deployments. Increasing the number of sectors, increases the directivity of the antenna and the cell range, reduces the interference and increases the over all capacity of a base station. However, in mobile scenarios a multitude of sectors will burden the system with a large number of handovers. The typical number of sectors in mobile systems is three. Within a single network, one can find a variety of cell configurations, depending on the specific zone. In mobile networks one can find a hierarchy of base stations which cover the same region but for different purposes.

- Umbrella cells, also called macrocells, installed on high masts. Those cells carry the traffic of high-speed users along roads and streets. Their long range and wide coverage reduces the need for handovers for the fast moving mobiles.

- Microcells, installed on roof tops or below rooftop levels. They are designed to carry the traffic of pedestrian users, and indoor users reachable by indoor penetration.

- Picocells, installed outdoor below rooftop levels or indoor, mainly within commercial buildings, such as shopping malls, airports, bus stations etc. The purpose of those cells is to cover hot spots and provide indoor coverage.

- Femtocells, installed indoor within residences, with the goal to provide indoor coverage for a small number of users and provide cellular access to the fixed users.

20.2.2.2 Site Location

Base station site location is critical in wireless networks deployments. Base station sites are one of the main cost factors of the network. It is not only the cost of equipment but also the cost of acquisition or rental, licensing, environmental implications, building and construction, electricity and air conditioning, backhaul and transmission network and long-term maintenance. All of these add up to form one of the major sources for capital and operation expanses. A planning tool that enables network optimization and deployment with a minimal number of sites will provide a very handsome return on investment on the tool cost. The classical method of planning (Lee, 1986, McDonald, 1979) was based on the well-known hexagonal grid pattern. The hexagon, being the polygon with the largest number of edges which can cover a plane, is the best approximation for a circle, which reflects the fact that the signal strength is highly dependent on range. After determining the cell range, the grid of hexagons is overlaid over the area to be covered. Actual locations of base stations are selected as close as possible to cell centers, no further than a quarter of the cell radius.

The problem of base station placement in a realistic environment, taking into account the user density, has attracted a lot of attention in the literature. Direct search, genetic algorithms, simulated annealing, vector quantization and heuristic approaches have all been presented for macrocellular, microcellular and indoor environments. (Aguado-Agelet *et al.* (2002), Anderson and McGeehan (1994), Chavez-Santiago *et al.* (2007), Huang *et al.* (2000), Hurley (2002), Molina *et al.* (2000), Yang and Ephremides (1997) are only a small sample of the literature available.)

Sophisticated algorithms are aiming to find a minimal number of base station sites such that the coverage and capacity constraints are satisfied. More complex external constraints can be considered, for example, to enable the deployment of a microwave backbone network, that will require, on top of meeting adequate coverage and signal-to-noise levels, an

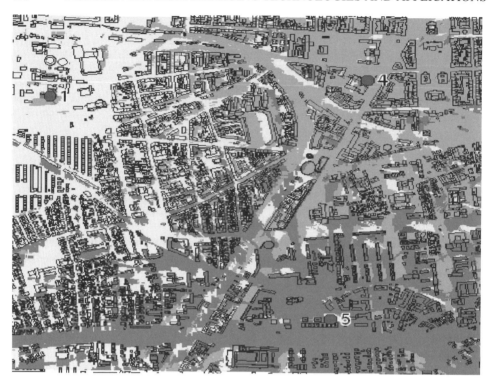

Figure 20.3 Site connectivity map (by highest signal power) showing noncontiguous cells. Each grey level indicates connectivity to a certain cell.

appropriate site-to-site propagation environment that ensures Radiofrequency (RF) visibility. More elaborate cost functions, which take into account sites' cost and even possible profit, can also be implemented.

20.2.2.3 Sites Connectivity and Coverage

Once the sites are in place, the coverage area of each cell (a cell is typically referred to as a sector within a base station) can be determined. A cell is said to cover a reference point or a bin if its signal strength at that point is above a given level that enables the base station to serve a SS at that point. In a fixed wireless environment, the coverage area, or the set of possible SS locations, is partitioned among the cells with no overlap. In the case of a mobile system, there is an inherent overlap, which is essential to provide the ability of handover, but a too large overlap indicates a high overload on the network and possible interference source.

An important point to emphasize is that the coverage zone of a realistic cell is neither circular nor hexagonal. As a matter of fact it is not even contiguous. Figure 20.3 shows a realistic scenario with three cells, showing the largest received signal strength in each bin. Figure 20.3 shows how the land cover, buildings in this case, affects the coverage area of each cell. In the figure, each grey level indicates which base station's signal is the strongest at the particular point. The coverage area of a cell is not contiguous and interlaces with that of its neighbors.

20.2.2.4 Required Cell Capacity

Once the coverage area per cell is determined, the total traffic capacity to be carried by each cell can be readily calculated, summing over the traffic nodes and the estimated handover load. Accordingly the number of RF channels/physical resource blocks needed for each cell may be determined (in terms of WiMAX this would be the number of physical symbols per frame or an equivalent measure). However, it should be noted that this number may vary according to the actual channel conditions and interference of each SS. Adaptive modulation, spatial multiplexing and other techniques may enable more traffic to be carried per physical resource block, and reduce the requirement per channel. So, a rough evaluation is taken at the first stage, and later corrected when more accurate estimate of the channel condition per user is made.

20.2.2.5 Frequency and System Configuration Planning

Given the number of RF channels per cell, it is now possible to plan the frequency of each RF channel. In a mobile WiMAX system this stage includes much more, as elaborated in the next section. Classical frequency planning algorithms (e.g. Lee (1986) and McDonald (1979)) are based on the reuse factor principle. Based on the hexagonal grid, frequencies are assigned such that co-channel frequencies are reused in cells as far away as possible. Modern frequency planning algorithms use graph coloring techniques, simulated annealing, genetic algorithms and other approaches; Niessen (1997) and Struzak (1982) provide examples of solutions to this problem.

20.2.2.6 Performance Evaluation and Plan Optimization

Now that the base station sites and their configurations are determined, the system performance can be evaluated. This involves the calculation, for each reference point or bin, the RSS, the interference and the resulting SINR value, which determines the operation conditions at that reference point. The actual channel nature can also be evaluated and, together with the SINR estimate, the achievable throughput can be estimated. Performance evaluation will most likely make it necessary for the planner to return to previous stages, improve and optimize the plan. This may include recalculation of cells load, replanning of frequency and system configuration, changing the cells configuration such as antenna azimuths and tilts, adjust handover thresholds, etc. As systems become more complex, simulation, which involves traffic and scheduler simulation, might also be used. Still, as it is most often a time-consuming process, simulation is usually made at the last stage for final verification.

20.2.2.7 The Planning Process Output

The output of the process we have described includes:

- base station sites;

- sector loads;

- base station antenna configurations;

- cells frequencies and other parameters;

- performance maps including SINR maps, expected rate and throughput;

- neighbor list.

20.2.3   Planning Verification and Update

The planning process is actually a closed-loop process. Planning should be verified and updated according to feedback from the field, received along the lifetime of the system.

Before the actual deployment, field surveys are performed to verify the data used for planning. Radio surveys are also performed wherein measurements of test links are made to verify the suitability of the propagation model.

The current art of prediction verification is to perform drive tests. During system deployment and operation a set of drive tests and walk tests are made, aiming to check the actual coverage and compare it with the predicted coverage. During those tests the signal strength of the cells are measured at each point along the test track. The test equipment is typically equipped with navigation equipment, such as a Global Positioning System (GPS) receiver and provides a list of coordinates and the corresponding signal strength from the base stations received at that point.

A typical usage of drive tests is to calibrate the propagation models. In the calibration process a set of parameters of the models are adjusted to yield a minimal deviation from the measurements. The calibrated models are assumed to provide more accurate predictions for points where measurements were not made. Given a large enough set of drive tests, calibration can be specific to a given area and take into account local effects. Sophisticated tools provide calibrated models specific to a part of a cell.

However, with the accumulated experience in cellular network operation, most of the operators agree that a drive test is a very limited technique to collect information on the network coverage: it is often compared with searching for a coin under a lamp post. Drive tests provide coverage information on the roads and, to a very limited extent, on public areas where walk tests can be performed. Today, most of the traffic, 70% to 80%, is performed from the customer premises, away from the road. Furthermore, data services such as Web browsing and video steaming, will significantly increase the amount of indoor traffic. A common concern that operators state today is if, after performing drive tests, one learns that down-tilting sector A will decrease the amount of interference it generates to sector B, how can one be sure that such an antenna down tilt will not damage our customers in the second, third and higher floors?

An acceptable solution must provide an efficient and accurate means to collect the coverage information at the customer premises. Regular mobile stations' reports can also be used for this purpose. However, since most of the reports are not accompanied by coordinates reading, a location technique should be applied, be it GPS, assisted GPS, triangulation, trilateration, finger printing or others (see Hata and Nagatsu (1980), Ott (1977), Riter and McCoy (1977) and Wigren (2007)). With such a technique mobile stations' measurements are turned into a Virtual Drive Test (VDT), which extends the capability of drive tests in time and place and, most importantly, reflects the actual usage of the system by its users and not by an artificial measurement equipment.

In many wireless systems such mobile station reports are sent to indicate problems, such as dropped calls. Locating a dropped call report and comparing the measurements with the predicted coverage database at that point enables the planner to fix the problem. This is an example where an integration of three sources of information: the predicted coverage, drive test measurements and mobile station measurements provides the ability to obtain a complete picture of the network status. While predictions are ubiquitous and can be used for what-if analyses, they are not so accurate. Drive tests are accurate but are limited to the time and place they were performed. Launching a drive test requires time and manpower resources. Mobile Station readings are not as accurate as drive test equipment, but they provide actual status of real users. They cannot be used for analyses of areas where the coverage does not exist. Hence, the integration of the three sources provides the most complete picture.

In addition to the signal strength information it is very important to measure the actual traffic distribution. Aggregate traffic data can be read of base stations and cells. The distribution of that traffic within the cells coverage area is not always straightforward. A common technique is to distribute it by clutter, namely land use data.

Updated coverage and traffic information can be used to optimize the network. The optimization process can be seen as repeating the planning process, using the measured data, with constraints imposed by the existing network deployment and parameters.

It is very common to use Key Performance Indicators (KPIs) to monitor network performance. These KPIs can include accessibility indicators such as service availability, service access success ratio, etc., retainability indicators such as drop ratios, and integrity indicators such as service access time, round trip time and more. Mapping the indicator on the coverage database brings the insight of the wireless planning process into those indicators and provides means to optimize the network accordingly.

20.3 The Impact of WiMAX on Network Planning

20.3.1 Flexibility of WiMAX Deployment

From the network operator point of view, WiMAX brings to the table, first and foremost, flexibility. There are many aspects to this flexibility.

- *Trade-off throughput with robustness.* Mobile stations which enjoy ideal channel conditions can use the full rate of the channel using spatial multiplexing, together with high bandwidth and very high level modulation and coding. This could bring the spectral efficiency of the link to be in an order of magnitude of tens of bits per second (bps) per Hertz. On the other hand, the WiMAX system has at its disposal a set of tools to support Mobile Stations that do not have good reception conditions. These include space diversity, robust modulation and coding, decrease in the allocated bandwidth, Automatic Request (ARQ) and Hybrid Automatic Request (HARQ) mechanisms. All of these techniques increase robustness and hence increase operation range and enable operation in difficult reception conditions, such as indoors, at the street level or under high interference conditions. To obtain the increased robustness one must trade-off throughput.

- *Trade-off in frequency allocation.* In traditional cellular systems, channel allocation was rigidly made per cell. The introduction of CDMA techniques enabled a trade-off between system load and interference. Practically it enables operation of a whole system with multiple cells, using a single wide frequency channel. Thanks to the CDMA technique, each transmission interferes very lightly with other transmissions.

Only the aggregation of interference caused by many users may inhibit the operation of the system. In a WiMAX system, which uses OFDMA as the basic modulation scheme, the subcarriers comprising the OFDMA symbols can be allocated orthogonally like a traditional system or, similarly to CDMA, the same subcarriers can be allocated to several transmitters thus creating interference. However, thanks to the statistical nature of the use of the common channel and to the coding scheme that is able to correct errors caused by interference, the transmissions involved can be successfully decoded. The interference is a function of the system load and, as in the CDMA case, there is a trade-off between load and interference. In an extreme situation WiMAX can be operated using a 'universal reuse 1', namely all of the cells are allocated a single RF channel.

WiMAX offers another degree of flexibility: the concept of zones, in which the frame is divided into time intervals, each with its own set of parameters. This allows the frequency assignment plan to change within a basic frame. Thus, interference free planning can be applied to part of the frame, in which nonoverlapping sets of subcarriers are allocated to different cells, then, in another zone, the whole bandwidth can be applied to all of the cells, and the interference between them is mitigated thanks to statistical nature of the load.

- *Flexibility in traffic assignment.* WiMAX basic profile is Time Division Duplex (TDD), in which transmission time is divided into base station transmission and terminal station transmission, to enable bidirectional communications. This mode allows flexible allocation of resources for downlink and uplink traffic.

- *Advantage in multi-user environments.* WiMAX offers the ability to enhance the overall system performance in case of multi-user environments. The first mechanism is that of a localized allocation, which uses the selective nature of the channel to allocate to each user the best set of subcarriers, as it is most likely that those subcarriers would not be identical for all users, it allows the best usage of the system subcarriers. Another mechanism is of collaborative spatial multiplexing, in which two terminals share the same time and frequency resources, but thanks to the different spatial channel, the base station can separate the transmissions. A similar technique, spatial multiplexing from two base stations, can be applied to a single terminal station, if it is equipped with several antennas.

- *Variety of user equipment and user services.* WiMAX provides a lot of possibilities to the operator to offer a variety of services within a large set of cost and pricing schemes. WiMAX can support fixed wireless applications, with outdoor antennas or self-install indoor antennas. It can support mobile applications with a variety of user equipment. WiMAX also supports quality of service, allowing the operator to slice the market and provide a cost-effective service according to the customers needs.

20.3.2   WiMAX Network Planning

Let us now follow the network planning process described above, and discuss the impact that WiMAX might have on it.

20.3.2.1   Data Collection

The main difference between a deployment of a WiMAX system and a non-WiMAX system is in the variety of services enabled by the system. In view of this the market data, market penetration, service offers and pricing schemes should be prepared and fit a viable business plan. In addition, one of the key factors in the deployment is the spectrum allocation. As WiMAX is capable of a 'reuse 1' deployment, it could be envisaged that only a single frequency channel is allocated for the deployment.

20.3.2.2   Network Planning

The impact of the new WiMAX features is the greatest at the stage of estimating the required cell capacity and frequency planning. Obviously, as WiMAX trades off interference with capacity, the two stages are highly linked and should be treated together. Frequency planning in WiMAX is not limited to the assignment of frequency channels to the radios. Actually this is only the first stage in the planning, and in some deployments, where only a single frequency channel is allocated, this is a trivial stage. In fact, at this stage a large set of parameters has to be planned. A better name for this stage would be cell configuration phase and it comprises of the following parts:

1. planning of the segmented zones;

2. planning of the full usage zones;

3. throughput estimation for the users covered by each cell;

4. frame planning;

5. power planning.

The segmented zones are the zones in which the cells use only a part of channel bandwidth. Cells that are allocated different segments of the spectrum do not interfere with each other. In the full usage zones there is interference between cells, but it is controlled, and is a function of the load. Full usage zones enable basically higher capacity, as resources in those zones can be reused by neighboring cells. For each cell, one must determine the segment each base station will take when operating in a segmented zone, the preamble sequence, the permutation bases for the downlink and uplink directions and the sequence used for pilot modulation and data randomization, etc.

The throughput a user can enjoy is highly dependent on the SINR, channel conditions and equipment capability. The average resource requirements per frame (in terms of time–frequency physical slots) are a function of the throughput and the user demand. At the planning stage one cannot know exactly the users' demand, equipment capability and operation conditions by location, however, assuming some user distribution the global user demand and equipment capability can be translated into specific user distribution per cell, which, in conjunction with throughput estimation of the bins within the cell coverage area, can be used to estimate the resource requirement per cell.

This resource requirement is needed for frame design. It is well known that in TDD systems, the downlink and uplink intervals should be switched simultaneously in all of the cells, to avoid base station to base station and mobile station to mobile station interference.

The same is true with the segmented usage and full usage zones. Those zones should be allocated exactly the same time intervals in all cells, to avoid interference between full usage zones and segmented zones. The process of frame design is basically location of the switching point between the zones, and its goal is to provide adequate resources to each zone according to the demand. It is an iterative process whereby the user demand per cell is partitioned between the zones. Allocation of a user to the segmented zone will enable its operation with less interference, but it will require more time slots as the base station is limited to a smaller number of subcarriers. Moving the demand to the full usage zone will enable to reduce the number of required time slots but it might expose the receiver to interference. It is then expected that users at the cell edge will be allocated to the segmented zone, while those which are not susceptible to interference will be allocated to the full usage zone.

Another dimension of planning would be power adjustment. Basically it is the scheduler function; however, at the planning stage it is important to assess the effect of power adjustment. The procedure can be performed by reducing the power allocated to users for which the allocated power is above the threshold needed for the particular modulation and coding state they can operate with. In the downlink this power can be allocated to other users, without affecting the interference to other cells, or decrease the total transmission power of the base station, reducing the interference to other cells. In the uplink, reducing the power can only reduce the interference caused to other base station receivers. Reduction of the interference will increase the possible supported rate.

All of those processes are inter-related and dependent on each other. The planning process is usually iterative; starting from a certain frequency and zone plan, the SINR distribution can be calculated, and out of this, and the channel conditions, the throughput can be estimated, and the resources required per cell evaluated. Following frame design and power allocation planning, a new frequency and segment planning can be made.

20.4   Planning of Future WiMAX Networks

Browsing this book, one cannot fail to be impressed by the potential and extent of new developments that can clearly extend the capabilities of future WiMAX systems and the new services and applications it can support. We have chosen to concentrate on the impact on network planning of five of those developments:

- advanced spatial techniques;

- mesh, relays and femtocells;

- self-configuring networks;

- cooperative WiMAX;

- cognitive radios.

20.4.1   Advanced Spatial Techniques

Advanced spatial techniques, such as beamforming, interference cancellation, diversity enhancement and spatial multiplexing are not techniques of the future but are very much present-day and currently implemented. Those techniques make use of the fact that multiple antennas are present in the base station, mobile station or in both. A very brief list of those techniques is given below, with the impact they have on network planning.

- *Receive diversity and transmit diversity*: improves link performance in fading conditions and hence reduces the fade margin that needs to be taken.

- *Beamforming*: increases the effective antenna gain towards a wanted user and decreases the power radiated in other directions. The impact on network planning is that some statistics have to be applied for the beam direction, in order to estimate the interference caused by an interfering base station to a victim mobile, as the transmission direction is not always the same.

- *Interference cancellation*: helps avoid energy being transmitted in a direction where a victim is found, and in suppressing the reception of an interfering signal coming from an unwanted direction. The impact of such a technique on planning is highly dependent on the specific implementation.

- *SDMA*: enables a number of users to be served simultaneously, using the same frequency resources, by using beamforming to a wanted user and interference cancellation to unwanted users. The effect on network planning is by increasing the number of users a base station is capable of supporting.

- *Spatial multiplexing*: using multi-user detection techniques, a receiver with multiple antennas can separate the transmission arriving at it from several different source antennas, as long as the number of sources is less than or equal to that of the receiver antennas and the medium is scattering rich, namely it provides enough independent paths such that the separation is possible. The source antennas can either belong to the same transmitter, thus increasing the throughput of a link, or belong to different transmitters thus enabling a better use of the time and frequency resource by increasing the number of users. A network planning tool has to be able to predict, in addition to the propagation loss and interference strength, if the channel can support spatial multiplexing, and also estimate the link, cell and system capacities when those techniques are used.

20.4.2 Relays, Femtocells and Mesh Networks

Relays, femtocells and mesh networks shift the paradigm of the point-to-multipoint cellular network into a distributed network and extend the range of solutions for network deployment. Relays extend the coverage of a cell beyond the reach of a base station. The relay should be a device that is simpler and much less expensive than a base station. There are many types of relays, but we should distinguish between in-band relays and out-of-band relays, The in-band relay uses the resources of the cell both for the base station to relay links and for the access link between the relay and the mobile station. The out-of-band relay uses other means (microwave point-to-point link, optical fiber, etc.) for the base station to relay link. From the point of view of the wireless network the relay transforms the point of transmission closer to the mobile stations, thus allowing the transmission to overcome obstacles, reducing the transmission power needed both on the relay side and on the mobile station side and thus it may reduce interference. An in-band relay requires that resources be allocated also for the base station to relay link, thus reducing the total available capacity.

Relays can be considered as an additional means for the planner, replacing costly base stations with relays, while loading the traffic they cover on the source base station, as long as it is possible.

Femtocells offer the possibility for private users to install a small base station at their premises, thus providing the counterpart of fixed service by a mobile operator, bundled with smooth handover to the cellular network when on the move. With such a solution available, a lot of the load is taken off the outdoor wireless network, which can now focus on outdoor coverage. The indoor coverage and indoor penetration problem is relieved.

Planning a system with femtocells could be made similarly to the way conventional planning is made, with the difference that the exact location and often the number of the femtocells are not known. System dimensioning, in terms of traffic and system capacity, can be performed assuming a certain penetration of femtocells. As for possible interference, it should be emphasized that although femtocells are supposed to be low-power devices, with the expected density of those devices, the interference cannot be taken as negligible. Means have to be taken to control the interference between the femtocells and the outdoor network and between a femtocell and its neighbors. The first problem can be handled by allocation of different resource (e.g. other frequency channels) to the femtocells or, by the planner consideration, take the random interference into account by requiring higher carrier to interference ratio and assuming the interference results in some uniform noise rise. The extent of the second problem is more difficult to predict as no previous experience with femtocells exist. Judging from the spread of Wireless Local Area Networks (WLANs) access points nowadays, any single mobile station may be exposed to receiving tens of femtocell signals in an urban environment. The femtocells should be made to coexist in such a dense environment.

Mesh networks use SSs to relay information to other users. Information can travel in those networks in any path: between users or through the base stations. Many network structures exist. In one extreme the base stations serve only as points of presence of the Internet and the outside world. In other schemes, the base station serves as a controller for the traffic of the SSs within its coverage zone. In other cases traffic must be routed through the base station in a tree-like topology, where the base station is the root of the tree.

Planning of conventional cellular networks reflects the centralized nature of those systems. Deployment planning of distributed systems, such as mesh networks, should take this fact into consideration. The planner cannot initially know the traffic routes, and hence is not able to tell the load, interference patterns or link performance and throughput. In order to dimension the system, the network capacity should be evaluated taking into account the fact that a terminal is not only a traffic source but also provides capacity to the system. For a green-field deployment the planner has to provide the infrastructure to enable basic operation, and can start with a coverage only deployment. As more users join the network, each may provide additional traffic routes that provide additional capacity to the system.

20.4.3 Cognitive Radios, Self-configuring and Cooperative Networks

With the development of computing power and processing capability, it is only natural to find more and more intelligence on the front-end of the communication link, namely at the base stations and terminal stations.

Cognitive radios are defined (IEEE, 2007) as a type of radio in which communication systems are aware of their environment and internal state and can make decisions about their radio operating behavior based on that information and predefined objectives. Such a radio may recognize interference and perform a set of actions to avoid, mitigate or suppress it. It can also sense the traffic, assess and analyze it and determine its internal configuration to optimize its performance. When a group of such radios cooperate in order to achieve common predefined objectives, we are now looking at a self-configuring and cooperative network, which can share the load among the radios, configure and coordinate their operation to avoid mutual interference, provide diversity and spatial multiplexing, as described in other chapters in this book.

With this intelligent and environment awareness, many of the functionalities can now be implemented within the base stations themselves, with the capability of performing them online and adaptively. It seems as if the planning tool of tomorrow will probably be a module within the base station.

However, the following points should be considered.

- A tool is still needed for green-field deployments.

- The optimization performed by a local entity is only local, while performance objective are global in nature.

- The planning tool database contains a large amount of valuable information, such as geographical information, prediction information, etc., which is needed for optimization.

The future planning tool should be envisioned as a part of a global optimization system that fuses network measurements, dedicated measurements, geographical information and predictions; functionalities that can be found today in separate planning, management, drive test analysis, network probes and optimization tools. The operation of such a tool is based on the initial database created by the planning tool. This database includes the geographical information, traffic load estimate information and network configuration. The tool can create a set of predictions for the entire coverage area, including indoor on ground and top floors. In urban areas, using ray tracing, a set of rays can be defined for each point, which could be valuable information for a base station to set a link to a user at that point. Measurements made by drive tests, real or virtual, and other dedicated measurement devices can be used to update and correct the predictions. This central database can be used by the base stations and other network elements for self-configuration. As the network develops, the inherent intelligence of each base station can be used for local optimization, while the central system provides global optimization solutions.

20.5 Modeling: the Key to Integration of Planning Information

In order to achieve the integration of the planning information described above, into a single solution, it is essential to monitor and track the network behavior along its lifetime. Wireless networks in general are susceptible to coupling between various parts of the network; a site

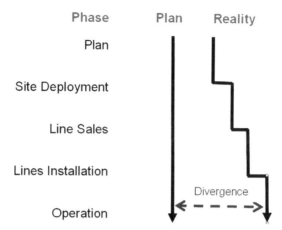

Figure 20.4 Divergence of the plan accuracy during deployment phase.

in one area can cause interference to distant sites. In addition, equipment malfunctions are indistinguishable from intra-system interference

Knowing the location of the terminal stations is of paramount importance for both fixed and mobile networks. Fixed wireless networks in general are characterized by known terminal station locations that should be served with high quality of service, base station configurations that are highly dependent on customer distributions and usage pattern and inherent inflexibility in the network growth. In mobile networks, the location of a call can improve significantly the performance of optimization and coverage assessment procedures.

20.5.1   The Problem

The process going from drawing board to a real operating wireless network may be described as follows:

- planning phase, in which the network is planned;

- site deployment phase, where base stations are deployed;

- line sale phase, in which users are joining the network;

- line installation phase, which may exist in case of fixed network deployment;

- operation.

The main impediment for an operator is a constantly growing discrepancy between the plan and the reality, that is, between the existing database, and the actual traffic behavior and propagation environment, as depicted schematically in Figure 20.4.

There are several sources for the accumulated gap between plan and reality:

- inaccuracy in the geographical database;

- actual sites location, structure and configuration;

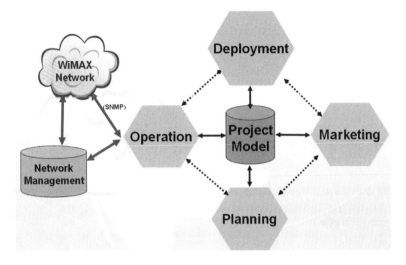

Figure 20.5 Company-wise integration solution.

- actual propagation environment;

- actual customer distribution and used demand.

On each of these phases the operator has to take operational decisions based on the available information: site locations, derived coverage and supplied capacity, frequency planning and fault location.

Consequently, the resulted network has an improvable network quality; it suffers from under utilization of resources, it is hard to control, it has high maintenance costs and complex fault location process.

20.5.2   Suggested Solutions

The solutions for the growing discrepancy between the plan and reality rely on the implementation of feedback from the network and integration of information, which may be found in different departments of the operator organization.

20.5.2.1   Company-wise Integration

The suggested solution is to create and maintain a model of the project, that is, a constantly updated geographical database, calibrated propagation calculations and modified user-demand models. The information is to be gathered across the operators departments via integrated software suite comprising four components as depicted in Figure 20.5.

The Network Management System (NMS) is the native system developed by the equipment vendor or by an independent vendor. Other software packages are included in an operator organization, for example, the customer database software and enterprise resource planning software. Note that the suggested solution is not designed to replace these software tools but to connect them to the model. The main idea behind this concept is to turn the model into common knowledge across the company. Deployment teams, which deploy base stations and, in the fixed network case, terminal stations, update the project model. The marketing team consults the project model for coverage information and updates the planning department for current and future user demand information. The operation team analyzes measurements and reports from the NMS and from the network elements via SNMP and interacts with the model by updating the coverage information and performing optimization.

The model enhances the whole operator's organization. It facilitates coordination of operator activities: planning, deployment, sale and operation. It constantly monitors the actual structure and performance of the network and provides a geographical Graphical User Interface (GUI). The main benefit is that the model allows online network optimization: constantly monitoring the frequency plan and the traffic pattern and provide appropriate alarms to the network manager, testing network manipulation on an updated model, providing platform for fault location expert system, calculating new cell service policy and guiding the sales/marketing efforts.

20.5.2.2 Calls Positioning: VDTs

The information integration solution described above is not sufficient for a mobile operator, as one of the key components: the terminal station location is not known *a priori*. Figure 20.6 describes a proposed solution. The solution is based on analysis of events (e.g. normal call, dropped calls, transfer of calls from one technology to another) collected by probes, which trace the various interfaces. Using the probe data observations the tool performs VDT which accurately locates calls on a geographical three-dimensional map and uses the measured signal strength information for the located event to assess the coverage in that area as well as any other applicable information such as distance measure and timing information. The next stage would be to apply automatic optimization to the network, which includes:

- cell configuration (e.g. azimuth, tilt);

- frequency and RF parameters planning:

    - measurements-based impact matrix;

    - accurate balance of full and partial frequency schemes;

- neighbor list;

- load balancing via handover parameters optimization.

At this stage the optimization tool automatically identifies and reports problematic spots, coverage holes, etc.

20.6   Conclusions

Network planning is the primary stage before any wireless system deployment. The network planning tools is essential for any operator for dimensioning the network, assess vendor proposals and provide a deployment plan that provides the necessary capacity and coverage

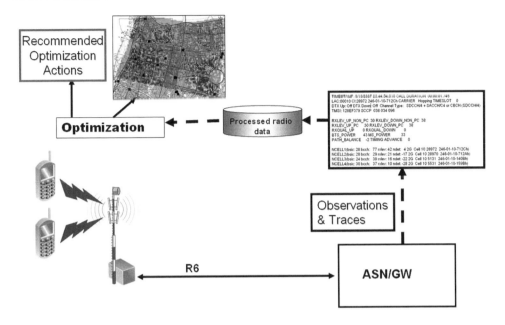

Figure 20.6 A mobile operator system feedback solution.

with the most cost-effective set of sites. The network planning process is actually an on-going process that takes place along the whole lifecycle of a wireless network, and should be reiterated to answer new demands, extend the coverage or improve network performance.

WiMAX is a broadband wireless access system that can provide high-speed data connections to fixed and mobile applications. From the network planner point of view, WiMAX provides a degree of flexibility and scalability that enable planning a system with a more stringent set of capacity, coverage and spectrum constraints than before, including the usage of spatial techniques, fractional frequency reuse, zone planning and more. However, to make use of those possibilities, the planning process has more parameters to plan and configure than traditional systems.

Future networks will include additional means to the planner to provide better and more cost-effective deployment solutions. Such means include relays, femtocells, mesh network solutions, etc. Future networks will also include processes and tools that will allow better modeling of the dynamic nature of the network, mainly the demand pattern, customer's distribution and resources used.

In the near future intelligence will find its way into the network system elements. Self-configuring networks, self-organizing networks, cooperative network elements and cognitive radios are some of the techniques that enable intelligent network behavior. Some of the functionalities found nowadays in the planning and optimization tools will be found within the network elements themselves, which in turn will interact with a central entity that will provide it with the geographical and coverage information needed for intelligent decisions.

WiMAX technology provides mechanisms to improve network performance via careful planning and optimization, in order to benefit from these mechanisms the optimization process has to take into account the information available in many shapes and forms across the network operator organization. All information gathering solutions should have the ability to translate feedback and measurements into a geographical information system format. This can be done using VDTs, which transform measurements made by subscriber terminals into a drive test associating them with a specific location.

Thus, the future network will witness the integration of a variety of support systems: network management, customer database, enterprise resource planning and radio planning and optimization systems, together with the inherent intelligence embedded within the network elements themselves to provide a better solution for wireless network operation.

References

Aguado-Agelet, F., Varela, A., Alvarez-Vazquez, L. *et al.* (2002) Optimization methods for optimal transmitter locations in a mobile wireless system. *IEEE Transactions on Vehicular Technology*, **51**(6), 1316–1321.

Anderson, H. and McGeehan, J. (1994) Optimizing microcell base station locations using simulated annealing techniques. *Proceedings of the IEEE Semiannual Vehicular Technology Conference*, Vol. 2, pp. 858–862.

Andrews, J. Ghosh, A. and Muhamed, R. (2007) *Fundamentals of WiMAX*. Prentice Hall, Englewood Cliffs, NJ.

Blaunstein, N. (1999) *Radio Propagation in Cellular Systems*. Artech House, Boston, MA.

Chavez-Santiago, R., Raymond, A. and Lyandres, V. (2007) Enhanced efficiency and frequency assignment by optimizing the base stations location in a mobile radio network. *Wireless Networks Online First*.

COST (1999) Digital Mobile Radio Towards Future Generation Systems, *COST Action 231*.

Eklund, C., Marks, R., Ponnuswamy, S. and Stanwood, K. (2006) *WirelessMAN: Inside the IEEE 802.16 Standard for Wireless Metropolitan Area Networks*. IEEE Press, New York.

Hata, M. and Nagatsu, T. (1980) Mobile location using signal strength measurements in a cellular system. *IEEE Transactions on Vehicular Technology*, **29**(2), 245–252.

Huang, X., Behr, U. and Wiesbeck, W. (2000) Automatic base station placement and dimensioning for mobile network planning. *Proceedings of the 52nd IEEE Semiannual Vehicular Technology Conference*, Vol. 4, pp. 1544–1549.

Hurley, S. (2002) Planning effective cellular mobile radio networks. *IEEE Transactions on Vehicular Technology*, **51**(2), 48–56.

ITU (2007a) Method for point-to-area predictions for terrestrial services in the frequency range 30 MHz to 3000 MHz. *Technical Report 1546 IRRP*, ITU.

ITU (2007b) A path-specific propagation prediction method for point-to-area terrestrial services in the vhf and uhf bands. *Technical Report 1812 IRRP*, ITU.

IEEE (2004) Part 16: Air Interface for Fixed Broadband Wireless Access Systems, *IEEE Standard 802.16-2004*.

IEEE (2005) Part 16: Air Interface for Fixed and Mobile Broadband Wireless Access Systems Amendment 2: Physical and Medium Access Control Layers for Combined Fixed and Mobile Operation in Licensed Bands and Corrigendum 1. *IEEE Standard 802.16e 2005*.

IEEE (2007) Draft Standard Definitions and Concepts for Dynamic Spectrum Access: Terminology Relating to Emerging Wireless Networks, System Functionality, and Spectrum Management. *IEEE, Draft Standard P1900.1/D2.0 1*.

Laiho, J. Wacker, A. and Novosad, T. (eds) (2007) *Radio Network Planning and Optimisation for UMTS*. John Wiley & Sons Ltd, Chichester.

Lee, W.C.Y. (1986) Elements of cellular mobile radio systems. *IEEE Transactions on Vehicular Technology*, **35**(2), 48–56.

McDonald, V.H. (1979) The cellular concept. *Bell Systems Technical Journal*, **58**(1), 15–41.

Mishra, A.R. (ed.) (2007) *Advanced Cellular Network Planning and Optimisation: 2G/2.5G/3G. . . Evolution to 4G*. John Wiley & Sons Ltd, Chichester.

Molina, A., Nix, A. and Athanasiadou, G. (2000) Optimised base-station location algorithm for next generation microcellular networks. *Electronics Letters*, **36**(7), 68–669.

Niessen, T. (1997) Optimal channel allocation for several types of radio networks. *Discrete Applied Mathematics*, **79**(1–3), 155–170.

Ott, G. (1977) Vehicle location in cellular mobile radio systems. *IEEE Transactions on Vehicular Technology*, **26**(1), 43–46.

Parsons, J.D. (1992) *The Mobile Radio Propagation Channel*. Pentech Press, London.

Riter, S. and McCoy, J. (1977) Automatic vehicle location – an overview. *IEEE Transactions on Vehicular Technology*, **26**(1), 7–11.

Struzak, R.G. (1982) Optimum frequency planning: a new concept. *Telecommunication Journal*, **49**, 29–36.

WiMAX Forum (2006) WiMAX Forum mobile system profile, http://www.wimaxforum.org/technology/documents/wimax_forum_mobile_system_profile_v1_40.pdf.

Wigren, T. (2007) Adaptive enhanced cell-ID fingerprinting localization by clustering of precise position measurements. *IEEE Transactions on Vehicular Technology*, **56**(5), 3199–3209.

Yang, S. and Ephremides, A. (1997) Optimal network design: the base station placement problem. *Proceedings of the 36th IEEE Conference on Decision and Control*, Vol. 3, pp. 2381–2386. IEEE, Press, Piscataway, NJ.

21

WiMAX Network Automation: Neighbor Discovery, Capabilities Negotiation, Auto-configuration and Network Topology Learning

Alexander Bachmutsky

21.1 Introduction

Mobile networks frequently have very complex hierarchical architectures (for example, Node B–RNC–SGSN–GGSN). WiMAX networks are not much different with Access Service Networks (ASNs) consisting of Base Stations (BSs) and ASN Gateways (ASN-GWs), and Connectivity Service Networks (CSNs) consisting of Home Agents (HAs), Authentication, Authorization and Accounting (AAA) proxies and servers, Dynamic Host Configuration Protocol (DHCP) proxies, relays and servers, IP Multimedia Subsystem (IMS) infrastructure and more. The deployment and management of such networks is a huge task and one of the biggest investments from both Telecommunications Equipment Manufacturers (TEMs) and operators. Based on the experience of many different projects from many TEMs, the development of Operations and Management (O&M) infrastructure and products counts for as much as a half of all development costs.

At the same time standards do not really help to ease the deployment and management of mobile networks. Of course, there is a Simple Network Management Protocol (SNMP) that is used for some management tasks, but it is very basic and cannot be used for many complex management jobs, and definitely not the deployment ones. The WiMAX Forum does not

WiMAX Evolution: Emerging Technologies and Applications   Edited by Marcos D. Katz and Frank H.P. Fitzek
© 2009 John Wiley & Sons, Ltd

pay much attention to the problem either: it concentrates more on the terminal management (over-the-air provisioning), not on the network management.

In addition, any network automation should take into account the dynamic nature of the network, which is affected by upgrades and downgrades, failures and recoveries, overload situations and mobility changing a processing chain per subscriber.

In this chapter we try to describe some related topics and to at least ignite a discussion of the WiMAX network automation helping all involved parties to build better, more reliable and lower cost solutions.

21.2   WiMAX Network Elements Auto-discovery

Performing the relatively simple task of adding one BS to the existing network requires configuring this BS with all neighboring BSs, configuring all neighboring BSs to include the new one, configuring the new BS with all logically connected ASN-GWs and configuring all of these ASN-GWs to include the new BS. Taking into account that the operator needs to perform such operations for every deployed BS, it is understandable why they require so complex and sophisticated O&M tools.

The first step in network automation is to discover the neighbors instead of a manual provisioning. BSs can discover their neighbor BSs from their radio resource management, either by themselves or with the help of a terminal. In the latter case the terminals report BSs that they can hear, and such information can be processed and saved by all BSs. Operators can 'train' the network during the deployment stage by driving around with a few terminals (they do this anyway in many cases for various reasons). On the other side, such learning will only provide information about BS identification, but will not provide the networking parameters (for example, Internet Protocol (IP) addresses). It is also possible to apply techniques and basic learning principles from mesh networking and Self-Organized Networks (SONs) that can help to automatically build a list of neighboring devices.

There are multiple ways to automatically learn a mapping between the WiMAX information (WiMAX Network Element (NE) identifications) and the transport information (Media Access Control (MAC) and IP addresses).

The first option is similar to a scheme that allows IPv4 routers to auto-discover each other (Deering, 1991). The proposal is to reuse ICMP Router Discovery from RFC 1256 in *WiMAX Network Element Advertisements* (WNEAs) by including the *WiMAX Network Element Advertisement Extension*. The presented mechanism covers only an IPv4 advertisement, but it can be easily extended to the IPv6 case.

WNEAs are sent using the network element IP address as a source and either 'all systems on this link' multicast address (224.0.0.1) or broadcast address (255.255.255.255) as a destination. In addition, WNEAs can be sent with a unicast address if such an address is known *a priori*, for example, from a previous WNEA. Usually, a new NE would advertise itself to the network using a multicast or broadcast IP address, while existing NEs would advertise themselves to this new one using its unicast address taken from a previously received multicast/broadcast packet. To eliminate a need for solicitation request, it is mandated to send the WNEA if another WNEA was received from a previously unknown network element.

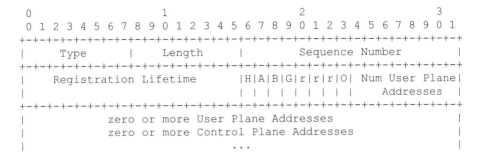

Figure 21.1 WNEA extension.

WiMAX NEs can have separate user and control planes with separate IP addresses. To accommodate such a case it is possible to include all user plane IP addresses in a standard Internet Control Message Protocol (ICMP) portion of the packet in the 'Router Address(es)' field. If the NE does not have a user plane component, the ICMP message 'Num User Plane Addresses' field shall be set to zero; otherwise it is set to the number of addresses provided in the 'Router Address(es)' field. All control plane IP addresses will be specified using the WNEA Extension (shown in Figure 21.1 in a form used frequently in IETF documents).

- *Registration Lifetime* is the time in seconds that this element should be kept as a valid WiMAX NE in tables of other NEs without receiving an updated WNEA. A value of 0xFFFF indicates infinity. Registration lifetime is the same for all functions advertised in the message. If there is a need for different registration lifetimes for every function, separate WNEA messages should be sent.

- *H*: HA functionality is included in this NE if set to 1.

- *A*: AAA server functionality is included in this NE if set to 1.

  B: BS functionality is included in this NE if set to 1.

- *G*: ASN-GW functionality is included in this NE if set to 1.

- *r*: Reserved bit.

- *O*: overload/busy flag; setting this to 1 means that this NE should not be used for any new connections until the end of a registration lifetime or new advertisement without an 'O' flag being set. This flag is applicable to all advertised functions (H, A, B and/or G). This flag is not relevant purely for a neighbor discovery, but shows that it is possible to perform both discovery and a status exchange in the same message.

- *User Plane Addresses*: IP addresses used for the transport of end-user packets. All addresses will apply to all advertised functions.

- *Control Plane Addresses*: IP addresses used for a WiMAX control/signaling plane. All addresses will apply to all advertised functions.

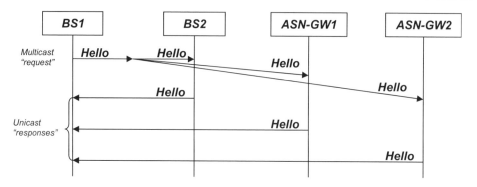

Figure 21.2 Exchange of *Hello* messages.

The above mechanism allows the NE to learn all IP addresses of its neighbors and their role in the WiMAX network. To create a mapping between WiMAX identifications (BSID, Authenticator ID, etc.), we can either add them to the WNEA Extension or use the described further *Hello* exchange with now known unicast destination address.

An alternative advertisement solution can be a WiMAX control plane *Hello* message sent to a known unicast or to a pre-configured multicast destination IP address (see Figure 21.2). As any other WiMAX message, *Hello* would be using the User Datagram Protocol (UDP) as a transport protocol with well-known WiMAX source and destination port values set to 2231. One of problems is that the existing WiMAX transport protocol does not support multicast transactions when a single request message can cause many replies, and the best way would be to include multicast transactions handling the standard. Unfortunately, it is not that simple, because the current transaction layer in WiMAX is very limited and requires a transaction numbering based on peers' information. In the case of multicast, the request has a multicast destination IP address, the response has specific IP address of the NE; therefore, it would be very hard to match the response to its request. Multicast transaction support practically asks for a unique identifier or the addition of an original multicast address in the response. Another option is to make *Hello* messages a one-way transaction (sent multiple times to minimize potential packet loss effect), where from the transaction management point of view the response becomes a new 'independent' *Hello* transaction initiated in the opposite direction.

In any case, a *Hello* transaction should be the first transaction sent after the NE (re-)start, but it also can be sent at any point of time if needed.

One of *Hello* message components is a WNEA that can include the following information.

- *Network Element Type*: equivalent to H, A, B and G above. Advertises HA, AAA server, BS or ASN-GW functionality.

- *Registration Lifetime*: similar to the field with the same name described above. Here it refers to the specific advertised NE ID.

- *Control Plane Addresses*: advertises all relevant control plane addresses for the corresponding NE (some implementations have a single control plane address, some have multiple). Here we would include all types of addresses by having the address type specified. For that purpose the WNEA would consist of *Number of Addresses* field and an *Address Description* for every advertised address with *Address Length*, *Address Type* (Ethernet MAC, IPv4, IPv6, etc.), *Address Value* and *Address Status*. The last element can be used for signaling health conditions (*OK*, *OVERLOAD*, *FAIL*) of this particular address. The reason for the status being set per address is to cover a bladed (or similar) architecture, where different IP addresses are assigned to different instances for internal load balancing purposes; the status would point to the health of a particular instance. For example, if an address becomes unavailable for any reason (the blade crashed or was removed, address removed by configuration), it has to be included in the WNEA with *Address Status* set to *FAIL*. There is no need to advertise every address in every *Hello* message, only changes have to be advertised; exceptions would include the first message right after its own restart or responding to the NE after the restart. However, if *Hello* is implemented as a one-way transaction, there is a need to advertise every change for the pre-configured amount of times to ensure that a single message loss does not affect the entire network behavior.

For the purpose of mapping between WiMAX identification(s) and IP addresses we will also add one or more WiMAX Identification Information Element(s) (IEs) with *Identification Type* (BS, Authenticator, etc.), *Identification Length* and *Identification Value*.

- *User Plane Addresses*: similar to the above, but related to user plane addresses. There is no need for *WiMAX Identification* for user plane addresses.

There are other ways to discover the network. Each WiMAX NE can register in some server (for example, the Lightweight Directory Access Protocol (LDAP) is used here, but it can be DNS-SD, IETF Service Location Protocol described in Veizades (1997), and others) and make inquiries about the network from the same server. The difference from other schemes is that there is no built-in information about neighbors: the NE would know all other NEs, but it would not know its immediate neighbors. In some cases it actually does not matter, for example, ASN-GWs are connected in a logical full mesh anyway. A huge advantage is that the scheme allows the entire list of all devices in the network to be obtained at once. Frequently, the solution for neighborhood information is some kind of a domain configuration. One could configure BSs and some ASN-GWs as belonging to the 'California Bay Area' domain (each NE can be included in multiple domains). With such a configuration a newly deployed BS would advertise itself to the server as being a part of that domain, inquire about all ASN-GWs in the same domain and select one or more for R6 connectivity.

The diagram in Figure 21.3 shows the procedure of adding a new BS to the 'California Bay Area' domain with already existing ASN-GW1, ASN-GW2 and ASN-GW3. It assumes that LDAP implementation supports notification of new registration (the LDAP v3 feature added in Wahl *et al.* (1997)), but it can be replaced by a LDAP inquiry to obtain information about the domain members.

The scheme still requires a domain configuration, but it is definitely simpler than what is happening today.

As a result of one of the above procedures, WiMAX NEs will be aware of other WiMAX NEs in their network and also their reachability information.

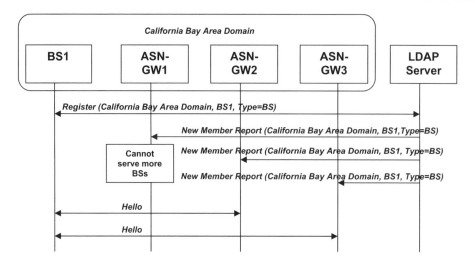

Figure 21.3 LDAP-based NE registration and discovery.

21.3   Automatic Learning of the WiMAX Network Topology

Knowing the immediate WiMAX neighbors opens a way to the entire WiMAX network topology learning. The learning principle can be taken directly from routing protocols, and one scheme described below is even utilizing such protocols. The idea is to advertise your neighbor information to the rest of the network.

One way is to introduce a capability to relay *Hello* messages with the addition of a new *Originator Identifier* as a part of the *Network Element Advertisement*. It is similar to the existing *Source Identifier*; and it provides the identity of the NE that initiated the neighborhood information. Relaying *Hello* messages can provide information about BSs connected to other ASN-GWs (required for the cases when the Anchor or Target ASN-GW is not co-located with the Serving one, and also helps in the topology-aware paging to wake-up idle terminals), BSs that can be connected to multiple ASN-GWs simultaneously (multiple ASN-GWs had reported the same BS as 'belonging' to them), BSs under the same ASN-GW for an optimized handover decisions, and many more uses. Figure 21.4 shows such *Hello* relay function with a table of WiMAX NEs and their WiMAX next hops.

The second possible implementation to learn network topology is by using real routing protocols. The idea is based on the fact that many WiMAX NEs are running routing protocols (Open Shortest Path First (OSPF), Border Gateway Protocol (BGP)) anyway, so why not make use of them? The only difference from 'regular' IP routing sessions is that we will run them over WiMAX interfaces (reference points): R3/R4/R6/R8. Routing protocols can, of course, provide information that some WiMAX NE has an IP address X, but can it do more than that? In general, the answer is 'yes', and there is no need to add some proprietary extensions: routing protocols can stay without modifications and without awareness of the special functionality that they are actually performing.

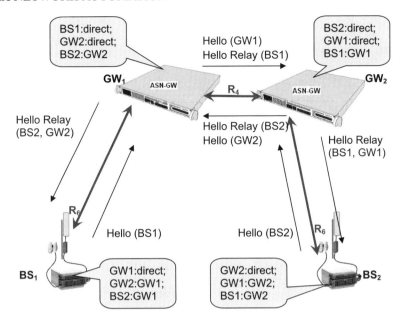

Figure 21.4 Network topology learning based on *Hello* message relay.

One very powerful enabler for this is a virtual routing, which was designed to separate the address space of multiple networks and allows overlapping addresses. It is easy to separate WiMAX network into a WiMAX virtual routing domain where only the addresses of WiMAX NEs appear. It is possible to go further and introduce virtual routing domains for WiMAX BSs, WiMAX ASN-GWs and so on; so not only will the addresses be advertised, but also the type of NE based on the corresponding Virtual Router (VR) ID. Routing protocols also support routing distribution policies to enable one-way or two-way address distribution between VRs, for example, learn BSs by ASN-GWs using the distribution of addresses from the BS VR domain into the ASN-GW VR domain; at the same time the policy can either allow or prevent learning in the opposite direction.

Routing protocols include also a route metric that is also called the route cost. Different costs can be assigned to different VRs or different WiMAX reference points. Using WiMAX route metrics can change the current method of topology aware decisions. To give a preference for R6 handover over R4 handover, it is necessary to assign R6 a lower cost compared with R4 and choose the lowest WiMAX route cost instead of a search in the database that contains the list of connected ASN-GWs and their corresponding BSs. It is just taking advantage of the fact that WiMAX network is in fact an overlay over a regular IP network. In Figure 21.5 describing the use of the routing protocol, it is assumed that all BSs and ASN-GWs are capable of running routing protocols.

Another example for cost-based decision making benefits is health management and overload control. If one R6 link becomes too loaded for any reason or fails, that can affect the link cost or even cause withdrawal of the corresponding BS IP address from the table. As a result this BS will be excluded from the list of potential target BSs even when the radio

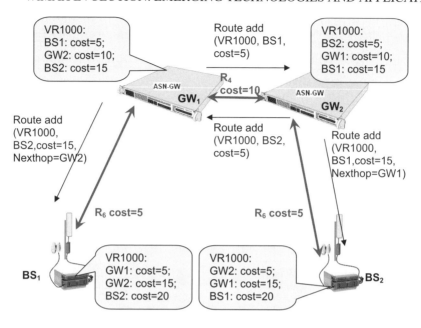

Figure 21.5 Network topology learning based on routing protocol advertisements.

link of this BS (established potentially through a different R8 path) has no problems. One possibility is to include the WiMAX NE load as part of the advertised cost, but it is necessary to be careful not to cause huge amount of routing updates just because of a NE load change. One option is to change the cost with every $X\%$ load change (an arbitrary number selected based on the fact that the load change by that number does not happen very frequently) and an additional hysteresis-based smoothing algorithm to prevent frequent updates when the load is bouncing around the reporting value.

An overload control and even classic WiMAX decisions for handovers or NE selection can benefit from the WiMAX topology learning, and they become a by-product of a routing-based topology discovery.

Routing protocols can efficiently distribute IP addresses, but what can be done to distribute WiMAX IDs that are different from IP addresses? For example, how can we distribute BSIDs that are usually created without taking into account any IP addressing? One 'simple' way is to generate an IP address from the WiMAX ID: if the ID is up to 4 bytes, an IPv4 address can be used; otherwise (the majority of cases) an IPv6 address can be used; it is similar to the IPv4-compatible IPv6 address representation by using WiMAX ID as Least Significant Bits (LSBs) in the address and setting Most Significant Bits (MSBs) to zero or even encode MSBs using WiMAX NE type. WiMAX ID can be set as an IPv6 interface identifier and the rest of bits can be encoded with network ID, operator ID, NE type (BS, ASN-GW, etc.) and so on. VRs can be utilized to prevent overlap of such artificially generated IP addresses with real IPv4 or IPv6 address space, or (especially in the IPv6 case) bits left after WiMAX ID can be made WiMAX-specific. Continuing the above example, BSID can be represented as WIMAX:x:y:z:BS-TYPE:48-bit-BSID (of course, this is only an example, there are many different ways to encode the same information). It is obvious that WiMAX NEs would need support for IPv6 addresses in routing protocols, but that becomes an independent requirement from many operators anyway as preparation for the next generation IPv6-based transport and services.

Whatever the chosen mechanism, *Hello* message relay or the distribution of addresses and IDs through routing protocols, it will enable learning of the WiMAX network topology.

21.4 Capabilities Exchange

Some limited capabilities exchange exists in WiMAX networks today – one is on R3aaa reference point (introduced initially for Simple IP/Mobile IP negotiation), another is between Mobile Stations (MSs) and Mobile IP Foreign Agents (FAs) for reverse tunneling encapsulation negotiation.

There is, however, a need for a much more robust solution. It is especially true for WiMAX, which has a large number of optional features and a flexible functional architecture. WiMAX specification defines three distinct deployment Profiles, A, B and C, that are interoperable only over an open R4 interface making connection between a Profile A BS and Profile C ASN-GW generally impossible. These profiles also describe the location of each function, but they do not mandate that every NE has to include all possible functions. For example, Authenticator function in Profiles A and C has to be located in ASN-GW, but some ASN-GWs can be deployed without it. Also, one of functions can fail for any reason (software module crash or malfunctioning of the hardware running the function), and it would be unwise to bring down the entire NE as a result of the failure.

All of this practically provides a mandate for a capability exchange on all reference points. There are many possible ways to dynamically exchange capabilities, but since the previously described neighbor discovery has already introduced the *Hello* messages for other automation tasks, it becomes convenient to use the same *Hello* exchange also for the capabilities negotiation.

When a NE learns a new neighbor or detects a change in one of its capabilities, it advertises the change. Some capabilities are purely binary, Yes or No, while others can have more complex descriptions. As mentioned above, the Authenticator function capability can be described in a binary form, but mobile subscriber support can have related values, for example, maximum number of active and idle MSs. Non-binary capabilities would need some advertisement policies. One of mechanisms used in many protocols is a division of the entire range of possible values into multiple regions with defined threshold and hysteresis parameters. Each region can be associated with a corresponding action: informational only, not accepting new subscribers, accepting only subscribers for emergency services, stopped serving all subscribers, etc.

Advertised capabilities will definitely affect network behavior. If an ASN-GW advertises the lack of Authenticator function, it means that it cannot support the R3aaa interface, which can affect its selection as a serving or target ASN-GW for R4 handovers, and should definitely prevent any attempt for R3 handover to this ASN-GW. If BS is connected to multiple ASN-GWs simultaneously, the knowledge about ASN-GW capabilities can be used as a weight for the gateway selection to serve a particular MS. When all connected ASN-GWs cannot accept new subscribers, the corresponding target BS would reject any handover attempt. The AAA

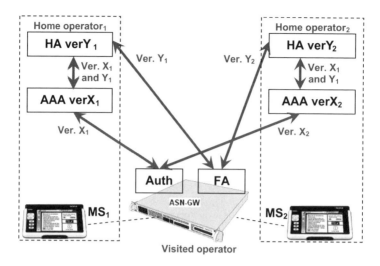

Figure 21.6 MS-dependent version handling.

server would have standard-based way to select the least-loaded HA for Mobile IP (MIP) subscriber. The list of possible smart network decisions enabled by the capabilities exchange can be very large.

21.5   Automatic WiMAX Version Management

Version management is a very important network automation feature for the operator. One can even call it critical because it cannot be replaced by any configuration.

The first challenge is to synchronize the version of all NEs. Let us assume for simplicity that an operator deploys his entire network to run a single WiMAX version X. Even this 'perfect' scenario has a potential risk, correct support for roaming subscribers. It is practically impossible to synchronize versions of all WiMAX NEs throughout the entire worldwide roaming domain. In some cases a visited operator will have to connect to a home AAA running a different version Y. The home AAA will also select a HA for MIP-based deployments (CMIP or PMIP), and that HA might run another version Z. This means that Authenticator and Charging functions in the visited network have to support version Y, and the visited network FA has to support version Z. Just add more subscribers roaming into their home networks running different WiMAX versions and it becomes obvious that the complexity is enormous with different functions running potentially multiple versions. In Figure 21.6 the Authenticator function is running versions X_1 and X_2, while the FA in the same ASN-GW is running versions X_2 and Y_2:

The first basic problem is to know all supported versions on both sides. Today there is a mechanism to advertise a single version X support in the Access Request message, and it is far from being as easy as shown in the above example. It definitely makes sense to advertise in our example all versions for the Authenticator and FA, and it would be possible to just define that the highest common version has to be selected, but an immediate issue is that the decision made by the home network can significantly complicate subscriber handling in the visited network when different home networks select different versions. There should be a way for a visited operator to influence the process, and one solution is to prioritize the supported version list.

The example touches one additional aspect of the version management: per function versions. The WiMAX standard is driven by a functional model where each function can be independent and not all functions have to be collocated. If the standard will allow different versions for different functions, the problem might become unmanageable because of added dimensions. The recommendation would be at least initially to force a single WiMAX version per NE, but even such a restriction does not fix the entire problem. In our example it means that the Authenticator and FA support the same version Y, but it would also mean that the AAA server and MIP HA have to support the same version Y.

Operators would require NEs to be interoperable with at least one common version, and this version is supported by all roaming partners. Is it possible to ensure that there is always a common version between any networks? The answer is 'yes' if all WiMAX devices support all versions starting from the very first. While initially such requirement does not seem to be a big deal, it will become a limiting factor with more and more new releases. The problem is not the development effort, but the verification complexity to test the entire functionality in all scenarios and all versions. A quick poll of a number of major TEMs confirms that existing products usually support no more than two or three previous versions. In this situation it is impossible to ensure that any network can talk to any network, moving the problem into the operator's domain of signing roaming agreements in a way that enables interoperability. The working assumption is that two or three previous versions would tolerate about 3 years difference between WiMAX versions supported by different operators. Otherwise, there will be cases when the network access for some subscribers is denied because of version incompatibility between home and visited networks.

TEMs and operators have another headache: minor versions. Each major WiMAX version can have multiple minor versions (for example, WiMAX version 1 has minor versions 1.0, 1.1, 1.2 and 1.3), and these minor versions might be not backwards compatible. With only three major versions and four minor versions every device has to support many different system behaviors simultaneously, again causing complex development, verification and field troubleshooting. This requires that the WiMAX Forum define only a single public potentially deployed minor version for every major version.

Until now the example for version management was relatively simple, because versions were static. The complexity significantly increases when operators perform upgrade or downgrade of some NEs. It is practically impossible to perform the version change for all devices in the network at once, in many cases it is impossible to do that at once even for specific type of NE, for example, all ASN-GWs. Therefore, there will be always a time period when multiple versions coexist in ASNs and CSNs. There is a possibility of different versions inside a single ASN, and Figure 21.7 can be considered as an example.

When the terminal was at BS_1, the version X was negotiated between all parties on R6, R4 and R3. If BS_2 was upgraded or downgraded to version Y, and then the mobile subscriber moves to the BS_2, there would be a mismatch between versions on R6 and R4. This is not very positive development, because BS_2 will communicate with ASN-GW$_1$ using version Y, but the messages will come through the R4 interface running version X.

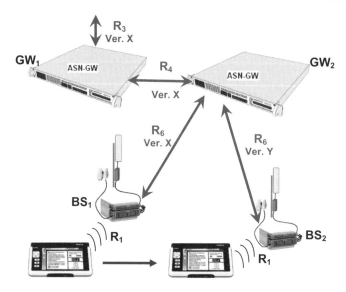

Figure 21.7 Different versions in a single ASN.

One possibility is to perform a version translation in ASN-GW$_2$ in both directions. It could be very complex functionality with multiple concurrent versions and significant differences between them. It would be unclear how to handle such translation in the case when BSID is changed somewhere between versions X and Y. Also, previously the relay ASN-GW$_2$ was not doing much except passing messages from one interface to another; now a potentially heavy translation is needed instead.

Another possibility is to introduce end-to-end version management, where ASN-GW$_2$ does not perform any translation for messages destined to ASN-GW$_1$ or BS$_2$. It becomes a kind of version tunneling: tunnel version Y inside version X. While the feature can be specified, it would solve only a problem of handling end-to-end messages. Unfortunately, many messages are not going end-to-end. For example, data path establishment messages do not fit that definition, and on the other side many informational elements are potentially copied between R6 and R4, and therefore they would have to be translated similarly to the first solution.

Yet another possibility is to re-negotiate the version per MS while in-service (in Figure 21.8 we have assumed that BS$_2$ cannot support version X because, for example, it was downgraded).

The very significant advantage of the dynamic in-service version renegotiation is that it has only a limited overhead compared with a version translation for every control plane message. It comes, however, with the disadvantage of the potential risk of version renegotiation failure when ASN-GW$_1$ or NEs in CSN (AAA server, HA) simply do not support the required ASN version. Practically, this scenario might not be realistic, but the handling is still needed as a negative use case. It is possible to prevent predictive handover that causes incompatible versions in the network by removing such incompatible BSs from the list of target BSs or even from the list of neighbor BSs. For an unpredictive handover with incompatible versions

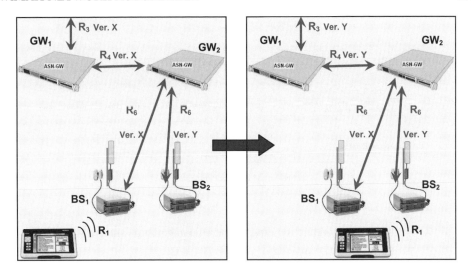

Figure 21.8 Dynamic version negotiation.

it is required to either deny a network access or at least to force a network re-entry and reset the version on the entire serving path similarly to the initial network entry.

In any case, it is our recommended procedure for version management from all of those described previously.

21.6 Automated Roaming

Roaming agreements are an additional area where practically everything is done today manually. A mobile subscriber can roam only when both home and visited operators have signed a corresponding roaming agreement. The process is very complex even for large operators, but becomes the bottleneck for small operators.

Fortunately, it does not have to be like that. A well-known work in the direction of automated roaming is the Ambient Network Project (Ahlgren *et al.*, 2005), which aims to establish connectivity and relationships between any heterogeneous networks.

Another solution for automated roaming is presented by Fu *et al.* (2007); While this paper is discussing heterogeneous networks, it is applicable also for pure WiMAX roaming. It proposes dynamic partnership negotiation via a special AAA entity called Partnership Management Application (PMA) (see Figure 21.9).

The paper also describes ways to establish trust relationship together with policy-based negotiation of the dynamic roaming agreement by means of Trusted Third Party (TTP) involvement.

Tuladhar (2007) raises concerns that the TTP concept brings significant bottleneck and creates a single point of failure for partnership negotiation. Tuladhar proposes an alternative solution that takes advantage of existing trust relationships to build new relationship similarly (but more securely thanks to the so-called proof-token exchange) to many current social networks: the friend of my friend can be my friend too (see Figure 21.10).

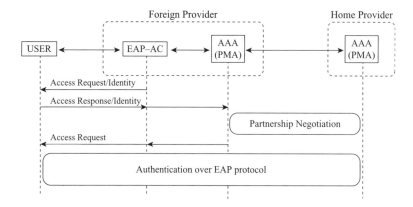

Figure 21.9 Dynamic roaming partnership negotiation using PMA.

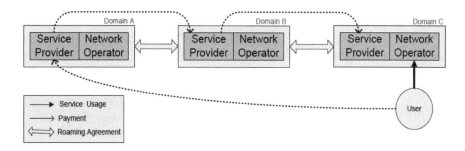

Figure 21.10 Dynamic roaming partnership negotiation using trust referrals.

In the described mechanism, roaming MSs can connect first to some large operator that has a trust relationship with their home operator, and after that use this trust to roam with smaller operators that have a trust relationship with the visited large operator instead of having to build a direct trust relationship with each of the small operators.

Without going into the details of all the above schemes, it is obvious that any automated roaming would need some level of modification in the authentication process (potentially some Extensible Authentication Protocol (EAP) and AAA functionality extensions), and therefore will need a standardization process. It would be great if the WiMAX Forum would start working on the solution as soon as possible, becoming the first major wireless standard to support roaming agreement automation.

21.7   Conclusion: Network Automation as a WiMAX Differentiator

Network automation enables much easier and more reliable network deployment, configuration and network management saving operators a lot of expense. It also can contribute to significantly better end-user experience because of more optimized signaling procedures and faster adaptation to dynamic network changes.

There are multiple broadband wireless technologies, each having some advantages and some disadvantages. Some operators are already choosing competing solutions, some are still 'sitting on the fence' and monitoring the status of such networks. Advanced WiMAX network automation can become a real differentiator between solutions and in some cases can potentially shift the balance towards the selection of WiMAX.

References

Ahlgren, B. *et al.* (2005) Ambient networks: bridging heterogeneous network domains. *Proceedings of the 16th IEEE Symposium on Personal Indoor and Mobile Radio Communications (PIMRC 2005)*, Berlin.

Deering, S. (ed.) (1991) ICMP Router Discovery Messages, *Request for Comments 1256*, IETF, http://www.ietf.org/rfc/rfc1256.txt.

Fu, Z.J. *et al.* (2007) AAA for spontaneous roaming agreements in heterogeneous wireless networks. *Proceedings of the ATC 2007*, Hong Kong, China.

Tuladhar, S.R. (2007) Inter-domain authentication for seamless roaming in heterogeneous wireless networks. *Master's Thesis*, University of Pittsburgh.

Veizades, J., Guttman, E., Perkins, C. and Kaplan, S. (1997) Service Location Protocol, *Request for Comments 2165*, IEFT, http://www.apps.ietf.org/rfc/rfc2165.html.

Wahl, M., Howes, T. and Kille, S. (1997) Light weight Directory Access Protocol (v3), *Request for Comments 2251*, IETF, http://www.apps.ietf.org/rfc/rfc2251.html.

22

An Overview of Next Generation Mobile WiMAX: Technology and Prospects

Sassan Ahmadi

22.1   Introduction

The growing demand for mobile Internet and wireless multimedia applications has motivated the development of broadband wireless access technologies in recent years. Mobile WiMAX was the first mobile broadband wireless access solution, based on IEEE 802.16e-2005 standard (IEEE, 2008d), that has enabled the convergence of mobile and fixed broadband networks through a common wide-area radio access technology and flexible network architecture. The mobile WiMAX air interface utilizes Orthogonal Frequency Division Multiple Access (OFDMA) as the preferred multiple access method in the Downlink (DL) and Uplink (UL) for improved multipath performance and bandwidth scalability.

Since January 2007, the IEEE 802.16 Working Group has embarked on the development of a new amendment of the IEEE 802.16 standard (i.e. IEEE 802.16m) in order to develop an advanced air interface to meet the requirements of ITU-R/IMT-Advanced for 4G systems as well as the next-generation mobile network operators.

Depending on the available bandwidth and antenna configuration, the next-generation mobile WiMAX will enable over-the-air data transfer rates in excess of 1 Gbps and will support a wide range of high-quality and high-capacity IP-based services and applications while maintaining full backward compatibility with the existing mobile WiMAX systems to preserve investments and continuing support for the first-generation products.

WiMAX Evolution: Emerging Technologies and Applications   Edited by Marcos D. Katz and Frank H.P. Fitzek
© 2009 John Wiley & Sons, Ltd

Mobile WiMAX and its evolution are facing serious challenges from other wireless broadband access technologies including 3GPP Long-Term Evolution (LTE), 3GPP2 Ultra-Mobile Broadband (UMB) and IEEE 802.20 that are expected to offer the same functionalities. Nevertheless, there are certain distinctive features and advantages that make mobile WiMAX and its evolution more attractive and more suitable for the realization of ubiquitous mobile Internet access. In this chapter we briefly describe the salient technical features of IEEE 802.16m and potentials for successful deployment of the next-generation of mobile WiMAX in 2011+.

The next-generation mobile WiMAX will build on the success of the existing WiMAX technology and its time-to-market advantage over other mobile broadband wireless access technologies. In fact, all Orthogonal Frequency Division Multiplex (OFDM)-based mobile broadband access technologies that have been developed lately exploit, enhance and expand fundamental concepts that were originally utilized in mobile WiMAX.

The IEEE 802.16m will be suitable for both green-field and mixed deployments with legacy Mobile Stations (MSs) and Base Stations (BSs). The backwards-compatibility feature will allow smooth upgrades and evolution paths for the existing deployments. It will enable roaming and seamless connectivity across IMT-Advanced and IMT-2000 systems through the use of appropriate interworking functions. The IEEE 802.16m systems can further utilize multi-hop transparent or nontransparent relay architectures for improved coverage and performance.

The active participation of a great number of companies from several countries across the globe adds more credibility to IEEE 802.16 efforts to create a worldwide radio access technology to satisfy 4G system and service requirements.

It should be noted that the IEEE 802.16m standard is currently under development and the features and functionalities described here are only proposals under consideration that have not been finalized and are subject to change (IEEE, 2008a).

22.2 Summary of IEEE 802.16m System Requirements

One of the most controversial requirements of IEEE 802.16m, during the development of its system requirement document (IEEE, 2008b), was the requirement for full backwards compatibility and interoperability with the legacy systems. However, the network operator has the ability to disable legacy support (i.e. green-field deployments). The reference system is defined as a system compliant with a subset of the IEEE 802.16e-2005 (IEEE, 2008d) features as specified by WiMAX Forum Mobile System Profile, Release 1.0 (WiMAX Forum, 2007).

The following items are the backwards compatibility requirements for IEEE 802.16m systems (IEEE, 2008b).

- An IEEE 802.16m MS shall be able to operate with a legacy BS, at a level of performance equivalent to that of a legacy MS.

- Systems based on IEEE 802.16m and the reference system shall be able to operate on the same Radiofrequency (RF) carrier, with the same channel bandwidth; and should be able to operate on the same RF carrier with different channel bandwidths.

- An IEEE 802.16m BS shall support a mix of IEEE 802.16m and legacy terminals when both are operating on the same RF carrier. The system performance with such a mix should improve with the fraction of IEEE 802.16m terminals attached to the BS.

- An IEEE 802.16m BS shall support handover of a legacy MS to and from a legacy BS and to and from an IEEE 802.16m BS, at a level of performance equivalent to handover between two legacy BSs.

- An IEEE 802.16m BS shall be able to support a legacy MS while also supporting IEEE 802.16m terminals on the same RF carrier, at a level of performance equivalent to that a legacy BS provides to a legacy MS.

The consideration and implementation of the above requirements ensure a smooth migration from the legacy to new systems without any significant impact on the performance of the legacy systems as long as they exist.

The requirements for IEEE 802.16m have been selected to ensure competitiveness with the emerging 4G radio access technologies while addressing and eliminating the perceived shortcomings of the reference system such as extreme L2 control/signaling overhead, unreliability and insufficient coverage of control and traffic channels at the cell edge, high air-link access latency due to long transmission time interval and Hybrid Automatic Repeat Request (HARQ) round-trip time of three radio frames, link budget deficiency in the UL, high MS power consumption due to several reasons including high UL Peak-to-Average-Power-Ratio (PAPR), frequent need to decode DL Medium Access Protocol (MAP), long scan latency and system entry/reentry time due to inefficient initialization procedures, extreme UL L1 overhead due to use of a subchannelization scheme with high pilot density in the UL, message-based signaling mechanism with extremely large Medium Access Control (MAC) header overhead, lack of support for Frequency Division Duplex (FDD) mode (this has been fixed in IEEE 802.16REV2 revision (IEEE, 2008d)), etc.

The IMT-Advanced requirements defined and approved by ITU-R/Working Party 5D and published as IMT.TECH (ITU, 2008b) are referred to as target requirements in IEEE 802.16m system requirement document and will be evaluated based on the methodology and guidelines specified by IMT.EVAL (ITU, 2008a). The baseline performance requirements will be evaluated according to the IEEE 802.16m evaluation methodology document (IEEE, 2008c). A careful examination of the IMT-Advanced requirements reveals that they are a subset of and less stringent than the IEEE 802.16m system requirements; therefore, the IEEE 802.16m standard will qualify as an IMT-Advanced technology.

Table 22.1 summarizes the IEEE 802.16m baseline system requirements. In the following sections we briefly discuss how these requirements can be met or exceeded. For the Voice over IP (VoIP) capacity, a 12.2 kbps codec with a 50% speech activity factor is assumed such that the percentage of users in outage is less than 2% where a user is defined to have experienced outage if less than 98% of the VoIP packets have been delivered successfully to the user within a one way radio access delay bound of 50 ms (IEEE, 2008b). It should be noted that the VoIP capacity is the minimum of the capacities calculated for the DL and UL.

Note that bidirectional VoIP capacity is measured in active users per megahertz per sector. The total number of active users on the DL and UL is divided by total bandwidth occupied by the system accounting for frequency reuse. For FDD configuration, the bandwidth is calculated as the sum of the UL and DL channel bandwidths. For a Time Division Duplex

Table 22.1 Summary of IEEE 802.16m baseline system requirements.

Metric	Downlink requirement	Uplink requirement
Peak spectral efficiency	15 bps per Hertz (4 × 4 MIMO)	6.75 bps per Hertz (2 × 4 MIMO)
Operating bandwidth	5 to 20 MHz (up to 100 MHz through band aggregation)	
Duplex scheme	TDD, FDD and H-FDD terminals	
Antenna configuration	2 × 2 (baseline), 2 × 4, 4 × 2, 4 × 4, 8 × 8	1 × 2 (baseline), 1 × 4, 2 × 4, 4 × 4
Data plane latency	<10 ms	
Control plane latency	<100 ms	
Handover interruption time	27.5 ms (intra-frequency) 40 ms (inter-frequency within spectrum)	
	60 ms (inter-frequency between spectrum)	
Average user throughput (using baseline antenna configuration)	> 2× (reference system)	> 2× (Reference System)
Cell-edge user throughput (using baseline antenna configuration)	0.26 bps per Hertz per sector > 2× (reference system)	0.13 bps per Hertz per sector > 2× (Reference System)
Sector throughput (using baseline antenna configuration)	0.09 bps per Hertz per sector > 2× (reference system)	0.05 bps per Hertz per sector > 2× (reference system)
VoIP capacity (using baseline antenna configuration)	2.6 bps per Hertz per sector > 30 active users per megahertz per sector	1.3 bps per Hertz per sector
Mobility	Up to 500 km/h with optimal performance up to 10 km/h Up to 100 km with optimal performance up to 5 km	
Cell range and coverage	4 bps per Hertz for ISD 0.5 km and 2 bps per Hertz for ISD 1.5 km	
Multicast and Broadcast Service	1.0 s (intra-frequency) 1.5 s (inter-frequency)	
MBS channel reselection interruption time		
Location-Based Service	Location determination latency < 30 s	
	MS-based position determination accuracy < 50 m	
	Network-based position determination accuracy < 100 m	

(TDD) configuration, the bandwidth is simply the channel bandwidth. Therefore, the VoIP capacity requirement for FDD and TDD systems is 60 and 30 active users per megahertz per sector, respectively.

Minimum performance requirements for enhanced multicast and broadcast are expressed in terms of spectral efficiency over 95% coverage area. The performance requirements apply to a wide-area multi-cell multicast broadcast single frequency network.

22.3 Areas of Improvement and Extension in Mobile WiMAX

Following a thorough gap analysis of the features, functionalities and performance of mobile WiMAX technology and further study of the 4G technology prospects and service requirements, we identified areas of improvement and extension to be pursued in IEEE 802.16m.

The following is a summary of key performance enhancement areas in IEEE 802.16m.

- Support for higher mobility.

    - Support for vehicular speeds up to 350 km h^{-1} (and up to 500 km h^{-1} depending on frequency band) is enabled through improved link adaptation and shorter link access delays and faster feedback mechanisms using subframe structure as well as faster and more reliable cell switching and handover.

- Higher spectral efficiency and peak data rates.

    - Downlink and uplink peak spectral efficiencies in the excess of 15 and 6.75 bps Hz^{-1} using 4×4 and 2×4 antenna configurations, respectively.

- Higher-order single-user and multi-user open-loop and closed-loop Multiple Input Multiple Output (MIMO) schemes with single-stream and multi-stream capability for each user.

- Lower overhead and increased efficiency that would translate into increased application capacity.

    - L1 overhead reduction through the use of new physical resource blocks and resource allocation, new pilot structure, etc.

    - L2 overhead reduction through new control/signaling channel design, compact single-user and multi-user MAC headers, etc.

- Lower latencies that would enable seamless connectivity and increased application performance and quality.

    - Air-link access latencies of less than 10 ms, inter-frequency and intra-frequency handover interruption times of 40–60 and 27.5 ms, respectively, and idle-state to active-state transition time of less than 100 ms have been targeted for IEEE 802.16m.

- Improved traffic and control channel coverage, improved link budget, and cell-edge performance

  - IEEE 802.16m is required to provide optimum performance for cell sizes up to 5 km and maintain functionality for cell sizes up to 100 km.

- Reduce MS power consumption

  - This is enabled through improved sleep and idle mode and paging protocols, enhanced initialization procedures, improved initial ranging and bandwidth request procedures, UL PAPR reduction, optimized Discontinuous Reception (DRX) protocol, etc.

We now provide a summary of the key functional enhancement areas in the next generation of mobile WiMAX.

- Higher flexibility for deployment through support of TDD and FDD duplexing schemes with maximum commonalities in MAC and Physical layer (PHY) and use of complementary scheduling to enable efficient Half-duplex FDD (H-FDD) terminal operation in FDD networks.

- Support for multi-hop relay architecture through properly classified end-to-end and hop-by-hop functionalities and unified access and relay links.

- Support of different IMT band classes (from 450 to 3600 MHz) through the use of multiple OFDM numerologies for performance optimization and support of wider RF channel bandwidths up to 100 MHz to meet IMT-Advanced requirements.

- Support of multi-carrier operation and RF bandwidths up to 100 MHz through aggregation of contiguous and/or noncontiguous RF channels using a single MAC instantiation.

- Provision for coexistence of colocated multi-radios on the same user terminal to minimize inter-radio interference and service disruptions.

- Inter-frequency and Inter-access-technology handover.

- Improved application and service performance.

  - Enhanced Quality of Service (QoS) classes to support delay-sensitive applications such as interactive gaming and VoIP.
  - Enhanced and competitive VoIP, video-streaming, multicast and broadcast services.
    * Support for more than 60 active users per megahertz per sector for FDD mode (i.e. more than 1200 users per sector at 20 MHz and baseline antenna configuration (WiMAX Forum, 2007).
    * Support for a minimum spectral efficiency of 4 bps Hz^{-1} for multicast and broadcast service with inter-site distance of 0.5 km.

- Enhanced support for location-based services with improved location determination latency and position accuracy to meet E911 Phase II requirements (WiMAX Forum, 2007).

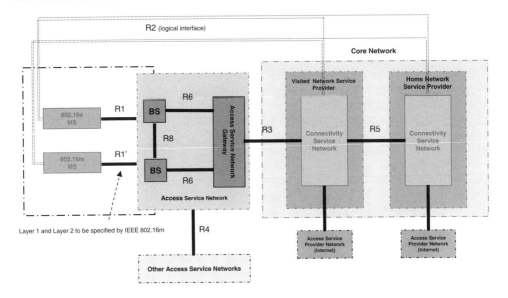

Figure 22.1 Mobile WiMAX overall network architecture; R_i reference points are specified in WiMAX Forum (2008).

22.4 IEEE 802.16m Architecture and Protocol Structure

One of the distinctive features of IEEE 802.16m is the support for a mobile-aware multi-hop relay architecture with unified relay and access links. The design of a new and enhanced air interface would allow more flexibility in design of relay functions/protocols that can be configured as a simple intermediate relay node to a sophisticated BS providing all functionalities that a regular BS would otherwise provide.

The WiMAX Network Architecture release 1.0 (WiMAX Forum, 2008) specifies a nonhierarchical end-to-end network reference model (shown in Figure 22.1) for mobile WiMAX that can be expanded to further include optional relay entities (to be specified by IEEE 802.16m standard) for coverage and performance enhancement. It is expected that the future releases of WiMAX network architecture will specify the reference points between the BS and Relay Station (RS) and between two RSs in a multi-hop network. The IEEE 802.16 standard describes both MAC and PHY for fixed and mobile broadband wireless access systems. The MAC and PHY functions can be classified into three categories, namely data plane, control plane and management plane. The data plane comprises functions in the data processing path such as header compression as well as MAC and PHY data packet processing functions.

A set of L2 control functions is needed to support various radio resource configuration, coordination, signaling and management. This set of functions are collectively referred to as control plane functions. A management plane is also defined for external management and system configuration. Therefore, all management entities fall into the management plane category.

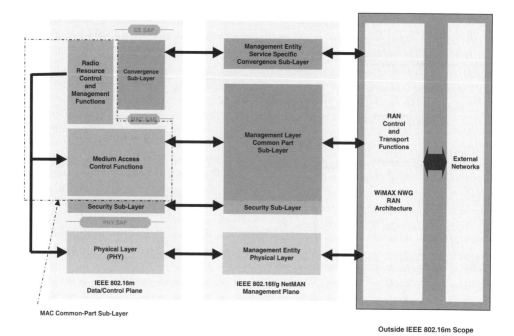

Figure 22.2 IEEE 802.16m reference model.

The IEEE 802.16e-2005 MAC layer is composed of two sublayers: Convergence Sublayer (CS) and MAC Common Part Sublayer (MAC CPS) (IEEE, 2008d). For convenience, we logically classify MAC CPS functions into two groups based on their characteristics as shown in Figure 22.2. The upper and lower classes are called resource control and management functional group and MAC functional group, respectively. The control plane functions and data plane functions are also classified separately. This would allow more organized, more efficient, more structured methods for specifying the MAC services in IEEE 802.16m standard specification. As shown in Figure 22.3, the resource control and management functional group comprises several functional blocks including the following.

- *The radio resource management block* adjusts radio network parameters related to the traffic load, and also includes the functions of load control (load balancing), admission control and interference control.

- *The mobility management block* scans neighbor BSs and decides whether the MS should perform a handover operation.

- *The network-entry management block* controls initialization and access procedures and generates management messages during initialization and access procedures.

- *The location management block* supports Location-Based Service (LBS), generates messages including the LBS information, and manages location update operations during idle mode.

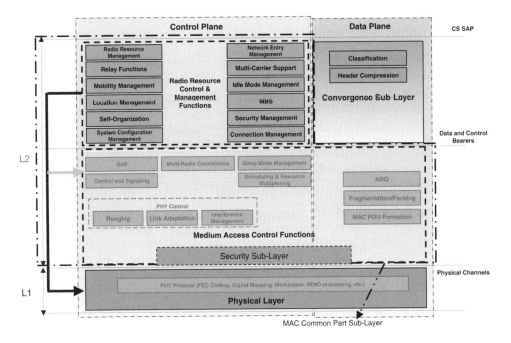

Figure 22.3 IEEE 802.16m protocol stack.

- *The idle mode management block* controls idle mode operation, and generates the paging advertisement message based on paging message from paging controller in the core network.

- *The security management block* performs key management for secure communication. Using managed key, traffic encryption/decryption and authentication are performed.

- *The system configuration management block* manages system configuration parameters, and generates broadcast control messages such as DL/UL Channel Descriptor (DCD/UCD).

- *The Multicast and Broadcast Service (MBS) block* controls and generates management messages and data associated with MBS.

- *Connection management block* allocates Connection Identifiers (CIDs) during initialization/handover service flow creation procedures, interacts with convergence sublayer to classify MAC Service Data Units (MSDUs) from upper layers, and maps MSDUs into a particular transport connection.

The medium access control functional group includes functional blocks which are related to physical layer and link controls such as:

- *The PHY control block* handles PHY signaling such as ranging, Channel Quality measurement/feedback (CQI), and HARQ Acknowledgment (ACK) or Negative

Acknowledgment (NACK) signaling. Based on CQI and HARQ ACK/NACK signals, PHY control block estimates channel environment of MS and performs link adaptation via adjusting Modulation and Coding Scheme (MCS) or power level.

- The control signaling block generates resource allocation messages such as DL/UL MAP as well as specific control signaling messages, and other signaling messages not in the form of general MAC messages, for example, a DL Frame Control Header (FCH).

- *The sleep mode management block* handles sleep mode operation and generates management messages related to sleep operation, and may communicate with the scheduler block in order to operate properly according to the sleep period.

- *The QoS block* performs rate control based on QoS input parameters from connection management function for each connection. The scheduler operates based on the input from the QoS block in order to satisfy QoS requirements.

- *The scheduling and resource and multiplexing block* schedules and multiplexes packets based on the properties of connections. In order to reflect the properties of connections, the scheduling and resource and multiplexing block receives QoS information from QoS block for each connection.

- *The Automatic Repeat Request (ARQ) block* performs MAC ARQ function. For ARQ-enabled connections, the ARQ block logically splits MSDUs and sequences logical ARQ blocks. The ARQ block may also generate ARQ management messages such as a feedback message (ACK/NACK information).

- *The fragmentation/packing block* performs fragmentation or packing of MSDUs based on input from the scheduler block.

- *The MAC Protocol Data Unit (PDU) formation block* constructs MAC PDUs so that BSs/MSs can transmit user traffic or management messages into PHY channels. MAC PDU formation block may add subheaders or extended subheaders. MAC PDU formation block may also add MAC Cyclic Redundancy Checks (CRCs) if necessary, and add a generic MAC header.

IEEE 802.16m protocol structure is expected to be similar to that of IEEE 802.16e-2005 with some additional functional blocks for newly proposed features. In the proposed protocol structure for IEEE 802.16m, the following additional functional blocks are included.

- *Routing (relay) functions* to enable relay functionalities and packet routing.

- *Self-organization and self-optimization functions* to enable home BS for femtocells and plug-and-play form of operation for indoor BS.

- *Multi-carrier functions* to enable control and operation of a number of adjacent or nonadjacent RF carriers (virtual wideband operation) where the RF carriers can be assigned to unicast and/or multicast and broadcast services. A single MAC instantiation will be used to control several physical layers. The mobile terminal is

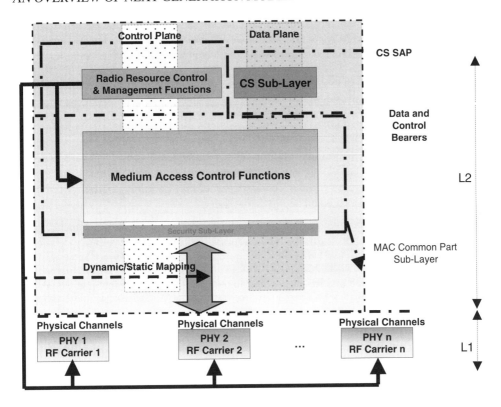

Figure 22.4 IEEE 802.16m protocol stack for multi-carrier operations. The traffic and control channels are statically or dynamically mapped to the PHY channels corresponding to different RF carriers.

not required to support multi-carrier operating. However, if it does support multi-carrier operations it may receive control and signaling, broadcast and synchronization channels through a primary carrier and traffic assignments (or services) may be made on the secondary carriers.

A generalization of the protocol structure to multi-carrier support using a single MAC instantiation is shown in Figure 22.4. The load-balancing functions and RF carrier mapping and control are performed via radio resource control and management functional class.

The carriers utilized in a multi-carrier system, from perspective of a MS can be divided into two categories.

- A primary RF carrier is the carrier that is used by the BS and the MS to exchange traffic and full PHY/MAC control information. The primary carrier delivers control information for proper MS operation, such as network entry. Each MS acquires only one primary carrier in a cell.

  – A secondary RF carrier is an additional carrier which the BS may use for traffic
    allocations for MSs capable of multi-carrier support. The secondary carrier may
    also include dedicated control signaling to support multi-carrier operation.

Based on the primary and/or secondary usage, the carriers of a multi-carriers system
may be configured differently as follows.

  – *Fully configured carrier*: a carrier for which all control channels including syn-
    chronization, broadcast, multicast and unicast control signaling are configured.
    Further, information and parameters regarding multi-carrier operation and the
    other carriers can also be included in the control channels. A primary carrier shall
    be fully configured while a secondary carrier may be fully or partially configured
    depending on usage and deployment model.

  – *Partially configured carrier*: a carrier with only essential control channel config-
    uration to support traffic exchanges during multi-carrier operation.

In the event that the user terminal RF front end and/or its baseband is not capable of
processing more than one RF carrier simultaneously, the user terminal may be allowed
during certain intervals to monitor secondary RF carriers (as shown in Figure 22.5) and
resume monitoring of the primary carrier prior to transmission of the synchronization,
broadcast, and common control channels. This condition ensures that user terminals
will remain synchronized and will receive essential system information at all times
regardless of their bandwidth capabilities that may dynamically change over time.
An example multi-carrier configuration is shown in Figure 22.5. In this example the
primary carrier is fully configured and the secondary carriers are partially configured.

• *Multi-radio coexistence functions* to allow nondisruptive operation of multiple radios
  on a user terminal by coordinating the operation of those radios to minimize inter-
  system interference.

Figure 22.6 illustrates the data plane and control plane protocol stack terminations in the
BS, RS or MS when relay functionality is enabled in the network. Certain radio resource
control and management functions may not exist in the RSs depending whether those
functions are performed in a centralized or a distributed mode as well as whether the RSs are
deployed with full functionalities of a BS. Furthermore, in order to ensure that the security
of the network will not be compromised by untrusted entities, having access to indoor BSs or
femtocell access points, security functions may be limited in the nodes that are outside of the
direct control of the network operator. The new function in the control plane protocol stack
is the security sublayer that would enable ciphering certain management messages.

Note that the access and relay links are defined by IEEE 802.16m standard. Depending
on the functionalities specified for the RSs, the termination points of certain protocols may
differ. The dotted lines in the figure indicate that those functions may be included.

22.5   IEEE 802.16m Mobile Station State Diagram

The IEEE 802.16e-2005 standard does not include an explicit mobile state diagram. However,
a mobile state diagram (i.e. a set of states and procedures between which the MSs transit

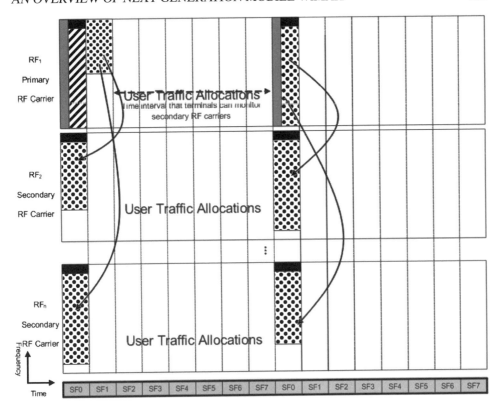

Figure 22.5 Multi-carrier scheme under consideration for IEEE 802.16m. The gray, black, striped and spotty areas in the figure denote synchronization channel, frame control header, super-frame header and common/dedicated control and signaling channels, respectively. The white areas designate user traffic allocations. The subframe/frame timing reference is shown at the bottom of the figure. The dedicated control channels may only reside on the primary carrier.

when operating in the system to receive and transmit data) for the reference system based on common understanding of its behavior can be established.

There are four states from the point of view of a MS when scanning and attaching to a BS in IEEE 802.16e-2005 (IEEE, 2008d), as follows.

1. *Initialization state.* Initialization is a state where a MS without any connection performs cell selection by scanning and synchronizing to a BS preamble and acquires the system configuration information through the broadcast channel before entering the access state. The MS can return back to the scanning step in the event it fails to perform the actions required in next step. During this state, if the MS cannot properly decode the broadcast channel information and cell selection, it should return to scanning and DL synchronization step.

IEEE 802.16m Data Plane Protocol Stack

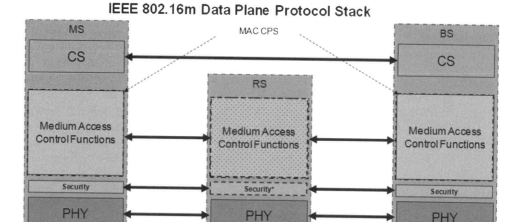

* Security functions may be limited in certain nodes of the network that are outside of direct control of the operator (also known as security zone)

IEEE 802.16m Control Plane Protocol Stack

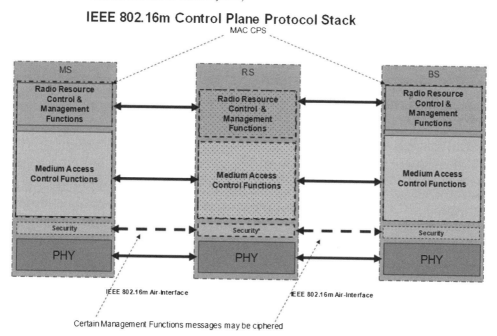

Certain Management Functions messages may be ciphered

Figure 22.6 IEEE 802.16m protocol stack and protocol terminations with unified relay and access links protocols. Dashed lines in the figures indicate that the existence of such links or functions may be optional and deployment dependent.

2. *Access state.* Access is a state where the MS performs network entry to the selected BS. The MS performs the initial ranging process (initial ranging code and RNG-REQ/RSP MAC message is used in the reference system) in order to obtain UL synchronization. Then the MS performs basic capability negotiation with the BS (SBC-REQ/RSP MAC message is used in the reference system). The MS later performs the authentication and authorization process. Next, the MS performs the registration process (REG-REQ/RSP MAC message is used in the reference system). The MS receives the 802.16m specific user identification as part of access state procedures. The IP address assignment may follow using appropriate procedures.

3. *Connected state:* The connected state consists of the following modes: (1) sleep mode, (2) active mode and (3) scanning mode. During the connected state, the MS maintains at least one connection as established during the access state, while the MS and BS may establish additional transport connections. In addition, in order to reduce power consumption of the MS, the MS or BS can request a transition to sleep mode. Also, the MS can scan neighbor BSs to reselect a cell which provides more robust and reliable services.

   - *Active mode*: where the MS and BS perform normal operations to exchange the DL or UL traffic. The MS can perform the fast network reentry procedures after handover. While in handover, the MS maintains any 802.16m specific user IDs required for handover and its IP address in accordance with upper layer protocols. Without going through access state, the MS may remain in connected state with the target BS.

   - *Sleep mode*: where the MS may enable power-saving techniques. The MS in active mode transitions to sleep mode through sleep mode MAC signaling management messages (MOB_SLP-REQ/RSP message is used in the reference system). The MS does not transmit and receive any traffic to/from its serving BS during the sleep interval. A MS can receive an indication message (MOB_TRF-IND message is used in the reference system) during listening interval and then based on the message content to decide whether it should transit to active mode or continue to stay in sleep mode. During the sleep interval, the MS may choose to transit to active mode.

   - *Scanning mode*: where the MS performs scanning operation and may temporarily be unavailable to the BS. While in active mode, the MS transitions to scanning mode via explicit MAC signaling (MOB_SCN-REQ/RSP message is used in the Reference System) or implicitly without scanning management messages generation. In scanning intervals, the MS is unavailable to the serving BS.

4. *Idle state:* The idle state consists of two separate modes, paging available mode and paging unavailable mode. During the idle state, the MS may attempt power saving by switching between paging available mode and paging unavailable mode. In the paging available mode, the MS may be paged by the BS (MOB_PAG-ADV message is used in the reference system). If the MS is paged, it shall transition to the access state for its network reentry. The MS may perform location update procedure during idle state. In the paging unavailable mode, the MS does not need to monitor the DL channel in

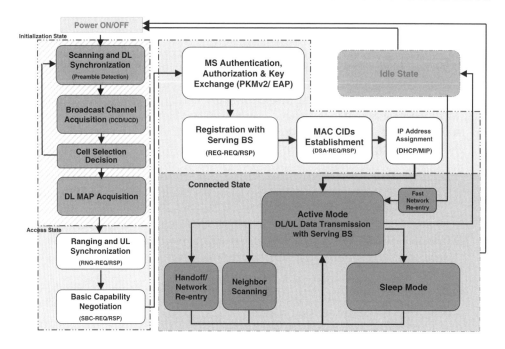

Figure 22.7 IEEE 802.16m MS state diagram, the MAC message acronyms are defined in (IEEE, 2008d).

order to reduce its power consumption. While in this mode, the MS can also transition to access state if required.

In the proposed MS state diagram which is under consideration for IEEE 802.16m, there are four states similar to that of the reference system with the exception that initialization state is simplified to reduce the scan latency and to enable fast cell selection or reselection.

If the location of the system configuration information (FCH and DCD/UCD messages) is fixed so that upon successful DL synchronization and preamble detection, the broadcast channel containing the system configuration information can be acquired (as shown in Figure 22.7), this would enable the MS to make decisions for attachment to the BS without acquiring and decoding the DL MAP and then waiting for the DCD/UCD arrival (DCD/UCD messages are transmitted every few hundred milliseconds). This modification would further result in power saving in the MS due to shortening and simplification of the initialization procedure.

22.6   IEEE 802.16m Physical Layer

In order to achieve the performance targets required by the IEEE 802.16m SRD (IEEE, 2008b), some basic modifications in key aspects of the mobile WiMAX technology such as frame structure, HARQ operation, synchronization and broadcast channel structures, CQI measurement and reporting mechanism, etc., are required. These modifications will enable

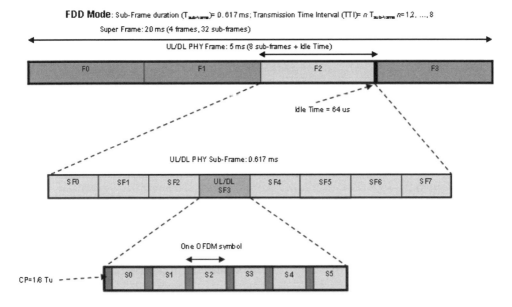

Figure 22.8 IEEE 802.16m FDD basic frame structure for CP length of 1/8 of OFDMA symbol useful time.

faster HARQ retransmissions for improved application performance and higher capacity, fast cell-selection, mobile-aware relay operation, multi-user MIMO and multi-carrier operation.

The IEEE 802.16m uses OFDMA as the DL and UL multiple access scheme same as the reference system. This would reduce the complexity due to use of heterogeneous multiple access schemes for the DL and UL as well as maximizes the commonalities with the legacy system. The OFDMA parameters also remain the same for the new and legacy systems. However, to enable deployment of IEEE 802.16m in new frequency bands such as 700 MHz and 3.6 GHz where large delay spread or large Doppler spread effects are more pronounced, respectively, other OFDMA parameters such as larger cyclic prefixes or larger sub-carrier spacings can be utilized to overcome those effects.

The new frame structure is shown in Figures 22.8 and 22.9 for FDD and TDD modes, respectively. Aside from some special considerations for TDD and FDD duplex schemes, the proposed frame structure equally applies to both duplex schemes, resulting in maximal baseband processing commonalities in both duplex schemes (which further includes H-FDD) that are highly desirable from implementation perspective.

The super-frame is a new concept introduced to IEEE 802.16m where a super-frame is a collection of consecutive equally-sized radio frames whose beginning is marked with a super-frame header. The super-frame header carries short-term and long-term system configuration information or collectively the broadcast channel and may further carry the new synchronization channel. It is desirable to design the synchronization, broadcast and common control and signaling channels such that they only occupy the minimum bandwidth and be detectable by all terminals regardless of their bandwidth.

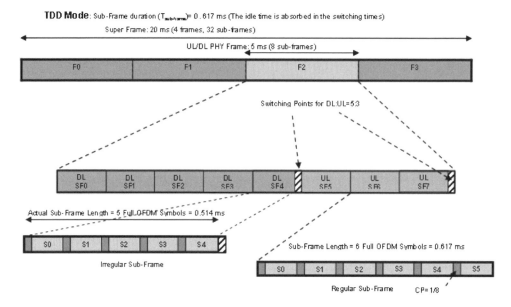

Figure 22.9   IEEE 802.16m TDD basic frame structure for CP length of 1/8 of OFDMA symbol useful time.

In order to decrease the air-link access latency, the radio frames are further divided into a number of subframes where each subframe comprises an integer number of OFDMA symbols. The Transmission Time Interval (TTI) is defined as the basic PHY layer transmission latency over the air-link and is equal to a multiple of subframe length (default of one subframe). Considering the DL/UL switching intervals within a TDD radio frame, there are two types of subframes defined for this mode. As shown in Figure 22.9, the regular subframes consist of six OFDMA symbols whereas the irregular subframes (e.g. those frames that are immediately preceding a switching point in TDD mode) may comprise less than six OFDMA symbols. In that case, the unused OFDMA symbol is an idle symbol. It is understood that no pilot, data and control bits are allocated on the idle symbol in the irregular subframes.

In the basic frame structure, superframe length is 20 ms (comprising four radio frames), radio frame size is 5 ms (comprising eight subframes), and subframe length is 0.617 ms. The use of the subframe concept with the latter parameter set would reduce the one-way air-link access latency from 18.5 ms (corresponding to the reference system) to less than 5 ms.

To accommodate the operation of the new and legacy systems and at the same time enable design and development of improved schemes such as new subchannelization, resource allocation, pilot structure, etc., for the IEEE 802.16m systems, the concept of time zones is introduced that is equally applied to TDD and FDD systems. The new and legacy time zones use the Time-Division Multiplex (TDM) approach across the time domain for the DL. For UL transmissions both TDM and Frequency-Division Multiplex (FDM) approaches are supported for multiplexing of legacy and new terminals. The nonbackwards-compatible improvements and features are restricted to the new zones. All backwards-compatible features and functions are used in the legacy zones. In the absence of any legacy system, the legacy zones will disappear and the entire frame will be allocated to the new zones.

There are several ways where the control and user data blocks can be multiplexed over time and frequency (and code that is, Code Division Multiplex (CDM)). The DL control and data blocks are defined within certain subframes where control and data blocks are mapped to one-dimensional resource blocks (i.e. 18 sub-carriers × 6 symbols physical or logical resource units). Note that in IEEE 802.16m frame structure there are six OFDMA symbols in each subframe. In the case of hybrid TDM/FDM, the control channel is limited within the boundaries of a subframe and the number of subcarriers in the frequency domain is an integer multiple of the number of subcarriers in a one-dimensional resource block. The results of the studies suggest that a combination of the TDM and FDM over the extent of subframe provides the advantages of both TDM and FDM schemes. To efficiently utilize the radio resources and to reduce complexity, the control information are allocated in the units of one-dimensional physical resource blocks. Therefore, unused physical resource blocks in the subframes that contain the control channel can be used for scheduling user data.

There is a growing demand from prominent mobile operators for support of user terminals with various bandwidth capabilities in a radio access network. The challenge will be further aggravated with the increase of operating bandwidth (bandwidths in excess of 20 MHz in the BS) in 4G radio access systems. The user terminal cost, complexity, power consumption and form factor will unjustifiably increase, if the user terminals are all required to support the system bandwidth. The system bandwidth refers to the maximum RF bandwidth supported by a BS. This bandwidth could be a single contiguous RF band or aggregation of small contiguous and/or noncontiguous RF bands.

The multi-bandwidth terminal support will enable operation of user terminals with various bandwidths (with a minimum bandwidth supported by all user terminals) in a broadband wireless access network, and in particular, the IEEE 802.16m standard which is currently under development. The scheme would enable a wide range of IEEE 802.16m-compliant products with different bandwidth capabilities and form factors targeted for various geographical, business or usage models to roam across IEEE 802.16m networks and receive service.

The multi-bandwidth support scheme is based on the assumption that all user terminals have as a minimum (which is required by the standard) the capability to receive and transmit over the minimum bandwidth. Therefore, if the synchronization, broadcast and common control and signaling channels occupy the minimum bandwidth (usually at the center band), all terminals regardless of their bandwidth capability are capable of acquiring the essential system information and DL synchronization.

The use of new subframe structure combined with new one-dimensional physical resource blocks with a low-density and scalable pilot structure and an efficiently-structured localized and distributed resource allocation scheme address some of the inefficiencies of the reference system PHY layer. Other physical layer features include the open-loop and closed-loop multi-user MIMO (MU-MIMO) (see Figure 22.10) that can collapse to single-user MIMO as a special case, support of larger bandwidths through aggregation of multiple RF carriers with a single MAC instantiation, asynchronous HARQ in the downlink and synchronous HARQ in the uplink, FDD mode, rate matching for more efficient mapping of data blocks to the physical resource blocks, CQI feedback with adaptive granularity, etc. Note that IEEE 802.16e-2005 does not include MU-MIMO feature. The shorter TTI helps reduce the

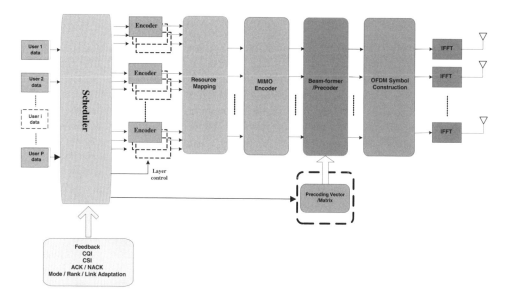

Figure 22.10 Functional block diagram of unified single-user/multi-user MIMO structure, the functions encircled by dashed lines are new functions.

feedback delay in the closed-loop MIMO schemes, resulting in improved throughput. Also note that IEEE 802.16e-2005 does not support synchronous HARQ that can reduce the L2 signaling overhead.

22.7   IEEE 802.16m MAC Layer

The concepts of compact MAC header (where the MAC header and trailer size is reduced from 10 to 4 octets) and multi-user MAC header and scheduling (where the size of the MAC header and trailer is reduced to two 1/4 octets and a group of users with similar channel conditions are scheduled simultaneously in the DL) are introduced to reduce the MAC header overhead and significantly increase capacity of the small-payload applications. In order to reduce the MAC header overhead for small payload applications, the 16-bit CID in the legacy system is split into two parts (1) a user identifier (User-ID) and (2) a user connection identifier (User-Connection-ID), where the User-Connection-ID is a small portion of the CID. The User-ID is assigned following the completion of the access state and the User-Connection-IDs are assigned for the application-specific or management connections that are established for each user.

As was mentioned in the previous section, the use of the subframe concept would help reduce the data and control planes latency. It can further reduce the handover interruption time, that is, the time interval during which a MS does not transmit/receive data packets to/from any neighboring BS, enabling seamless service connectivity when roaming across different cells throughout network. IEEE 802.16e-2005 supports various handover schemes including hard handover, fast BS switching and macro diversity handover, among which only

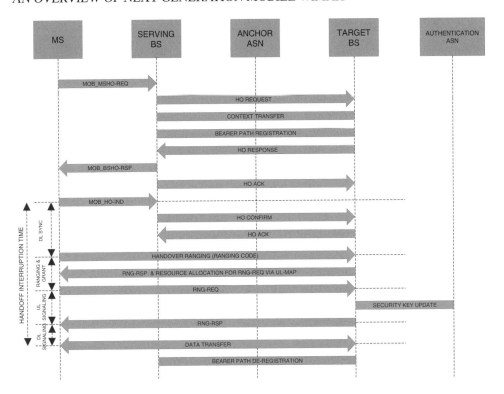

Figure 22.11 Handover procedure in mobile WiMAX (illustration of handover interruption time).

hard handover was made mandatory in the mobile WiMAX profile. The procedures involved in handover interruption time are illustrated in Figure 22.11. The improvements in neighbor scanning and acquisition of target BS, uplink random access, etc. would enable meeting the requirements for handover in IEEE 802.16m to be met. Also some optimizations are considered for hard handover scheme that would improve the overall handover performance. Another distinctive feature of IEEE 802.16m is the requirement for L2 handover capability to/from nonhomogeneous radio access networks, for example, 3GPP LTE, IEEE 802.11, etc. This functionality is an extension of the existing mobility management function in the radio resource control and management functional class (see Figure 22.3).

Other MAC improvements include security protocol enhancement and use of pseudo-identity for active terminals, ciphering of certain management messages for increased security, extension of radio resource management function to enable load balancing across multiple RF carriers, introduction of routing function to facilitate multi-hop relay operation, and self-organization, improve idle and sleep mode protocols to allow MS power saving and self-optimization function to enable plug-and-play BS installation for femtocell and picocell installation and operation. These MAC functions are categorized and shown in Figure 22.3.

22.8   Conclusions

The IEEE 802.16m standard will build on the success of mobile WiMAX to provide the state-of-the-art broadband wireless access in the next decade and to satisfy the growing demand for advanced wireless multimedia applications and services. The standardization of IEEE 802.16m and release 2.0 of mobile WiMAX profile are expected to complete by the end of 2010.

Multi-hop relay architecture, self-configuration, advanced single-user/multi-user multi-antenna schemes and interference mitigation techniques, enhanced multicast broadcast service, increased VoIP capacity, improved cell-edge user throughput, support of vehicular speeds up to 500 km h^{-1}, etc. are among the most prominent features that would make IEEE 802.16m one of the most successful and advanced broadband wireless access systems in the next decade.

References

IEEE (2008a) IEEE 802.16m System Description Document, 80216m-08-003r4.

IEEE (2008b) IEEE 802.16m System Requirements Document, IEEE 802.16m-07/002r5.

IEEE (2008c) IEEE 802.16m Evaluation Methodology Document, IEEE 802.16m-08/004r2.

IEEE (2008d) PART 16: Air Interface For Broadband Wireless Access Systems, P802.16Rev2/D6, Revision of IEEE Std 802.16-2004 and consolidates material from IEEE Std 802.16e-2005,IEEE Std 802.16-2004/Cor1-2005,IEEE Std 802.16f-2005 and IEEE Std802.16g-2007.

ITU (2008a) Guidelines for evaluation of radio interface technologies for IMT-Advanced [IMT.EVAL]. Draft New Report ITU-R M.[IMT.EVAL]

ITU (2008b) Requirements related to technical system performance for IMT-Advanced radio interface(s) [IMT.TECH]. Draft new Report ITU-R M.[IMT.TECH]

WiMAX Forum (2007) WiMAX Forum Mobile System Profile, Release 1.0 Approved Specification.

WiMAX Forum (2008) WiMAX Forum Network Architecture Stage 2–3: Release 1, Version 1.2.

Index

3G, 62
4G, 48

Absolute Continuous Rating (ACR), 230
Access Service Network (ASN), 88, 188, 210,
        215, 266, 281, 425
Access Service Network Gateway
        (ASN-GW), 165, 267, 425
Adaptive Modulation and Coding (AMC), 134
Adaptive Power Distribution (APD), 339
Additive White Gaussian Noise (AWGN), 244
Admission Control (AC), 228, 238, 279
        Measurement Based Admission Control
                (MBAC), 228
        speech quality aware, 237
advantage factor, 233
Application Programming Interface (API),
        165
application session
        conversation flow, 268
asymptotic throughput scaling laws, 328
Asynchronous Transfer Mode (ATM), 133
authentication domain, 173
Authentication, Authorization and Accounting
        (AAA), 138, 173, 203, 210
Automatic Repeat Request (ARQ), 118
availability, 163

Base Station (BS), 168, 266, 425
        backhaul, 394
        cluster, 178
beamforming, 415
        post-FFT, 374
        pre-FFT, 373
Best Effort (BE), 204, 281, 283
Border Gateway Protocol (BGP), 166, 274,
        430
Broadband Wireless Access (BWA), 133
broadcast address, 426

capabilities exchange, 433
capacity limited deployment, 406
Capital Expenditure (CAPEX), 69–71, 182
Care-of-Address (CoA), 275
carrier-grade, 163
        Linux (CGL), 164
Cellular-Controlled Peer-to-Peer (CCP2P),
        106
centralized scheduling, 147
channel model, 369
        3GPP/3GPP2, 369
        analytical, 371
        COST 259/273, 369
        SUI, 370
        WINNER, 370
Cipher Block Chaining (CBC), 138
Client Mobile IP (CMIP), 173, 434
coding
        cost, 155
        gain, 154
        management module, 155
cognitive radios, 416
commercial off-the-shelf (COTS), 51
Common Part Sublayer (CPS), 201, 202
compact MAC header, 460
Connection Identifier (CID), 201, 204
Connectivity Services Network (CSN), 210,
        266
        anchored mobility, 276
context identifier (CID), 187
control plane, 447
control plane IP address, 427
Control Service Access Point (C-SAP), 203,
        205–207
Convergence Sublayer (CS), 155, 201–203
cooperative networks, 416
cooperative principles, 105
coordinated beamforming, 320